Groundwater Development and Management

Pradip K. Sikdar
Editor

Groundwater Development and Management

Issues and Challenges in South Asia

Springer

CP

Editor
Pradip K. Sikdar
Department of Environment Management
Indian Institute of Social Welfare &
Business Management
Kolkata, India

Co-published by Springer International Publishing, Cham, Switzerland, with Capital Publishing Company, New Delhi, India.

Sold and distributed in North, Central and South America by Springer, 233 Spring Street, New York 10013, USA.

In all other countries, except SAARC countries—Afghanistan, Bangladesh, Bhutan, India, Maldives, Nepal, Pakistan and Sri Lanka—sold and distributed by Springer, Tiergartenstr. 15, 69121 Heidelberg, Germany.

In SAARC countries—Afghanistan, Bangladesh, Bhutan, India, Maldives, Nepal, Pakistan and Sri Lanka—printed book sold and distributed by Capital Publishing Company, 7/28, Mahaveer Street, Ansari Road, Daryaganj, New Delhi 110 002, India.

ISBN 978-3-319-75114-6 ISBN 978-3-319-75115-3 (eBook)
https://doi.org/10.1007/978-3-319-75115-3

Library of Congress Control Number: 2018939756

Printed on acid-free paper

This Springer imprint is published by the registered company Springer International Publishing AG part of Springer Nature.
The registered company address is: Gewerbestrasse 11, 6330 Cham, Switzerland

Preface

Groundwater is the most preferred source of water in different user sectors in South Asia due to its near-universal availability, dependability, and low capital cost. The increasing dependence on groundwater as a reliable source of water has resulted in indiscriminate extraction in various parts of South Asia without taking into account the recharging capacities of aquifers and other environmental factors. The irrigation sector remains the major consumer of groundwater in India, accounting for 92% of its annual withdrawal. The development of groundwater in India is highly uneven and shows considerable variations from place to place. Meeting the growing demand for water is further constrained by the deteriorating groundwater quality. In the past, a major advantage of using groundwater was that it normally required little or no treatment, but this is no longer the case. High arsenic, fluoride, chloride, nitrate, etc. concentrations, primarily as a result of increased agricultural production since the 1960s, affect many groundwater supplies. The same is true for pesticides that are widely used for weed control in agriculture, on roads, and railways, and that are also used to control pests in agriculture and industry. Perhaps the greatest threat to South Asia and India's water resources, and as a consequence also water supply and the environment, may come from the changing climate. However, there is great uncertainty about what the effects on water resources will be from climate change. From a groundwater perspective, it could cause a long-term decline in aquifer storage, increased frequency, and severity of droughts and floods as well as the mobilization of pollutants due to seasonally high water tables and saline intrusion in coastal aquifers.

Management of groundwater resources in South Asia is an extremely complex challenge. The highly uneven distribution of groundwater and its utilization make it impossible to have a single management strategy for the region as a whole. Any strategy for scientific management of groundwater resources should involve a combination of supply-side and demand-side measures depending on the regional setting. The likely adverse impacts of global climate change on the availability and quality of groundwater also demand significant political attention at the international and national levels. Groundwater management also calls for a paradigm shift from development strategy to management strategy against the backdrop of several million operating groundwater abstraction structures in South Asia to ensure food and energy security in the region.

This book is organized in twenty-three chapters, which not only highlight the problems of groundwater management in South Asia, in general, and in India, in particular, but also provides solutions using both traditional and modern techniques. The first two chapters cover the problems and challenges in groundwater management in South Asia and India. The next five chapters deal with the various tools that are used in understanding the geology, structure, 3D configuration of the aquifer systems, and groundwater flow and pollutant transport, such as geophysical techniques, environmental isotopes, geostatistics, remote sensing, geographical information system, and numerical modeling. Chapters 8 and 9 describe the development and management of the coastal aquifer systems and springs in hilly regions for community water supply. Drilling, construction, design, and development of wells are very important for a sustainable supply of groundwater from different geological formations; these are covered in Chap. 10. Chapter 11 deals with pumping tests and a field method to analyze pumping test data to determine the cardinal aquifer parameters, such as hydraulic conductivity, transmissivity, and storage coefficient; and safe spacing between two pumping wells. The next five chapters cover the quality of groundwater with respect to arsenic and fluoride, their impact on the food chain and human health, and the methods of treatment of arsenic and fluoride contaminated groundwater. The likely adverse impacts of global climate change on the availability and quality of groundwater also invite significant political attention at the international and national levels, which are covered in Chap. 17. Aspects related to rainwater harvesting, artificial recharging, and development and management of baseflow for public water supply in semi-arid areas are included in the next two chapters. The subject of groundwater governance has been addressed in Chaps. 20, 21, and 22 encompassing economics, pricing, jurisdiction, legislation, policy and regulations, legal framework, and community participation. The final chapter of this book depicts a road-map to evolve a paradigm to ensure safety and security of groundwater-based water supply.

This book addresses the various technical aspects of groundwater development and management and offers a meaningful and feasible guidance for better managing the stressed groundwater resources in South Asia including India. The book will serve as a guide to those who are concerned with various aspects of groundwater science like researchers, academics, professionals, and students in diverse fields like geology, geophysics, hydrology, environmental science, environmental engineering, environmental management, civil engineering, and irrigation engineering. Others who will benefit from this book are administrators, policy makers, and economists who are also concerned with the formulation and evaluation of plans for the development and management of groundwater resources.

The editor conveys his appreciation to all the authors of this book for their invaluable contributions. Without their contributions and support, this book would not have been possible. The editor is thankful to the Director of the Indian Institute of

Social Welfare and Business Management for providing the infrastructural facilities. The editor also thanks the Publisher Capital Publishing Company and Co-publisher Springer for their kind support and great efforts in bringing the book to fruition.

Kolkata, India

Pradip K. Sikdar

Contents

Contributors

Achiransu Acharyya Department of Economics, Visva Bharati (A Central University), Santiniketan, India

Bimalendu B. Bhattacharya S.N. Bose National Centre for Basic Sciences, Salt Lake, Kolkata, India

Indian School of Mines, Dhanbad, India

Ayan Biswas Asia Hub, Akvo, Amsterdam, The Netherlands

Jared Buono Eco-Hydrologist and Independent Consultant, Ruby, NY, USA

Ashis Chakraborty Central Ground Water Board, Faridabad, India

Surajit Chakraborty Department of Environment Management, Indian Institute of Social Welfare and Business Management, Kolkata, India

Rajib Chattopadhyay Indian Institute of Tropical Meteorology, Pune, India

Saheb Das Centre for Ground Water Studies, Kolkata, India

Shouri Dutta Department of Environment Management, Indian Institute of Social Welfare and Business Management, Kolkata, India

Utsab Ghosal Department of Environment Management, Indian Institute of Social Welfare and Business Management, Kolkata, India

Gautam Ghosh Geological Survey of India, Kolkata, India

Subrata Halder State Water Investigation Directorate, Department of Water Resources Investigation & Development, Government of West Bengal, Kolkata, West Bengal, India

Sudhir Kumar National Institute of Hydrology, Roorkee, Uttarakhand, India

Arunabha Majumder School of Water Resources Engineering, Jadavpur University, Kolkata, India

Arsenic & Fluoride Task Force, Government of West Bengal, Kolkata, India

All India Institute of Hygiene & Public Health, Government of India, Kolkata, India

John M. McArthur Department of Earth Science, University College London, London WC1E 6BT, UK

Ajoy Kumar Misra Central Ground Water Board, Kolkata, India

Dipak K. Mukhopadhyay Geological Survey of India, Kolkata, India

Ranjan K. Ray Central Ground Water Board, Faridabad, India

Tapan K. Ray State Water Investigation Directorate, Department of Water Resources Investigation & Development, Government of West Bengal, Kolkata, West Bengal, India

Dipankar Saha Central Ground Water Board, Faridabad, India

Atul K. Sahai Indian Institute of Tropical Meteorology, Pune, India

Paulami Sahu School of Environment & Sustainable Development, Central University of Gujarat, Gandhinagar, India

Saroj K. Sanyal Bidhan Chandra Krishi Viswavidyalaya, Mohanpur, Nadia, West Bengal, India

Bhabesh C. Sarkar Department of Applied Geology, Indian Institute of Technology (ISM), Dhanbad, India

Pradip K. Sikdar Department of Environment Management, Indian Institute of Social Welfare and Business Management, Kolkata, India

About the Editor

Professor Pradip K. Sikdar received his education at the University of Calcutta (Geology/Hydrogeology) and the University of Newcastle-upon-Tyne, UK (Integrated Coastal Zone Management). A teacher by profession and a hydrogeologist by education, he was a Senior Scientist and Executive Secretary at the Centre for Study of Man and Environment, Kolkata, and a Lecturer at the Department of Applied Geology, Indian School of Mines, Dhanbad, before joining the Indian Institute of Social Welfare and Business Management at the Department of Environment Management in 2000. Currently as a Professor at the Department of Environment Management, he is actively engaged in research and teaching of groundwater hydrology and environmental impact assessment. He has over 28 years of experience in project management and implementation, field and research work in groundwater systems including the application of computer models in the scientific evaluation of hydrogeologic systems, and groundwater resources development and protection. He also provides technical and management advice on a wide range of subjects dealing with watershed management: natural resources development; and environmental management issues to industries, consultant groups, and NGOs. He has carried out extensive research on arsenic contamination in groundwater of the Bengal Basin, and his primary interest in arsenic is to explain the spatial distribution of As-pollution in aquifers of the Bengal Basin. His present research also deals with the sustainability of water supply in the fluoride affected and semi-arid regions of West Bengal and the design of groundwater abstraction structures.

Professor Sikdar supervises several students for Master, Doctoral, and Post-doctoral research work. Up until now, five research students have been conferred the Ph.D. degree under his guidance. He is also Visiting Faculty at the Departments of Geology and Environmental Science of the University of Calcutta and at the Department of Geology of the Presidency University, where he teaches Hydrogeology and Water Resource Management. He has delivered several lectures in national and international conferences, seminars, workshops, UGC-sponsored refresher courses, etc. He is serving as a reviewer for many national and international journals. He is a member of several national and international professional institutions and societies, and is also on the Board of Editors of some leading national journals. He was a Member of the CGWB State-Level Technical Evaluation Committee and West Bengal Fact-Finding Commission on Environment. He has published more than 100 research papers in international and national journals, conference proceedings, books, and 35 project reports. He has authored the book "Groundwater of West Bengal: Assessment, Development and Management" (in Bengali) and co-authored the book "A Text Book of Environment". He has also edited two books on "Interlinking of Indian Rivers." He is the recipient of the 22nd Dewang Mehta Business School Award 2014 for the "Best Professor in Water Resources Management" and Groundwater Science Excellence Award in 2015 by the International Association of Hydrogeologists – India Chapter in appreciation of his achievements in the field of groundwater in India.

Chapter 1
Problems and Challenges for Groundwater Management in South Asia

Pradip K. Sikdar

1 Introduction

South Asia represents the southern region of the Asian continent, which comprises Afghanistan, Bangladesh, Bhutan, Maldives, Nepal, India, Pakistan and Sri Lanka. Topographically, it is dominated by the Indian Plate, which rises above sea level as Nepal and northern parts of India situated south of the Himalayas and the Hindu Kush. South Asia is bounded on the south by the Indian Ocean and on land by West Asia, Central Asia, East Asia, and Southeast Asia. South Asia covers about 5.1 million km^2, which is 11.51% of the Asian continent or 3.4% of the world's land surface area. The region is home to about 39.5% of Asia's population and over 24% of the world's population, making it both the most populous and the most densely populated geographical region in the world. The important rivers of South Asia are Ganges, Indus and Brahmaputra. These rivers have contributed to the rise and prosperity of some of the earliest civilizations in history and today are the source of livelihood for millions. The South Asian river basins, most of which have their source in the Himalayas, support rich ecosystems and irrigate millions of hectares of fields, thereby supporting some of the highest population densities in the world.

The climate of this vast region varies considerably from area to area from tropical monsoon in the south to temperate in the north. The variety is influenced by not only the altitude, but also by factors such as proximity to the sea coast and the seasonal impact of the monsoons. Southern parts are mostly hot in summers and receive rain during monsoon periods. The northern belt of Indo-Gangetic plains also is hot in summer, but cooler in winter. The mountainous north is colder and receives snowfall at higher altitudes of Himalayan ranges. As the Himalayas block the north-Asian

P. K. Sikdar (✉)
Department of Environment Management, Indian Institute of Social Welfare and Business Management, Kolkata, India
e-mail: p_sikdar@hotmail.com

P. K. Sikdar (ed.), *Groundwater Development and Management*,
https://doi.org/10.1007/978-3-319-75115-3_1

bitter cold winds, the temperatures are considerably moderate in the plains. South Asia is largely divided into four broad climate zones (Olive 2005). They are (i) dry subtropical continental climate between the northern Indian edge and northern Pakistani uplands, (ii) equatorial climate between the far south of India and southwest Sri Lanka, (iii) tropical climate in most of the peninsula with variations such as hot sub-tropical climate in northwest India, cool winter hot tropical climate in Bangladesh and tropical semi-arid climate in the centre, and (iv) alpine climate in the Himalayas.

Maximum relative humidity of over 80% has been recorded in Khasi and Jaintia Hills and Sri Lanka, while the area adjoining Pakistan and western India records lower than 20%. Climate of South Asia is largely characterized by monsoons. Two monsoon systems exist in the region (Tyson 2002). They are (i) summer monsoon when wind blows from southwest to most of parts of the region and accounts for 70%–90% of the annual precipitation, and (ii) winter monsoon when wind blows from northeast. The warmest period of the year (March to mid-June) precedes the monsoon season. In the summer the low pressures are centered over the Indus-Gangetic Plain and high wind from the Indian Ocean blows towards the centre. The monsoon season is relatively cooler because of cloud cover and rain. In early June the jet streams vanish above the Tibetan Plateau, low pressure over the Indus Valley deepens and the Inter-tropical Convergence Zone (ITCZ) moves in. This change results in depressions in the Bay of Bengal which brings in rain from June to September (Olive 2005).

2 Hydrogeological Framework

The behaviour of groundwater in South Asia is highly complicated due to the occurrence of diversified geological formations with considerable lithological and chronological variations, complex tectonic framework, climatological dissimilarities and various hydrochemical conditions. Studies carried out over the years have revealed that aquifer groups in alluvial/soft rocks even transcend the surface basin boundaries. The hydrogeological framework of the countries of South Asia is briefly described below.

2.1 India

In India groundwater occurs broadly in two groups of rock formations depending on characteristically different hydrogeological and hydrodynamical conditions. They are porous formations (unconsolidated and semi-consolidated) and fissured formations.

The unconsolidated formations are alluvial sediments of river basins, coastal and deltaic tracts where groundwater occurs in primary porosity. These are by far the most significant groundwater reservoirs for large scale and extensive development.

The hydrogeological environment and groundwater regime conditions in the Ganga-Brahmaputra basin indicate the existence of potential aquifers having enormous fresh groundwater resources. Bestowed with high incidence of rainfall and covered by a thick pile of porous sediments, these groundwater reservoirs get replenished every year and are being used heavily. The semi-consolidated formations normally occur in narrow valleys or structurally faulted basins. The Gondwanas, Lathis, Tipams, Cuddalore sandstones and their equivalents are the most extensive productive aquifers. Under favourable situations, these formations give rise to free flowing wells. In select tracts of northeastern India, these water-bearing formations are quite productive. The Upper Gondwanas, which are generally arenaceous, constitute prolific aquifers.

The fisssured formations occupy almost two-third of the country. Groundwater in these formations occurs in secondary porosity except vesicular volcanic rocks where it occurs in primary porosity. From the hydrogeological point of view, fissured rocks are broadly classified into four types. They are (i) igneous and metamorphic rocks excluding volcanic and carbonate rocks, (ii) volcanic rocks, (iii) consolidated sedimentary rocks and (iv) carbonate rocks. The first type consists of granites, gneisses, charnockites, khondalites, quartzites, schists and associated phyllites, slates, etc. These rocks possess negligible primary porosity but develop secondary porosity and hydraulic conductivity due to fracturing and weathering. Groundwater yield depends on rock type, grade of metamorphism and degree of weathering and fracturing. The predominant types of the volcanic rocks are the basaltic lava flows of Deccan Plateau. The Deccan Traps have usually poor to moderate hydraulic conductivity depending on the presence of primary and secondary porosity. The consolidated sedimentary rocks occur in Cuddapahs, Vindhyans and their equivalents. The formations consist of conglomerates, sandstones, shales, slates and quartzites. The presence of bedding planes, joints, contact zones and fractures control the groundwater occurrence, movement and yield. The carbonate rocks consist of mainly limestones and dolomites in the Cuddapah, Vindhyan and Bijawar group of rocks. In carbonate rocks, the circulation of water creates solution cavities. The solution activity leads to widely contrasting permeability within short distances through which turbulent flow of groundwater takes place.

The estimated replenishable groundwater resources in India as estimated by the Central Ground Water Board is 432 billion cubic metres (BCM) of which nearly 60% of the resources has already been developed. Major utilization of groundwater resources, nearly 80%, is in the agriculture sector.

2.2 *Pakistan*

The Indus Plain in Pakistan was formed by sediment deposits from the River Indus and its tributaries, and it is underlain by a highly transmissive unconfined aquifer. In the Punjab, most groundwater supplies are fresh. The main exceptions are the areas of saline groundwater in the centre of the interfluviums ('doab'), particularly those between Multan and Faisalabad, around Sargodha, and in the south-eastern part of the province. In the southern part of the lndus Plain, in Sindh, groundwater supplies are more problematic. With the exception of a small strip along the River Indus, groundwater supplies are highly saline. The discharge from the aquifers in Sindh is generally less than that in Punjab (Survey of Pakistan 1989). Supplies are even less in the eastern deserts and the western limestone ridges bordering the lndus Plain. Balochistan and North West Frontier Province are geologically more complex. In Balochistan, land formations consist of limestone, sandstone, and shale formations alternating with sand, silt, and gravel deposits. The aquifers are confined and generally yield only limited quantities of water. The quality of groundwater is unsuitable for agriculture in parts of Balochistan, particularly in some of the coastal areas, in patches in the large, alluvial Kacchi plain in the east, and in several valleys in the uplands. Although the picture for North West Frontier Province resembles that of Balochistan, irrigation with groundwater in North West Frontier Province is more recent and, until recently, less extensive. This is due to the less arid conditions and ample surface water resources. Another difference is in the areas around Bannu and Mardan, where there are thick and extensive aquifers. In Bannu, however, the fresh water is overlain by saline water.

Total groundwater resources in Pakistan has been estimated to be 57 BCM out of which 30 to 40% is considered unsuitable for agriculture. Nearly 1.1 million tubewells (agricultural wells) are among the major causes of rapidly depleting groundwater levels. The over-exploitation of the confined aquifer has caused such an alarming situation that in the forseeable future the supply from groundwater may dry up in Quetta, the capital of Baluchistan. Large scale groundwater abstraction has posed considerable threat to soil and groundwater quality.

2.3 Bangladesh

The deposits of thick unconsolidated Pliestocene and Holocene alluvial sediments of the Ganga-Brahmaputra-Meghna (GBM) delta system form one of the most productive aquifer systems in the world. The aquifer gets fully recharged each year by monsoonal rains and floods. Deeper aquifers below the shallow zones of saline water intrusion are exploited in the coastal regions. Jones (1985) suggested that fresh water may also be available from older Tertiary rocks down to depths of 1800 m. The main aquifers in Bangladesh are (i) Late Pliestocene to Holocene coarse sands, gravels and cobbles of the Tista and Brahmaputra mega fans and basal fan delta gravels along the incised Brahmaputra channel (MMP 1977; UNDP 1982; MMP 1983), (ii) Late Pliestocene to Holocene braided river coarse sand and gravels deposited along the incised palaeo-Ganga, lower Brahmaputra and Meghna main channels (UNDP 1982; MMP 1983; Davies et al. 1988; Davies and Exley 1992; MMI 1992), (iii) Early to Middle Pliestocene stacked fluvial main channel made of medium to coarse sand at >150 m depth in the Khulna, Noakhali, Jessore/Kushtia and western moribund Ganga delta areas in the subsiding delta basin (UNDP 1982), (iv) Early to Middle

Pliestocene red-brown medium to fine sands that underlie grey Holocene medium to fine sands in the Old Brahmaputra and Chandina areas (UNDP 1982; MMP 1983; MMI 1992; Davies and Exley 1992), and (v) Early to Middle Pliestocene coarse to fine fluvial sands of the Dupi Tila Formation that underlie the Madhupur and Barind Tracts, capped by deposits of Madhupur Clay Residuum (Welsh 1966).

In most of the groundwater studies undertaken in Bangladesh, the aquifer system has not been divided stratigraphically. Conceptual models of hydrogeological conditions based on simple lithological units have been used to assess the hydraulic properties of aquifers and tubewell designs to depth of about 150 m. The aquifers have been divided into two groups according to colour and degree of weathering (Clark and Lea 1992). They are (i) the grey sediments mainly deposited during the last 20 ka, and (ii) the red brown sediments mainly older than 100 ka with iron oxides cements and grey smectitic clays. Hydraulic conductivities determined for grey sediments are estimated to be in the range 0.4–100 m/day. Those for red brown sediments are in the range 0.2–50 m/day (UNDP 1982). These give a ratio of hydraulic conductivities of 2:1 for grey-brown sediments. In general, the regional groundwater flow in the aquifers of Bangladesh is from north to south with local variation near major rivers. National Water Plan Phase-II estimated average groundwater as 21 cubic kilometres in 1991 (Sengupta et al. 2012). More than 60 per cent of the groundwater in Bangladesh contains naturally occurring arsenic, with concentration levels often significantly exceeding World Health Organisation's (WHO) standard of 10 μg/L.

2.4 Afghanistan

According to United Nations Department of Technical Cooperation for Development (United Nations 1986) the aquifers in Afghanistan can be divided into three categories. They are (i) alluvial and colluvial unconsolidated to semi-consolidated aquifers comprising about 20% of the total mapped aquifers and containing about 70% of the available groundwater resources, (ii) limestone and dolomite aquifers making up only 15% of the total mapped aquifers and containing about 20% of the available groundwater resources, and (iii) the remaining 65% of aquifers are low permeability units that contain about 10% of the available groundwater resources.

The three hydrogeological regions in Afghanistan are the Great Southern Plain (Siestan Basin) in the south, the Central Highland Region including the Hindu Kush mountain range and its associated ranges, and the Northern Plain (Amu Darya Basin) (United Nations 1986).

The intermontane stream basins of the Central Highland Region are hydrogeologically most significant (Gellasch 2014). These basins are fault controlled and filled with a variety of unconsolidated materials ranging from alluvial, colluvial, lacustrine and glacial deposits. The aquifers in these basins contain good amount of fresh water. The most important of the intermontane basins include those near the cities of Ghazni, Khowst (Khost), Jalalabad and Kabul (United Nations

1986). Of these basins, the Kabul Basin has been studied most extensively since 2001 (Broshears et al. 2005; Akbari et al. 2007; Lashkaripour and Hussaini 2008; Houben et al. 2009a, b; Mack et al. 2010). The Kabul Basin geology consists of consolidated rocks in the mountains surrounding the basin with unconsolidated sediments in the basin serving as the principal aquifer system. There are four aquifers in the Kabul Basin consisting mainly of sand and gravel which become slightly cemented with increasing depth. The Paghman-Darulaman basin has two aquifers lying along the course of River Paghman and the upper course of River Kabul. The other two aquifers are located in Logar Basin and the southern part of the Kabul Basin and follow the course of River Logar and the lower course of River Kabul.

The general groundwater flow direction is from western or south-western basin margin, through the basin centre, to the eastern basin margin. Locally the thickness of the aquifer can be up to 80 m (Bockh 1971). The Kabul, Paghman and Logar aquifers provide most of the drinking water to the residents of Kabul. The hydraulic conductivity of the aquifers varies from 2.3×10^{-5} to 1.3×10^{-3} m/s and can be classified as permeable to very permeable (Himmelsbach et al. 2005). Each of these three aquifers near Kabul is capped by a loess layer that varies between 1 and 5 m in thickness and helps to protect groundwater from contaminants migrating downward from the surface. The loess also inhibits infiltration and impacts recharge. The Kunar River valley in eastern Afghanistan is another basin that contains an important aquifer system. According to Banks and Soldal (2002), these intermontane basins are similar to two other thoroughly studied locations: the intermontane trough between the Greater and Lesser Caucasus in Azerbaijan and the intermontane trough of the Altiplano, between the Cordilleras Oriental and Occidental of Bolivia. The total groundwater recharge has been estimated to be $10{,}650 \times 10^6$ m^3/year and the total groundwater usage is 2800×10^6 m^3/year (Uhl 2003).

2.5 Nepal

Nepal is among the richest in terms of water resource availability. Water resources are abundant throughout the country in the form of snow covers, rivers, springs, lakes and groundwater. The total renewable water resource of the country is estimated to be 237 km^3/year (225 km^3/year for surface water sources and 12 km^3/year for groundwater sources) and per capita water availability for 2001 was 9600 m^3/capita/year (http://www.wepa-db.net/policies/state/nepal/state.htm, accessed on February 2017).

Groundwater is abundant in the aquifers of the Terai and the Kathmandu Valley. About 50% of the water used in the city of Kathmandu is derived from groundwater. Groundwater availability is more limited in the populated hilly regions because of the lower permeability of the indurated and crystalline rock types. In the Kathmandu Valley (area around 500 square kilometres), groundwater is abstracted from two main aquifers within the thick alluvial sediment sequence. A shallow unconfined aquifer occurs at >10 m depth and a deep confined aquifer occurs at around 310–370 m (Khadka 1993). The transmissivity value ranges between 163 and 1056 m^2/day for the shallow aquifer and 22.5 and 737 m^2/day for the deep aquifer (Pandey and Kazma 2011) indicating that the shallow aquifer has a better capacity to transmit water. Exploitation of these aquifers, especially the shallow aquifer, has increased rapidly in recent years as a result of the increasing urbanisation of the region. Recent abstraction of groundwater from the deep aquifer has led to a decrease in the groundwater level by around 15–20 m since the mid 1980s (Khadka 1993). Below the alluvial sediments in the Kathmandu Valley, karstic limestone aquifers also exist. A small number of natural springs issue from these and are used for water supply in the southern part of the valley. The limestone aquifer has not been developed and has received little hydrogeological attention. Shallow and deep aquifers are also present in the young alluvial sediments throughout most of the Terai region (Jacobson 1996). The shallow aquifer appears to be unconfined and well developed in most areas, although it is thin or absent in Kapilvastu and Nawalparasi (Upadhyay 1993). The deep aquifer of the Terai region is reported to be artesian (Basnyat 2001).

2.6 Sri Lanka

Six main types of groundwater aquifers have been identified and characterized in Sri Lanka. They are (i) shallow karstic aquifer of Jaffna Peninsula, (ii) deep confined aquifer, (iii) coastal sand aquifers, (iv) alluvial aquifers, (v) shallow regolith aquifer of the hard rock region, and (vi) south western lateritic (Cabook) aquifer. In 1985, the internal renewable groundwater resources were estimated at 7.8 km^3, most (estimated at 7 km^3/year) returning to the river systems and being included in the surface water resources estimate of 50 km^3/year (http://www.eoearth.org/view/article/156991/, accessed on February 2017).

The shallow karstic aquifer of the Jaffna peninsula is composed of Miocene limestone and is intensively used. The aquifer is 100 to 150 m thick, distinctly bedded and well jointed. The shallow groundwater forms mounds or lenses floating over the saline water. These water mounds or lenses reach their peak during the monsoon rains of November–December (Panabokke and Perera 2005).

The deep confined aquifers occur within the sedimentary limestone and sandstone formations of the northwest coastal plain and are least utilized. These aquifers are more than 60 m deep and have a relatively high recharge rate. The deep aquifer is highly faulted and it separates the aquifer into a series of isolated blocks thus forming seven distinct groundwater basins. The shallow coastal sand aquifer occurs on the coastal beaches and spits in the Kalpitiya Peninsula and the Mannar Island in the north west of Sri Lanka and on raised beaches of Nilaveli-Kuchchaweli, Pulmoddai and Kalkuda in the north-eastern region. The shallow coastal aquifers occupy about 1250 sq. km and are intensively used for drinking, domestic, agriculture and tourism industry.

The alluvial aquifers occur on both coastal and inland flood plains, inland river valleys of varying size, and old buried river beds and is unconfined in nature. The deeper and larger alluvial aquifers occur along the lower reaches of the major rivers that cut across the various coastal plains. Rivers such as Mahaweli Ganga, Kelani Ganga, Deduru Oya, Mi Oya, Kirindi Oya and Malwathu Oya have broad and deep alluvial beds of variable texture and gravel content in their lower reaches. Old buried riverbeds with high groundwater yield are present in the lower part of River Kelani. The alluvial formations of these larger rivers may vary between 10 and 35 m thickness, and may extend to several hundreds of meters on either side of the riverbeds. A reliable volume of groundwater can be extracted from these alluvial aquifers throughout the year and are intensively used at present.

Groundwater potential in the hard rock region of Sri Lanka is limited because of the low groundwater storage capacity and transmissivity of the underlying crystalline basement hard rock (Sirimanne 1952). Groundwater is found in the weathered rock zone, or the regolith, as well as in the deeper fracture zone. The weathered zone generally ranges from 2 to 10 m in thickness, while the fracture zone is located at depths of more than 30–40 m (Panabokke 2003). This shallow regolith aquifer is mainly confined to a narrow belt along the inland valley systems of this undulating mantled plain landscape, and despite its low yield and transmisivity, it has provided the basic minimum water needs for village settlements. The average thickness of the regolith is not more than 10 m in this region. Recent developments in agro-well farming in the north central provinces of the country are wholly dependant on this shallow groundwater (Karunaratne and Pathmarajah 2002).

The laterite or cabook, aquifers which occur in southwestern Sri Lanka have considerable water holding capacity depending on the depth of the Cabook Formation and is unconfined in nature. Due to the highly dissected nature of the macro landscape in this region, the aquifer is highly fragmented into a number of discreet, low mounds, within the residual landscape which is separated from each other by intervening valley floors. This aquifer has been highly exploited due to rapid expansion of industrial estates, urban housing schemes and bottled water projects especially in the area of outer Colombo and adjacent districts.

2.7 *Maldives*

Groundwater in Maldives is found in freshwater lenses underlying the atolls and floating on top of the saline water. Heavy abstraction of this as the main source of drinking water has depleted the freshwater lenses, especially in the capital city of Male, causing salt water intrusion. Groundwater is recharged by rainfall but becomes contaminated while percolating through the soil, which is generally polluted with organic and human wastes. A rough estimate of the groundwater resources, based on an assumed 0.1 m/year recharge throughout the country (300 km^2), is 0.03 km^3/year, which would be the only renewable resource of Maldives, though hardly exploitable because of seawater intrusion and pollution (http://www.eoearth.org/view/article/156968/, accessed on February 2017).

2.8 *Bhutan*

Very little hydrogeological work has been done in Bhutan. A reconnaissance hydrogeological work was carried out by Sikdar (2014) in Samdrup Jongkhar town. The town shares border with Indian state of Assam in the South, Arunachal Pradesh in the East, Pemagatshel Dzongkhag of Bhutan in the west and Trashigang Dzongkhag of Bhutan in the north. The area consists of low to moderate altitude denudation structural hills and is characterised by high run off, low infiltration to the groundwater body and development of springs. Groundwater occurs under unconfined condition within the weathered residuum of the underlying semi-consolidated formation consisting of claystone/siltstone/sandstone. Groundwater also occurs under semi-confined conditions within the fractures, joints and bedding planes of the semi-consolidated formation. Geophysical investigation up to a depth of 60 m by PRCS and ATWMC in 2010 reveals that in the northern part of the town in the Pinchina area the aquifer comprising fine to coarse sand mixed with gravels, pebbles and boulders (weathered/fractured sandstone?) extends up to a depth of 30 to 40 m followed by a low permeability bed made of clay (claystone or siltstone?). In the industrial area of the town the aquifer extends up to a depth of 60 m. In the southern part of the town the aquifer is thinner, extending up to a depth of 20 m followed by a low permeability bed. At Dredulthang the aquifer has not been recorded and the low permeability bed occurs at a depth of 3.5 m only. Central Ground Water Board of India has reported the existence of two to three promising aquifer zones down to the depth of maximum 200 m below ground level and the aquifer shows various degree of lateral and vertical variation in the adjoining area of Assam. It is expected that similar hydrogeological condition prevails in Samdrup Jongkhar town. The master slope of the land surface is towards south and hence the regional groundwater flow is also towards south.

3 Problems of Groundwater

South Asia is home to about a quarter of the global population, but has less than 5% of the world's annual renewable water resources. Low groundwater recharge (Fig. 1.1), low per capita water availability, coupled with a very high relative level of water use (dominated by irrigation), makes South Asia one of the most water scarce regions of the world, and a region where scarcity impacts on economic development (Fig. 1.2).

Since the 1970s, groundwater extraction has increased greatly in South Asia especially for food security. For example, in India, groundwater irrigated areas witnessed a spectacular increase from around 11.9 million hectare in 1970–1971 to 33.1 million hectare in 1998–1999, an increase of over 178%. The number of groundwater extraction mechanism rose from less than one million in 1960 to almost 26–28 million in 2002. In Pakistan Punjab, the number of mechanized wells and tube

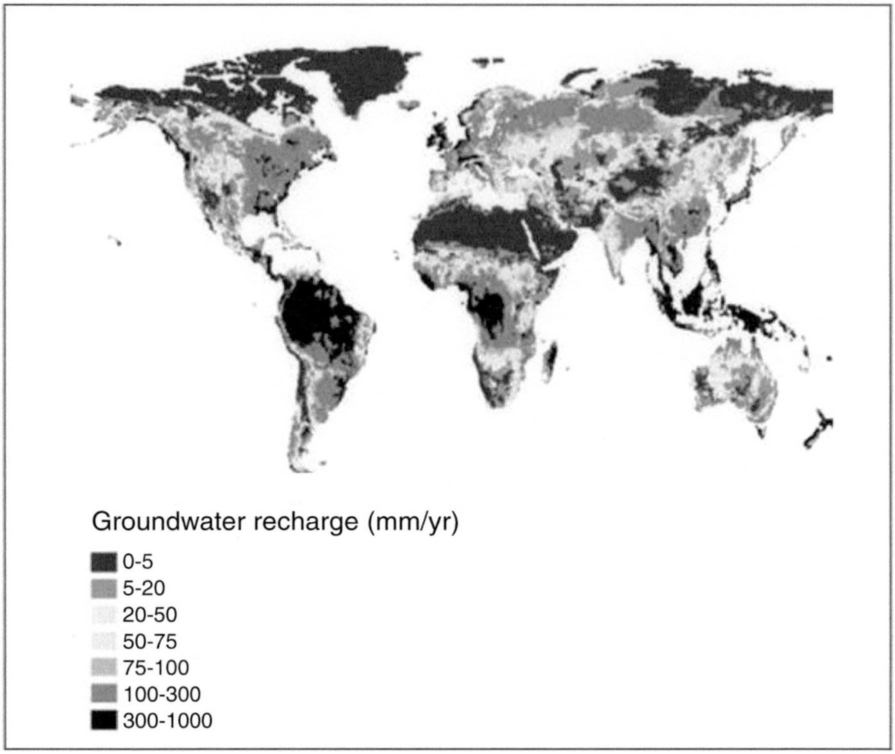

Fig. 1.1 Long-term average groundwater recharge. (*Source:* Döll et al. 2002)

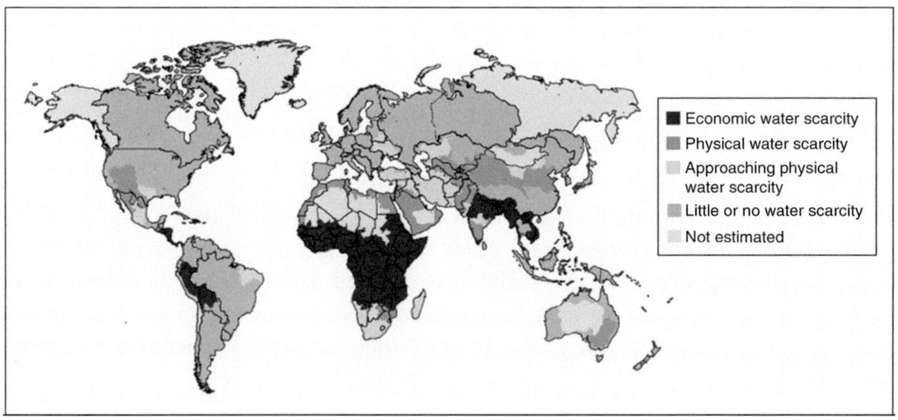

Fig. 1.2 Global water stress. (*Source:* UNWWDR 2012)

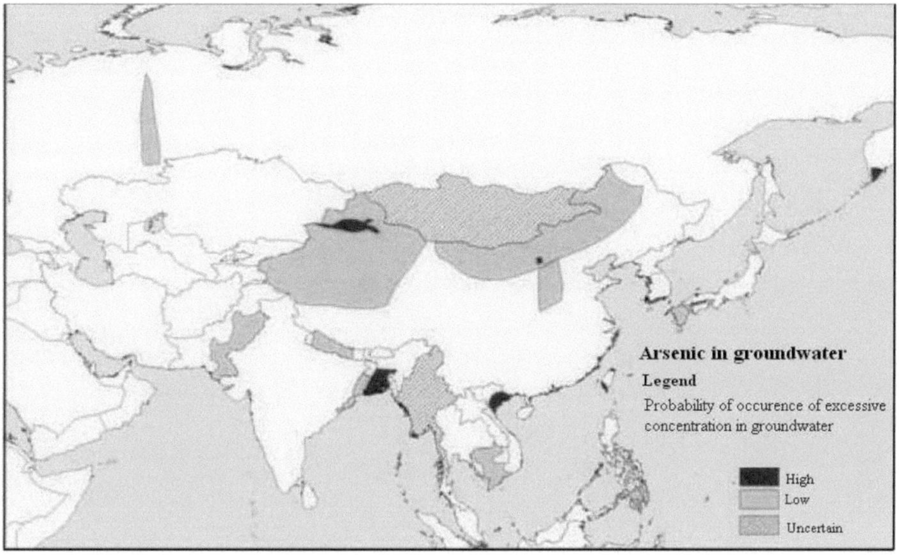

Fig. 1.3 Distribution of arsenic in groundwater in Asia. (*Source:* Brunt et al. 2004a)

wells increased from barely a few thousand in 1960 to 500 thousands in 2000. Bangladesh saw an increase in the number of tube wells, from 93,000 in 1982–1983 to almost 800,000 in 1999–2000. The beneficial impacts of groundwater use are increased productivity, food security, increased recharge, decreased flood intensity, job creation, livelihood diversification, poverty alleviation and general economic and social improvement. But in the long run, the impact of groundwater extraction might be negative especially in over-exploitation situation, such as permanent lowering of the water table, deterioration of water quality, saline intrusion in coastal areas, etc.

3.1 Groundwater Quality

Groundwater quality in South Asia has progressively deteriorated due to increasing withdrawals for various uses, increased use of agrochemicals, discharge of untreated domestic sewage and industrial effluents. In parts of India, Bangladesh, Nepal and Pakistan, high concentration of arsenic in groundwater is a menacing problem and estimated 500 million people are at risk of being exposed to arsenic poisoning through drinking water (Fig. 1.3). Long-term exposure to low levels of arsenic in food and water produces adverse effects on human health that are often described by the term arsenicosis. Early symptoms are non-specific effects such as muscular weakness, lassitude and mild psychological effects. These are followed by characteristic skin ailments such as changes in skin pigmentation and progressively painful

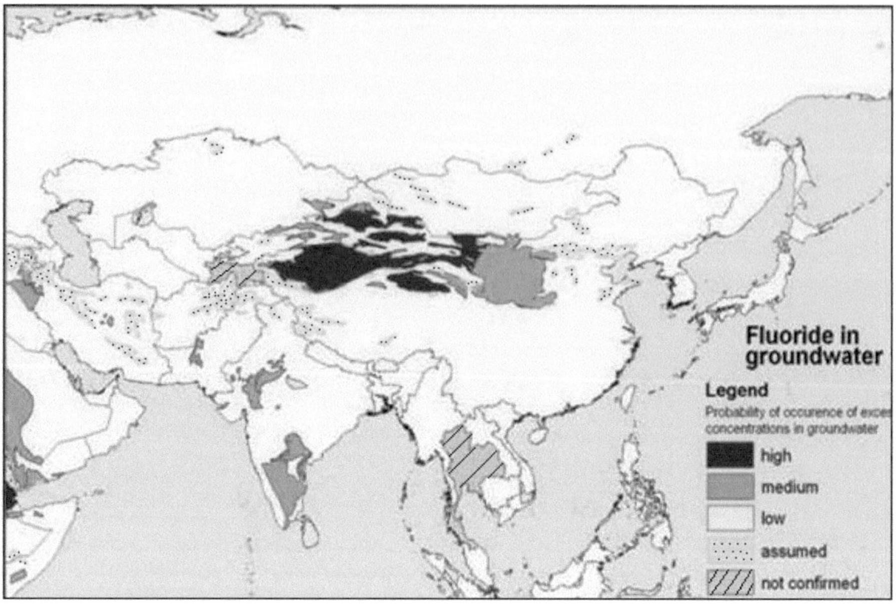

Fig. 1.4 Distribution of fluoride in groundwater in Asia. (*Source:* Brunt et al. 2004b)

skin lesions, known as keratosis. At the same time, arsenic causes a wide range of other effects on health, including diseases of the liver and kidney, cardio-vascular and peripheral vascular diseases, neurological effects, diabetes and chronic and acute lung disease. Continued exposure to arsenic can lead to gangrene, cancers of the skin, lung, liver, kidney and bladder, and thereby to death. High fluoride in groundwater is another problem in India and Pakistan (Fig. 1.4). Consumption of water having excess fluoride over a prolonged period leads to a chronic ailment known as fluorosis. Dental fluorosis, also called "mottled enamel", occurs when the fluoride level in drinking water is marginally above 1.0 mg/L. Typical manifestations of dental fluorosis are loss of shining and development of horizontal yellow streaks on teeth. Since this is caused by high fluoride in or adjacent to developing enamel, dental fluorosis develops in children born and brought up in endemic areas of fluorosis. Skeletal fluorosis affects both adults and children and is generally manifested after consumption of water with fluoride levels exceeding 3 mg/L. Typical symptoms of skeletal fluorosis are pain in the joints and backbone. In severe cases this can result in crippling the patient. Recent studies have shown that excess intake of fluoride can also have certain non-skeletal health impacts such as gastro-intestinal problems, allergies, anemia and urinary tract problems. Nutritional deficiencies can enhance the undesirable effects of arsenic and fluoride. Anthropogenic contamination in the form of heavy metals such as Cr, Pb, Mn and bacteriological parameters has already made the groundwater unacceptable at many places.

3.2 Over-Exploitation of Groundwater

Groundwater over-exploitation has occurred in most countries of South Asia which has resulted in fall in groundwater levels, reduction of well yields, sea water intrusion in coastal aquifers, land subsidence and transport of polluted water into the aquifer. Generally, the fall of groundwater levels results in increased cost of groundwater owing to the expenditure involved in deepening the wells and pumping up water from the correspondingly increased depths. In some cases, over-exploitation can lower the water table to such depths that the existing wells have to be abandoned. Countries which are facing problems related to excessive withdrawal of groundwater in certain locations include India, Bangladesh, Pakistan, Maldives and Sri Lanka. For example, in Quetta, the state capital of Baluchistan over-exploitation due to large scale development through private tube wells of the confined aquifer has caused such an alarming situation that in the foreseeable future even the supply from deep groundwater may dry up. In Kolkata, India (Sikdar et al. 2001) and Dhaka, Bangladesh (Hoque et al. 2007) because of over-abstraction of groundwater from deep aquifers water level fails to recover fully due to monsoonal recharge resulting in a long-term water level decline in the post-monsoon period and change in groundwater flow pattern.

3.3 Seawater Intrusion

Seawater intrusion into coastal fresh water aquifers is another serious groundwater problem. Since a large portion of South Asia's population is located along the coasts there are many problems of this kind in this region. When groundwater is pumped from aquifers that are in hydraulic connection with the sea, the gradients that are set up may induce a flow of salt water from the sea towards the well. The migration of salt water into freshwater aquifers under the influence of groundwater development is known as seawater intrusion. There is a tendency to indicate occurrence of any saline or brackish water along the coastal formations to sea water intrusion. Although seawater intrusion is a slow process, in an area where pumping is continuous, it tends to be an irreversible process. As groundwater is extracted from the wells, the salt water slowly moves through the water-bearing formations in the direction of the wells and, unless corrective measures are taken, the salt water will ultimately begin to contaminate the fresh water in the wells. Such contamination manifests itself in a gradual increase in the salt content of the water being pumped. In India, sea water intrusion is observed along the coastal areas of Gujarat and Tamil Nadu.

3.4 Land Subsidence

Land subsidence occurs when large amounts of groundwater have been withdrawn from certain types of rocks, such as fine-grained sediments. The sediment compacts because the water is partly responsible for holding up the ground. Decline of water table or piezometric surface results in vertical compression of the subsurface materials (Bouwer 1977). Along with vertical compression, lateral compression may also take place due to initiation or acceleration of lateral flow of groundwater. This lateral movement also results in subsidence of the land surface. Any flow or overdraft of groundwater in unconsolidated material should produce some movement of the land surface. This movement is generally small, but may become very significant where subsurface materials are thick and/or compressible and the groundwater level declines appreciably (Sikdar et al. 1996). Land subsidence may not be noticeable because it can occur over large areas rather than in a small spot. Over-exploitation of groundwater in India, Pakistan and Bangladesh requires to be controlled so that possibilities of land subsidence, a vulnerable environmental threat, can be avoided.

4 Groundwater Legislation

Groundwater legislation is concerned with the provision for the quantification, planning, allocation and conservation of groundwater resources, including water abstraction and use rights. It is also concerned with providing cooperative interaction between water administrators and water users.

In India, groundwater is treated as a state subject as the Constitution of India does not empower Central Government to directly deal with its management. The Government of India prepared a Model Bill for regulation of groundwater in 1970 and circulated it to all states for implementation which advocated the view that government has the right to regulate the extraction of groundwater which, therefore, should not be regarded as private property like land. The bill was revised in 1992, 1996 and 2005. The main thrust of all the versions of the Model Bill is to constitute a state groundwater authority which would identify the critical areas and would notify them for regulation. Eleven states (Andhra Pradesh, Goa, Tamil Nadu, Kerala, West Bengal, Himachal Pradesh, Bihar, Jammu-Kashmir, Karnataka, Assam and Maharahtra) and three Union Territories (Lakshadweep, Pondicherry and Dadra-Nagar Haveli) have so far enacted legislation for regulation of groundwater in their states. A Central Ground Water Authority has also been constituted as a statutory body under Environment (Protection) Act, 1986 from January 1997 to regulate and control development and management of groundwater resources in the country. The Authority has been given wide powers including power to impose penalty on any person, company, Government Department etc. The Authority has notified 162 places/blocks/mandals/talukas for control and regulation of groundwater.

In Bangladesh the first effort towards management of groundwater was in the form of 'Groundwater Management Ordinance 1985'. The ordinance was primarily meant for management of groundwater used for irrigation purpose. In 2013 Water Act 2013 has been enacted and is designed for integrated development, management, extraction, distribution, usage, protection and conservation of water resources in Bangladesh. As per Water Act 2013 groundwater in Bangladesh belong to the government and a permit or a license is required for any large scale withdrawal of water by individuals and organizations beyond domestic use.

In Pakistan, Balochistan is the only provincial government to issue legislation to control groundwater mining. In 1978 the Groundwater Rights Administration Ordinance was announced. The objective of the Ordinance was 'to regulate the use of groundwater and to administer the rights of the various persons therein.' The Ordinance established the procedures and framework within the district civil administration to issue permits for the development of new karezes, dug wells and tubewells.

In Maldives, as per the Maldives Tourism Act (Law No. 2/99) groundwater cannot be extracted for the purpose of construction in an island or land leased for the development of tourism. In Sri Lanka, Nepal, Afghanisthan groundwater has practically remained an unregulated resource.

5 Challenges for Groundwater Management

The present system of groundwater management is based on political and administrative boundaries. This approach may be faulty as flow of water ignores political boundaries. Therefore, the first challenge for effective groundwater management is to advocate the importance of managing water at catchment or river basin scales. Database on groundwater systems are either poor or haphazardly maintained. Hence, the second challenge is to improve and share basic data and generate and transfer scientific knowledge. In many parts of the region the aquifer system is not well understood. Therefore, the third challenge is to map the aquifer at river basin scale which should also include the quality aspect. A nationwide programme of regular water-level and water-quality monitoring of the stressed aquifer systems would both characterize the state of water quality in the aquifer and serve as an early-warning system for the impending arrival of contaminants in water supply wells. Therefore, developing a framework of monitoring of the groundwater level and quality is the fourth challenge for effective groundwater management. The fifth challenge is to evolve an effective legal and institutional framework for groundwater governance which would include well-defined groundwater use rights, water pricing and energy pricing. The sixth and the final challenge is to train a group of hydrogeologists into 'social hydrogeologists' or 'ecological hydrogeologists' who will have the capacities to comprehend a sustainable system that combines not only the technical, but also financial, social and environmental aspects and their impacts.

Therefore, challenges for groundwater management in South Asia are huge, and perhaps, more importantly, that there is an urgent need for all professionals working with water-related management to realize their personal responsibilities in creating an accountable and conscious stewardship of water for present and future generations.

6 Conclusion

A review of hydrogeological situation in South Asia reveals several alarming trends of over-exploitation of groundwater causing long-term decline of groundwater level, ingress of saline water and deterioration of groundwater quality. This calls for a new institutional framework for groundwater management. There should be a clear shift from focusing on the unsustainable exploitation of groundwater towards an approach centered on providing incentives to different stakeholders for better and more equitable management of groundwater resource. Providing incentives may not be enough. An empowering environment and a fine-tuned combination of top-down and bottom-up approaches are also required. This new groundwater management framework will focus not only on groundwater use rights and groundwater pricing but also on the users themselves including low income users and other vulnerable groups, and to develop a group of 'social hydrogeologists' or 'ecological hydrogeologists' who would have the skills to better understand not only the technical, but also financial, social and environmental aspects and their impacts. There is dire need to evolve workable methods and approaches to synchronize the demand and supply gap. In order to improve water supply in urban areas, the installation of water meters need to be encouraged. Building a near social framework including community participation at all levels of water system is necessary. The community participation in water pumping policies, incentives of efficient use, affordability by low income users and other vulnerable groups, water awareness, especially among women and children are prime factors for success of any domestic water project.

References

Akbari, M.A., Tahir, M., Litke, D.W. and Chornack, M.P. (2007). Ground-Water Levels in the Kabul Basin, Afghanistan, 2004–2007. U.S. Geological Survey Open-File Report 2007–1294, 46 p.

Banks, D. and Soldal, O. (2002). Towards a policy for sustainable use of groundwater by non-governmental organizations in Afghanistan. *Hydrogeol J*, **10:** 377–392. doi: https://doi.org/10.1007/s10040-002-0203-y.

Basnyat, K. (2001). Sustainable manageable models essential. *The Rising Nepal* (Daily), Kathmandu.

Bockh, E.G. (1971). Report on Groundwater Resources of the city of Kabul – Report for Bundesanstalt fur Geowissenschaften und Rohstoffe (unpublished), BGR File Number: 0021016.

Bouwer, H. (1977). Land subsidence and cracking due to groundwater depletion. *Groundwater*, **15 (5):** 358–364.

Broshears, R.E., Akbari, M.A., Chornack, M.P., Mueller, D.K. and Ruddy, B.C. (2005). Inventory of Ground-Water Resources in the Kabul Basin, Afghanistan. U.S. Geological Survey Scientific Investigations Report 2005-5090, 34 p.

Brunt, R., Vasak, L. and Griffioen, J. (2004a). Arsenic in groundwater: Probability of occurrence of excessive concentration on global scale. International Groundwater Resources Assessment Centre (IGRAC), Report nr. SP 2004-1.

Brunt, R., Vasak, L. and Griffioen, J. (2004b). Fluoride in groundwater: Probability of occurrence of excessive concentration on global scale International Groundwater Resources Assessment Centre (IGRAC), Report nr. SP 2004-2.

Clark, P.V. and Lea, P.D. (1992). The last inter-glacial transition in North America. The Geological Society of America, Special Paper 270.

Davies, J., Herbert, R., Nuruzzaman, M.D., Shedlock, S.L., Marks, R.J. and Barker, J. (1988). The pilot study into optium well design: IDA 4000 Deep Tubewell II Project. Volume 1 Fieldwork – Results. Main volume with six data volumes. British Geological Survey Technical Report WD/88/21.

Davies, J. and Exley, C. (1992). Short Term BGS Pilot Project to Assess the "Hydrochemical character of the main aquifer units of Central and North-eastern Bangladesh and Possible Toxicity of Groundwater to Fish and Humans". Final Report. British Geological Survey Technical Report WD/92/43R.

Doll, P., Lehner, B. and Kaspar, F. (2002). Global modeling of groundwater recharge. *In:* Schmitz, G.H. (ed.). Proceedings of Third International Conference on Water Resources and the Environment Research, (I). Technical University of Dresden, Germany, pp 27–31.

Gellasch, C.A. (2014). Hydrogeology of Afghanistan and its impact on military operations. *In:* The Geological Society of America Reviews in Engineering Geology XXII, pp 69– 81.

Himmelsbach, T., Tunnermeier, T. and Houben, G. (2005). Hydrogeology of the Kabul Basin Part I: Geology, Aquifer characteristics, climate and hydrography. Federal Institute for Geosciences and Natural Resources (BGR), Hannover, Germany, 46 p.

Hoque, M.A., Hoque, M. and Ahmed, K.M. (2007). Declining groundwater level and aquifer dewatering in Dhaka metropolitan area, Bangladesh: Causes and quantification. *Hydrogeol J*, **15:** 1523–1534.

Houben, G., Niard, N., Tunnermeier, T. and Himmelsbach, T. (2009a). Hydrogeology of the Kabul Basin (Afghanistan), Part I: Aquifers and Hydrology. *Hydrogeol J*, **17:** 665–677. doi: https://doi.org/10.1007/s10040-008-0377-z.

Houben, G., Tunnermeier, T., Eqrar, N. and Himmelsbach, T. (2009b). Hydrogeology of the Kabul Basin (Afghanistan), Part II: Groundwater Geochemistry. *Hydrogeol J*, **17:** 935–948, doi: https://doi.org/10.1007/s10040-008-0375-1.

Jacobson, G. (1996). Urban groundwater database. AGSO Report, Australia, http://www.clw.csiro . au/UGD/DB/Kathmandu/Kathmandu.html.

Jones, P.H. (1985). Geology and Groundwater resources of Bangladesh. Hydrogeology Inc., Baton Rouge, Louisiana, USA.

Karunaratne, A.D.M. and Pathmarajah, S. (2002). Groundwater development through introduction of agro-wells and micro-irrigation in Sri Lanka. Symposium on use of groundwater for agriculture in Sri Lanka. AESL/PGIA.

Khadka, M.S. (1993). The groundwater quality situation in alluvial aquifers of the Kathmandu Valley, Nepal. AGSO. *J Austra Geol & Geophy*, **14:** 207–211.

Lashkaripour, G.R. and Hussaini, S.A. (2008). Water resource management in Kabul river basin, eastern Afghanistan. *The Environmentalist*, **28:** 253–260. doi: https://doi.org/10.1007/s10669-007-9136-2.

Mack, T.J., Akbari, M.A., Ashnoor, M.H., Chornack, M.P., Coplen, T.B., Emerson, D.G., Hubbard, B.E., Litke, D.W., Mitchell, R.L., Plummer, L.N., Rezai, M.T., Senay, G.B., Verdin, J.P. and Verstraeten, I.M. (2010). Conceptual Model of Water Resources in the Kabul Basin, Afghanistan: U.S. Geological Survey Scientific Investigations Report 2009-5262 (p. 240).

MMI (1992). Deep Tubewell II Project, Final Report. Mott MacDonald International in association with Hunting Technical Services. Report prepared for the Bangladesh Agricultural Development Corporation, Dhaka, Bangladesh.

MMP (1977). BADC/IDA Tubewell Project, Consultants Report, Volume III: Groundwater. Sir M MacDonald and Partners for United Nations and Bangladesh Water Development Corporation.

MMP (1983). Water balance studies, Bangladesh: Final report – Report II: Groundwater. Sir M MacDonald and Partners for United Nations and Bangladesh Water Development Board.

Olive, J.E. (2005). The Encyclopedia of World Climatology, pp. 115–117, Springer.

Panabokke, C.R. 2003. Sustainable use of groundwater in the NCP, NWP and EP. Trop. Agri. Ruhunu University: 8–13.

Panabokke, C.R., Perera, A.P.G.R.L (2005). Groundwater Resources of Sri Lanka. Water Resources Board, Colombo, 28 p.

Pandey, V.P. and Kazma, F. (2011). Hydrogeologic characteristics of groundwater aquifers in Kathmundu Valley Nepal. *Environ Earth Sci*, **62:** 1723–1732.

Sengupta, S., Kang, A. and Jacob, N. (2012). Water Wealth: A Briefing Paper on the State of Groundwater Management in Bangladesh, http://www.cseindia.org/userfiles/groundwater_management_bangladesh.pdf (Accessed online on February 2016).

Sikdar, P.K. (2014). A Report on Reconnaissance Hydrogeological Appraisal of Samdrup Jongkhar town, Bhutan. Unpublished report submitted to STUP Pvt. Ltd. p. 25.

Sikdar, P.K., Biswas, A.B. and Saha, A.K. (1996). A study on the possible land subsidence in Calcutta and Howrah cities due to groundwater overdraft. *Ind J Geol*, **68(3):** 193–200.

Sikdar, P.K., Sarkar, S.S. and Palchoudhury, S. (2001). Geochemical evolution of groundwater in Quaternary aquifers of Calcutta and Howrah, India. *J Asian Erth Sc*, **19(5):** 579–594.

Sirimanne, C.H.L. (1952). Geology for water supply. Presidential address. Sec. D.CAAS: 87-118.

Survey of Pakistan/Directorate General Hydrogeology/EUAD (1989). Hydrogeological Map of Pakistan, 1:2.000.000.

Tyson, P.D. (2002). Global-Regional Linkages in the Earth System. Springer.

Uhl, V.U. (2003). Afghanistan: An Overview of Groundwater Resources and Challenges. A Report of Uhl, Baron, Rana Associates, Inc. Washington Crossing, PA, USA, http://www.vuawater.com/Case-Study-Files/Afghanistan/Afghanistan_Overview_of_GW_Resources_Study-2003.pdf (Accessed online on February 2017).

UNWWDR (2012). Managing Water under Uncertainty and Risk. Report of United Nations World Water Development Report 1(4), 887 p.

UNDP (1982). The Hydrogeological Condition of Bangladesh. Groundwater Survey Technical Report DP/UN/BGD-74-0071, United Nation Development Programme, New York.

United Nations (1986). Ground Water in Continental Asia (Central, Eastern, Southern, South-Eastern Asia): New York, United Nations, p. 391.

Upadhyay, S.K. (1993). Use of groundwater resources to alleviate poverty in Nepal: Policy issues. *In:* Groundwater Irrigation and the Rural Poor: Options for Development in the Gangetic Basin. Kahnert, F. and Levine, G. (eds). World Bank, Washington DC, USA.

Welsh, J.L. (1966). Report on Dhaka Groundwater Supply, East Pakistan, Parsons Corporation.

Chapter 2
Groundwater Resources of India: Potential, Challenges and Management

Dipankar Saha and Ranjan K. Ray

1 Introduction

Recent research has revealed that two-thirds of the world's population currently lives in areas that experience water scarcity for at least 1 month a year. Noteworthy is that about 50% of the people facing this level of water scarcity live in China and India (UN-Water 2017). Though India receives a copious annual precipitation of around 4000×10^9 m^3, only around one fourth (1123×10^9 m^3) of it is utilizable. A country is considered to be under regular water stress when the renewable water supplies drop below 1700 m^3 per capita per year and it faces chronic water scarcity when the water supplies drop below 1000 m^3 per capita per year (Falkenmark and Widstrand 1992). The per capita average water availability in India in the year 2001 was 1816 m^3 which is likely to reduce to 1140 m^3 in 2050 (MoWR 2015). In the recent past, major share of the increased demand for water has been met from aquifers and groundwater has steadily emerged as the backbone of India's agriculture and drinking water security (Vijay Shankar et al. 2011). Today, contribution of groundwater is ~62% in irrigation, ~85% in rural water supply and ~45% in urban water consumption. High dependence on groundwater resources has led to stressed conditions in various parts of the country. This calls for holistic understanding of the aquifer systems and management of this precious natural resource in a sustainable manner.

D. Saha (✉) · R. K. Ray
Central Ground Water Board, Faridabad, India
e-mail: dsaha002@yahoo.com

P. K. Sikdar (ed.), *Groundwater Development and Management*,
https://doi.org/10.1007/978-3-319-75115-3_2

2 Hydrogeological Setup

Hydrogeological setup is the primary control of occurrence and movement of groundwater. The groundwater resources map of Asia brought out by the World-wide Hydrogeological Mapping and Assessment Programme (WHYMAP; UNESCO 2008) shows that India holds more potential aquifer systems in comparison to most of her neighbours (Fig. 2.1). In this map, except the Indo-Ganga-Bramhaputra plains and tracts along large rivers, major part of India has been designated as areas with 'complex hydrogeological structure' (UNESCO 2008). Taylor (1959), based on geology and terrains, delineated eight broad groundwater provinces. This classification of groundwater provinces of India has been in wide use in literature (Karanth 2003). The National Atlas on Aquifer Systems of India (CGWB 2012a) has adopted a two-tier classification of aquifer systems of India. The aquifers are classified into 14 principal aquifer systems, which in turn have been sub-divided into 42 major aquifer systems. Kulkarni et al. (2015) have divided the country into six hydrogeological settings (aquifer typology).

The geological units in India can be grouped into two broad categories based on groundwater storage and transmissive properties: (i) soft rock and (ii) hard rock. The soft rocks characterized by the predominance of primary porosity can further be divided into unconsolidated sediments and semi-consolidated rocks. The hard rock formations, characterized by predominance of secondary porosity like fractures and joints can be broadly grouped into Precambrian sedimentaries, basaltic aquifers,

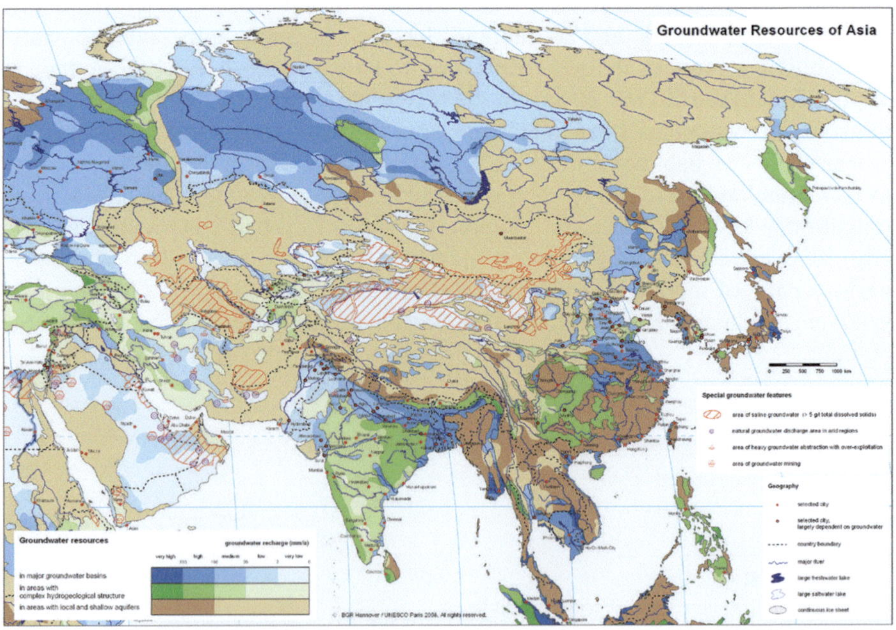

Fig. 2.1 Groundwater resources of Asia. (*Source*: UNESCO 2008)

crystalline aquifers and the carbonate aquifers. Aquifers in the hilly areas (Himalayan Terrain) are complex and discontinuous. Similarly, the aquifers and the groundwater dynamics in the islands are unique and are described separately. Geographical distribution of various hydrogeologic units as described above is shown in Fig. 2.2.

2.1 *Hard Rock Aquifers*

The hard rock aquifers, as a group, cover major part of the geographical area of the country, of which the crystalline aquifers are the most predominant type. Most

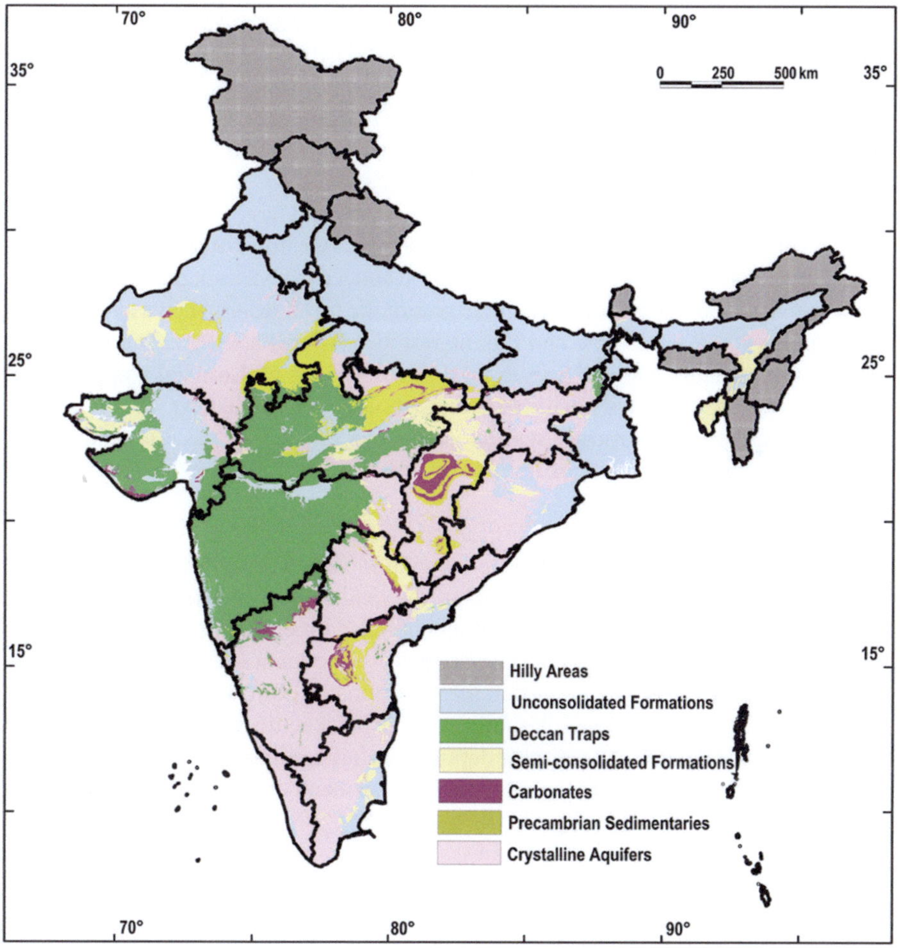

Fig. 2.2 Major hydrogeological units in India. (Adapted from the Aquifer Atlas of India; CGWB, 2012)

common rock types categorized under crystalline aquifers are granites, gneisses, charnockites, khondalites, quartzites, schist and associated phyllite, slate, etc. These rocks possess negligible primary porosity but are rendered porous and permeable due to weathering and due to the presence of fractures and joints. The crystalline aquifers cover major parts of the States of Andhra Pradesh, Chhattisgarh, Karnataka, Kerala, Odisha, Tamilnadu, Telangana and Jharkhand. The upper weathered part and the fractured zone below form the aquifers. The fracture zones are connected along some alignments (lineaments) and when they are well-connected with the top weathered zone form potential aquifers (Saha et al. 2013). Potential aquifers are restricted mostly up to a depth of 150 m. Transmissivity of these aquifers remain mostly within 100 m^2/day, though occasional values up to 600 m^2/day has also been reported. Though of limited potential, these aquifers occur widely and act as the principal source of freshwater in the country.

The Precambrian sedimentaries are encountered mostly in parts of Andhra Pradesh, Chhattisgarh, Madhya Pradesh and Rajasthan. Major rock formations include sandstones and shale. Usually the Precambrian sandstones are highly silicified and the shales are compact (Ray et al. 2017). Both sandstones and shale have very limited aquifer potential with specific yield value of around 0.0038 (Ray et al. 2014).

Basaltic aquifers occupy most part of Maharashtra and large geographical areas in Gujarat and Madhya Pradesh. They comprise multiple flows (traps). Each flow is marked by a potential vescicular zone at the top and a massive rock unit at the bottom (Saha and Agrawal 2006). The weathered part of the top flow and the vescicular zone of the successive flows below and the intertrappeans form aquifers. The weathered zone mostly remains within 15 m though in some parts of Karnataka and Gujarat 40 m thick weathered zones have also been reported (CGWB 2012a). Transmissivity of these aquifers remain mostly within 70 m^2/day. Patches of basaltic aquifers are also reported from other parts of India like Chhattisgarh, Rajmahal Traps in Eastern parts of Jharkhand etc.

Carbonates ranging in age from Precambrian (as in Chhattisgarh State) to Tertiary (as in Rajasthan) make potential aquifers owing to various degrees of karstification. Most prominent carbonate aquifers are in the central part of Chhattisgarh. These carbonate units belong to the Precambrian Chhattisgarh Group of rocks (Mukherjee et al. 2014). Dar et al. (2014) provide a review of the karst (carbonate) aquifers in India. Potential water bearing zones in these carbonate aquifers are mostly restricted to 80 m below ground level. Usually, degree of karstification and consequent groundwater potential decrease with depth. Rainfall Infiltration factor and specific yield of the carbonate aquifers in Chhattisgarh has been estimated to be 4.5% (Ray et al. 2017) and 3.7% (Ray et al. 2014) respectively.

2.2 Soft Rock Aquifers

Soft rock aquifers can be further subdivided into unconsolidated sediments and semi-consolidated sedimentaries. The unconsolidated sediments in turn can be grouped into three broad categories: (i) alluvial deposits in Indo-Ganga-Bramhaputra Plains and along tracts of major rives, (ii) coastal deposits, most prominent along the east-coast, and (iii) aeolian deposits in the northwestern part. The Indo-Ganga-Brahmaputra plains hold one of the most potential soft rock aquifers in the world. The unconfined aquifers occurring at the top sometimes extend down to 125 m. Deeper aquifers are mostly leaky-confined/confined (Saha et al. 2007; Saha et al. 2011; Saha et al. 2013). The unconfined aquifers generally show storage coefficients between 5% and 25% (Saha et al. 2009). Transmissivity values vary widely from 1000 to 5000 m^2/day (CGWB 2012a). Transmissivity values of deeper aquifers may go up to 12,000 m^2/day (Saha et al. 2010). These aquifers may yield as high as 70 litres per second. The potential of alluvial aquifers along the peninsular rivers are rather moderate with yield up to 14 litres per second. But the alluvial deposits (~100 m thick) of Narmada, Tapi, Purna basins may yield up to 28 litres per second. The alluvial sequences in deltas of major rivers on the eastern coast and in Gujarat estuarine tracts have their hydrogeological potential limited by salinity hazards.

The aeolian deposits occurring in West Rajasthan, Gujarat, Haryana, Delhi and Punjab are well sorted and permeable and have moderate to high yield potentials. However, natural recharge is poor because of scanty rainfall in the area and water table is deep.

The semi-consolidated formations mainly comprise shales, sandstones and limestones. The sedimentary deposits belonging to Gondwana and Tertiary formations are included under this category. The sandstones form potential aquifers locally, particularly in Peninsular India, but at places they have only moderate potential. Under favourable situations, these sediments give rise to artesian conditions as in parts of Godavari Valley, Cambay Basin and parts of West Coast, Puducherry, Neyveli in Tamil Nadu and Tertiary belt in Tripura. Potential aquifers particularly those belonging to Gondwanas and Tertiaries have transmissivity values from 100 to 2300 m^2/day (CGWB 2012a).

2.3 Hills and Islands

Hydrogeological units in the Himalayan terrain (Fig. 2.2) are complex and have not yet been explored properly. The aquifers are discontinuous. Information about extent of the aquifers, their hydraulic properties, recharge areas, recharge mechanism, pathways of recharge etc. are scanty. The Himalayas occupy nearly 500,000 km^2 covering major parts of the States of Jammu and Kashmir, Uttarakhand, Himachal

Pradesh, Sikkim and the northeastern States. Valleys within the hilly terrain are the areas with significant groundwater potential. Because of highly undulating terrain and structurally complex nature owing to tectonic disturbances, groundwater often oozes out in the form of springs. These springs form potential sources of freshwater in this terrain.

Islands have unique hydrogeological characteristics, where fresh groundwater floats as a lens over saline water within the aquifers. Fresh groundwater lenses of small islands, in particular, are more vulnerable to external factors than continental coastal aquifers, and require additional attention (White and Falkland 2010; Ketabchi and Ataie-Ashtiani 2015). Two major islands in India are the Andaman and Nicobar Islands and the Lakshadweep Islands. Weathered and fractured ophiolites, calcareous marl and shell-coralline limestone are the major water baring formations in the Andaman and Nicobar Islands (CGWB 2012b). Coral sand and coral limestone are the main water bearing formations in Lakshadweep Islands (Najeeb and Vinaychandran 2011). In addition to the above two groups of islands (Archipelago), groundwater also form the major source of fresh water in the islands in other parts of India such as those in Andhra Pradesh, Assam, Daman and Diu, Goa, Gujarat, Karnataka, Kerala, Maharashtra, Odisha, Puducherry and West Bengal.

3 Groundwater Level Scenario

Central Ground Water Board (CGWB) periodically monitors groundwater levels through a network of 20,000 observation wells spread all over the country. Measurements of water levels are taken four times a year during the months of January, May, August and November. Besides this, the State Ground Water departments monitor approximately 60,000 wells. The monitoring wells include open dug wells as well as purpose-built piezometers. Detailed information regarding water level measurements done by CGWB is available on website (www.cgwb.gov.in) and its geoportal (www.india-wris.gov.in). Discussion on groundwater levels as presented here are based on water levels of the phreatic aquifers collected and archived by the Central Ground Water Board (CGWB 2016).

In general, annual deepest water levels are recorded during the month of May (Pre-monsoon) and the shallowest in August (Monsoon). As per the analysis of water levels for the year 2015, while nearly 30% of the wells show water levels shallower than 5 m below ground level (bgl) during the pre-monsoon period, percentage of such wells during monsoon period is as high as 60% (Fig. 2.3). Only less than 10% of the wells show water levels above 20 m bgl both during premonsoon as well as monsoon period. Spatial variation of water levels over the country during premonsoon period (May, 2015) is shown in Fig. 2.4 and that during monsoon period (August, 2015) is shown in Fig. 2.5.

During monsoon period (August), in major part of the country, water levels range between 5 and 10 m bgl. Very shallow water levels (<2 m bgl) are observed locally,

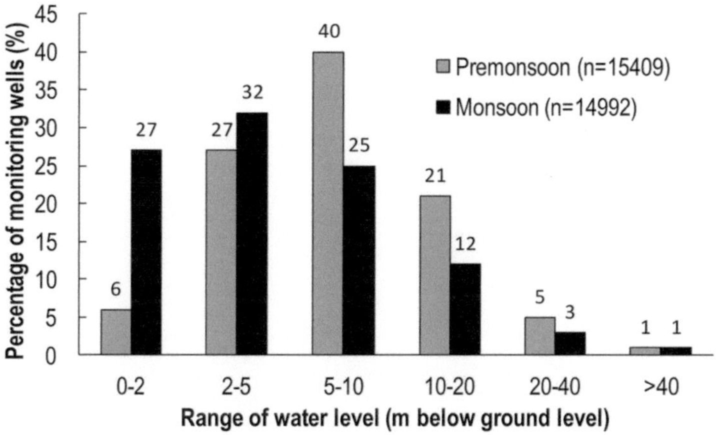

Fig. 2.3 Comparison of instances of water levels in different ranges during pre-monsoon and monsoon periods

in the states of Assam, Andhra Pradesh, Himachal Pradesh, and Tamil Nadu. The depth to water level generally varies from 2 to 5 m bgl in almost whole of Assam, coastal Tamil Nadu and in parts of Maharashtra, Andhra Pradesh, Uttar Pradesh, Bihar, Chhattisgarh, Odisha etc. In the western parts of the country, deeper water level is recorded in the depth range of 20–40 m bgl and more than 40 m bgl. In parts of Delhi and a major part of Rajasthan, water levels of more than 40 m bgl are recorded. Spatial pattern of water level variations over the country during the monsoon period is comparable to that in the premonsoon period albeit with shallower water levels (Figs 2.4 and 2.5).

Water levels show long-term (decadal) falling trends in many parts of the country. Areas, where falling trend is recorded during either pre-monsoon or post-monsoon period are shown in Fig. 2.6. Most of the areas with falling trends are in the States of Rajasthan, Punjab, Haryana, Delhi, Karnataka and Tamilnadu (Fig. 2.6).

4 Groundwater Resource Availability

Available groundwater resources can be defined as the volume of annual groundwater recharge from all sources that is available for use after taking into account the natural discharges that goes out from the aquifers (MoWR 2009). Rainfall is the major source of recharge, other sources being return flow from irrigation, seepage from surface water bodies, transboundary flows etc. While there have been many attempts to estimate the recharge at local and regional scales (Rangarajan and Athavale 2000), the country level groundwater assessments are done by CGWB in association with the State groundwater departments (CGWB 2014a; Chatterjee and Purohit 2009).

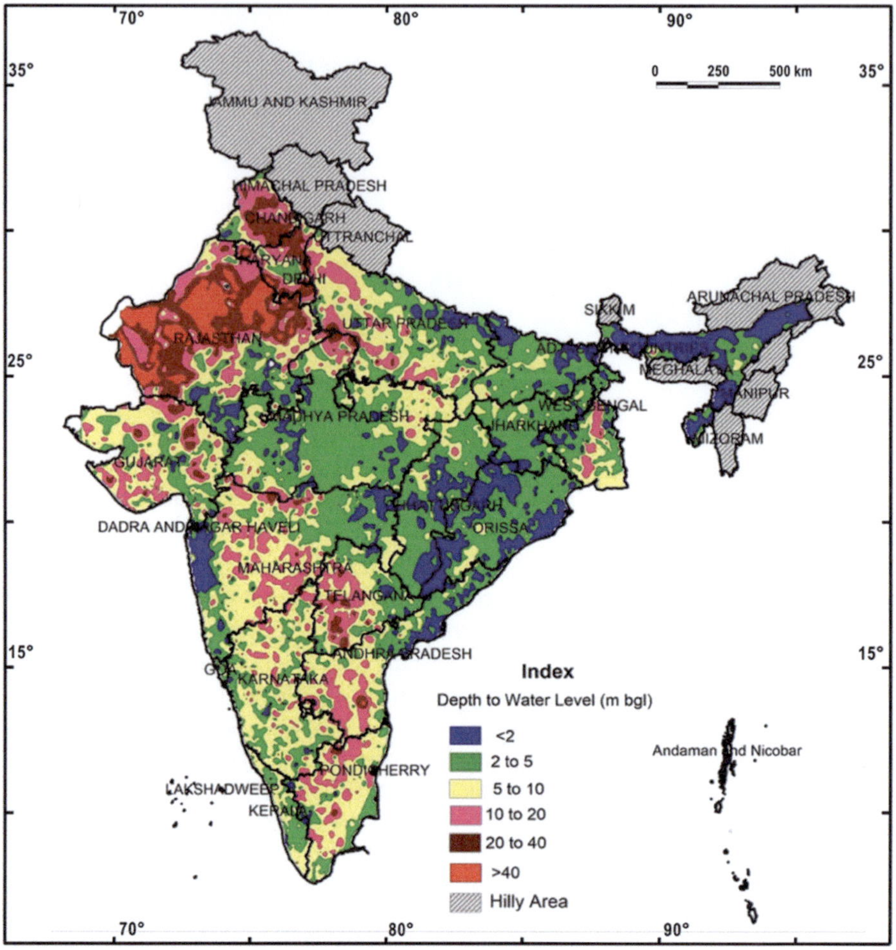

Fig. 2.4 Pre-monsoon (May, 2015) water levels over India

Groundwater resource assessment for the year 2011 (CGWB 2014a) records that total estimated annual availability of groundwater resources of the phreatic aquifers in the country is 398×10^9 m^3 of which recharge from rainfall accounts for nearly 70% and other sources contribute nearly 30%. Annual available resources in depth terms (mm) are shown in Fig. 2.7. Availability is much higher in the Indo-Ganga-Bramhaputra plains (>200 mm) in comparison to the peninsular part (mostly 100–200 mm). Annual availability is very poor (<50 mm) in almost entire Rajasthan, which receives scanty rainfall. Lower availability (<50 mm) as shown in parts of Gujarat is because it is the Rann area, where groundwater is mostly saline. Similarly, in south-eastern part of West Bengal also availability has been shown as <50 mm as in this part fresh water availability in the phreatic aquifer is poor and deeper aquifers are the principal sources of fresh groundwater.

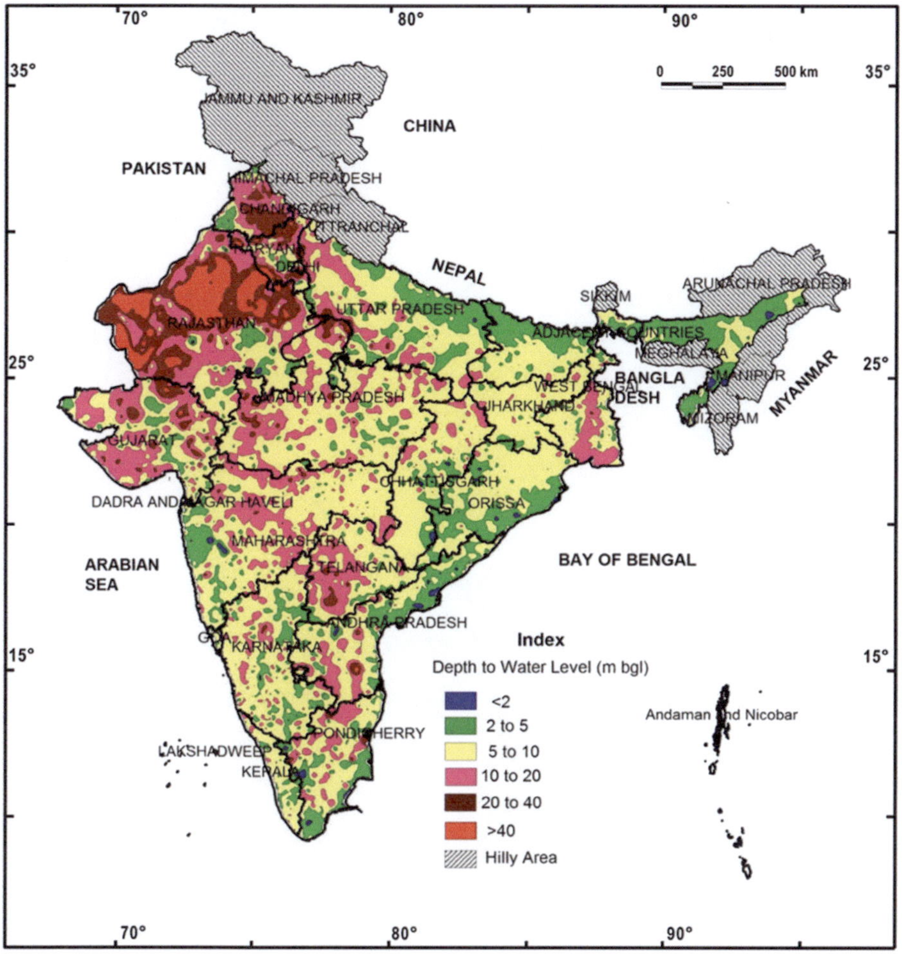

Fig. 2.5 Water levels during monsoon period (August, 2015)

5 Groundwater Extraction, Stage of Development and Categorisation

As per the estimates of the Government of India (CGWB 2014a), annual groundwater extraction (draft) in the country is 245×10^9 m^3, of which about 90% (222×10^9 m^3) is extracted for irrigation purposes. Remaining extraction of less than 10% (23×10^9 m^3) is for domestic and industrial purposes. Groundwater draft per unit area is much higher in the States of Punjab, Haryana, Uttar Pradesh and West Bengal in comparison to the other parts of the country (Fig. 2.8) as groundwater-based irrigation is intense in these areas. Draft per unit area is the lowest in Rajasthan commensurate with poor availability.

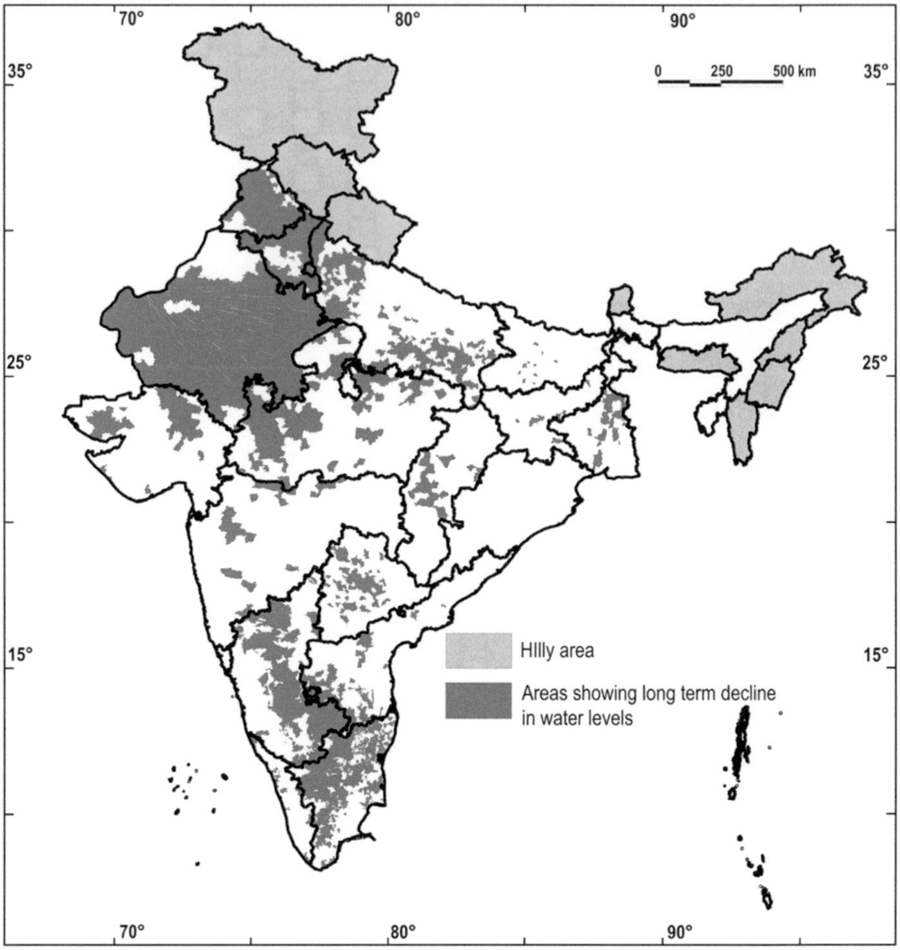

Fig. 2.6 Areas showing long-term decline in water levels during either of the one or both pre-monsoon and post-monsoon periods. (Adapted from CGWB, 2014)

'Stage of groundwater development' is a measure of extent of use of groundwater relative to annual availability (Eq. 2.1; MoWR 2009).

$$\text{Stage of Development} = \frac{\text{Annual Groundwater Draft}}{\text{Annual Available Resources}} \times 100 \qquad (2.1)$$

A stage of development of more than 100% means that annual groundwater extraction in the concerned area is more than annual groundwater availability. Stage of development for the country as a whole is 61% (CGWB 2014a), though there is wide variation over the country (Fig. 2.9). Out of 6607 assessment units (CGWB 2014a), those with more than 100% stage of development are mostly concentrated in the States of Rajasthan, Haryana, Punjab, Tamilnadu, Delhi and

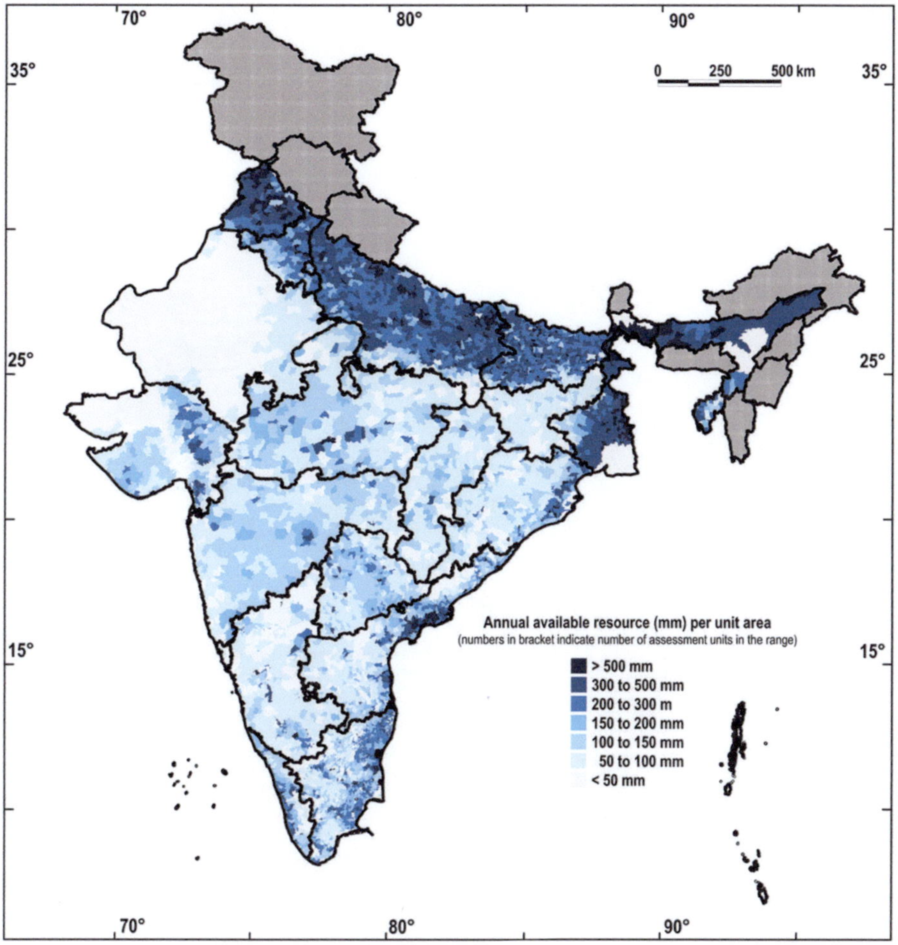

Fig. 2.7 Annual available groundwater resources in India. (Adapted from CGWB, 2014)

Karnataka (Fig. 2.10). Stage of groundwater development is low (<30%) mostly in the eastern part of the country covering parts of Bihar, Jharkhand, Chhattisgarh, West Bengal, Assam and eastern parts of Maharashtra and Madhya Pradesh.

For prioritization of areas for groundwater management, the assessment units (*firka* in Tamilnadu, *mandal* in Andhra Pradesh, *taluka* in Karnataka and blocks in most other states) have been grouped into four categories: 'over-exploited', 'critical', 'semi-critical' and 'safe' (CGWB 2014a). The categorisation is based on stages of groundwater development and long-term water level trends (MoWR 2009). Groundwater extraction in the overexploited units exceeds annual recharge and the water levels show falling trends. In critical units, extraction is more than 90% of the annual availability and in 'semi-critical' units extraction is more than 70% of the annual available resources. While immediate management interventions are recommended

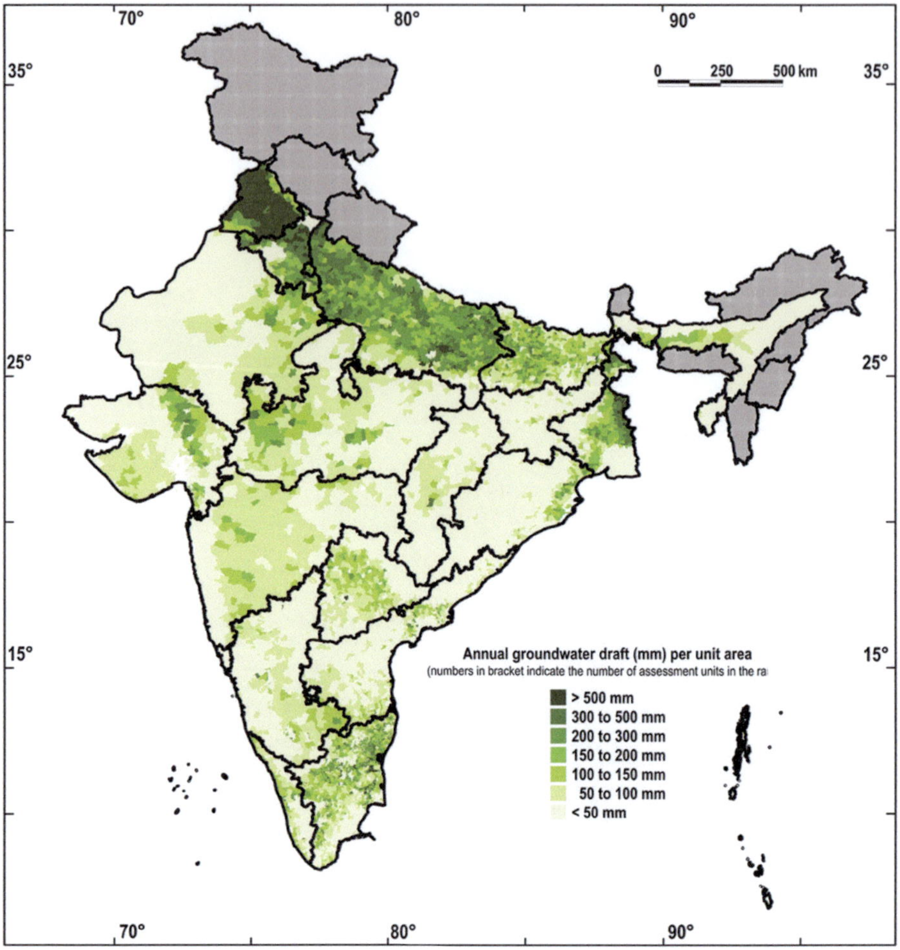

Fig. 2.8 Annual groundwater extraction (draft) in depth terms (mm)

for 'over-exploited' and 'critical' blocks, caution should be exercised in 'semi-critical units'. Out of 6607 numbers of assessed administrative units, 1071 are 'over-exploited', 217 are 'critical', 697 are 'semi-critical' and 4530 units are 'safe'. The remaining 92 assessment units are classified as 'saline' as major part of the phreatic aquifers in these units is saline (Fig. 2.10). Over-exploited blocks are concentrated in the north-western, western and southern Peninsular part of the country. Number of 'over-exploited' and 'critical' administrative units are significantly higher (more than 15% of the total assessed units) in Delhi, Haryana, Karnataka, Punjab, Rajasthan, Tamil Nadu, Uttar Pradesh and also in the Union Territories of Daman & Diu and Puducherry.

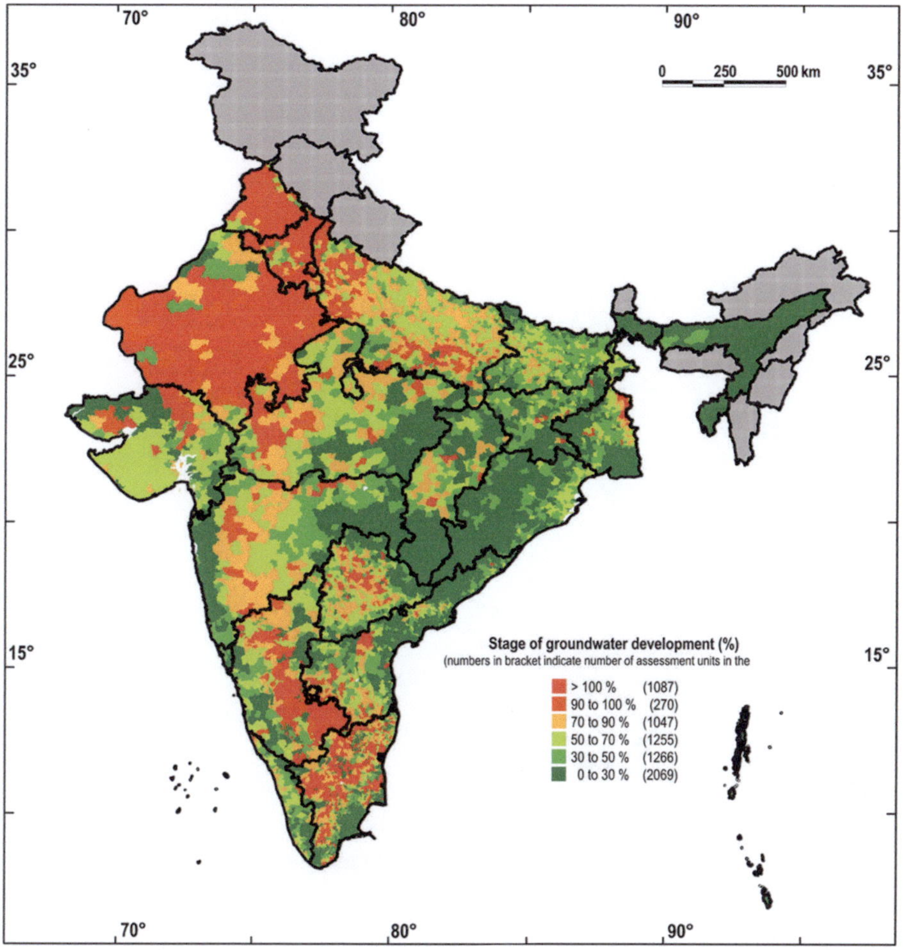

Fig. 2.9 Assessment unit wise stage of groundwater development (Adapted from CGWB, 2014)

6 Groundwater Contamination

Groundwater contamination, both from geogenic and anthropogenic sources, are reported from various parts of the country. Geogenic contamination, particularly that caused by high fluoride and high arsenic is widely reported. Besides this, salinity and high iron, though they are not as acute health hazard as arsenic and fluoride, have been reported from various parts. Anthropogenic contamination on the other hand is originated from industrial effluents, untreated sewage, leachates from landfills etc. The anthropogenic contamination is localized in geographical extent. The Integrated Management Information System (www.indiawater.gov.in) of the Ministry of drinking water and sanitation provide habitation wise status of different contaminants in drinking water (in most of the cases it is groundwater). CGWB has compiled the

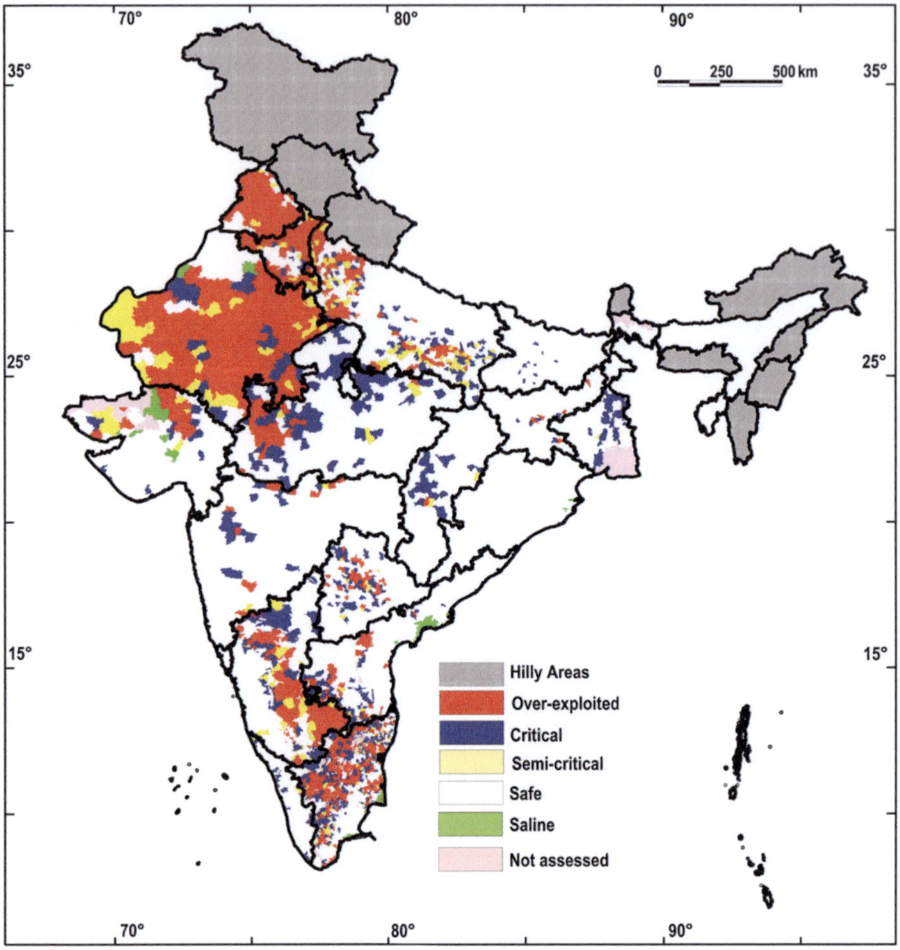

Fig. 2.10 Categorisation of assessment units as per groundwater resource assessments done by CGWB in association with State Ground Water Departments

contamination of groundwater in two documents (CGWB 2010, 2014b), which provide detailed account of groundwater quality in India. A summary of constituent wise maximum permissible limits in drinking water (Bureau of Indian Standard: IS 10500:2012) and States from where they were reported is provided in Table 2.1.

High concentration (>1.5 mg/L) of fluoride in groundwater is widely prevalent across the length and breadth of the country. However, it is more prevalent in the states of Andhra Pradesh, Tamil Nadu, Uttar Pradesh, Gujarat and Rajasthan, where 50–100% of the districts have drinking water sources containing excess level of fluoride. As per an estimate (FRRDF 1999) about 66 million people in India are consuming water with fluoride level beyond the permissible limit.

Table 2.1 Permissible limits in drinking water and State-wise instances of occurrence of fluoride (F), arsenic (As), salinity, iron (Fe) and nitrate (NO_3)

Constituent	Maximum Permissible Limit in Drinking Water (BIS 2012, 2015)	Number of States	States from where instances of contamination have been reported
Fluoride	1.5 mg/L	20	Andhra Pradesh, Assam, Bihar, Chhattisgarh, Delhi, Gujarat, Haryana, Jammu and Kashmir, Jharkhand, Karnataka, Kerala, Madhya Pradesh, Maharashtra, Odisha, Punjab, Rajasthan, Tamilnadu, Telangana, Uttar Pradesh, West Bengal
Arsenic	0.01 mg/L	21	Andhra Pradesh, Assam, Bihar, Chhattisgarh, Delhi, Gujarat, Haryana, Himachal Pradesh, Jammu and Kashmir, Jharkhand, Karnataka, Madhya Pradesh, Manipur, Odisha, Punjab, Rajasthan, Tamilnadu, Telangana, Uttar Pradesh, West Bengal, Daman and Diu
Salinity	2000 (Total Dissolved Solids) mg/L	16	Andhra Pradesh, Chhattisgarh, Delhi, Gujarat, Haryana, Karnataka, Kerala, Madhya Pradesh, Maharashtra, Odisha, Punjab, Rajasthan, Tamilnadu, Telangana, Uttar Pradesh, West Bengal
Iron	0.3 mg/L	26	Andhra Pradesh, Assam, Arunachal Pradesh, Bihar, Chhattisgarh, Goa, Gujarat, Haryana, Jammu and Kashmir, Jharkhand, Karnataka, Kerala, Madhya Pradesh, Maharashtra, Manipur, Meghalaya, Nagaland, Odisha, Punjab, Rajasthan, Tamilnadu, Telangana, Tripura, Uttar Pradesh, West Bengal, Andaman and Nicobar Islands
Nitrate	45 mg/L	21	Andhra Pradesh, Bihar, Chhattisgarh, Delhi, Gujarat, Haryana, Himachal Pradesh, Jammu and Kashmir, Jharkhand, Karnataka, Kerala, Madhya Pradesh, Maharashtra, Odisha, Punjab, Rajasthan, Tamilnadu, Telangana, Uttar Pradesh, Uttarakhand, West Bengal

Source: Loksabha (2017a)

Elevated concentration of arsenic in groundwater is widely recognised as a global threat to human health and poses a challenge to drinking water supply in the affected areas. As per the BIS standards (IS 10500: 2012) maximum permissible limit of arsenic in drinking water was 50 μg/L, which vide an amendment in June 2015 has been changed to 10 μg/L (http://cgwb.gov.in/Documents/WQ-standards.pdf). According to the studies by CGWB and state government organisations, arsenic concentration in groundwater in excess of 10 μg/L has been reported from parts of

21 States in India (Loksabha 2017a). However, the contamination is mostly confined to the Indo-Ganga-Bramhaputra alluvial terrain covering parts of West Bengal, Uttar Pradesh, Bihar, Punjab, Haryana, Assam and Manipur (Saha 2009; Saha et al. 2009; CGWB 2014b; Sahu and Saha 2015). Significant instances of elevated arsenic have also been reported from the mineralised zones in hard rock areas of Chhattisgarh (Mukherjee et al. 2014) and Karnataka. In addition to above, isolated instances of arsenic contamination have also been reported from parts of another 12 States (Loksabha 2017a) as summarised in Table 2.1.

Salinity does not cause serious health effects as compared to arsenic or fluoride. Based on its origin, salinity could be broadly divided into inland salinity caused by rock-water interaction and coastal salinity caused by saline water intrusion. Inland salinity has been reported mainly from western, north-western and southern parts of India. The states largely affected by it are Rajasthan, Gujarat and Haryana. Inland salinity caused by high sulphate in ground water has also been reported from some parts of the country like central part of Chhattisgarh (CGWB 2006). Salinity originated from saline-water intrusion in coastal areas due to over pumping has been reported from the states of Tamil Nadu, West Bengal, Odisha, Gujarat and Andhra Pradesh (CGWB 2014b, c). Besides, percolation from spread sea-water during high tide or storm event also contributes to groundwater salinity in low lying coastal areas.

Excessive iron in groundwater has been reported from almost all states. Iron is so abundant in the natural environment that with favourable hydrochemical condition almost every geological formation can act as a source of iron in groundwater. Iron has been reported from parts of 26 States (Table 2.1).

Among anthropogenic contaminants, nitrate is the most widespread. It is the most ubiquitous chemical contaminant in the world's aquifers and the concentration levels are increasing (CGWB 2014b). Nitrogenous fertilizers and domestic wastes are the most important sources of nitrate in groundwater. Nitrate has been reported both from hard-rock as well as alluvial aquifers particularly from areas under intense cultivation (Saha and Alam 2014). Instances of elevated concentrations of nitrate have been reported from parts of 21 States (Table 2.1). Besides nitrate, zinc (Zn), mercury (Hg), manganese (Mn) have also been reported from some pockets (CGWB 2014b). Pesticides have also been reported from some pockets (Ghose et al. 2009).

7 Groundwater Issues and Challenges

7.1 Unequal Spatial and Temporal Distribution

There is a wide spatial variation in annual availabilities of groundwater resources within the country. Replenishable annual groundwater availability in India is 398×10^9 m^3, which translates into 120 mm for the country as whole, whereas it varies from more than 500 mm in the northern alluvial areas to less than 50 mm in the western part of the country (Fig. 2.2). More than 60% of the available resources are restricted to the northern alluvial terrains that occupy around 20% of the geographical area. In addition to unequal spatial distribution, intra- annual availabilities of the resources also vary within wide ranges. Reduced groundwater availabilities during the lean period (April–June) is apparent in many parts especially the Peninsular part covered by the hard rocks. Recent instance of drinking water scarcity in Latur district of Maharashtra is an example of reduced groundwater availability during lean period.

7.2 Dependence on Groundwater

While low availability in itself is a challenge in many parts of the country, high dependence on groundwater makes the situation worse. The United Nations (UN) led Drinking Water and Sanitation Decade (1981–1990) promoted construction of low cost bore wells especially in the hard rock areas around the world (Foster 2012). Though the drive was for drinking water, the introduction of fast drilling rigs resulted in the increase in construction of wells for irrigation as well. Introduction of energy efficient pumps and State subsidies on electricity used for pumping has also given a boost to groundwater extraction. In the last couple of decades groundwater accounted for 80% addition to net irrigated area in the country (Vijay Shankar et al. 2011). The increased dependence on groundwater has put the sustainability of groundwater resources at risk.

7.3 Over-Exploitation

Annual extraction of groundwater in India is the highest in the world, which even supersedes those of USA and China put together (NGWA 2016). Out of 6607 assessment units, there are nearly 2000 assessment units in which groundwater extraction has already exceeded 70% of the annual availability (Fig. 2.3). Around 16% (1071) of the assessment units have been categorised as 'over-exploited'. Signs of unsustainable extractions are already apparent in terms of declining water levels, drying up of wells, reduced flow in the rivers etc.

7.4 Low Groundwater Development

Ironically, alongside over-exploitation, there are also areas where groundwater extraction is sub-optimal. Most of these areas are in the central, east central and north eastern parts of the country. Most of these areas are also irrigation deprived. The Government of India is contemplating a plan to promote groundwater based irrigation in these irrigation deprived areas where groundwater development is sub-optimal and utilizable resource is available.

7.5 Groundwater Contamination

Groundwater contamination has emerged as a major challenge in ensuring safe drinking water supply as nearly 85% of rural drinking water requirements and 50% of urban drinking water requirements are met from groundwater. Open dug wells used to be the main sources of rural drinking water supply in India in the 1970s. Pathogenic contamination of drinking water was widespread. In an attempt to overcome this, Governments promoted borewells as drinking water sources. This, in turn, gave rise to a new set of issues. Contaminants like arsenic, fluoride, nitrate, iron, salinity etc. have been reported from groundwater in many parts of the country.

8 Groundwater Management

8.1 Supply and Demand Side Interventions

Groundwater management refers to managing the resources in a sustainable manner. The core idea is that the utilisation should remain largely within the dynamic resources on a long-term basis. Off course, the dynamic or replenishable resource tends to increase as the resource is exploited more, which is reflected in higher fluctuation of pre and post monsoon water levels. Further, management objective is also to prevent or mitigate any long-term decline of water levels.

The management interventions can be grouped into two broad categories: supply management and demand management. The supply side intervention aims at enhancing the resource availability by rainwater harvesting, artificial recharge, recycling of wastewater or resorting to alternative sources. The demand management, on the other hand, envisages enhancing water use efficiency by adopting pressure irrigation, crop diversification, plugging unnecessary losses by seepage and evaporation between well head and irrigation field etc. The other important demand management tool is to address the groundwater-energy nexus. High rate of subsidy—even free electricity—is being provided to the farmers in most of the States and Union Territories. This issue is to be addressed with a rational approach like introducing appropriate pricing mechanism, incentivizing power/water conservation efforts, separating the electricity lines for use in pumps and regulating power supply for irrigation etc.

One essential link in groundwater management is mass awareness and public participation. Given the socio-economic condition in India, participatory groundwater management is advocated to be the most effective mode of popularizing efficient use of groundwater and implementing management interventions.

8.2 Groundwater Quality Management

Groundwater contamination is an impediment in ensuring safe drinking water as a majority of the populace in India is dependent on groundwater resources for meeting their drinking water needs. Tapping contamination-free aquifers, wherever available, is by far the most common way of dealing with groundwater quality. In the arsenic affected areas of West Bengal, Uttar Pradesh and Bihar, CGWB has constructed wells by sealing the upper arsenic contaminated aquifers and tapping the arsenic safe deeper aquifers (CGWB 2014b). Arsenic contamination reported from the hard-rock areas of Chhattisgarh has a strong geological control and CGWB has constructed water wells tapping water bearing zones, which are arsenic safe (Mukherjee et al. 2009). Post extraction (ex-situ) treatment of groundwater is another common mode of dealing with contaminated groundwater for drinking water supply, as in-situ treatment techniques are still in development stage (CGWB 2014b). Awareness regarding available safe sources, water-borne diseases etc. can greatly reduce the vulnerability to groundwater contamination.

With increasing instances of anthropogenic pollution, prevention of contamination has emerged as an important issue. Proper identification of waste dumping sites, reducing excessive use of fertilizers and pesticides, treatment of domestic and industrial wastes and protecting recharge areas are some of the ways through which man-made groundwater contamination can be prevented. While there are laws, public awareness and participation is vital to prevent groundwater contamination.

8.3 Groundwater Legislation

Regulating groundwater use is one of the ways to manage demand and it is a huge challenge for India as the rights of groundwater extraction is tied to the land ownership. As per the existing legal framework in India, groundwater does not seem to have a legal existence separate from the land (Koonan 2016). As per Sect. 7 of the India Easement Act, 1882 every owner of land has the right to collect and dispose within his own limits of all water under the land which does not pass in a defined channel. This legal proposition is still in force in India owing to Article 372 of the Constitution of India that keeps pre-Constitution laws in force until they are changed or repealed through subsequent laws.

However, alongside the existing law on ownership, the evolving statutory framework focuses on regulation of groundwater use. As per article 246 (read with seventh schedule List II), State Governments are entrusted with the power to adopt groundwater law under the Constitution. The necessity for regulation of groundwater withdrawal was felt in the 1970s, when signs of groundwater depletion started emerging due to extensive withdrawal. The Union Government circulated a Model

Bill in 1970 to all the States and Union Territories to enable them to enact suitable legislation for adopting control and regulatory measures in groundwater development. The model bill was subsequently revised in 1992 and 1996 (MoWR, RD, & GR 2016).

Though model bills were circulated since 1970, no noteworthy progress could be achieved until 1997. In response to a Public Interest Litigation (PIL) regarding alarming decline in groundwater levels due to overexploitation filed in 1985, the Hon'ble Supreme Court of India ordered Government of India to constitute an authority to regulate groundwater use and development under the Environmental (Protection) Act, 1986. As per the directions of the Hon'ble Supreme Court of India, Central Ground Water Authority (CGWA) was constituted in 1997. With creation of CGWA with clear statutory mandate and over-riding power for controlling and regulating the development and management of groundwater resources and equipped with the National Water Policy (MoWR 2002), the then Ministry of Water Resources (now it is Ministry of Water Resources, River Development and Ganga Rejuvenation) revised the model bill (MoWR 2005) and circulated it to the States for promulgation of State groundwater act and constitution of State Ground Water Authority, which would regulate and control the development and management of groundwater in the State. So far, 15 States/UTs have adopted and implemented the ground water legislation on the lines of Model bill 2005 (Loksabha 2017b). The States where separate groundwater law does not exist, regulation is implemented by the Central Ground Water Authority (CGWA). Subsequently, the model bill was again revised in 2011 (MoWR 2011) and 2016 (MoWR, RD, & GR 2016).

The groundwater laws in effect mainly envisage three regulatory tools (Koonan 2016): (i) Notification of areas in the state where groundwater situation requires regulatory interventions; (ii) Users in the notified areas are required to seek permission from the groundwater authority constituted under the groundwater law; (iii) Drilling agencies are required to register their machinery and are bound by the instructions issued by the groundwater authority. CGWA has notified 162 areas for regulation of groundwater development and management. Under the CGWA guidelines, in notified areas, no permission is accorded to extract groundwater through any energized means for any purpose other than drinking water. However, for non-notified areas, groundwater withdrawal by industries is regulated by means of guidelines/criteria as specified by CGWA (CGWA 2015).

Appropriate legislation is an important tool for prevention and control of groundwater pollution as well. Under the ambit of 'environment', the Environment (Protection) Act of 1986 of Government of India has provisions for protecting water resources from pollution. The Model Bill 2016 (MoWR, RD, & GR 2016), inter alia, envisages to 'reduce and prevent pollution and degradation of groundwater'. The model bill also stipulates that whoever contravenes the provisions of the Act by polluting or contaminating groundwater shall be strictly liable for any groundwater pollution they cause and shall be responsible for the cost of its remediation.

9 Conclusions

India, host to 16% of total population of world, has only 4% of total freshwater resources of the world. There is huge inequality in distribution of water resources within the country. Groundwater, which is the main source of fresh water in the country, is unevenly distributed. 60% of groundwater resources are restricted to the Indo-Ganga-Bramhaputra plains, which account for only 20% of the geographical area. Out of 6607 groundwater assessment units in the country 1071 have been categorized as 'over exploited', where annual extractions exceed annual recharge thereby threatening the sustainability of the sources. While over-exploitation is a major issue in the country, there are also areas where enough scope exists for further groundwater development. Adding to the woes is contamination of sources. Contaminants like arsenic, fluoride, nitrate, iron and salinity have been reported from groundwater in many parts of the country. Further, climate change is feared to contribute towards increased vagaries of monsoon affecting intra annual availabilities. Climate change induced sea level rise may increase saline ingress and increase in salinity of coastal aquifers. Groundwater is very much susceptible to deterioration in terms of quantity and quality due to unplanned and uniformed anthropogenic interventions and the results usually do not show until it is too late.

Artificial recharge, rainwater harvesting, treatment and recycling of wastewater are some of the supply side interventions that can help in augmenting resources. Reducing demand of groundwater through increasing water use efficiency, choosing less water intensive crops and regulating extractions are also effective management tools. Management interventions are a big challenge especially in view of the fact that more than 25 million abstraction structures are in operation in the country and most of them are private owned. Water is a 'state subject' in India and management interventions are implemented primarily by the respective State Governments. The Union Government plays a supportive role. The National Water Policy of India recognizes that water is a scarce natural resource and is fundamental to life, livelihood, food security and sustainable development. It proposes a framework for creation of a system of laws and institutions and for a plan of action with unified national perspective. The most important link in ensuring effective management is awareness and community participation.

References

BIS (2012). Indian standard-drinking water specification (second revision) IS 10500:2012, Bureau of Indian standards, Govt. of India (http://cgwb.gov.in/Documents/WQ-Standards.pdf)

BIS (2015). Amendment No. 1 to IS 10500:2012 drinking water specification (second revision), Bureau of Indian standards, Govt. of India (http://cgwb.gov.in/Documents/WQ-Standards.pdf)

CGWA (2015). Guidelines/Criteria for evaluation of proposals/requests for ground water abstraction (With effect from 16.11.2015). Central Ground Water Authority (CGWA). 27 p. http://www.cgwb.gov.in/CGWA/Documents/Revised%20guidelines_12112015.pdf accessed 15th May 2017.

CGWB (2006). State Report: Hydrogeology of Chhattisgarh. Unpublished Report, CGWB, NCCR, Raipur, India, 184 p.

CGWB (2010). Ground water quality in shallow aquifers of India. Central Ground Water Board, MoWR, RD&GR, Govt of India. http://www.cgwb.gov.in/WQ/gw_quality_in_shallow_aquifers.pdf. Accessed 15th May 2017.

CGWB (2012a). Aquifer systems of India. Central Ground Water Board, MoWR, RD&GR, Govt of India. http://www.cgwb.gov.in/AQM/India.pdf

CGWB (2012b). Water management challenges in islands of India. Central Ground Water Board, MoWR, RD&GR, Govt of India, 163 p.

CGWB (2014a). Dynamic ground water resources of India (As on 31st March 2011) Faridabad July 2014, http://www.cgwb.gov.in/Documents/Dynamic-GW-Resources-2011.pdf

CGWB (2014b). Concept Note On geogenic contamination of ground water in India with a special note on Nitrate. Central Ground Water Board, MoWR, RD&GR, Govt of India. http://www.cgwb.gov.in/WQ/Geogenic%20Final.pdf

CGWB (2014c). Status of ground water quality in coastal aquifers of India. Central Ground Water Board, MoWR, RD&GR, Govt of India http://www.cgwb.gov.in/WQ/Costal%20Report.pdf

CGWB (2016). Ground water year book of India 2014-15, CGWB, Ministry of WR, RD&GR, Govt of India. http://cgwb.gov.in/Ground Water/Ground%20Water%20Year%20Book%202014-15.pdf

Chatterjee, R. and Purohit, R. (2009). Estimation of Replenishable groundwater resources of India and their status of utilization. *Curr. Sci.*, **96(12)**, doi: https://doi.org/10.1007/s12665-009-0064-yCGWB2016.

Dar, F.A., Perrin, J., Ahmed, S. and Narayana, A.C. (2014). Review: Carbonate aquifers and future perspectives of karst hydrogeology in India. *Hydrogeol J*, **22(7):** 1493–1506.

Falkenmark, M. and Widstrand, C. (1992). Population and water resources: A delicate balance. Population Bulletin No. 3. Population Reference Bureau, Washington DC. http://www.ircwash.org/sites/default/files/276-92PO-10997.pdf.

Foster, S. (2012). Hard-rock aquifers in tropical regions: Using science to inform development and management policy. *Hydrogeol J*, **20(4):** 659-672, doi: https://doi.org/10.1007/s10040-011-0828-9.

FRRDF (1999). State-of-art report on the extent of fluoride in drinking water and the resulting endemicity in India. Fluorosis Research and Rural Development Foundation, New Delhi.

Ghose, N.C., Saha, D. and Gupta, A. (2009). Synthetic detergents (surfactants) and organochlorine pesticide signatures in surface water and groundwater of greater Kolkata, India. *Journal of Water Resources and Protection*. doi: https://doi.org/10.4236/jwarp.2009.14036.

Karanth, K.R. (2003). Ground Water Assessment Development and Management. Tata McGraw-Hill, New Delhi.

Ketabchi, H. and Ataie-Ashtiani, B. (2015). Review: Coastal groundwater optimization—advances, challenges, and practical solutions. *J Hydrol*, **23:** 1129–1154. DOI: https://doi.org/10.1007/s10040-015-1254-1

Koonan, S. (2016). Revamping the groundwater legal regime in India: Towards ensuring equity and sustainability. *Socio-Legal Review*, **12(2):** 45–73.

Kulkarni, H., Shah, M. and Vijay Shankar, P.S. (2015). Shaping the contours of groundwater governance in India. *Journal of Hydrology: Regional Studies*, **4(A):** 172–192. https://doi.org/10.1016/j.ejrh.2014.11.004

Loksabha2017a. Written reply to the question no. 3663 regarding contamination of ground water tabled in the Loksabha on 23.03.2017 http://164.100.47.194/Loksabha/Questions/QResult15.aspx?qref=51044&lsno=16

Loksabha2017b. Written reply to the question no. 4712 regarding rainwater harvesting tabled in the Loksabha on 30.03.2017. http://164.100.47.190/loksabhaquestions/annex/11/AU4712.pdf

MoWR (2002). National Water Policy – 2002. Ministry of Water Resources, Govt of India. http://wrmin.nic.in/writereaddata/NationalWaterPolicy/nwp20025617515534.pdf

MoWR (2005). Model bill 2005: Model Bill to regulate and control the development and management of ground water, 2005 http://cgwb.gov.in/NEW/CGWA/Documents/Model-bill.pdf

MoWR (2009). Report of the ground water resource estimation committee. Ministry of Water Resources, Govt. of India. 107 p. http://cgwb.gov.in/documents/gec97.pdf . Cited 18 November 2016

MoWR (2011). Model bill 2011: Model bill for the protection, conservation, management and regulation of groundwater. Draft of 18 October 2011 http://www.planningcommission.gov.in/aboutus/committee/wrkgrp12/wr/mb_wtrgrndsor_181011.pdf

MoWR (2015). A year of Inclusive Development in Water Resources sector. National Water Mission, Ministry of WR, RD&GR, Govt of India. http://www.nationalwatermission.gov.in/sites/default/files/COP%2021%20MoWR%20Booklet.pdf (accessed 17 January 2017).

MoWR, RD & GR (2016). Model bill for the conservation, protection, regulation and management of groundwater, 2016. Draft of 17 May 2016, Ministry of Water Resources, River Development and Ganga Rejuvenation, Govt of India. http://wrmin.nic.in/writereaddata/Model_Bill_Ground water_May_2016.pdf

Mukherjee, A., Ray, R.K., Tewari, D., Ingle, V.K., Sahoo, B.K. and Khan, M.W.Y. (2014). Revisiting the stratigraphy of the Mesoproterozoic Chhattisgarh Supergroup, Bastar craton, India based on subsurface litho information. *J Earth Syst Sci*, **123(3):** 617–632.

Mukherjee, A., Tewari, D., Verma, J.R., Subramanian, S., Ray, R.K. and Devangan, R. (2009). Geogenic arsenic contamination to ground water in parts of Ambagarh Chowki Block, Rajnandgaon District, Chhattisgarh. *Bhujal News*, **24(2):** 40–58.

Najeeb, K. Md. and Vinaychandran, N. (2011). Groundwater Scenario in Lakshadweep Islands. *Journal Geological Society of India*, **78(4)**, doi: https://doi.org/10.1007/s12594-011-0095-3.

NGWA (2016). Facts about global groundwater usage compiled by NGWA, the groundwater association. http://www.ngwa.org/Fundamentals/use/Documents/global-groundwater-use-fact-sheet.pdf Cited 12 June 2016.

Rangarajan, R. and Athavale, R.N. (2000). Annual replenishable groundwater potential of India—An estimate based on injected tritium studies. *J Hydrol*, **234:** 38–53.

Ray, R.K., Mukherjee, A. and Mukherjee, R. (2014). Estimation of specific yields of individual litho-units in a terrain with multiple litho-units: A water balance approach. *J Geol Soc India*, **84 (2):** 221–225, doi: https://doi.org/10.1007/s12594-014-0126-y.

Ray, R.K., Syed, T.H., Saha, D., Sarkar, B.C. and Reddy, D.V. (2017). Recharge mechanism and processes controlling groundwater chemistry in a Precambrian sedimentary terrain: A case study from Central India. *Environ Earth Sci*, **76:** 136, doi: https://doi.org/10.1007/s12665-017-6435-x.

Saha, D. and Agrawal, A.K. (2006). Determination of specific yield using a water balance approach: Case study of Torla Odha watershed in Deccan Trap province, Maharashtra State, India. *Hydrogeol J*, **14:** 625–635, doi: https://doi.org/10.1007/s10040-005-0471-4.

Saha, D., Upadhyay, S., Dhar, Y.R. and Singh, R. (2007). The aquifer system and evaluation of its hydraulic parameters in parts of South Ganga Plain, Bihar, India. *J Geol Soc Ind*, **69:** 1031–1041.

Saha, D., Sinha, U.K. and Dwivedi, S.N. (2011). Characterization of recharge processes in shallow and deeper aquifers using isotopic signatures and geochemical behavior of groundwater in an arsenic-enriched part of the Ganga Plain. *Appl Geochem*, doi: https://doi.org/10.1016/j.apgeochem.2011.01.003.

Saha, D. and Alam, F. (2014). Groundwater vulnerability assessment using DRASTIC and Pesticide DRASTIC models in intense agriculture area of the Gangetic plains, India. *Environ Monit Assess*, doi: https://doi.org/10.1007/s10661-014-4041-x.

Saha, D., Dwivedi, S.N. and Singh, R.K. (2013). Aquifer system response to intensive pumping in urban areas of the Gangetic plains, India: The case study of Patna Urban area. *Environ Erth Sci*, doi: https://doi.org/10.1007/s12665-013-2577-7.

Saha, D. (2009). Arsenic groundwater contamination in parts of Middle Ganga Plain, Bihar.*Current Sci*, **97:** 753–755.

Saha, D., Sreehari, S.M.S., Dwivedi, S.N. and Bhartariya, K.G. (2009). Evaluation of hydrogeochemical processes in arsenic contaminated alluvial aquifers in parts of MidGanga Basin, Bihar, Eastern India. *Environ. Earth Sci.*, doi: https://doi.org/10.1007/s12665-009-0392-y.

Saha, D., Sahoo, S. and Chandra, P.C. (2010). Arsenic safe alternate aquifers and their hydraulic characteristics in contaminated areas of Middle Ganga Plain, Eastern India. *Environ Assess and Monit*, doi: https://doi.org/10.1007/s10661-010-1535-z.

Sahu, S. and Saha, D. (2015). Role of shallow alluvial stratigraphy and Holocene geomorphology on groundwater arsenic contamination in the Middle Ganga Plain, India. *Environ Erth Sci*, **73 (7):** 3523–3536, doi: https://doi.org/10.1007/s12665-014-3637-3.

Taylor, G.C. Jr. (1959). Ground Water provinces of India. *Economic Geology*, **54(4):** 683–697.

UNESCO (2008). Groundwater Resources of Asia. BGR, Hannover/UNESCO, Paris, https://www.whymap.org/whymap/EN/Downloads/Continental_maps/contimaps_node_en.html

UN-Water (2017). Wastewater the untapped resource. The United Nations World Water Development Report 2017. United Nations Educational, Scientific and Cultural Organization, Paris.

Vijay Shankar, P.S., Kulkarni, H. and Krishnan, S. (2011). India's Groundwater Challenge and the Way Forward. *Economic and Political Weekly*, **XLVI(2)**, January 8.

White, I. and Falkland, T. (2010). Management of freshwater lenses on small Pacific islands. *Hydrogeol J*, **18(1):** 227–246.

Chapter 3
Application of Geophysical Techniques in Groundwater Management

Bimalendu B. Bhattacharya

1 Introduction

Geophysics is the application of the principles of physics to the study of the earth. The aim of Geophysics is to deduce the physical properties of the earth and its internal constitution from the physical phenomena associated with it. On the other hand, the object of Applied or Exploration Geophysics is concerned to investigate specific, relatively small scale and shallow features, which are presumed to exist within the earth's crust.

Geophysical techniques used in investigating the shallow features of the earth's crust vary in accordance with the physical properties of which these features are composed, but broadly speaking they fall into four classes: *static, dynamic, relaxation* and *integrated effect* methods. In the *static* methods the distortions of a static physical field are detected and measured accurately to delineate the features producing them. The static fields are geomagnetic, gravity, geothermal, self-potential (SP) and telluric etc. In the *dynamic* methods, on the other hand, signals are sent in the ground, and returning signals are detected and their strengths and times are measured at suitable points, e.g., resistivity, seismic, electromagnetic (EM) methods. There are methods which lie in between the two methods already mentioned. These may be called *relaxation* methods, e.g., induced polarization (IP) and transient electromagnetic (TEM) methods. Their feature is that the dimension of time appears in them. Finally there is what we may call *integrated effect* methods, in which the detected signals are statistical averages over a given area or within a given volume. The radioactive method falls in this category.

B. B. Bhattacharya (✉)
S.N. Bose National Centre for Basic Sciences, Salt Lake, Kolkata, India

Indian School of Mines, Dhanbad, India
e-mail: bimalism@gmail.com

P. K. Sikdar (ed.), *Groundwater Development and Management*,
https://doi.org/10.1007/978-3-319-75115-3_3

This chapter will be confined only to the geophysical techniques used extensively for groundwater exploration, i.e., resistivity imaging and sounding, refraction seismic, ground penetrating radar (GPR), induced polarization and well logging methods.

2 Resistivity Methods

In the resistivity method mostly a direct current (dc) or at times low-frequency alternating current is injected into the ground by means of two electrodes, i.e., metal stakes connected to the terminals of a portable power supply source. The resulting potential distribution on the ground, mapped by means of two non-polarizable electrodes, provides information on the distribution electrical resistivity below the surface.

2.1 Ohm's Law

Ohm established that electric current I in a conducting wire is proportional to the potential difference V across it. The linear relationship is expressed by the equation

$$V = IR \tag{3.1}$$

where R is the resistance of the conductor. The unit of resistance is ohm (Ω). The inverse of resistance is called conductance. The SI unit of conductance is Siemens (S).

For a given material the resistance is proportional to the length L and inversely proportional to the cross-sectional area A of the conductor (Fig. 3.1). The relationship is expressed by the equation

$$R = \rho \frac{L}{A} \tag{3.2}$$

The proportionality constant ρ is the resistivity of the conductor, which expresses its ability to oppose a flow of charge. The inverse of ρ is called the conductivity σ of the material. The unit of resistivity is ohm-metre (Ω-m and also written as Ωm). The SI unit of conductivity is Siemens per metre (S/m).

Substituting Eq. (3.2) for R in Eq. (3.1) and rearranging the terms, one gets

$$\frac{V}{L} = \rho \frac{\mathbf{I}}{\mathbf{A}} \tag{3.3}$$

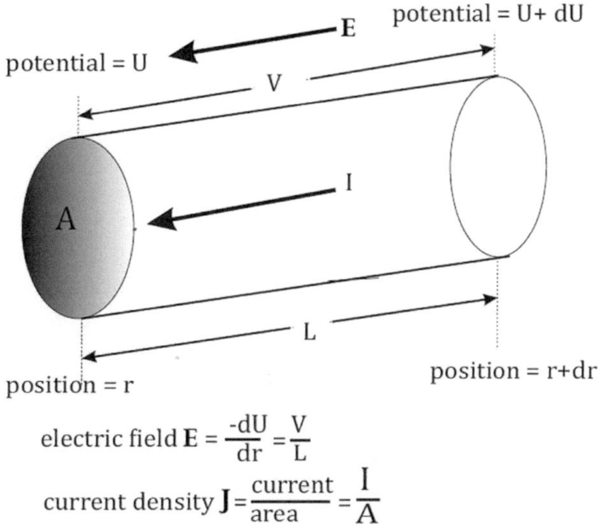

Fig. 3.1 Parameters explaining Ohm's law for a straight conductor

This expression, i.e., the Ohm's law using the vector terms, can be rewritten in the form of electric field $\boldsymbol{E}$ (=V/L) and current density $\boldsymbol{J}$ (=I/A), i.e.,

$$\boldsymbol{E} = \rho J \text{ or } \boldsymbol{J} = \sigma \boldsymbol{E} \tag{3.4}$$

It may be noted that irrespective of the formula used, the quantities measured are V and I.

2.1.1 Types of Electrical Conduction

Electric current passes through a material by one of the three different modes: *electronic, dielectric* and *electrolytic conduction*. *Electronic conduction* occurs in metals and crystals, *dielectric conduction* in insulators and *electrolytic conduction* in liquids. *Electronic conduction* is typical of metals and generally not of relevance to exploration geophysicists. *Dielectric conduction* occurs in insulators, which contain no free electron, unlike metal. When the electric field is applied to such a medium the atom or ions acquire an electric polarization and acts like an electric dipole. The net effect is to change the permittivity of the material ε from the physical constant $\boldsymbol{\varepsilon}_0$, commonly called the vacuum permittivity, permittivity of free space—an ideal, physical constant, which is the value of the absolute dielectric permittivity of classical vacuum. It has an exactly defined value

$$\varepsilon_0 = 8.854187817\ldots \times 10^{-12}\,\text{Fm}^{-1}\,(\text{farads per metre}).$$

Table 3.1 Dielectric constant of different rock types

Rock type	Dielectric constant (k)
Sandstones	5–12
Granite	3–19
Diorite	6
Basalt	12
Water	81

The relationship betweenε and $\boldsymbol{\varepsilon}_0$ is given by e = ke_0 where k is the *dielectric constant* or *permittivity* of the material. It is dimensionless and has a value commonly in the range of 3–81. Examples of k for some rock types and water are shown in Table 3.1.

Dielectric effects are not important in resistivity method. However, ground penetrating radar (GPR) uses frequencies in the MHz and GHz range and dielectric or permittivity contrast plays a significant role. *Electrolytic conduction* occurs in aqueous solutions that contain free ions. For example, in a saline solution the molecule of sodium chloride (NaCl) dissociates into separate Na^+ and Cl^- ions. The solution is called *electrolyte*. The ions in the electrolyte are mobilized by an electric field, which causes current to flow. Electric charge is transported by positive ions in the direction of the field and negative ions in the opposite direction. Ionic conduction is slower than electronic conduction.

2.2 *Resistivity for Rocks*

The important physical property of rocks for resistivity surveying is resistivity (or conductivity). The resistivity method will be applicable only when there is resistivity (or conductivity) contrast either laterally or vertically. The resistivity of rocks is an extremely variable property ranging from about 10^{-6} Ωm for minerals such as graphite to more than 10^{12} Ωm for dry quartzitic rocks. Most rocks and minerals are insulators in the dry state. In nature they almost always hold some interstitial water with dissolved salts and therefore acquire an ionic conductivity which then depends upon the moisture content, the nature of the electrolytes and the degree to which the open spaces in a rock (pores, micro fissures, cracks, fractures etc.) are saturated with water. The form of pores in a rock plays a subordinate role in determining the conductivity but the degree of pore interconnection is very important. In rocks like basalt or granite, for example, the resistivity can vary enormously depending upon the degree of pore interconnection or lack of it, from thousands of Ωm in compact, non-porous formation to as low as 10–50 Ωm in rocks like porous, water-bearing basalts in most tropical areas. Water-bearing fracture zones in rocks usually have fairly low resistivities while the resistivity of clays depends on proportion of clay minerals and other material. Some minerals, notably graphite, pyrite, chalcopyrite, galena and magnetite are relatively good electronic conductors.

Table 3.2 Resistivities of igneous, metamorphic, sedimentary rocks and sediments

Rock type	Resistivity range (ohm-m)
Igneous rocks	
Andesite	1.7×10^2 to 4.5×10^4
Basalt	10 to 1.3×10^7
Dacite	2×10^4
Diorite	10^4 to 10^5
Gabbro	10^3 to 10^6
Granite	3 x 10^2 to 10^6
Lavas	10^2 to 5×10^4
Olivine norite	10^3 to 6×10^4
Peridotite	3×10^3 to 6.5×10^3
Metamorphic rocks	
Gneiss	6.8×10^4 to 3×10^6
Graphitic schists	10 to 100
Marble	10^2 to 10^{12}
Quartzite	10 to 10^8
Schists	20 to 10^4
Skarn	2.5×10^2 to 2.5×10^6
Slates	6×10^2 to 4×10^7
Tuffs	2×10^3 to 10^5
Sedimentary rocks and sediments	
Argillites	10 to 8×10^2
Dolomite	3.5×10^2 to 5×10^3
Limestone	50 to 10^7
Marls	3 to 70
Sandstones	1 to 6.4×10^8
Shale (consolidated)	20 to 2×10^3
Sediments	
Alluvium and sands	10 to 800
Clays	1 to 120
Clay (unconsolidated and wet)	20
Conglomerates	2×10^3 to 10^4
Moraine	8–4000

Modified from Bhattacharya and Shalivahan (2016)

Igneous rocks that contain no water can have very high resistivity. The resistivity range of any given rock type is wide and overlaps with other rock types (Table 3.2).

The resistivity of the rocks is strongly influenced by the presence of groundwater, which acts as an electrolyte. This is especially important in porous rocks sediments and sedimentary rocks. The minerals that form the matrix of a rock are generally poorer conductor than groundwater, so the conductivity of sediment increases with the amount of groundwater it contains. This depends on the fraction of the rock that consists of pore spaces (*the porosity,* φ), and the fraction of the pore volume that is water-filled (the *water saturation, S*). The conductivity of the groundwater, which is quite variable because it depends on the concentration and type of dissolved minerals and salts it contains. These observations are summarized in an empirical formula, called Archie's for the *resistivity* ρ of the rock,

$$\rho = \frac{a}{\varphi^m S^n} \rho_w \tag{3.5}$$

By definition the φ and S are fractions between 0 and 1, ρ_w is the resistivity of groundwater, and parameters a, m and n are empirical constants that have to be determined for each case. Generally, $0.5 \leq a \leq 2.5$, $1.3 \leq m \leq 2.5$ and $n \approx 2$.

2.3 *Natural Potentials and Currents*

Electrical investigations of natural electrical properties are based on measurements of the voltage between a pair of electrodes implanted on the ground.

2.3.1 Self-Potential (or Spontaneous Polarization)

The potentials that originate spontaneously in the ground are self-potential (SP) or spontaneous polarization. These are the result of coupling between electric and non-electric flows and forces in the earth. The forces and fluxes are: electric potential gradients and electric current density (Ohm's Law), hydraulic gradients and fluid flux (Darcy's Law), chemical gradients and solute flux (Fick's Law), and thermal gradients and heat flow (Fourier's Law) (Minsley 2007; Bhattacharya and Shalivahan 2016).

Most of the self-potentials have an electrochemical origin. For example, if the ionic concentration in an electrolyte varies with location, the ions tend to diffuse through the electrolyte so as to equalize the concentration. The diffusion is driven by an electric *diffusion potential* which depends on the temperature as well as the difference in ionic concentration. A metallic electrode inserted into the ground reacts electrochemically with the electrolyte (i.e., groundwater), giving rise to contact potential. If two identical electrodes are inserted into the ground, variations in concentration of the electrolyte cause different electrochemical reactions at each electrode. A potential difference arises, called the *Nernst potential.* The combined diffusion and Nernst potentials are called *electrochemical* self-potential. It is temperature sensitive and may be either positive or negative, and generally of the order of a few tens of millivolts.

The self-potentials originating due to above mechanisms are attracting increased attention in tackling the near-surface problems like environmental and shallow engineering problems. However, in the mineral exploration problems, these are often much smaller than the potentials associated with electronically conducting ore bodies like sulphide, graphite etc. and treated as 'background potential'. The

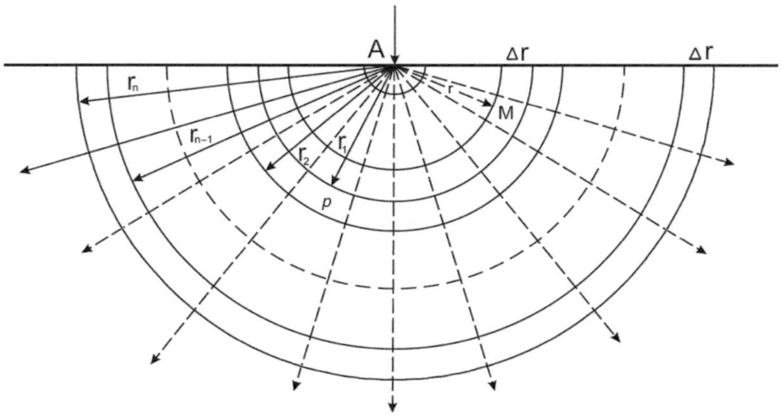

Fig. 3.2 Potential and current lines for a point source of current: continuous lines – equipotential lines; discontinuous lines with arrows – current directions

potentials associated with the ore bodies are called *mineralization potential* and the background potentials are noise in such cases.

2.4 Basic Theory of Resistivity Method

2.4.1 Potential Due to a Single Electrode

Consider the flow of a current around an electrode that introduces a current I at the surface of a minimum half-space (Fig. 3.2). The point of contact acts as a current source from which current radiates outward. The electric field lines are parallel to the current flow and normal to the equipotential surfaces, which are hemispherical in shape. The current density $\boldsymbol{J}$ is equal to current I divided by the surface area, which is $2\pi r^2$ for a hemisphere of radius r. The electric field $\boldsymbol{E}$ obtained from Ohm's law (Eq. 3.4) at a distance r from the current electrode A is given by

$$\boldsymbol{E} = \rho \boldsymbol{J} = \rho \frac{I}{2\pi r^2} \tag{3.6}$$

We know that,

$$\boldsymbol{E} = -\frac{dV}{dr} \tag{3.7}$$

Substituting Eq. (3.6) in Eq. (3.7), one gets the electric potential V at a distance r from the point current electrode A

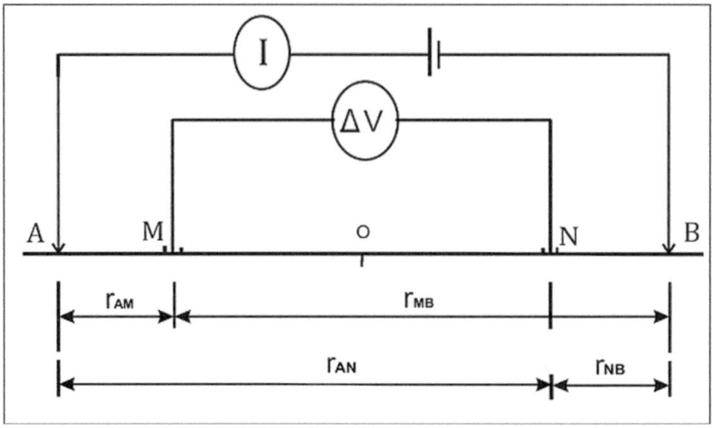

Fig. 3.3a Generalized set up of collinear four-electrode arrangement

$$\frac{dV}{dr} = -\rho\frac{I}{2\pi r^2} \tag{3.8}$$

that is,

$$V = \frac{\rho I}{2\pi r^2} \tag{3.9}$$

The equipotential around a source or sink electrodes are hemispheres, if we regard the electrode in isolation. The potential around a source is positive and diminishes as $1/r$ with increasing distance. The sign of I is negative at a sink, where the current flows out of the ground. Thus around a sink the potential is negative and increases, i.e., becomes less negative, as $1/r$ with increasing distance from the sink. These observations can be used to calculate the potential difference between a second pair of electrode at a known distance from the source and sink.

2.4.2 Collinear Four-Electrode Method

Consider a generalized arrangement of electrodes in a collinear, i.e., arranged along a line, pattern consisting of a pair of current electrodes and a pair of potential electrodes (Fig. 3.3a). The current electrodes A and B act as a source and sink (also termed sometime as transmitter), respectively. The potentials measured at M and N are receiving potential electrodes (also termed sometime as receiver) measuring the potential difference at these two points. The potential at M due to source A is $+\rho I/2\pi r_{AM}$, while the potential due to sink B is $-\rho I/2\pi r_{BM}$. The combined potential (V_M) at M is

$$V_M = \frac{\rho I}{2\pi}\left(\frac{1}{r_{AM}} - \frac{1}{r_{BM}}\right) \tag{3.10}$$

Similarly the resultant potential (V_1N) at N is

$$V_N = \frac{\rho I}{2\pi}\left(\frac{1}{r_{AN}} - \frac{1}{r_{BN}}\right) \tag{3.11}$$

The potential difference (V) between M and N is

$$V = \frac{\rho I}{2\pi}\left[\left(\frac{1}{r_{AM}} - \frac{1}{r_{BM}}\right) - \left(\frac{1}{r_{AN}} - \frac{1}{r_{BN}}\right)\right] \tag{3.12}$$

All the quantities in Eq. (3.12) can be measured on the ground surface except the resistivity ρ, which is given by,

$$\rho = 2\pi\frac{V}{I}\left[\frac{1}{\left(\frac{1}{r_{AM}} - \frac{1}{r_{BM}}\right) - \left(\frac{1}{r_{AN}} - \frac{1}{r_{BN}}\right)}\right] \tag{3.13}$$

$$\rho = K\frac{V}{I} \tag{3.13a}$$

where

$$K = \frac{2\pi}{\left(\frac{1}{r_{AM}} - \frac{1}{r_{BM}} - \frac{1}{r_{AN}} + \frac{1}{r_{BN}}\right)} \tag{3.13b}$$

The constant K, referred also as geometrical constant in literature, is determined from the distances between the electrode positions and is known as geometrical constant. According to the Eq. (3.13), the dimension of K is the dimension of length. Equation (3.13) obtained over the homogenous medium may be used for the measurements with four electrodes placed on the surface of the inhomogeneous earth. The resistivity measured over the inhomogeneous earth is termed as apparent resistivity ρ_a. Thus, in such cases Eq. (3.13) reduces to

$$\rho_a = K\frac{V}{I}\text{i.e.}\rho_a = K\frac{V}{I} = KR \tag{3.13c}$$

where R is the resistance offered by the ground.

Equation (3.13c) shows that apparent resistivity ρ_a, in addition to the function of K, is also a function of current I injected into the ground, which in turn gives rise to a potential difference (V) and it depends on the geoelectric sections underneath the electrode configuration. The unit of apparent resistivity is ohm-m and at times presented in literature as Ω-m or Ω.m or Ωm.

2.4.3 Special Electrode Configuration

The general formula for the resistivity measured by a four-electrode method is simpler for some special geometry of current and potential electrodes. The most commonly used configurations are Wenner, Schlumberger and dipole-dipole set ups. In each configuration the four electrodes are collinear but the geometries and spacing are different in the electrode arrays.

In *Wenner configuration* (Fig. 3.3b) the current and potential electrode pairs have a common mid-point and the distances between adjacent electrodes are equal, so that $r_{AM} = r_{BN} = a$ and $r_{BM} = r_{AN} = 2a$. Using these values in Eq. (3.13), one gets

$$\rho = 2\pi\frac{V}{I}\left[\frac{1}{\left(\frac{1}{a}-\frac{1}{2a}\right)-\left(\frac{1}{2a}-\frac{1}{a}\right)}\right] \tag{3.14}$$

$$\rho = 2\pi a\frac{V}{I} \tag{3.15}$$

In *Schlumberger configuration* (Fig. 3.3c) the current and potential pairs of electrodes also have a common mid-point, but the distances between adjacent electrodes differ. Let the separations between current and potential electrodes be L and a, respectively. The $r_{AM} = r_{BN} = (L - a)/2$ and $r_{AN} = r_{BM} = ((L + a))/2$. Substituting in the general formula, one gets

$$\rho = 2\pi\frac{V}{I}\left[\frac{1}{\left(\frac{2}{L-a}-\frac{2}{L+a}\right)-\left(\frac{2}{L+a}-\frac{2}{L-a}\right)}\right] \tag{3.16}$$

$$\rho = \frac{\pi}{4}\frac{V}{I}\left(\frac{L^2-a^2}{a}\right) \tag{3.17}$$

Schlumberger array measures gradient of potential i.e., electric field unlike Wenner array where potential is measured. If $L \geq 5a$, then for all practical purposes, the condition of measuring the electric field at the mid-point O is satisfied.

In the dipole-dipole configuration (Fig. 3.3d) the spacing of the electrodes in each pair is a, i.e., $AB = MN = a$; distance between BM is na where $n = 1, 2, 3, 4$ etc. In this case L is the distance between the midpoints of the transmitting (AB) and receiving (MN) dipoles. Therefore, $r_{AM} = r_{BN} = L\left(\frac{a}{2}+na\right) = \frac{a(2n+1)}{2}$, $r_{BM} = (L-a) = na$, $r_{AN} = (L + a) = na + a = a(n + 1)$

The expression for resistivity in this case reduces to

$$\rho = 2\pi\frac{V}{I}\left[\frac{1}{\left(\frac{1}{L}-\frac{1}{L-a}\right)-\left(\frac{1}{L+a}-\frac{1}{L}\right)}\right]$$

or

$$\rho = 2\pi \frac{V}{I} \left[\frac{1}{\frac{1}{\left(\frac{1}{\left(\frac{a}{2}+na\right)} - \frac{1}{na} \right) - \left(\frac{1}{a(n+1)} - \frac{1}{\left(\frac{a}{2}+na\right)} \right)}} \right] \tag{3.18}$$

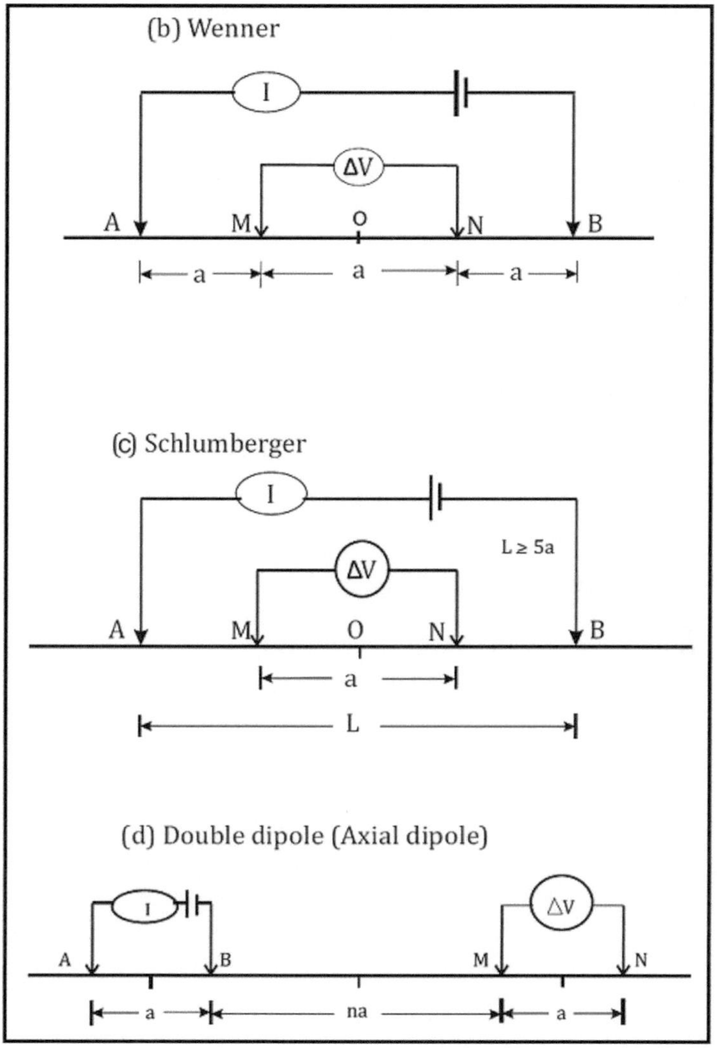

Fig. 3.3 (**b–d**) Special geometries of current and potential electrodes: (**b**) Wenner, (**c**) Schlumberger and (**d**) dipole-dipole

$$\rho = \pi \frac{V}{I} \left[\frac{L(L^2 - a^2)}{a^2} \right] \quad \text{or} \quad \rho = \pi \{ n(n+1)(n+2)a \} \frac{V}{I} \tag{3.19}$$

In addition to these three commonly used configurations (Parasnis 1986; Lowrie 1997; Bhattacharya and Shalivahan 2016), the arrays like two-electrode or pole-pole, three-electrode and pole-dipole electrode are also used in geoelectrical surveys (Bhattacharya and Shalivahan 2016).

Two modes of investigatiton can be used with each electrode configuration: profiling and sounding. The assemblage of four electrodes is displaced stepwise along a profile while maintaining constant values of the inter-electrode distances corresponding to the configuration employed to study the lateral variation of resistivity up to a certain depth. The separation of the current electrodes is chosen so that the current flow is maximized in depths where lateral resistivity contrasts are expected. Wenner method is best suited for profiling. Results from a number of profiles may be compiled in a resistivity map of the region of interest. The resistivity map so prepared reveals the horizontal variations in resistivity within an area up to a particular depth. Geological features like contacts, dikes and mineralized veins etc. are delineated in such maps.

The vertical variation in resistivity is detected by sounding and it is known as *vertical electrical sounding* (VES) – a widely used method in groundwater exploration and engineering geology problems. The technique is best adapted to determine depth and resistivity of horizontal and near-horizontal layered formations, such as sedimentary beds and depth to the water table. Schlumberger configuration is most commonly used for VES studies. The mid-point of the array is kept fixed while the distance between the current electrodes is progressively increased. This causes the current lines to penetrate to ever greater depths, depending on the vertical distribution of electrical conductivity (or resistivity).

Let us consider a case of horizontally layered structure where a layer of thickness h and resistivity ρ_1 overlies a layer of resistivity ρ_2 (Fig. 3.4a). If the current electrodes AB ($= L$) are close together (i.e., $L \ll h$), then most of the current flows in the upper layer (Fig. 3.4b – left) and the observed apparent resistivity is close to the true resistivity of the top layer ρ_1, i.e., $\rho_a \to \rho_1$. With increasing current electrode separation current lines flow deeper and proportionally more current flows in the second layer (Fig. 3.4b – right). If the second layer is conducting (i.e., $\rho_2 < \rho_1$) then apparent resistivity ρ_a decreases with increasing electrode separation. When the electrode separation is much larger than the thickness of the upper layer (i.e., $L \gg h$), the measured apparent resistivity ρ_a is close to the resistivity of the second layer i.e., $\rho_a \to \rho_2$. Between these two extreme situations the apparent resistivity (ρa) decreases if the second layer is conducting (Fig. 3.4c – left) and conversely increases if the second layer is resistive (Fig. 3.4c – right). The apparent resistivity observed between these two extreme situations is not related in a simple way to the true resistivity of these two layers.

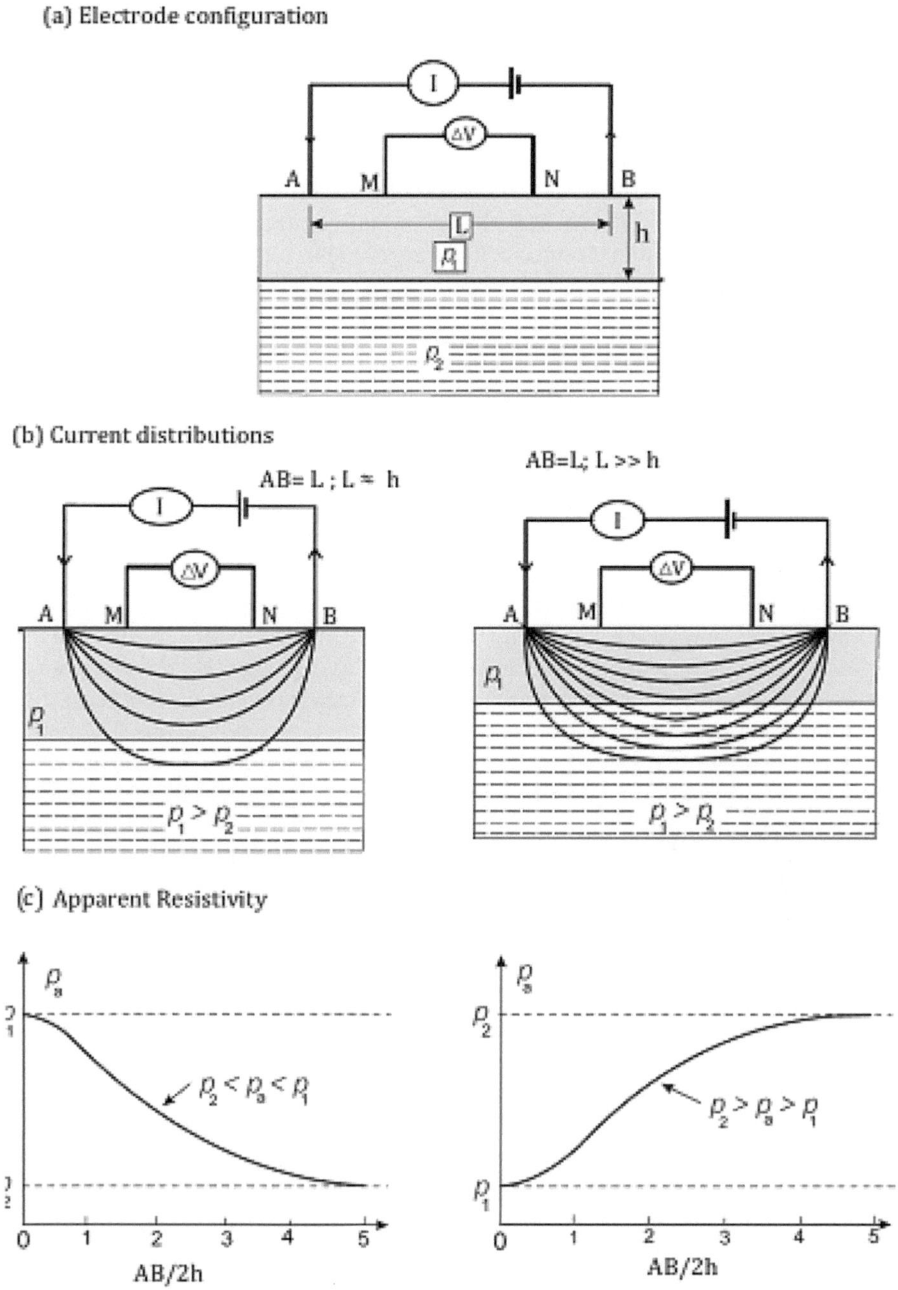

Fig. 3.4 Distribution of current lines over a two-layered earth for a generalized four-electrode set up (**a**); (**b**) for the model $\rho_2 < \rho_1$(Left); and (**c**) trend of apparent resistivity curve as the electrode spacing is increased for the two cases (i) $\rho_2 < \rho_1$ (left), and (ii) $\rho_2 > \rho_1$ (right)

The resistivity contrast between the layers is expressed by a *k*-factor defined as

$$k = \frac{\rho_2 - \rho_1}{\rho_2 + \rho_1} \tag{3.20}$$

The *k*-factor ranges between −1 and +1 as the resistivity ratio ρ_2/ρ_1 for infinitely conducting and resistive bottom layer respectively, varies between 0 and ∞. The apparent resistivity field curves are plotted on double logarithmic plot to cover a wide range of values of current electrode separations ranging from, say 1 m to 1000 m, and resistivity values generally changing by five to six orders.

Modern VES analyses take advantage of flexibility offered by computers with graphic outputs on which the apparent resistivity curves can be visually assessed. The first step in the analysis consists of classifying the shape of the vertical sounding curve. Two-layer VES curves fall broadly under two categories: ascending or descending type of curves. The trend of rise or fall of the various curves will depend on the *k*-factor.

The apparent resistivity curve for three-layer earth model (Parasnis 1986; Lowrie 1997; Bhattacharya and Shalivahan 2016) having resistivities of $\rho_{1,}$ ρ_2 and $\rho_{3,}$ generally has one of the four typical shapes, determined by the vertical sequence of resistivities in the layers (Fig. 3.5). The type K curve ($\rho_1<\rho_2> \rho_3$) rises to a maximum and then decreases indicating that the intermediate layer has higher resistivity than the top and bottom layers (Fig. 3.5a). The type H curve ($\rho_1>\rho_2<\rho_3$) shows opposite effect. It falls to a minimum and increases again due to intermediate layer that is better conductor than the top and bottom layers (Fig. 3.5b). The type A curve ($\rho_1<\rho_2<\rho_3$) may show some changes of gradient but the apparent resistivity generally increases continuously with increasing electrode separation, indicating the true resistivities with depth from layer to layer (Fig. 3.5c). A careful qualitative analysis will clearly distinguish between A type curves from two-layer ascending type curve. The Q type curve ($\rho_{1>}\rho_{2>}\rho_3$) exhibits the opposite effect. It decreases continuously along with a progressive decrease of resistivity with depth (Fig. 3.5d). In this case also a careful qualitative analysis will clearly distinguish between Q type curves from two-layer descending type curve.

2.4.4 Multilayer Curves

There are eight types of four layer curves: KH, QH, HA, AA, HK, AK, KQ, QQ (Bhattacharya and Shalivahan 2016). The first four types of curves are for substratum showing maximum value of resistivity whereas the last four are for the substratum showing least value of resistivity. In ground water exploration VES curves of 3–5 layers are generally encountered. In cases where the resistivity of the fourth layer is very high ($\rho_4 \rightarrow \infty$) then it may screen the effect of the fifth layer and in such cases VES curve may appear like a four-layer only.

The interpretation of VES data initially is done qualitatively to determine the possible number of layer of the earth model. The next step is the one-dimensional

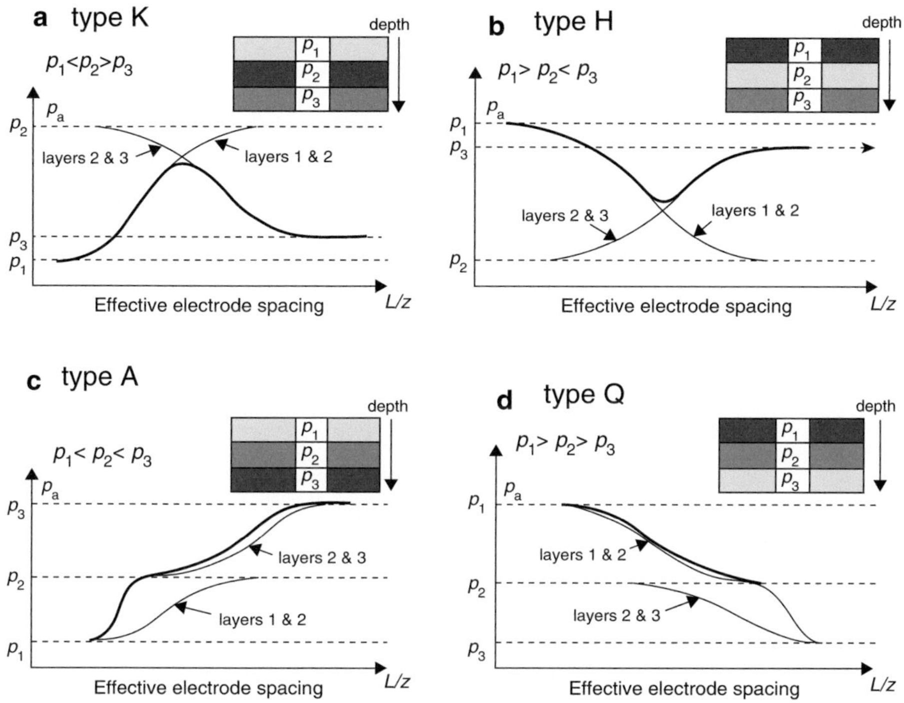

Fig. 3.5 Three layer VES curve types: (**a**) K ($\rho_1 < \rho_2 > \rho_3$); (**b**) H ($\rho_1 > \rho_2 < \rho_3$); (**c**) A ($\rho_1 < \rho_2 < \rho_3$) and (**d**) Q ($\rho_1 > \rho_2 > \rho_3$)

inversion of field data. It is an iterative procedure which is quickly performed now-a-days with the help of portable fast computers. The method assumes the equations for the theoretical response of a multi-layered earth. The layers are characterized by their thickness and resistivity which are to be determined. An initial guess of these parameters is made for each layer and the predicted curve of apparent resistivity versus electrode spacing is computed. The discrepancies between the observed and theoretical curves are then determined point by point. The layer parameters used in the governing equations are next adjusted and the calculation is repeated with corrected values, giving a new predicted curve to match with the field curve. The procedures are reiterated several times until the discrepancies are smaller enough to be accepted as the correct layer parameters. One-dimensional inversion is an automatic matching technique. The vertical layer earth has an analytical solution and many efficient inversion algorithms have been designed and widely used. In recent years procedures have been designed that take into account the lateral inhomogeneities also. The response of two- or three-dimensional structures must be approximated by a numerical solution, based mainly on finite-difference or finite element techniques. In two- or three-dimensional inversions number of unknown parameters increase, thereby increasing the computational difficulties as well as computational time considerably.

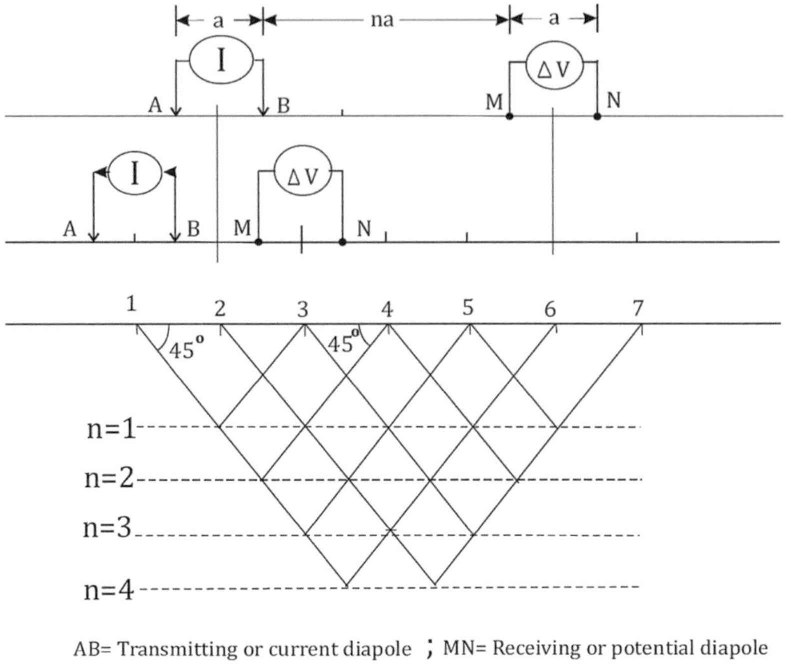

Fig. 3.6 Layout of pseudo section

2.4.5 Pseudo Section

It is an alternative display of data where both lateral profiling and vertical sounding are combined to plot resistivity data to illustrate the effect of variation of electrode spacing. A collinear dipole-dipole array is widely used though the other electrode arrays can be used as well to prepare the pseudo section. If the transmitter and receiver dipoles are of length '*a*' then the distance between the nearest electrodes of the transmitter and receiver dipoles is a multiple *na* of the electrode spacing '*a*'. Apparent resistivities are plotted on vertical section (Fig. 3.6). The observed values are plotted at the point of intersection of lines making an angle of 45° with the line representing the profile drawn from the mid-points of transmitter and receiver dipoles in opposite directions. Measurements are made at several discrete positions as the receiver pair is moved incrementally away from the fixed transmitter position. The transmitter pair is then shifted by one increment along the same profile and the procedure is repeated for $n = 2, 3, 4, 5$ etc. As 'n' increases the information is obtained from increasingly greater depths. Thus the measurements for $n = 2$ in such a section will appear along a line below the line $n = 1$ and $n = 3$ will be plotted along a further deeper line and so on. Finally the contours of equal apparent resistivity values are plotted at some regular intervals. The plot is called pseudo section because measurements along a profile are indicative of lateral resistivity variations. The observations with larger values of 'n' are expected to provide information from deeper parts of the subsurface.

Table 3.3 Broad correlation of resistivity of geoelectric layers and lithology

Layer	Resistivity range (ohm-m)	Lithology from boreholes
1	< 30	Wet top soil
	30–700	Dry top soil
2	10–150	Highly weathered/weathered/semi-weathered layer
3	150–600	Partially fractured/fractured rock
4	> 600	Massive rock

After Bhattacharya and Shalivahan (2016)

3 Groundwater Exploration in Drought Prone Hard Rock Areas

Systematic VES should be carried out in the hard rock terrain with stations separated from each other by reasonable distance. The lineament map of the area based on Landsat imagery indicating sets of prominent lineaments representing fracture zones in the bed rock and their trend should be plotted for the area. Lineament density contour map can be prepared by dividing the entire area in the pixel of say, 25 sq. km and lineament density calculated in the unit of km/25 sq. km.

Fence diagram using most of the VES data in the area of investigation should be prepared. Fence diagram is a three-dimensional depiction of an area showing the locations and relationship of geoelectric horizons from VES data. The diagram is constructed from several geoelectric sections drawn in position corresponding to their actual locations and the geoelectric layers of similar range of values are joined. It gives an overall idea of the variation in thickness of weathered and fractured system overlying the massive rock for the entire area of geophysical investigations.

VES must also be carried out in the vicinity of available boreholes in the area for comparing the geoelectric and geologic sections. A typical example of nature of broad correlation between resistivity values of geoelectric layers and lithology from available boreholes in drought prone area of Kalahandi district in Odisha, India is shown in Table 3.3 (Bhattacharya and Shalivahan 2016).

The presence of weathering in geologic section below the top soil cover is generally likely to be reflected in the entire geoelectric sections. However, in a few boreholes the total effect of the closed fractures reduces the range of resistivity from that of the massive rock. In some cases the fracture zones could broadly be identified in correlation to boreholes. However, very thin fractures are not reflected in most of the VES.

3.1 Groundwater Management Plan of the Hard Rock Area

Results of VES survey along with other relevant features like pre- and post-monsoon depth to water table, short-and long-term water table fluctuation, lineament density, pre- and post-monsoon flow direction and lineament density axes, spatial distribution of total dissolved solids, total hardness, chloride, fluoride, chemical sensitivity maps etc. can be collated in the area of investigation for groundwater management plan. This would serve, in general, as a broad guide line for groundwater management plan for hard rock areas under investigation.

The availability of groundwater in the draught prone hard rock districts is problematic because of factors like (i) unsuitable geomorphological condition, (ii) poor aquifer characteristics, (iii) high run-off in general and low infiltration (sometime only 3% of the total rainfall infiltrates into the groundwater system), (iv) extensive deforestation, (v) long duration of hot-dry season, and (vi) unscientific methods used in locating spot sources for dug wells and tube wells.

The suggested steps to increase the availability of groundwater resources in such areas are (i) demarcating groundwater potential zones, (ii) reducing run off and thereby increase the storage of water in the aquifer by artificial recharge, and (iii) construction of efficient water yielding spot sources in different potential zones.

3.1.1 Groundwater Potential Zones

The technique of thematic overlays using several sensitivity maps can be used to demarcate the groundwater potential zones (Bhattacharya and Shalivahan 2016). The major components of this analysis should be (i) depth to massive rock (DMR), (ii) aquifer transmissivity (AT), (iii) lineament density (LD), (iv) chemical quality (CQ), (v) depth to water table (DWT), and (vi) post-monsoon long-term water table fluctuation (PLWTF).

Bhattacharya and Shalivahan (2016) have explained in great detail the procedure. Each of these components has a number of factors of influence which are visualized in a decision tree by breaking up the above components into sub-categories to measurable parameter or indicator. The combined effect of various sub-categories of each component has been categorized into high, medium and low potential zones by overlay procedure explained through matrices.

4 2D Resistivity Imaging

The multi-electrode resistivity technique uses a multi-core cable connecting large number of electrodes (viz., 24 or its multiple) (Barker 1981; Griffiths et al. 1990; Griffiths and Barker 1993; Xu and Noel 1993; Beresnev et al. 2002 and Bhattacharya and Shalivahan 2016) where references to several other relevant papers have been cited) placed on the ground at a fixed spacing, every 5 m for instance (Fig. 3.7). In this type of measurements the measuring device, i.e., resistivity meter contains relays which ensure the switching of preselected electrodes according to a sequence of readings predefined and stored in an internal memory of the device. The various combinations of transmitting (A, B) and receiving (M, N) pairs of electrodes construct the pseudosection i.e., a combination of profiling and sounding data set, with depth of investigation decided by the total length of the cable. Various types of electrode set ups are used, such as Wenner, Schlumberger, Dipole-Dipole, Wenner-Schlumberger, Pole-Pole and Pole-Dipole arrays etc. The measuring device generally consists of internal microprocessor controlled circuit and electronic switching system to automatically select the relevant four electrodes of the array decided for the survey.

Figure 3.7 shows a typical field Wenner setup for a 2-D survey with a 20 number at a constant inter- electrode spacing of '*a*' along a traverse attached to a multi-core cable which in turn is attached to the resistivity measuring device. Normally a constant spacing between adjacent electrodes is used. First of all the data is acquired for data level 1 i.e., for $n = a$ and in such a situation for station 1 the current electrodes (*A*, *B*) are at positions 1 and 4 whereas the potential electrodes (*M*, *N*) at 2 and 3 respectively. Then for station 2 the position of each electrode is shifted by '*a*' and the electrodes occupy stations 2, 3, 4 and 5 are now used. This process is repeated till the last electrode 20 is occupied by current electrode B. The level 1 measurements is complete with the electrodes (*A*, *M*, *N*, *B*) occupying the stations 17, 18, 19 and 20 respectively. The number of possible stations for level *'n'* is equal to the number of last electrode station – ($n \times 3$). With $n = 1$ and the last electrode number 20, the total number of measurements at this level is ($20 - 1 \times 3) = 17$. After completing the data acquisition for level 1 the procedure commences for level 2. The first electrode positions of *A*, *M*, *N*, *B* at this level occupies the stations 1, 3, 5 and 7 respectively for electrode spacing of '2*a*'. This level is complete when the electrodes occupy the stations 14, 16, 18 and 20 respectively and the total number of possible measurements is 14, i.e., ($20–2 \times 3) = 14$. Subsequently measurements at all the levels are made for n = '3*a*', '4*a*', '5*a*' etc. The idea is to acquire all possible data for better model from inversion of the measured data. The roll-along method, a system with a limited number of electrodes, is frequently used to increase the lateral extension of the survey area. After completing the entire sequence of measurements, the cable is moved past one end of the line by several unit electrode spacing. All the measurements that involve the electrodes on part of the cable that do not overlap the original end of the survey line are repeated (Fig. 3.8) (Loke 2000). A final field map is obtained by contouring the data in a pseudo section.

4.1 Comparison of Different Electrode Arrays

The arrays that are most commonly used for 2-D imaging surveys are the collinear arrays like (a) Wenner, (b) dipole-dipole, (c) Wenner-Schlumberger, (d) pole-pole and (e) pole-dipole. The deciding factors in favour of a particular array are: (i) the depth of investigation, (ii) the sensitivity of the array to vertical and horizontal changes in the subsurface resistivity, (iii) the horizontal data coverage and (iv) the signal strength. The first of these factors are determined from the sensitivity function of the array for a homogeneous model that depicts to the extent the measured

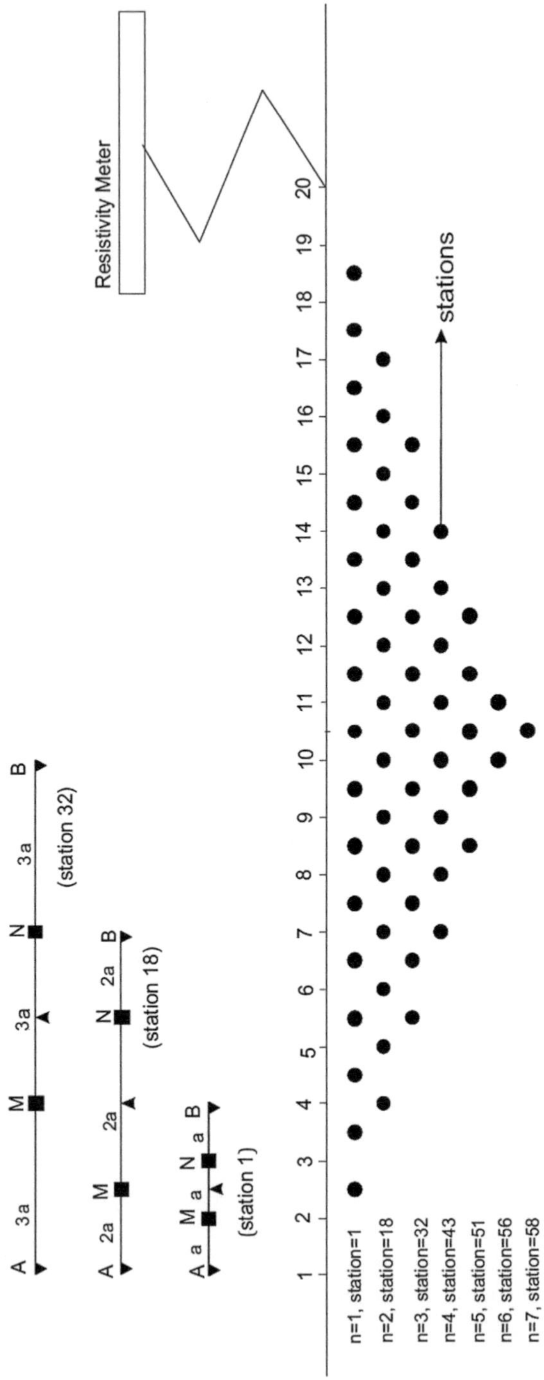

Fig. 3.7 Electrode set up for 2-D electrical survey and the sequence of measurements for the pseudo section

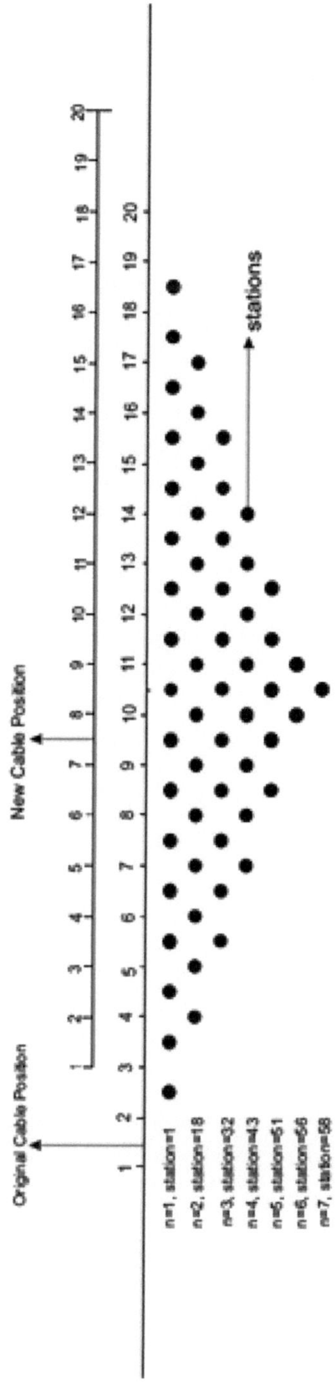

Fig. 3.8 The roll-along method for extension of the area covered by 2-D survey

potentials are influenced by the change in the resistivity. The higher the value of the sensitivity function, the greater is the influence of the subsurface region on the measurement.

5 Seismic Refraction Method

VES and Resistivity Imaging are widely used geophysical technique in groundwater exploration. Seismic refraction method is as much employed in hard rock terrain to distinguish highly weathered, less weathered, fractured and massive rock or basement. In Deccan trap areas, though generally treated as hard rock terrain, the strategy for geophysical surveys for groundwater will also look for vesicular lava and intertrappean beds as well for mapping the aquifer zones.

The basis of seismic methods is the theory of elasticity. The elastic properties are characterized by elastic modulii or constants specifying the relation between 'stress' and 'strain'. Seismic wave is a transfer of energy by way of particle motion. Different types of seismic waves are characterized by their particle motion. These are compressional (*P*), shear (*S*) which are known as body wave and surface, i.e., Love and Raleigh waves. Only body waves are of interest in seismic refraction.

In the compressional, longitudinal or *P* waves the motion of the particles in the medium is in the same direction as the direction of wave propagation. Its velocity (V_P) is given by

$$V_P = \left(\frac{k + \frac{4\mu}{3}}{\rho} \right)^{1/2} \tag{3.21}$$

where k, μ and ρ are bulk modulus, shear modulus and density of the medium respectively. In the shear, transverse or *S* waves the particles of the medium move perpendicular to the direction of wave propagation and the velocity (V_S) is given by

$$V_S = \left(\frac{\mu}{\rho} \right)^{1/2} \tag{3.22}$$

It is clear from the Eqs. (3.21) and (3.22) that $V_P > V_S$. As the fluid (liquid and gas) do not possess shear modulus (m) the *S* waves do not propagate through liquids and gases, i.e., $V_S = 0$.

If a medium has a free surface, there are also surface waves in addition to the above two types of body waves. Rayleigh wave is the main surface wave relevant in seismic exploration. In this case the particles describe elliptical motion in the vertical plane containing the direction of propagation of the wave.

The frequency range of body waves in the earth generally varies from 15 Hz to about 100 Hz. The surface waves are mostly below 15 Hz. In exploration domain the *P* waves are mainly important. *S* waves at times can also be used. Surface waves are

Table 3.4 Seismic velocities of air, water and some rocks

Materials	Compressional velocity (m/s)	Shear velocity (m/s)
Air	330	–
Water	1450	–
Loose sand	300–1500	100–450
Clay	1000–2500	500–1500
Glacial moraine	1500–2700	900–1300
Granite	4500–7000	2500–4000
Compact Basalt	4000–6500	2300–3800
Compact limestone and dolomite	3500–6500	1800–3800
Soft limestone	1500–4000	950–2600
Compact deep seated rocks	4000–6500	1900–3800

After Parasnis (1986), Dobrin and Savit (1988), Telford et al. (1988) and Lowrie (1997)

not capable of giving information regarding structures at depth and generally not applied in geophysical exploration programmes. However, in recent years surface waves have found wider applications in shallow geotechnical problems. In exploration seismology the surface waves are unwanted signal which need to be suppressed to enhance the desired signal.

Some typical values of the velocity of *P* and *S* waves in some rocks are given in Table 3.4. In general the velocities are greater in igneous, crystalline and compact rocks than the sedimentary and weathered rocks. Seismic velocities can be measured in the field as well as in the laboratory on samples of rocks. Well velocities surveys are also commonly used to obtain the velocity of rock formations.

5.1 Seismic Wave Propagation

A seismic disturbance is transmitted by periodic elastic displacements of particles of the medium. The surfaces in a medium on which the wave motion is the same at all points are known as wavefronts. If it is plane, it is termed as plane wave. The normal to a wavefront at any point is the direction of a ray and is the direction of wave propagation at that point. The rays are straight lines in a medium of constant seismic velocity. The rays are curved in inhomogeneous media. Quite often the wave propagations are described by rays. When the rays are intercepted by an interface between two media, it may be reflected or refracted.

The passage of a wave through a medium and across interfaces between adjacent media has been explained by the Dutch mathematician and physicist Huygens. Huygens' Principle states that every point on a wavefront may be considered to be a secondary source that emits waves travelling radially outward from the point. Reflection, refraction and Snell's law can be derived from Huygens' principle. The behaviour of seismic ray paths at an interface is explained by Fermat's principle – formulated by the French mathematician Pierre de Fermat. It states that, of the many possible paths between two points A and B, seismic ray follows the path that gives the shortest travel-time between these two points.

The principle difference between the geometry of the refraction and that of the reflection methods is in the interaction that takes place between the seismic waves and the lithological boundaries they encounter during propagation. The waves reflected by the boundaries travel along paths which are easy to visualize. Refracted wave, with which this section is mainly concerned, follows a somewhat complicated trajectory. Refraction paths cross boundaries between media having different velocities in such a way that energy travel from source to receiver in the shortest possible time, as per the Fermat's principle. In refraction prospecting the waves propagating through the formations with velocities must be substantially greater than that of overlying formations. The velocities and depths of such layers are from the times required for the refracted waves to travel between sources and the receivers both placed on the surface of the earth. This distance between the source and the receiver is always several times greater than the thicknesses and depths of the formations. In refraction work, in most of the surveys, only the initial or first arrival of seismic wave is recorded.

Refraction survey has been traditionally used in groundwater exploration work in hard rock terrain. In such a terrain general trend of the lithology is top soil cover overlying highly weathered rock formations, followed sequentially by less weathered formations, semi-weathered formations with cracks, fissures and joints and hard and compact basement. These layers show gradually higher velocity contrasts as is the essential requirement for refraction of seismic waves according to the various laws of physics discussed earlier.

Let us consider a two-layered earth model, upper having a thickness z. The longitudinal velocity of the upper layer is V_O and the lower layer velocity is V_1 with $V_1 > V_0$ (Fig. 3.9). Seismic wave is generated at the source point S on the surface and propagates in hemispherical wavefronts. A detector (geophone) is placed at a point D on the surface at a distance x from S. For small distances of x the first arrival of the seismic wave at D will be the one that travels horizontally with a velocity of V_0. For greater distances the wave takes an indirect path, propagating down from source S, then travels along the interface and finally upward from bottom layer with velocity V_1 will arrive first because the time is gained while travelling through higher velocity interface in spite of wave travelling a longer path.

The spherical wavefronts from S striking the boundary layer undergoing a velocity change are refracted in the lower medium according to Snell's law. The ray SA strikes the interface at a critical angle and further right of A the wavefronts below the interface travel faster than that propagating above. The wave originating at a point B in the lower medium will travel a distance BC while for the same time the wavefront spreading out at in the upper medium will attain a radius of BE. The resultant wavefront above the interface will follow the line CE, which makes an angle i_c with interface. Thus

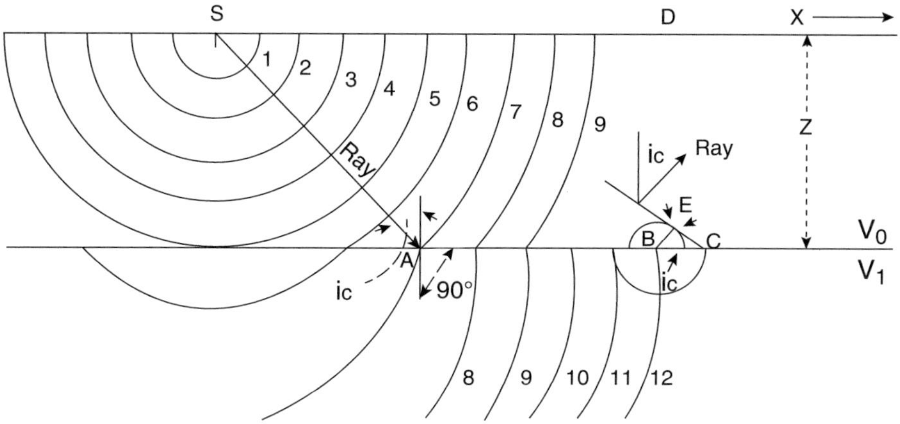

Fig. 3.9 Mechanism for transmission of refracted waves in two-layered earth

$$\sin i_c = \frac{BE}{BC} = \frac{V_0 t}{V_1 t} = \frac{V_0}{V_1} \tag{3.23}$$

The angle that the wavefront makes with the horizontal is the same as the ray perpendicular to it makes with the vertical, so that the ray returns to the surface at the critical angle $\langle \sin^{-1} V_0/V_1 \rangle$ with a line perpendicular to the interface.

Refraction data are represented by time-distance curve by plotting the first-arrival time T versus the shot-detector distance x. Figure 3.10 shows the basic appearance of time-distance graph for a simple two-layer earth. In this diagram A is the source point of generation of seismic waves caused by detonator, weight drop or explosives. D is the locations of the detectors which are moved along a profile, passing through the source point, at predetermined locations.

The direct wave travels from shot to detector on the surface with a velocity of V_0, and therefore, $T = x/V_0$. It is a straight line, in the T versus x plot, passing through the origin with a slope of $1/V_0$ (Fig. 3.10). The wave refracted along the interface at depth z, reaching and leaving it at critical angle i_c, takes a path consisting of three segments: AB, BC and CD. The total time T for the refracted path $ABCD$ is

$$T = T_{AB} + T_{BC} + T_{CD} \tag{3.24}$$

It can be rewritten as

$$T = \frac{2z}{V_0 \cos i_c} - \frac{2z \sin i_c}{V_1 \cos i_c} + \frac{x}{V_1} \tag{3.25}$$

Finally it can be shown that

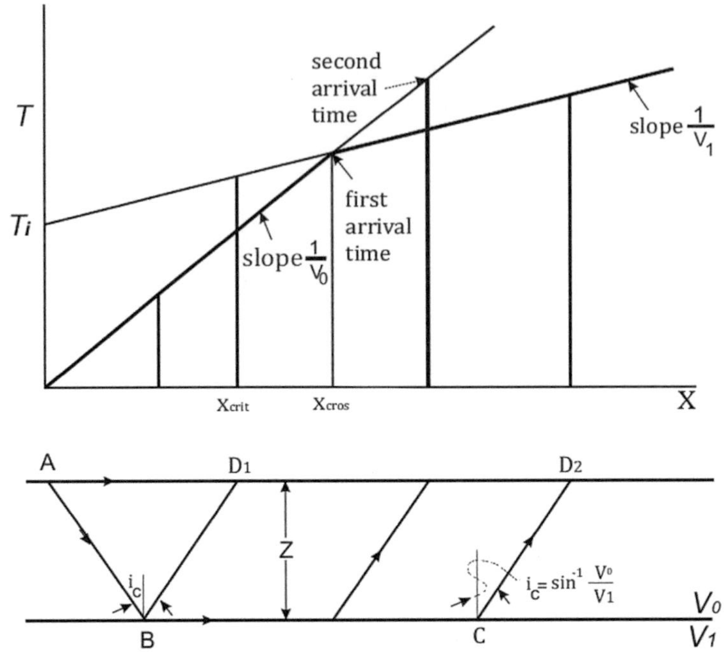

Fig. 3.10 Time-distance curve for a two-layer earth. x_{crit} and x_{cros} are the critical distance and crossover distance respectively

$$T = \frac{x}{V_1} + \frac{2z\sqrt{V_1^2 - V_0^2}}{V_1 V_0} \tag{3.26}$$

In a time (T) – distance (x) plot this is a straight line with a slope of $1/V_1$ which intercepts T axis ($x = 0$) at a time T_i, known as intercept time, given by

$$T_i = 2z\frac{\sqrt{V_1^2 - V_0^2}}{V_1 V_0} \tag{3.27}$$

The intercept time $T_i = T - x/V_1$ may be considered to be consisting of two *delay times*: D_1 associated with the shot end of the trajectory and D_2 with the detector end.

At a distance x_{cros} (Fig. 3.10) two linear segment cross. At a distance less than this, the direct wave travelling along the surface, having velocity, V_0 reaches the detector first. At greater distances the waves refracted by the interface arrives earlier than the direct wave. It is referred as crossover distance (x_{cros}).

Using the Eq. (3.27) the depth z can be determined and is expressed as

$$z = \frac{T_i}{2} \frac{V_1 V_0}{\sqrt{V_1^2 - V_0^2}} \tag{3.28}$$

T_i can be determined from the time-distance curve (Fig. 3.10) or numerically from the relation (Eqs. 3.26 and 3.28), i.e., $T_i = T - x/V_1$.

The depth can also be determined from crossover distance (x_{cros}) by using the relationship that $T_0 = \frac{x}{V_0}$ and $T_1 = \frac{x}{V_1} + 2z\frac{\sqrt{V_1^2 - V_0^2}}{V_1 V_0}$ are equal at crossover point. Therefore,

$$\frac{x_{cros}}{V_0} = \frac{x_{cros}}{V_1} + \frac{2z\sqrt{V_1^2 - V_0^2}}{V_1 V_0} \tag{3.29}$$

Thus the depth z can be expressed as

$$z = \frac{1}{2} \frac{V_0 V_1 x_{cros}}{\sqrt{V_1^2 - V_0^2}} \left(\frac{1}{V_0} - \frac{1}{V_1} \right) \tag{3.30}$$

Simplifying Eq. (3.30), the depth z can be expressed as

$$z = \frac{1}{2} x_{cros} \sqrt{\frac{V_1 - V_0}{V_1 + V_0}} \tag{3.31}$$

The three-layer formations, say with velocities V_0, V_1 and V_2 following the condition $V_2 > V_1 > V_0$, in somewhat complicated way, can be treated similarly. The time-depth relations for multilayer earth can also be extrapolated from the relations derived for two-, and three-layer earth models provided the velocity in each layer is greater than the overlying layer.

5.2 Blind Zone

If any bed in the multilayer earth sequence has lower velocity than the overlying layer, then it will not be detected in refraction method as ray in such a layer ray gets deflected in downward direction and cannot travel horizontally as it happens in refraction. This will cause erroneous calculations of depths of all interfaces below the low velocity layer. If the thickness of a sandwiched layer is small and/or the velocity contrast is small with that of the underlying layer then this intermediate layer will not get detected. This type of intermediate layer is known as *blind zone*. If a high velocity layer, like that of basalt, limestone etc., on the surface or at shallower depths masks the deeper layers of equal or slightly greater velocity, then the application of refraction method becomes difficult. But if the shallow high velocity layer is thin compared to the wavelength of the seismic pulse, then it is transparent to seismic energy and will not be able to mask the seismic waves to travel in the deeper horizons. This problem can be overcome in many cases by operating a large shot (source) – geophone (detector) distance. For dipping bed formations the angle of dip can be mathematically determined in refraction method.

5.3 Field Setup

In field layout of refraction surveys, the arrangements for shots and detectors are decided by a number of factors like the nature of geologic problems, terrain and the facilities available. Some of the widely used field set ups are: profile shooting, broadside shooting etc. Profile shooting is the most widely used of all the field set ups in refraction work where the shots and detectors are spread out in long profiles. Successive shots are located at regular intervals along each line and the successive detector spreads are shifted about the same distance as the corresponding shot so that the shot-detector distances remain same for all the shots. In each spread shots are taken in reverse directions. The distance is chosen on the basis of first arrivals refracted from the layer or formation of interest and is guided by the time-distance curve plot (see Fig. 3.10) on the basis of experimental shooting. The main problem of profile shooting is the conversion of intercept times into delay times, representing respective depths to the refractor under shot and receiver. Several techniques have been developed for the same. Some of these methods are: Barthelmes, Wyrobek-Gardner, Slotnick, Wavefront, Tarrant and Hales.

In the areas broadside shooting is resorted to where, for some reason or the other, profile shooting technique is not possible to determine the structure across strike. Whenever possible, the broadsides should be tied to an in-line profile where delay times and depths have been determined at the subsurface tie point by any of the profiling method.

5.4 Elevation and Weathering Corrections

In refraction method the times must be corrected for elevation and for changes in the thickness of the weathered layer (Parasnis 1986; Dobrin and Savit 1988; Telford et al. 1988). The elevation correction takes care to remove the differences in travel times only due to the variations in the surface elevation of the shots and geophone stations. The common approach is to put both the shot and the geophone on the same imaginary datum plane. The time is subtracted if the wave travels from the datum to the respective shot or detector locations if they are higher or added if they are lower.

The weathering corrections, on the other hand, remove the differences in travel times through the near-surface zone of unconsolidated, low-speed sediments which may vary in thickness from place to place. It is necessary to correct for weathering by shooting with short shot-geophone distances.

5.5 *Geophysical Exploration Strategy for Groundwater in Hard Rock Terrain*

Geophysical exploration for groundwater in hard rock terrain along with seismic refraction work, resistivity sounding (VES) and/or resistivity imaging should simultaneously be carried out as the sequences from the top like highly weathered, less weathered, moderately fractured and massive rock or basement show definite electrical and seismic velocity contrasts and two together provide better subsurface information. In addition to these methods some of the other methods discussed below depending on the nature of the problem may also be utilized.

6 Other Relevant Geophysical Surveys

In hard rock terrain electromagnetic method like Very Low Frequency (VLF) method is very useful in detecting near-surface fractures, fissures, lineaments and faults etc. VLF operates mostly in the range of 15 kHz to 25 kHz. In this method only a receiver is required to map the near-surface geological features in the region of influence of a powerful VLF transmitter. Moving transmitter-receiver set up using extra low frequency in the range of 5 kHz to 100 Hz is also used to map somewhat relatively deeper fracture and faults. Ground probe radar (GPR) is also commonly used now-a-days to map the near-surface features, up to depth of about 20–25 m using very high frequency transmitting signals in the range of 200 MHz to a few GHz.

In sedimentary formations the localization of clay in aquifer can be better mapped by induced polarization (IP) method as clay shows relatively higher chargeability than sand formations due to membrane polarization characteristics of clay. IP profiling and sounding will enable to map such a region.

7 Geophysical Logging

In groundwater exploration geophysical logging has not been given its due that it deserves in many of the programmes. Porosity is effective parameter to decide the quantity of water in an aquifer. Porosity is the fractional portion of rock volume which is occupied by pore space, expressed as percent. It is, however, not enough to ensure that high porosity signifies better quality of aquifer in terms of water recovery. Besides porosity, an equally important property is the degree to which the pores are interconnected, that is permeability. Permeability is measured in darcy or subunit millidarcy. It has the dimensions (length)^2 and 1 darcy $= 0.987 \times 10^{-12}$ m ^{2}. The formations with high porosity but poor permeability do not make good aquifer because these do not lead to desirable water recovery. For good water well recovery it is essential that the formation be permeable to allow the flow of water under a hydraulic pressure gradient. Sandy layers, sandstone formations etc. have both high porosity and permeability and make good aquifer. Clay and shales have good porosity but poor permeability and, therefore, are not candidate for aquifer.

The resistivity of porous, water bearing rocks, free of clay minerals, follows Archie's law stated in Eq. (3.5). Using the notations of the logging industry the Eq. (3.5) can be rewritten as

$$R_t = \frac{R_w}{\varphi^m S_w^n} \tag{3.32}$$

where R_t is the true resistivity of the geologic formation, R_w is the resistivity of water in the pores and S_w is the fraction of the pore volume filled with water. It is thus very clear from Eq. (3.32) that there are basically three logs that we must consider in groundwater exploration activity, namely permeability zone log, resistivity and porosity logs.

7.1 *Permeability Zone Log*

In permeability zone log the self-potential (SP) and gamma-ray logging play the main role. In SP log the self-potential value is generally constant in the same formation but jumps suddenly to another level on crossing the boundary between two formations. In gamma-ray logging the technique utilizes natural radioactivity of rocks and is useful in correlating formations like clay, shale which show very high radioactivity.

7.2 *Resistivity Log*

There are several types of resistivity of logs used in petroleum industry and some of these are relevant in groundwater exploration work as well. The various types of resistivity logs are single electrode, normal arrangement using short and long normal, lateral arrangement, focused-current logs etc. Single electrode is obsolete now as the apparent resistivity is considerably different from the true value.

In normal arrangement one current and one potential electrode on the logging sonde are closely spaced – 16 inches apart for the short normal and 64 inches for long normal. The other two electrodes are placed on the surface at the top of the hole. Focussed current logs are used for varieties of problems in oil industry and some of these may not always be relevant for groundwater logging problems.

7.3 Porosity Logs

Three types of logs play very important role in determining porosity and these are density, neutron and sonic or acoustic logs. If d_M, d_F and d_B are the rock matrix density of a clean formation, density of the fluid filling and bulk density respectively, then

$$d_B = \varphi d_F + (1 - \varphi) d_M$$

and finally the expression for porosity, φ, can be expressed as

$$\varphi = \frac{(d_M - d_B)}{(d_M - d_F)} \tag{3.33}$$

In the above expression d_B is obtained by lowering a gamma-ray source in the hole and counting the number of high energy gamma rays arriving at fixed distance when the source radiation is scattered by collisions with the electrons in the rock matrix. It essentially uses the well known Compton scattering phenomena. Dual-detector density log, also known as compensated density log (CDL), is used for density logging. Subsequent development is the litho-density (LDT) logging. In this case low-energy gamma rays within a narrow band are registered in addition to the high-energy gamma rays in the CDL. Absorption of low-energy gamma rays is a photoelectric phenomenon and not a loss of energy as in Compton scattering.

In neutron logging, two of the methods employing neutron are neutron-gamma and neutron-neutron logging. In rocks, hydrogen nuclei are present in water (and also in hydrocarbons) and the resulting gamma radiation can be detected by lowering a neutron source just ahead of a gamma ray counter. In the neutron-neutron method the intensity of the neutrons scattered by the hydrogen nuclei, rather than the intensity of the gamma radiation due to their capture is detected. Presently compensated neutron log (CNL) is used instead of neutron-gamma and the single detector neutron-neutron logs. It is the general experience that neutron log indicates a somewhat higher porosity than the density log. Empirically the true porosity quite likely is the mean of these two values.

The velocity of sound in a porous rock depends upon the velocity in the rock matrix as well as that in the fluid filling the pore spaces and sonic log observes these parameters to determine the porosity of the formation. Wyllie's formula for the travel time is given by

$$t = \varphi t_F + (1 - \varphi) t_M \tag{3.34}$$

where t_F and t_M are travel times through the fluid and rock matrix respectively. Equation (3.34) can be rewritten as

$$\varphi = \frac{t - t_M}{t_F - t_M} \tag{3.35}$$

To determine the porosity, j correctly, t_M of different rock types must be known accurately. Standard recommended values for t_M for sandstone, limestone and dolomite are 177, 144 and 161 μsm^{-1} respectively.

To model any aquifer system, which ultimately must be the goal to obtain the aquifer characteristics, the essential requirements are the determination of parameters like thickness, porosity and permeability that play the most important role and these can be reliably estimated only from quality geophysical logs.

8 Conclusion

Applications of geophysical exploration techniques can broadly be related to types of areas: (i) alluvial or sedimentary formations and (ii) hard rock terrain. In the alluvial types of areas mainly vertical electrical sounding (VES) and resistivity imaging are employed. Induced Polarization (IP) method can play an important role in isolating the clay enriched zones in the aquifer horizons. Seismic refraction method is hardly employed in such areas as the seismic velocity of sand and clay horizons are generally not favourably disposed in nature. In hard rock terrains VES, resistivity imaging and seismic refraction methods are mainly employed. Deccan trap areas, grouped as hard rock region, vesicular lava and intertrappean beds also should be mapped for finding the aquifer horizons. The other geophysical methods like self-potential, time-and frequency domain electromagnetic methods, and very low frequency (VLF) methods may be gainfully employed in groundwater exploration under various types of geological conditions.

References

Barker, R. (1981). The offset system of electrical resistivity sounding and its use with a multicore cable. Geophys Prospec, 29: 128–143.

Beresnev, I.A., Hruby, C.E. and Davis, C.A. (2002). The use of multi-electrode resistivity imaging in gravel prospecting. J Appl Geophy, 49: 245–254.

Bhattacharya, Bimalendu B. and Shalivahan (2016). Geoelectric Methods: Theory and Applications: McGraw Hill Education (India) Private Limited, 735 p.
Dobrin, Milton B. and Savit, Carl H. (1988). Introduction to Geophysical Prospecting (International Edition). McGraw Hill, 867 p.
Griffiths, D.H., Turnbull, J. and Olayinka, A.J. (1990). Two-dimensional resistivity mapping with a computer-controlled array. First Break, 8**:** 120–129.
Griffiths, D.H. and Barker, R.D. (1993). Two-dimensional resistivity imaging and modelling in areas of complex geology. J Appl Geophy, 29**:** 211–226.
Loke, M.H. (2000). Tutorial: 2-D and 3-D electrical imaging surveys.
Lowrie, William (1997). Fundamentals of Geophysics. Cambridge University Press, 354 p.
Minsley, B.J. (2007). Modeling and inversion of self-potential data. Ph D thesis. Massachusetts Institute of Technology, 251 p.
Parasnis, D.S. (1986). Principles of Applied Geophysics (Fourth Edition). Chapman and Hall, 402 p.
Telford, T.M., Geldart, L.P., Sheriff, R.E. and Keys, D.A. (1988). Applied Geophysics (First Indian Edition), Cambridge University Press, 860 p.
Xu, B. and Noel, M. (1993). On the completeness of data sets with multi-electrode systems for resistivity surveys. Geophys Prospec, 41**:** 791–801.

Chapter 4
Environmental Isotopes in Groundwater Applications

Sudhir Kumar

1 Introduction

The total amount of water on this earth is virtually constant but its distribution over time and space varies largely. Wherever people live, they must get a clean and continuous water supply as a primary requirement. The assessment of quality, supply and renewal of resources of water is a well known problem, but it is becoming critical with the growth of population and rapid industrialization.

Water resources management, to sustainably meet the current and future water demands, needs to have a comprehensive assessment of the quality and quantity of the resources. The assessment of both quality and quantity of water, though well recognised, is hampered due to non-availability of detailed hydrological information and understanding. Generally in our country, the available hydrological data/information is either incomplete or has wide gaps in it. These gaps are particularly acute with respect to groundwater resources.

It is estimated that more than 97% of the Earth's available water is saline and out of the rest 3% fresh water, 97% is located underground. Though groundwater is a vital resource, yet it is often poorly understood and poorly managed.

Stable and radioactive isotope techniques are cost effective tools in hydrological investigations and assessments, and are critical in supporting effective water management. Isotopes help in understanding various hydrological processes. Isotope techniques using "environmental isotopes" are commonly used in the developed countries by meteorologists, hydrologists and hydrogeologists in the study of water. The use of these techniques has greatly increased in our country, but still it requires momentum and training of the field persons in this subject. Study of the isotopes of oxygen and hydrogen in water, or of elements contained in dissolved salts, which

S. Kumar (✉)
National Institute of Hydrology, Roorkee, Uttarakhand, India
e-mail: sudhir.nih@gmail.com

P. K. Sikdar (ed.), *Groundwater Development and Management*,
https://doi.org/10.1007/978-3-319-75115-3_4

have the same behaviour as water, enable exact recording of phenomena affecting the occurrence and movement of water in all its forms. In the past few decades, sophisticated nuclear-hydrological instrumentation have been developed to measure accurately both radioactive as well as stable isotopes and accordingly various nuclear methods have been evolved. It is, therefore, now very easy to solve many hydrological problems related to planning of water resources, agriculture, industry and habitation using isotope techniques, which were very difficult, sometimes impossible to tackle in the past.

The use of isotopes in hydrology was introduced in early 1950's with the application of radiocarbon dating technique for determining the age of groundwater. After that, the application of isotopes is being successfully used to find the effective solutions of various hydrological problems in the developed countries. Later on the International Atomic Energy Agency (IAEA), Vienna, an independent intergovernmental organisation within the United Nations system, took a leading role in the development and use of isotope techniques in hydrology. Presently, isotope techniques are used frequently in the developed countries while their use in the developing counties is increasing slowly.

In groundwater, isotopes are commonly employed to investigate:

- Sources and mechanisms of groundwater recharge;
- Groundwater age and dynamics;
- Interconnections between aquifers;
- Interaction between surface water and groundwater;
- Effectiveness of artificial recharge measures;
- Groundwater salinization; and
- Groundwater pollution.

2 Isotopes

Isotopes are the atoms of an element having same atomic number (*Z*) but different atomic weight (*A*). In other words, the atoms of an element having different number of neutrons (N) but same number of protons or electrons are called isotopes. For example, hydrogen has three isotopes having the same atomic number of 1 but different atomic masses or weights of 1, 2 and 3 respectively i.e., ${}_1^1H_0$, ${}_1^2H_1$, and ${}_1^3H_2$ (Fig. 4.1). Similarly, oxygen has eleven isotopes, ^{12}O, ^{13}O, ^{14}O, ^{15}O, ^{16}O, ^{17}O, ^{18}O, ^{19}O, ^{20}O, ^{21}O and ^{22}O, and carbon has three isotopes ^{12}C, ^{13}C and ^{14}C.

There are two more terms i.e., isobars and isotones that are used to differentiate and distinguish the atoms of different elements showing similarities in physical and chemical properties. Isobars have same atomic weight ($Z + N$), i.e., $^{76}Ce_{32}$ and $^{76}Se_{34}$; or $^{58}Fe_{26}$ and $^{58}Ni_{27}$, but different atomic number. Isotones have same number of neutrons but different atomic number, i.e., $^{37}Cl_{17}$ and $^{39}K_{19}$ (both have 20 neutrons in the nuclei). Electronic configurations of atom of some light elements are shown in Fig. 4.2.

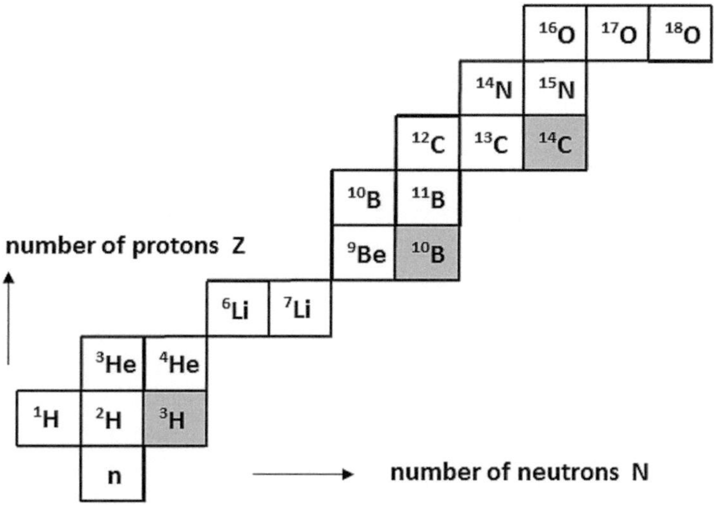

Fig. 4.1 The isotopes of an element (equal *Z*) are found in a horizontal row, isobars (equal *A*) along diagonal lines, isotones (equal *N*) in vertical columns. The natural radioactive isotopes of H, Be, and C are marked grey

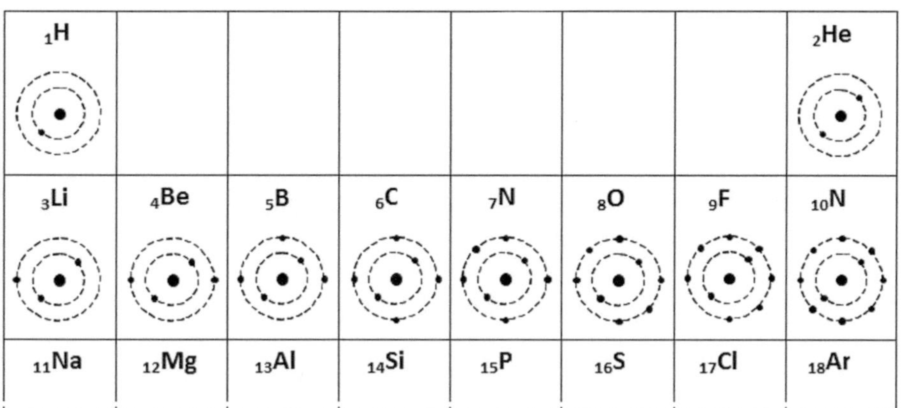

Fig. 4.2 Electronic configurations of atoms of various light elements

2.1 Classification of Isotopes

Isotopes are classified in two important categories: (i) stable isotopes and (ii) unstable isotopes.

Isotopes are also classified as natural and artificial isotopes, i.e., the isotopes that occur naturally are called natural isotopes while those produced in a reactor or laboratory under controlled conditions are known as artificial isotopes. Both, stable

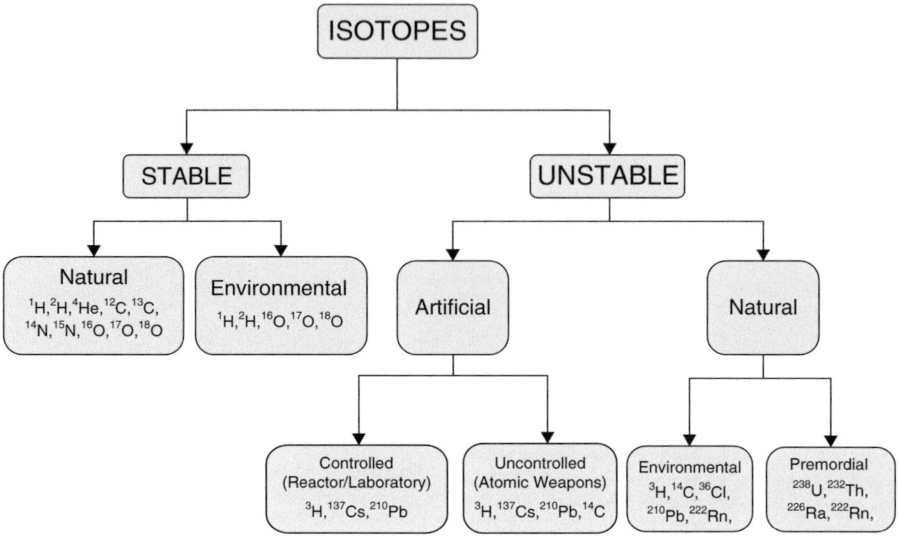

Fig. 4.3 Classification of isotopes

and radioactive isotopes occur naturally and some of the radioactive isotopes are produced artificially.

Another category of isotopes has been devised that is called environmental isotopes. These isotopes have different types of categories i.e. naturally occurring stable and radioactive isotopes and radioisotopes introduced into the atmosphere due to anthropogenic activities etc. The environmental radioisotopes whether naturally occurring due to cosmic ray interaction with various gaseous molecules or anthropogenically produced and becoming the part of hydrological cycle are safe in normal conditions and do not pose any threat to human health.

Classification scheme of various isotopes is shown in Fig. 4.3.

2.2 *Stable Isotopes*

Stable isotopes of an element are the atoms, which do not decay with time or take infinite time to decay. The atoms are satisfied with the present arrangement of proton, neutron and electron. Over 2000 isotopes of 92 naturally occurring elements have been identified out of which several hundred are stable isotopes. The stable isotopes commonly used in hydrology as water tracers are ^{2}H, ^{18}O and ^{13}C.

As water molecule is made up of two hydrogen atoms and one oxygen atom, therefore, isotopically 18 types of water are possible, out of which $^{1}H^{1}H^{16}O$, $^{1}H^{1}H^{18}$

Table 4.1 Stable isotopes with their natural abundance and reference standards used for ratio measurements

Isotope	Ratio	% natural abundance	Reference (abundance ratio)	Commonly measured phases
^{2}H	$^{2}H/^{1}H$	0.015	VSMOW (1.5575×10^{-4})	H_2O, CH_2O, CH_4, H_2, OH^- minerals
^{3}He	$^{3}He/^{4}He$	0.000138	Atmospheric He (1.3×10^{-6})	He in water or gas, crustal fluids, basalt
^{6}Li	$^{6}Li/^{7}Li$	7.5	L-SVEC (8.32×10^{-2})	Saline waters, rocks
^{11}B	$^{11}B/^{10}B$	80.1	NBS 951 (4.04362)	Saline waters, clays, borate, rocks
^{13}C	$^{13}C/^{12}C$	1.11	VPDB (1.1237×10^{-2})	CO_2, carbonate, DIC, CH_4, organics
^{15}N	$^{15}N/^{14}N$	0.366	AIR N_2 (3.677×10^{-3})	N^2, NH_4^+, NO_3^-, N- organics
^{18}O	$^{18}O/^{16}O$	0.204	VSMOW (2.0052×10^{-3}) VPDB (2.0672×10^{-3})	H_2O, CH_2O, CO_2, sulphates; NO_3^-, carbonates, silicates OH^- minerals
^{34}S	$^{34}S/^{32}S$	4.21	CDT (4.5005×10^{-3})	Sulphates, sulphides, H_2S, S-organics
^{37}Cl	$^{37}Cl/^{35}Cl$	24.23	SMOC (0.324)	Saline waters, rocks, evaporites, solvents
^{81}Br	$^{81}Br/^{79}Br$	49.31	SMOB	Developmental for saline waters
^{87}Sr	$^{87}Sr/^{86}Sr$	$^{87}Sr = 7.0$ $^{86}Sr = 9.86$	Absolute ratio measured	Water, carbonates, sulphates, feldspar

O, $^{1}HD^{16}O$, $^{1}HD^{18}O$, $^{1}H^{1}H^{17}O$ and $^{1}HD^{17}O$ are most abundant. The natural occurrence of very abundant types of water molecules is given below:

$$^{1}H_2{}^{16}O\tilde{\ }9,99,680 \text{ ppm } (99.9680\%)$$
$$^{1}HD^{16}O\tilde{\ }320 \text{ ppm } (0.032\%)$$
$$^{1}H_2{}^{18}O\tilde{\ }2,040 \text{ ppm } (0.204\%)$$

Natural abundance and reference standards of some important isotopes used in hydrological studies are shown in Table 4.1.

2.2.1 Isotopic Notations and Measurements

Stable isotopes are measured in terms of abundance ratios i.e. atomic mass of heavy atom to the atomic mass of light atom. For example, heavy water ($^{2}H_2{}^{16}O$) has a mass of 20 compared to normal water $^{1}H_2{}^{16}O$ which has a mass of 18. Similarly, heavier stable molecule of water $^{2}H_2{}^{18}O$ has a mass 22. This is because of the variation in the number of neutrons. The absolute abundance of isotopes is not

Table 4.2 Oxygen and hydrogen δ-values of the major water reference material

Name	Material	Status	Distribution	$\delta^{18}O$[‰]	$\delta^{2}H$[‰]
VSMOW	Water	CM	IAEA, NIST	0	0
SLAP	Water	CM	IAEA, NIST	−55.5	−428
GISP	Water	RM	IAEA, NIST	−24.78 ± 0.08 Gonfiantini et al. (1995)	−189.73 ± 0.87

usually measured in natural waters and also in other materials. Only the relative difference in the ratio of the heavy isotopes to the more abundant light isotope of the sample with respect to a reference is determined. The difference is designated by a Greek letter δ and is defined as follows:

$$\delta = \left(R_{\text{sample}} - R_{\text{reference}}\right)/R_{\text{reference}}$$

where R's are the ratios of the $^{18}O/^{16}O$ and D/H isotopes in case of water.

The difference between samples and references are usually quite small, δ values are, therefore, expressed in per mille differences (‰) i.e. per thousand, $\delta\,(‰) = \delta \times 1000$.

$$\delta(‰) = [(R_s - R_r)/R_r] \times 10^3 = [(R_s/R_r) - 1] \times 10^3$$

If the δ value is positive, it refers to the enrichment of the sample in the heavy-isotope species with respect to the reference and negative value corresponds to the sample depleted in the heavy-isotope species.

The reference standards normally considered are SMOW (Standard Mean Oceanic Water) and VSMOW (Vienna Standard Mean Ocean Water). VSMOW has the same ^{18}O content as defined in SMOW but its D-content is 0.2‰ lower. Other standards for calibration of oxygen and hydrogen isotopes are SLAP (Standard Light Antarctic Precipitation), and GISP (Greenland Ice Sheet Precipitation). The isotopic values of the standards are given in Table 4.2.

2.3 *Radioisotopes*

The unstable isotopes are the atoms of an element which do not satisfy with the present arrangement of atomic particles and disintegrate by giving out alpha (α), beta (β) particles and/or gamma (γ) radiation etc. and transform into another type of atom. This process continues till the stable nuclide (element) is formed. Because of disintegration or the property of giving out radiation, the unstable isotopes are also called radioactive isotopes.

In early days, the use of radioisotopes was in vogue. Mostly, the radioisotopes, artificially produced in reactor/laboratory were used as tracers. The radioisotope of

Table 4.3 Various radioisotopes with their half-life, decay mode and principal sources

Isotope	Half-life (years)	Decay model	Principal sources
^{3}H	12.43	β^{-}	Cosmogenic, weapons testing
^{14}C	5730	β^{-}	Cosmogenic, weapons testing
^{36}Cl	301,000	β^{-}	Cosmogenic and subsurface
^{39}Ar	269	β^{-}	Cosmogenic and subsurface
^{85}Kr	10.72	β^{-}	Nuclear fuel processing
^{81}Kr	2,10,000	ec	Cosmogenic and subsurface
^{129}I	1.6×10^{7}	β^{-}	Cosmogenic, subsurface, nuclear reactors
^{222}Rn	3.8 days	α	Daughter of ^{226}Rn in ^{238}U decay series
^{226}Ra	1600	α	Daughter of ^{230}Th in ^{238}U decay series
^{230}Th	75,400	α	Daughter of ^{234}U in ^{238}U decay series
^{234}U	2,46,000	α	Daughter of ^{234}Pa in ^{238}U decay series
^{238}U	4.47×10^{9}	α	Primordial

β^{-} – beta emission.; α – alpha emission.; ec – electron capture

hydrogen (tritium) in the form of water molecule ($^{3}H_2O$) and denoted by symbol ^{3}H or T is still widely used for various hydrological studies. There are other varieties of artificially produced radioisotopes like ^{60}Co, ^{82}Br, ^{131}I, ^{137}Cs, ^{198}Au, $^{226}Ra/^{241}Am$ etc. that are used for various hydrological investigations.

However, with the introduction of sophisticated instrumentation, the radioisotopes that occur in traces in the environment and are part of hydrological cycle, are used. This has reduced the use of artificial radioisotopes tremendously, as the radioisotopes are considered to be a health hazard both by the user as well as by the public.

The details of various radioisotopes with their half-lives, decay mode, principal sources etc. are given in Table 4.3.

2.4 *Environmental Isotopes*

Environmental isotopes, both stable and radioactive (unstable), occur in the Earth's environment in varying concentrations with respect to location and time over which the investigator has no direct control. Environmental isotopes are neither required to be purchased nor to be injected as these are freely available and automatically injected in the hydrological cycle. Earlier only artificially produced radioactive isotopes were used but with the better instrumentation facilities, now-a-days environmental isotopes are used more and more except in few cases where artificial radioisotopes can only be useful. The most commonly used environmental stable isotopes are deuterium (D), oxygen-18 (^{18}O), carbon-13 (^{13}C) and radioisotopes tritium (^{3}H) and carbon-14 (^{14}C), nitrogen-15 (^{15}N), chlorine-36 etc. Silicon-32 (^{32}Si), caesium-137 (^{137}Cs) and lead-210 (^{210}Pb) etc. are also used as environmental radioisotopes for few specific studies in hydrology. Silicon-32 (^{32}Si) is potentially attractive, because its half-life (100 year) is between that of ^{3}H and ^{14}C. Argon-39 (^{39} Ar) has also been investigated and research is still in progress, but the disadvantage of using both ^{32}Si and ^{39}Ar is that large amount of water (a few tons) is required to provide required amount of sample for measurement.

Environmental tritium is used to date the groundwater upto 50 years, while carbon-14 is used upto the age of 40,000 years.

3 Fractionation of Stable Isotopes

Among the different properties of isotopic substances, one that is of particular importance for hydrologists is their slightly different physico-chemical behaviours that lead to isotopic fractionation effects. Isotopic fractionation is the basis for its utilization in stable isotope geochemistry, isotope geology, biogeochemistry, paleo-oceanography and others. For instance, the analysis of the ratio of stable oxygen isotopes in calcium carbonate, secreted by organisms like belemnites, mollusks and foraminifera and buried in deep-sea sediments, has permitted the reconstruction of paleo-temperatures for the last 150 million years or so (McCrea 1950; Epstein et al. 1953; Emiliani 1966).

3.1 Isotope Fractionation

According to classical chemistry, the chemical characteristics of isotopes, or rather of molecules that contain different isotopes of the same element (such as $^{13}CO_2$ and $^{12}CO_2$) are same. Largely this is true; however, if sufficiently accurate measurement are made using modern mass spectrometers, tiny differences in chemical as well as physical behaviour of so-called *isotopic molecules* or *isotopic compounds* can be observed. Differences in chemical and physical properties arising from variations in atomic mass of an element are called "*isotope effects*". This phenomenon can be observed as a result of change in isotopic composition by transition of a compound from one state to another (liquid water to water vapour), or by conversion of one compound into another compound (carbon dioxide into plant organic carbon), or due to difference in isotopic composition between two compounds in chemical equilibrium (dissolved bicarbonate and carbon dioxide).

It is well known that the electronic structure of an atom of an element essentially determines its chemical behaviour, whereas the nucleus is more or less responsible for its physical properties. Because all isotopes of a given element contain the same number and arrangement of electrons, a far-reaching similarity in chemical behaviour is the logical consequence. However, this similarity is not unlimited; certain differences exist in physicochemical properties due to mass differences. The replacement of any atom in a molecule by one of its isotopes produces a very small change in chemical behaviour. The addition of one neutron can, for instance, depress the rate

Table 4.4 Characteristic of physical properties of $^{1}H_2{}^{16}O$, $^{2}H_2{}^{16}O$ and $^{1}H_2{}^{18}O$

S. No.	Property	$^{1}H_2{}^{16}O$	$^{2}H_2{}^{16}O$	$^{1}H_2{}^{18}O$
1	Density (20 °C in g cm^{-3})	0.997	1.1051	1.1106
2	Temperature of greatest density (°C)	3.980	11.240	4.300
3	Melting point (760 Torr, in °C)	0.00	3.810	0.280
4	Boiling point (760 Torr, in °C)	100.00	101.420	100.140
5	Vapour pressure (at 100 °C, in Torr)	760.00	721.600	
6	Viscosity (at 20 °C, in centipoise)	1.002	1.247	1.056

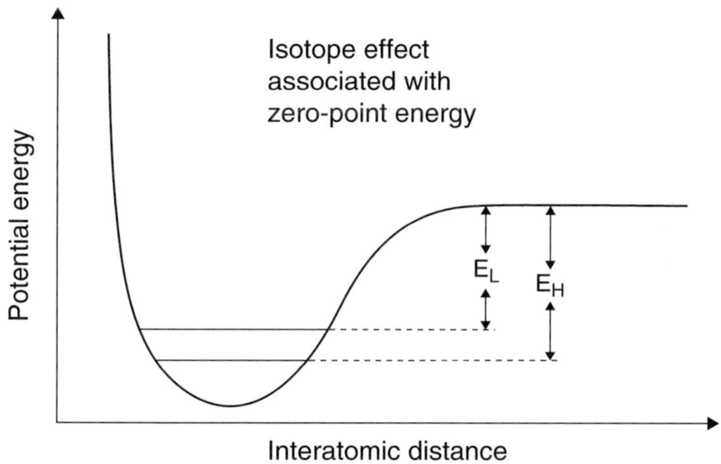

Fig. 4.4 Schematic potential energy curve for the interaction of two atoms in a stable molecule or between two molecules in a liquid or solid. (After Bigeleisen 1965)

of chemical reaction considerably. Furthermore, it may lead, for example, to a shift of the lines in the Raman and IR spectra. Such mass differences are most pronounced among the lightest elements. For example, some differences in physio-chemical properties of $^{1}H_2{}^{16}O$, $D_2{}^{16}O$, $^{1}H_2{}^{18}O$ are listed in Table 4.4. To summarize, the properties of molecules differing only in isotopic substitution are qualitatively the same, but quantitatively different.

Differences in the chemical properties of the isotopes of H, C, N, O, S and other elements are determined experimentally. These differences in the chemical properties can lead to considerable separation of the isotopes during chemical reactions.

The theory of isotope effects and a related isotope fractionation mechanism is discussed here briefly. Differences in the physicochemical properties of isotopes arise as a result of quantum mechanical effects. Figure 4.4 shows schematically the energy of a diatomic molecule as a function of the distance between the two atoms. In the figure, the upper horizontal line (E_L) represents the dissociation energy of the light molecule and the lower line (E_H) that of the heavy one is actually not a line, but an energy interval between the zero-point energy level and the "continuous" level.

This means that the bonds formed by the light isotope are weaker than bonds involving the heavy isotope. Thus, during a chemical reaction, molecules bearing the light isotope will, in general, react slightly more readily than those with the heavy isotope.

Isotopic composition of an element in a certain compound changes by the transition of the compound from one physical state or chemical composition to another.

Mass-dependent isotope fractionation takes place due to three processes, namely *thermodynamic* (in physical or chemical equilibrium systems), *kinetic* (in one-way (bio)chemical reactions) and *transport fractionation* during diffusive processes.

Within the hydrologic cycle, the variability in the isotope composition results primarily from mass-dependent isotope fractionation accompanying the phase transitions and transport processes in the cycle.

The differences in physical and chemical properties of isotopic compounds (i.e. chemical compounds consisting of molecules containing different isotopes of the same element) are brought about by mass differences of the atomic nuclei. The consequences of these mass differences are two-fold:

1. The heavier isotopic molecules have a lower mobility. The kinetic energy of a molecule is solely determined by temperature: $kT = \frac{1}{2}\,mv^2$ (k = Boltzmann constant, T = absolute temperature, m = molecular mass, v = average molecular velocity). Therefore, molecules have the same $\frac{1}{2}\,mv^2$, regardless of their isotope content. This means that the molecules with larger m necessarily have a smaller v. Some practical consequences are: (a) heavier molecules have a lower diffusion velocity; and (b) the collision frequency with other molecules – the primary condition for chemical reaction – is smaller for heavier molecules; this is one of the reasons why, as a rule, lighter molecules react faster.
2. The heavier molecules generally have higher binding energies.

Examples of this phenomenon are:

- ${}^{1}H_2{}^{18}O$ and ${}^{1}H^{2}H^{16}O$ have lower vapour pressures than ${}^{1}H_2{}^{16}O$; they also evaporate less easily, and
- in most chemical reactions the light isotopic species reacts faster than the heavy. For example, $Ca^{12}CO_3$ dissolves faster in an acid solution than does $Ca^{13}CO_3$.

In isotope equilibrium between two chemical compounds, the heavy isotope is generally concentrated in the compound, which has the largest molecular weight.

3.2 Equilibrium and Kinetic Fractionation

The isotope fractionation can occur due to physicochemical reactions under: (i) equilibrium condition; and (ii) non-equilibrium condition and also due to molecular diffusion. In the first case, reactant and product interact for sufficiently long duration so that isotopic equilibrium is established, whereas in the second case, sudden change in temperature or addition or removal of product or reactant prevents the isotopic equilibrium in a given physicochemical reaction. In the molecular diffusion, product under isotopic equilibrium slowly diffuses out of the product reservoir but the product reservoir may remain in isotopic equilibrium with large reservoir of reactant. Although these three types of fractionating processes are known to occur, for all practical purposes, fractionations are conveniently grouped into two major classes namely, equilibrium and kinetic. Isotopic fractionation can occur during: (i) equilibrium isotopic exchange reactions, and (ii) non-equilibrium (kinetic) processes.

3.2.1 Equilibrium Fractionation

Equilibrium exchange reactions involve thermodynamic equilibrium between the two phases during a phase change process during which redistribution of the isotopes between the two phases (products and reactants) takes place. When forward and backward rates of phase change reaction are equal, the thermodynamic equilibrium is said to have been attained. The equilibrium fractionation is primarily governed by the binding energy of the isotopologues in two phases (isotopologues are molecules of a substance that differ in their isotopic composition). Since the binding energies depend on temperature, the equilibrium fractionation depends on temperature. The heavier isotopologues get concentrated in the phase in which they have greater binding energy. As a 'rule of thumb', among different phases (vapour, liquid, solid) in which H_2O can exist, the denser the phase, the more it tends to be enriched in the heavier isotopes (D and ^{18}O).

The natural example of an equilibrium fractionation is vapour to liquid phase change in cloud, where liquid phase remains in contact with surrounding vapour and is believed to have attained the isotopic equilibrium with vapour phase, before it rains out from the cloud.

3.2.2 Non-equilibrium or Kinetic Fractionation

In systems out of equilibrium, forward and backward reaction rates are not identical, and isotope reactions become unidirectional and irreversible. This can happen due to sudden change in temperature or sudden addition or removal of product or reactant. Under these circumstances, relative amount of the reactant or product suddenly changes and hence the two cannot attain thermodynamic equilibrium with each other. Isotope fractionation due to such reactions is called non-equilibrium or kinetic fractionation. Although this is a non-equilibrium fractionation, it still strongly depends on temperature, like equilibrium fractionation.

Natural example of this type of fractionation is sudden uplift of warm vapour loaded air mass and resultant cloud burst; immediate removal of vapour emanating from water bodies (lakes, flowing rivers, surface water spread for irrigation etc.) due to strong winds etc.

3.2.3 Diffusive Fractionation

Fractionation due to differences in diffusive velocities is a variant of the non-equilibrium or kinetic fractionation. In this case, different isotopic molecules diffuse out of the bulk reservoir, with different diffusive velocities. Unlike the kinetic fractionation caused by sudden temperature change or sudden addition/ removal of mass, which has strong temperature dependence, the diffusive fractionation is strongly governed by diffusive velocities and has only slight temperature dependence, which for all practical purposes can be ignored in actual calculations of fractionation factors. The diffusive fractionation is also referred to as transport fractionation in some literature.

Natural scenarios where diffusive fractionation occurs are evaporation from open water body and diffusion of vapour from surface of water to air above. In isotope hydrology, the most accepted model for non-equilibrium evaporation from a water body involves diffusion of water vapour across a hypothetical microns thin boundary layer over the liquid water interface. The boundary layer has virtually 100% water saturation. This layer is in isotopic equilibrium with the underlying water column. Between the boundary layer and the mixed atmosphere above is a transition zone through which water vapour is transported in both directions by molecular diffusion. It is within the transition zone that non-equilibrium fractionation arises due to the fact that diffusive velocity of ${}^{1}H_2{}^{16}O$ in air is greater than that of ${}^{2}H^{1}H^{16}O$ or ${}^{1}H_2{}^{18}O$.

The vapour over ocean is isotopically depleted with respect to ocean water and the total fractionation is the sum of: (1) the equilibrium fractionation between ocean water and thin boundary layer; and (2) the diffusive fractionation between boundary layer and mixed atmospheric air above.

3.3 *Rayleigh Distillation*

The rain (or snow) is produced when an air mass containing vapour cools. The cooling occurs by (i) adiabatic (without loss of heat) expansion as the warm air rises to lower pressure; or (ii) radiative loss of heat. When the temperature of the air parcel drops below the dew point (temperature at which relative humidity is 100%) vapour condenses into liquid (rain) or snow (solid), in order to maintain the thermodynamic equilibrium at that temperature. The vapour and condensate remain in an intimate contact at a given temperature in the cloud. If the temperature drops further, condensation of vapour proceeds further and if temperature increases, the evaporation of liquid or snow occurs. The first condensates are tiny particles which float and remain in close contact with vapour under equilibrium. When tiny particles coalesce and sufficiently massive particle is formed, it begins to fall under gravity. This is how we receive rain on ground.

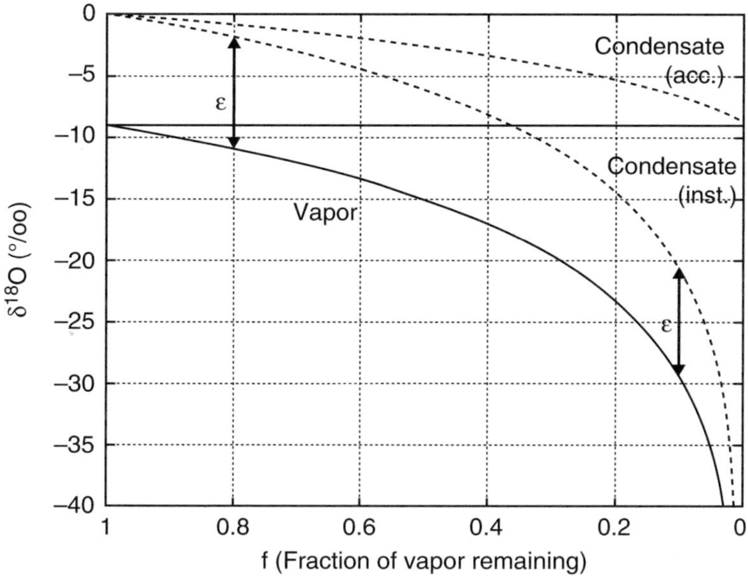

Fig. 4.5 Rayleigh fractionation process of water condensing from vapour. The oxygen isotope fractionation between reservoir (vapour) and the instantaneous product (condensate, inst.) is constant. When no vapour remains ($f = 0$), the accumulated product (condensate, acc.) has the same isotopic composition as the initial vapour since the vapour has completely condensed (horizontal line)

During equilibrium condensation at prevalent in-cloud temperature, isotopic molecular species differentially partition between vapour and liquid (or solid) phase such that more condensed phase is always enriched in heavier isotopes and vapour is depleted in heavier isotopes. Thus, with progressive rainout along the wind trajectory from marine vapour source region into continental interiors, the remaining vapour becomes progressively more and more depleted in heavier isotopes (Fig. 4.5). At every stage of rainout, the condensate is always enriched in heavier isotopes compared to vapour. Since vapour is progressively depleted at successive stage of rainout, the resultant rain at succesive stage is also isotopically depleted.

The effect of the admixture of evapotranspiration flux on the isotopic composition of the downwind atmospheric vapour and subsequent precipitation depends on the details of the evapotranspiration process. Transpiration returns precipitated water essentially un-fractionated back to the atmosphere, despite the complex fractionation in leaf water (Forstel 1982). Transpiration restores both the vapour mass and the heavy isotope depletion caused by the rainout in such a way that the next rainout event is not as depleted as it would have been without the transpiration flux. Under such circumstances, the change in the isotopic composition along the air mass trajectory is only due to the net loss of water from the air mass, rather than being a measure of the integrated total rainout. This causes apparent reduction in the downwind isotopic gradient. The evaporated water, on the other hand, usually gets depleted in heavy species relative to that of transpired vapour, thus restoring the vapour mass to the downwind cloud but reducing its isotopic composition. This may cause apparent increase in the downwind isotopic gradient.

3.4 Global Meteoric Water Line

Due to kinetic and equilibrium processes during evaporation from the ocean and subsequent condensation, δD and $\delta^{18}O$ in the air moisture and precipitation vary with temperature during condensation and with relative humidity during evaporation (Clark and Fritz 1997). The stable isotope ratios of air moisture reflect both the origin of the air-mass and the conditions under which condensation occurs.

Craig (1961) observed that the δD and $\delta^{18}O$ in the precipitation are linearly related, if it has not been evaporated. The relation between δD and $\delta^{18}O$ in precipitation is expressed by the equation:

$$\delta D = 8\delta^{18}O + 10‰(SMOW)$$

This equation, known as the "*Global Meteoric Water Line*" (GMWL), is based on precipitation data from locations around the globe, and has an $r^2 > 0.95$ (Fig. 4.6). This high correlation coefficient reflects the fact that the oxygen and hydrogen stable isotopes in water molecules are intimately associated.

Subsequent global monitoring of the stable isotopic composition of precipitation (IAEA Global Network for Isotopes in Precipitation – GNIP) has refined this relationship (Rozanski et al. 1993), as:

$$\delta D = 8.13\,\delta^{18}O + 10.8‰(VSMOW)$$

Craig's line is only global in application, and is actually an average of many local or regional meteoric water lines, which differ from the global line due to varying climatic and geographic parameters (Clark and Fritz 1997). Local lines may differ from the global line in both slope and deuterium intercept. Nonetheless, GMWL provides a reference for interpreting the provenance of groundwater.

The slope and intercept of the "Local Meteoric Water Line" (LMWL), which is the line derived from precipitation collected from a single site or set of "local" sites, can be significantly different from the GMWL. In general, most of these local lines have slopes of 8 ± 0.5, but slopes in the range of 5 and 9 are not uncommon.

Several processes cause waters to scatter away from the GMWL. Water that has evaporated or has mixed with evaporated water typically plots below the meteoric water line along lines that intersect the GMWL (or LMWL) at the location of the original un-evaporated composition of the water. In such waters, slopes in the range of 2 to 5 are common. Geothermal exchange also increases the ^{18}O content of waters

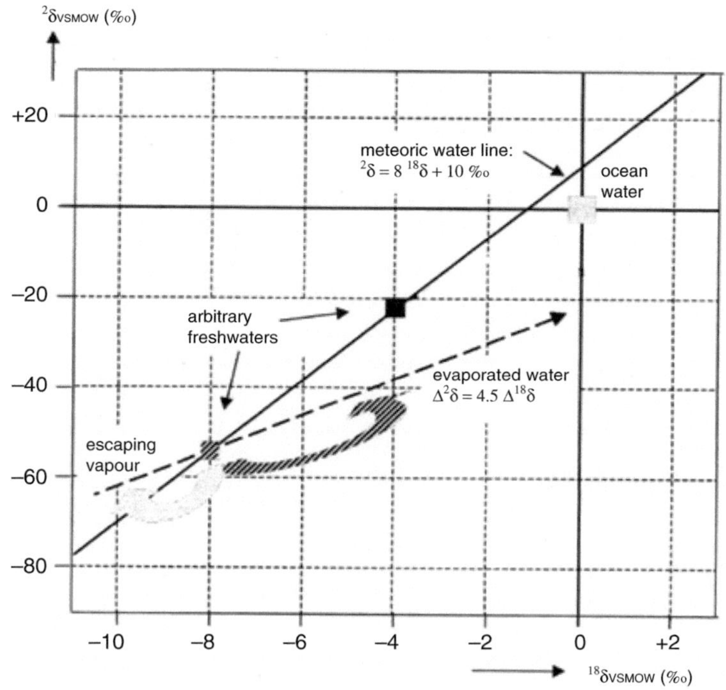

Fig. 4.6 Relation between $\delta^{18}O$ and δD for estuarine mixing and for evaporating surface water. Because the evaporation is a non-equilibrium process, isotope fractionations involved are not necessarily related by a factor of 8, as is the equilibrium condensation process

and decreases the ^{18}O content of rocks as the waters and rocks attempt to reach a new state of isotopic equilibrium at the elevated temperature. This causes a shift in the $\delta^{18}O$ values, but not the δD values of geothermal waters. Low temperature diagenetic reactions involving silicate hydrolysis can sometimes cause increases in the $\delta^{18}O$ and δD values of waters.

3.5 *Isotope Effects*

In principle, variations in $\delta^{18}O$ and δD are coupled under conditions of isotopic equilibrium with slope of 8, but under non-equilibrium (due to complicated kinetic process), the slope of GMWL deviates. The variation of isotopic composition in water/vapour is governed by various factors like latitude/annual temperature, altitude, season, distance from sea, amount of rain, etc. These are called as *isotope effects* and are described below, in brief.

3.5.1 Latitude/Annual Temperature Effect

As discussed earlier, the stable isotopic composition of precipitation on a global scale depends on two processes: (i) formation of atmospheric vapour by evaporation in regions with the highest surface ocean temperatures, and (ii) progressive condensation of the vapour during transport to higher latitudes with lower temperatures.

The progressive rainout process based on the Rayleigh fractionation/condensation model, results in a relation between the observed annually averaged δ^{18}O and δD values of the precipitation and the mean surface temperatures in reasonable agreement with the observed values from the GNIP data network.

Relations established by Daansgard (1964) and later by Yurtsever and Gat (1981) using annual average and monthly average temperatures are:

$$\delta^{18}\mathrm{O} = 0.695\,T_{\mathrm{annual}} - 13.6‰\mathrm{SMOW}\ \text{and,}\ \delta D = 5.6\,T_{\mathrm{annual}} - 100‰\mathrm{SMOW}$$

If monthly temperatures are used, then

$$\delta^{18}\mathrm{O} = (0.338 \pm 0.028)T_{\mathrm{monthly}} - 11.99‰\mathrm{VSMOW}$$

On an average, 1‰ decrease in average δ^{18}O corresponds to 1.1 to 1.7 °C decrease in the average annual temperature. As latitude increases, the temperature decreases, therefore isotopic composition depletes in precipitation. Polar Regions are located at the highest latitudes and at the end of Rayleigh rainout process; thus precipitation has maximum depleted values in heavier isotopic composition.

Thus, water vapours or precipitation depletes in heavier isotopes with the increase in latitude. In low latitudes water vapours depletes very less in heavier isotope species of water molecule. The variation of δ^{18}O is of the order of −0.6‰ per degree of latitude for continental stations of the North America, Europe and about −2‰ per degree latitude for the colder Antarctica stations.

3.5.2 Continental Effect

Precipitation depletes in heavier isotopes of water molecules as clouds move away from the coastal parts. On average, δ^{18}O depletes about −2‰ per 1000 km from seacoast. Global T-δ^{18}O relationship ($\delta^{18}\mathrm{O} = 0.695\ T_{\mathrm{annual}} - 13.6‰$ SMOW) changes significantly due to continental effect.

3.5.3 Altitude or Elevation Effect

Precipitation progressively depletes in δ-values with increase in altitude. This is mainly due to two reasons: (i) decrease in temperature with increase in altitude, and (ii) rainout process increases with increase in altitude due to orographic effect.

In general, $\delta^{18}O$ and δD depletes in the range of −0.15 to −0.5‰ and −1 to −4‰ respectively per 100 m increase in altitude. This effect is used in the identification of location/altitude and source of springs. Source of precipitation can also be identified by knowing the altitude and continental effects.

3.5.4 Seasonal Effects

Variation of δD and $\delta^{18}O$ due to change in season is called seasonal effects. Mainly two factors are responsible for the seasonal effects, i.e., (i) variation in temperature with respect to seasons, and (ii) change in amount of precipitation.

Evaporation and evapo-transpiration increases with increase in temperature. Local or regional water vapours mix with the water vapours originated from the sea and enrich the precipitation in $\delta^{18}O$ and δD. Increase in temperature increases the effect of evaporation in the falling raindrops and enriches the precipitation in δD and $\delta^{18}O$. This effect is least when precipitation occurs in large amount or with high intensity.

3.5.5 Amount Effect

Isotopic composition of the precipitation also depends on the amount of rain: heavier rain events, or greater monthly precipitation amounts, result in more negative δD and $\delta^{18}O$ values. Dansgaard (1964) proposed two major explanations for this amount effect, i.e., (i) lower ambient temperatures cause the formation of clouds with lighter isotopic composition (temperature effect) and also causes heavy precipitation; and (ii) falling raindrops undergo evaporation, enriching the falling rain in the heavy isotopes. This effect is less severe when ambient temperatures are low or when the amount of rain is large.

The amount of monthly rain varies during the year, causing a seasonal variation in the isotopic composition.

3.6 Distribution of Stable Isotopes

3.6.1 Natural Abundance of Stable Oxygen Isotopes

As mentioned earlier, oxygen has three stable isotopes, ^{16}O, ^{17}O and ^{18}O, with abundances of 99.76, 0.035 and 0.2%, respectively (Nier 1950). Variation in ^{18}O values in natural materials has a range of almost 100‰ (Fig. 4.7). ^{18}O is often enriched in (saline) lakes subjected to a high degree of evaporation, while high-altitude and cold-climate precipitation, especially in the Antarctic, is low in ^{18}O. Generally, in the hydrological cycle in temperate climates, values of ^{18}O do not exceed 30‰.

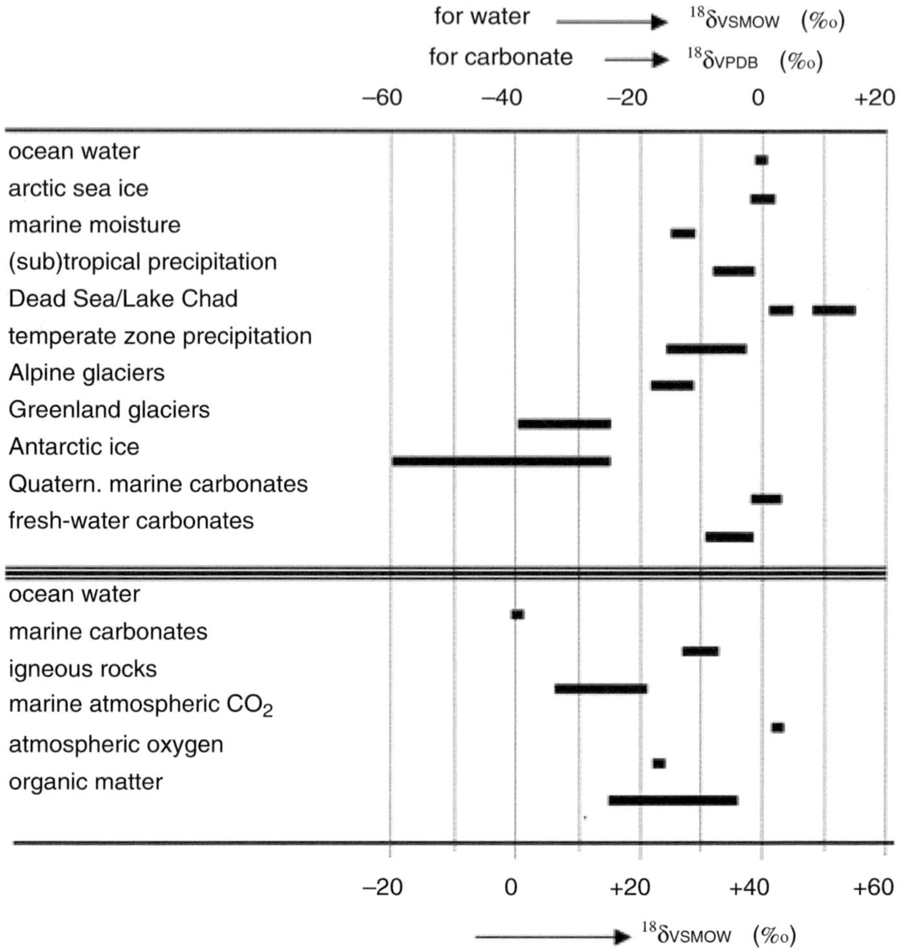

Fig. 4.7 Variation of ^{18}O in natural compounds

3.6.2 Natural Abundance of Stable Hydrogen Isotopes

The chemical element hydrogen consists of two stable isotopes, ^{1}H and ^{2}H (D or Deuterium), with an abundance of about 99.985 and 0.015% and an isotope ratio ^{2}H/^{1}H 0.00015 (Urey et al. 1932). This isotope ratio has a natural variation of about 250‰, higher than the ^{18}O variations, because of the relatively larger mass differences between the isotopes (Fig. 4.8). As with ^{18}O, high ^{2}H concentrations are observed in strongly evaporated surface waters, while low ^{2}H contents are found in polar ice. Variations of about 250‰ are present in the part of the hydrological cycle.

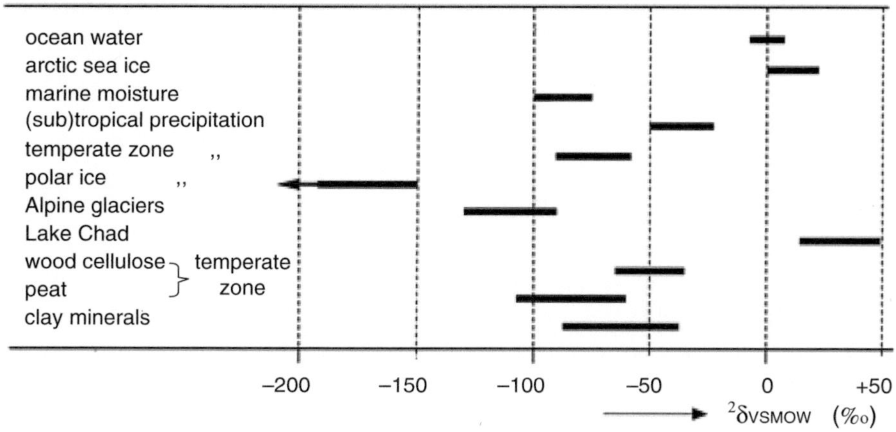

Fig. 4.8 Variation of δ^2H in natural compounds

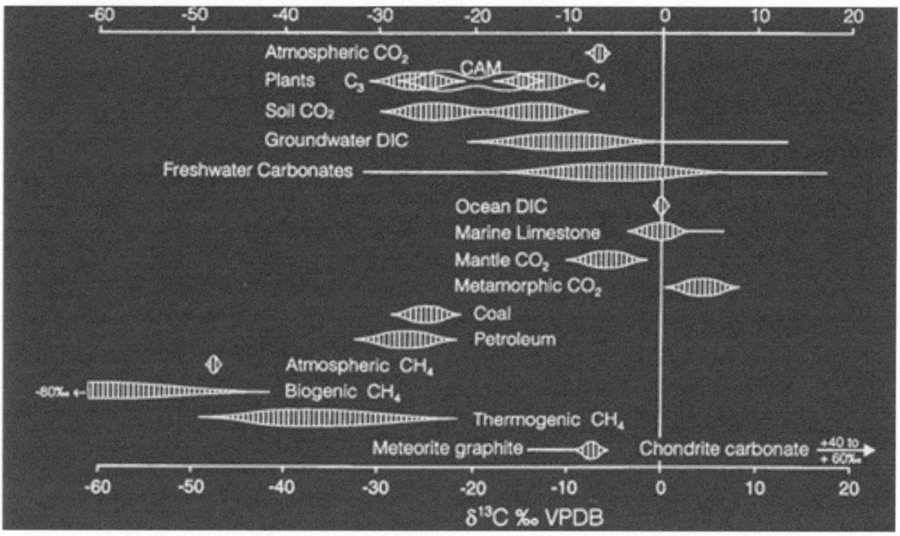

Fig. 4.9 Variation of ^{13}C in natural compounds

3.6.3 Natural Abundance of Stable Carbon Isotopes

There are two naturally occurring stable isotopes of carbon: 12 and 13, which occur in a natural proportion of approximately 99:1. Stable carbon isotopes are utilized differentially by plants during photosynthesis. Most plants in temperate climates follow C3 photosynthetic pathway that yield δ^{13}C values averaging about −26.5‰. Grasses in hot arid climates follow a C4 photosynthetic pathway that produces δ^{13}C values averaging about −12.5‰. The range of variation of 13C in natural materials is shown in Fig. 4.9.

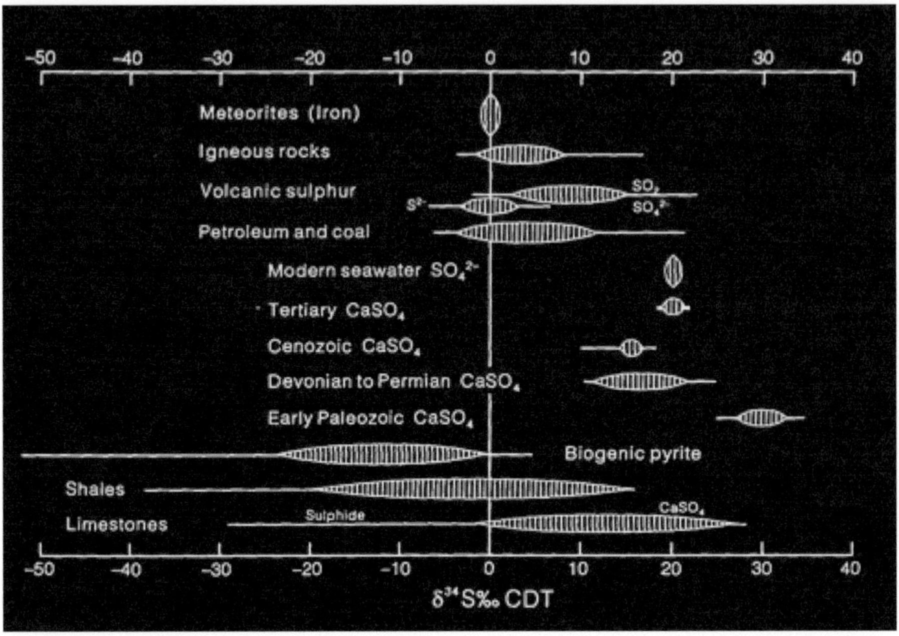

Fig. 4.10 Variation of ^{34}S in natural compounds

3.6.4 Natural Abundance of Stable Sulphur Isotopes

Sulphur (^{34}S) is a major element of sea water and marine sediments and is an essential nutrient for vegetation. A summary of the ranges for ^{34}S in natural materials is shown in Fig. 4.10. Meteorites and magmatic ^{34}sulphur are close to standard Canon Diablo Troilite (CDT), Values exceeding +20 are found associated with evaporates and limestone. Negative $\delta^{34}S$ values are typical of diagenetic environments where reduced sulphur compounds, such as pyrites in shales, are formed (Krouse 1980).

3.6.5 Natural Abundance of Stable Nitrogen Isotopes

Nitrogen (^{15}N), the stable isotope of nitrogen is being used for hydrological studies, particularly for tracing the source and pathways of NO_3 contamination in groundwaters. The combination of ^{15}N and ^{18}O in NO_3 provides a tool to distinguish between nitrates of different origins. The range for ^{15}N in natural materials is shown in Fig. 4.11.

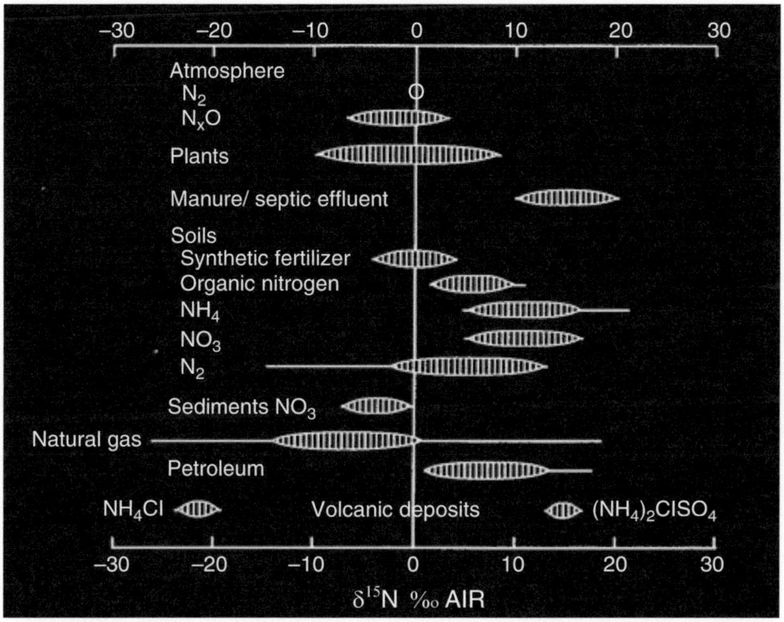

Fig. 4.11 Variation of ^{15}N in natural materials (Amberger and Schmidt 1987; Böttcher et al. 1990; Létolle 1980)

4 Groundwater Age Dating

Groundwater age is defined as the amount of time that has elapsed since a particular water molecule of interest was recharged into the subsurface environment system until this molecule reaches a specific location in the system where it is either sampled physically or studied theoretically for age-dating. On the other hand, groundwater *residence time* is the time taken by the water molecules to travel from the recharge area to the discharge area of the aquifer (Modica et al. 1998). It is the time interval between infiltration and exfiltration of water molecule in the subsurface media. This definition indicates that groundwater age is the groundwater residence time, but only at the discharge area.

Groundwater age is an intrinsic property of the groundwater molecule, like other parameters such as electrical conductivity, and temperature (Goode 1996; Etcheverry and Perrochet 2000; Bethke and Johnson 2002). Thus, age and groundwater are not two separable components, as the water starts aging from the instance it enters into the subsurface. Therefore, as soon as a water molecule enters the subsurface, it becomes groundwater and it has an age.

Dating method is the key criterion in differentiating and classifying groundwater age into three major groups: young, old and very old. Young groundwaters can be dated using techniques whose dating range extends from less than a year to about 50 or 60 years – post thermonuclear bombs or CFC-free groundwaters. Old groundwaters can be dated with methods whose range is between 60 and 50,000 years. Finally, very old groundwaters can be dated using techniques whose coverage ranges from 50,000 to 100,000 years to more than a few tens of millions of years.

The total volume of groundwater resources on the earth is about 23×10^6 km^3 (Gleick 1993). About 4–8 $\times$ 10^6 km^3 of this circulates continuously within the hydrologic cycle (Freeze and Cherry 1979; Gleick 1993) and is called as **active water**. The rest is sometimes referred to as **dead water**. Dead groundwater includes connate water (water entrapped in the sediments during deposition), magmatic water (water contained within magmas deep in the earth), metamorphic water (water produced as a result of recrystallization during metamorphism of minerals), and marine water (seawater intruded into coaster aquifer). Dead water also refers to stagnant waters in isolated envelopes in deep regional aquifers that are not in full hydraulic connection with the surrounding water (Mazor and Nativ 1992). With regard to groundwater dating, **active waters** are of greater interests because they are the main source of water used for various purposes.

A groundwater sample may contain waters that have originated from various recharge areas and input points. The result of such a situation is a sample that consists of many fractions with different ages. This phenomenon is referred to as **mixing** and represents a major challenge to age-dating practice. Similarly, age is subjected to the various processes governing mass transport in aquifers, such as advection, dispersion and mixing. When these processes are active, the age of a groundwater sample cannot be a single number, because the sampled mixture may consist of numerous fractions with different ages. Dispersion and transport of groundwater age fulfil the same function as mixing, but at a micro-scale level. The dispersion of groundwater ages (or residence times) is primarily due to heterogeneous groundwater velocities and hydrodynamic dispersion.

A groundwater sample contains billions of water molecules. In a well-mixed groundwater system, anyone of these molecules may have its own distinct particular age. **Mean age** or the age measured by isotopic and chemical methods, is practically the average of ages of all molecules in the sample. Statistically speaking, mean age is the first moment (i.e., the average) of the age distribution. Understanding the type of statistical age distribution of various ages in a groundwater sample is a must if one has to describe the age of a given sample. This can only be done through application of mathematical models. In situations where groundwater flow can be modelled accurately, the first moments (e.g., mean, variance, and skewness) of the age or residence time distributions can be simulated with temporal moments equations of the advective-dispersive type (e.g., Harvey and Gorelick 1995; Varni and Carrera 1998). Other recent mathematical approaches combining advective-dispersive equations and the reservoir theory (Etcheverry and Perrochet 2000; Cornaton and Perrochet 2005) yield the full age distributions at any point of an aquifer and the residence time distributions at its outlets.

4.1 Tools for Dating Groundwater

There are basically two different ways of estimating groundwater age at a given point in the aquifer: (i) by environmental tracers, and (ii) by groundwater flow modelling. A number of isotope methods can be used to assess mean residence times. The more routinely applied techniques are based on the decay of radionuclides. Those with a long half-life (^{14}C, ^{36}Cl, ^{39}Ar and ^{81}Kr) can be used to date paleo-groundwaters. Short-lived radioisotopes (^{3}H, ^{32}Si, ^{37}Ar, ^{85}Kr and ^{222}Rn) and those produced by man's nuclear activities over the past five decades (^{3}H, ^{36}Cl and ^{85} Kr) indicate modern recharge.

The "sub-modern" period in the dating range between modern waters and paleo-groundwaters is problematic. While this >45 to ~1000 years range can potentially be filled by ^{39}Ar dating ($t_{1/2} = 256$ yr), this method requires rather ideal aquifers, very large samples, complicated sample preparation techniques and special counting facilities. Very few laboratories can afford this and, therefore, ^{37}Ar dating has not developed into a routine tool.

4.2 Dating Young Groundwaters

To date young groundwaters (0–60 years old), ^{3}H, $^{3}H/^{3}He$, ^{4}He, ^{85}Kr, CFCs, SF_6 and ^{36}Cl techniques are used. Dating methods for young groundwaters are typically applicable to unconfined shallow aquifers only, while those methods for dating very old groundwaters are often for deep, confined aquifers. It is, therefore, safe to argue that only dating methods for old groundwaters can be used for both confined and unconfined aquifers.

4.2.1 Tritium Method

Hydrogen has three isotopes: ^{1}H (common hydrogen or protium, H); ^{2}H (deuterium, D); and ^{3}H (tritium, T). Protium has one proton, deuterium has one proton and one neutron, and tritium has one proton and two neutrons. Deuterium (also referred to as heavy stable isotope of hydrogen) and protium are stable, but tritium is radioactive with a half-life of 12.32 years. Tritium is produced in the atmosphere due to the interaction of the cosmic rays produced neutrons in the upper atmosphere with nitrogen atoms:

$$^{14}N + ^{1}n = ^{3}H + ^{12}C$$

Apart from the natural production, lot of tritium was added to the atmosphere due to testing of thermonuclear devices during 6th and 7th decades of the twentieth Century.

Tritium dating is used to trace water sources and to determine age of recent materials (maximum upto age of 50 ± 1 year). Tritium dating provides means to estimate the time since recharge to groundwater system occurred and susceptibility of the groundwater system to contamination. Sources directly fed by rainwater contain the same tritium levels as rainwater. Tritium values are reported in tritium units (TU).

$$1\,\mathrm{TU} = \frac{1(^{3}\mathrm{H})}{10^{18}(^{1}\mathrm{H})} = 7.1\,\mathrm{dpm}/l = 0.12\,\mathrm{Bq}$$

Tritium in the recharging water starts disintegrating into ^{3}He when it enters the subsurface environment:

$$^{2}_{1}\mathrm{H} \rightarrow ^{3}_{2}\mathrm{He} + \beta$$

Therefore, with the passage of time, the concentration of tritium in the groundwater decreases according to the decay law.

$$C = C_0 \ln e^{-\lambda t} \mathrm{or} ^{3}\mathrm{H} = ^{3}\mathrm{H}_0 \ln e^{-\lambda t}$$

where C or ^{3}H is the concentration of tritium in the water sample, C_0 or $^{3}\mathrm{H}_0$ is the concentration of tritium in the recharge water, or the initial value. λ is the decay constant of tritium, 0.056 yr.$^{-1}$. Tritium is of special value in detecting recent recharge because of its short half-life of 12.32 years and because of high levels of tritium in the atmosphere since the beginning of atmospheric testing of thermonuclear devices in 1952. Direct age estimation using tritium is difficult due to variable input of tritium, since 1952.

Tritium concentrations in environment are routinely measured (monthly averages) at various gauging stations fixed by IAEA/WMO since 1961. Tritium contents in the environment has reached to its normal value i.e., normally the natural level of tritium vary between 5 and 15 TU according to the geographical location of the area. The existence of tritium in a water sample is a definite proof of the presence of some components of modern recharge. Therefore, if tritium content is observed to be 10 TU, the groundwater may be of recent origin and if it is 5 TU, then the groundwater may be 12.32 years old (uncorrected).

In some systems, where the discharge represents a variable composite of current-year recharge and older water of low tritium content, it is possible to calculate the ratio of the two components on the basis of periodic tritium sampling.

Isotopic profiles in the unsaturated zone can help in the evaluation of the infiltrated water, although disturbances introduced in the profiles by exchange with the atmosphere need further investigation. The tritium vertical stratification study can be also extended to the saturated zone, but here disturbances introduced by the horizontal groundwater flow should be taken into account. Injection of artificial tritium or other radioisotopes should also be considered when the detection of environmental tritium peaks is difficult.

Major *advantages* of using this method are: (i) it is a well-established and a well-known method with plenty of references, (ii) laboratory facilities are worldwide and the cost of analysis is relatively small, (iii) it is the only tracer that is part of the water molecule, and (iv) tritium is still regarded as a supplementary dating method.

The *disadvantages* of this method are: (i) the method is approaching its expiry date, and (ii) due to the strong latitudinal variation, it is difficult to precisely determine the initial value.

4.2.2 Tritium Helium (^{3}H/He) Method

The fading of the tritium dating method has led to revive an old technique, namely ^{3}H/^{3}He, to replace it. By measuring ^{3}H together with its daughter ^{3}He, true ages can be determined through calculations that do not rely on complicated tritium input function. In ideal circumstances, the method is remarkably accurate for groundwater upto 40 years old.

There are four sources for ^{3}He in the groundwater, i.e., atmospheric, in-situ, tritium decay and from mantle.

The tritium-helium method measures the relative abundance of tritium and ^{3}He in a groundwater sample. The amount of ^{3}He from the decay of tritium is measured along with the amount of tritium remaining in the water. That sum is equal to the amount of tritium that was present at the time of recharge, or the initial value. Mathematically we write:

$$^3\mathrm{H}_0 = {}^3\mathrm{H} + {}^3\mathrm{He}_{\mathrm{tri}} \mathrm{and} {}^3\mathrm{H} = {}^3\mathrm{H}_0 \ln e^{-\lambda t}$$

Combining these two equations

$$\begin{aligned} {}^3\mathrm{H} &= ({}^3\mathrm{H} + {}^3\mathrm{He}_{\mathrm{tri}}) \ln e^{-\lambda t} \rightarrow \ln e^{-\lambda t} = {}^3\mathrm{H}/({}^3\mathrm{H} + {}^3\mathrm{He}_{\mathrm{tri}}) \rightarrow t \\ &= 1/\lambda \ln \left({}^3\mathrm{He}_{\mathrm{tri}}/{}^3\mathrm{H} + 1\right) \end{aligned}$$

It is clear from the equation that in order to measure the age of a groundwater sample, we simply need to measure its tritium and $^3\mathrm{He}_{\mathrm{tri}}$ simultaneously.

Main advantages of the method are (i) the resolution of this method is high in average situations (moderately thick unsaturated zone, limited sources of helium, etc.), (ii) data collected can be used for both ^{3}H/^{3}He and tritium methods, (iii) this method will be applicable for a long time, i.e., its effectiveness is not reduced in the future as is the case with methods like CFCs, tritium, etc., and (iv) this method does not need the initial value, a parameter that is fundamental and problematic for many of the dating methods.

Disadvantages of ^{3}H/^{3}He method are (i) sampling and analysis are expensive and laboratory facilities are not available worldwide, and (ii) it is difficult to separate tritiogenic helium from the other type of helium.

4.2.3 Krypton-85 (^{85}Kr) Method

Krypton (Kr) is a colourless and inert gas (so-called noble gas) belonging to the 8th group of the Mendeleyev's periodic table. Krypton is highly soluble in water, together with Xenon. Typically atmospheric Kr ranges from 7.61 to 12.57 × 10^{-8} cm^3 STP/g and from 2.26 to 3.80 × 10^{-8} cm^3 STP/g in groundwater and seawater respectively (Ozima and Podosek 2001).

Natural production of ^{85}Kr takes place in small amounts in the atmosphere by spallation and neutron activation of stable ^{84}Kr.

$$^{84}\mathrm{Kr}_{36} + n \rightarrow {}^{85}\mathrm{Kr}_{36} + \gamma$$

Manmade ^{85}Kr is produced by the fission of plutonium and uranium. The main anthropogenic sources of ^{85}Kr to the atmosphere are, therefore, nuclear weapon testing and nuclear reactors used for both commercial energy production and plutonium weapons production.

^{85}Kr disintegrates by beta decay to stable ^{85}Rb. In this respect, it can be used as a "clock" or a radioactive tracer.

$$^{84}\mathrm{Kr}_{36} \rightarrow {}^{85}\mathrm{Rb}_{37} + \beta^-$$

However, its use in age-dating is more based on its nearly linear increasing concentration in the atmosphere. Atmospheric ^{85}Kr is dissolved in rainwater and is carried out to the unsaturated and then saturated zones. The higher is the concentration of ^{85}Kr, the younger the groundwater age. Also, due to the short half-life (10.76 years) and minimal natural production in the Earth, the absence of ^{85}Kr verifies that groundwater is older than 1950. If ^{85}Kr is combined with an additional radioactive isotope with a similar half-life (such as ^{3}H), additional confidence in the results can be gained. ^{85}Kr is commonly used with tritium because the two tracers have similar half-lives but completely different input functions. Its short half-life and increasing concentrations in the atmosphere make ^{85}Kr a potential replacement for ^{3}H as tritium levels continue to decline.

Main advantages of the method are: (i) this method is much less sensitive to degassing than the other dating methods for young groundwaters, as this method is based on isotopic ratio (^{85}Kr/Kr), (ii) in anoxic environments and in CFCs and SF_6-contaminated areas, ^{85}Kr may prove to be superior because of low contamination possibility, (iii) the method may continue to be used in future also, because atmospheric concentration of ^{85}Kr is still on the rise, and (iv) geochemically ^{85}Kr is inert.

Disadvantages of ^{85}Kr method are: (i) large sample-size required, (ii) high costs due to the specialized measurement methods, (iii) in uranium-rich aquifers, some proportion of ^{85}Kr gets mixed and masks the atmospheric component of ^{85}Kr or make it difficult to be distinguished, and (iv) in aquifers with thick unsaturated zones, there is a significant time lag for transport of ^{85}Kr to the saturated zone (Cook and Solomon 1995), which may lead to overestimation of groundwater ages.

4.2.4 Chloroflourocarbons (CFCs)

Like tritium, CFCs, otherwise unwanted contaminants, resistant to degradation are now being utilized as a useful marker for modern groundwater. Atmospheric CFC concentrations have been increasing since 1940s, providing a characteristic input function. CFCs have been extensively used to trace oceanic circulation patterns over the past decade but some recent studies (Thompson and Hayes 1979) have documented their usefulness for dating young groundwaters.

Atmospheric CFCs, dissolved in percolating precipitation water, reach the groundwater system after passing through the unsaturated zone. Groundwater CFCs ages are obtained by converting measured CFCs concentrations in the groundwater sample to equivalent air concentrations using known solubility relationships developed for CFC-11 and CFC-12 (Warner and Weiss 1985), for CFC-13 (Bu and Warner 1995) and the recharge temperature (Cook and Herczeg 1998). The life time of CFC-11, CFC-12 and CFC-13 is 45, 100 and 85 years respectively.

Equivalent atmospheric concentration (or sometimes called "apparent atmospheric concentration") of CFCs can be computed by the equation given by Cook and Herczeg (1998), i.e.

$$\mathrm{EAC} = \mathrm{CFCs}_{\mathrm{gw}}/S \times MW$$

where EAC is the equivalent atmospheric concentration, $\mathrm{CFCs}_{\mathrm{gw}}$ is the concentration of CFCs in the groundwater, S is the solubility in $\mathrm{molkg}^{-1}\ \mathrm{atm}^{-1}$, and MW is the molecular weight of CFCs with unit of g/mol.

Main advantages of using the CFC method are: (i) presence of CFCs is a good indicator of post-1945 recharged groundwater. CFC-113 indicates post-1965 recharged groundwater, (ii) input function is relatively well known because spatial variations in atmospheric CFCs concentration are relatively moderate, (iii) it is possible to date the groundwater sample by EAC of one species and also by ratio of various species, (iv) concordant ages from various species may help to understand the geochemical processes in the aquifer, and (v) cost of analysis is cheap compared to all other methods.

Disadvantages of CFCs method are: (i) the method is losing its applicability (post-1990s), (ii) many parameters such as excess air, recharge temperature, degradation of CFCs, etc., can influence the accuracy of the ages, and (iii) great care is needed for sampling, and large errors may be introduced if proper guidelines are not followed.

4.2.5 Sulphur Hexafluoride (SF_6) Method

Sulphur (or sulfur) hexafluoride, SF_6, is a colourless and odourless gas used in the electric power industry, in the semiconductor industry, in the production of magnesium and aluminium for degassing melts of reactive metals, in blood products, in running shoes, and as intraocular gas tamponades for a wide range of complicated vitreoretinal diseases etc. SF_6 is primarily of anthropogenic origin but also occurs naturally in minerals, rocks, and volcanic and igneous fluids. It is the most potent greenhouse gas with an estimated atmospheric lifetime of 1935–3200 years. SF_6 is a conservative tracer for groundwater studies and behaves identically to bromide (Wilson and Mackay 1993). In addition to its use in dating young groundwaters, SF_6 is applied: (i) to estimate longitudinal dispersion coefficients in rivers, (ii) in the study of groundwater nitrate pollution, (iii) to estimate gas exchange rate in streams, and (iv) as a natural atmospheric tracer.

Atmospheric SF_6 is dissolved in rain and snow. Like other atmospheric-derived gas tracers, the amount of the dissolution is a function of the concentration of SF_6 in the air and the air temperature. Higher the atmospheric concentration, higher is the amount of the dissolution. Infiltrating rain- and snow water carry dissolved SF_6 into the unsaturated zone and finally to the saturated zone. Therefore, we can date a groundwater sample based on presence and concentration of its SF_6. The dating range of this method is *X*-1970, where *X* is the date the groundwater is sampled for dating; e.g., in year 2010, the dating range of the method would be 0–40 years.

Main advantages of the method are: (i) the concentration of SF_6 in the atmosphere continues to rise and the method is, therefore, going to hold effective until this trend is stopped or reversed, (ii) the atmospheric input is relatively well known, and the subsurface addition of SF_6 is thought to be insignificant, and (iii) the narrowness of the dating range is a plus in those situations where precise time scales are of interest.

Disadvantages of CFCs method are (i) groundwater ages obtained by the SF_6 method do not include the travel time of groundwater in the unsaturated zone. If this time is long, the ages obtained are substantially different from what is defined as groundwater age, (ii) the age range of this method is narrow, (iii) the knowledge about subsurface or natural production of SF_6, microbiological degradation, and other unfriendly causes is limited, and (iv) the main anthropogenic source of SF_6 is in the middle latitude of the Northern Hemisphere; hence the applicability of the methods in the other parts of the world is doubtful.

4.3 Dating Old and Very Old Groundwaters

Constraining the age of groundwaters that are clearly sub-modern or older can be important in establishing the long-term potential for aquifer recharge. For groundwater development and management policy, the question of renewability is most important. The methods to age-date old groundwaters (60–50,000 years old) include mostly ^{14}C, but less used and indirect methods such as ^{32}Si, ^{39}Ar, ^{18}O, ^{2}H, and conservative and reactive tracers are also being utilized.

4.3.1 Carbon Dating

Radiocarbon dating, or *carbon dating*, is a radiometric dating method that uses the naturally occurring radioisotope carbon-14 (^{14}C) with half-life 5730 ± 40 years to determine the age of carbonaceous materials up to about 50,000 years. Carbon-14 dating of groundwater is done by measuring ^{14}C activity in its dissolved inorganic carbon. ^{14}C is the leading tool in estimating the age of palaeo and fossil groundwaters. The method is based upon the incorporation of atmospherically derived ^{14}C from the decay of photosynthetically-fixed carbon in soil. Radiocarbon in the soil is taken into solution as dissolved inorganic carbon (DIC = $CO_{2(aq)} + HCO_3^- + CO_3^{2-}$) or as dissolved organic carbon (DOC).

Atmospheric ^{14}C is dissolved in the percolating rainwater and reaches the water table. In groundwater, ^{14}C starts decaying to nitrogen:

$$^{14}C_6 \rightarrow {}^{14}N_7 + \beta^-$$

If no further ^{14}C exchange occurs, measurement of the remaining ^{14}C atoms can be used to date groundwater following the first-order kinetic rate law for decay:

$$C = C_0 e^{-\lambda t}$$

where C_0 is the activity assuming no decay occurs (initial activity or activity at $t = 0$), and C is the observed or measured activity of the sample. The groundwater dating by ^{14}C ranges in age from 870 to 19,000 years if 10% and 90%, respectively of the original atoms are assumed to have decayed. However, ages up to 40,000 years or longer have been reported by this method.

Main advantages of the method are: (i) it is an old and well-established method that has been proved and developed by considerable research during the last half-century, (ii) sampling and analysis for this method are now routine and cheaper than the majority of the dating methods, (iii) it is the only popular method available to date old groundwaters and to fill the dating range between young and very old groundwaters, and (iv) the deficiencies, the principles, and the positive points of the methods are all well known.

Disadvantages of ^{14}C method are: (i) it is an extremely difficult task to determine the correct initial value due to the various processes that modify ^{14}C signature of the percolating rainwater, (ii) a large number of geochemical reactions modify the concentration of ^{14}C in the groundwater, and (iii) having pointed out the above two major obstacles, it is safe to argue that the ^{14}C method is often a semi-quantitative technique.

4.3.2 Chlorine-36 (^{36}Cl) Method

The long half-life of ^{36}Cl (301,000 years) and generally simple chemistry of Cl^- makes this radioisotope an interesting tool for dating very old groundwater. Interest in application of ^{36}Cl is now increasing in recharge studies, groundwater infiltration rates and rates of erosion. Groundwater dating is based on two fundamental methods: (1) decay of cosmogenic and epigenic ^{36}Cl over long periods of time in the subsurface, or (2) in growth of hypogenic^{36}Cl produced radiogenically in the subsurface.

Groundwater in recharge areas derives cosmogenic^{36}Cl from two sources: atmospheric production and epigenic or surface production. Atmospheric ^{36}Cl is produced in the upper stratosphere through the bombardment of argon gas by cosmic radiation, according to:

$$^{40}\mathrm{Ar} + \mathrm{p} \rightarrow {}^{36}\mathrm{Cl} + \mathrm{n} + \alpha(67\%)$$
$$^{36}\mathrm{Ar} + \mathrm{n} \rightarrow {}^{36}\mathrm{Cl} + \mathrm{p}(33\%)$$

where n = neutron, α = alpha particle and p = proton.

Atmospheric residence time is minimal and ^{36}Cl, together with stable Cl^-, is washed to the surface by precipitation or arrives as dry fallout. Common ^{35}Cl in the atmosphere can also be irradiated by cosmic flux to produce ^{36}Cl and gamma radiation.

$$^{35}\mathrm{Cl} + \mathrm{n} \rightarrow {}^{36}\mathrm{Cl} + \gamma$$

The principle of this method is simple. It is based on the radioactive decay of ^{36}Cl in the subsurface groundwater system. Above the Earth's surface, ^{36}Cl (atmospheric ^{36}Cl) with an initial value, $^{36}Cl_0$, enters groundwater by rainwater infiltration. After time t, it decays to ^{36}S and ^{36}Ar

$$^{36}\mathrm{Cl}_{17} \rightarrow {}^{36}\mathrm{Ar}_{18} + \beta^- + \nu$$
$$^{36}\mathrm{Cl}_{17} \rightarrow {}^{36}\mathrm{S}_{18} + \beta^- + \nu$$

and reaches a new concentration, ^{36}Cl, according to the decay equation

$$C = C_0 \mathrm{e}^{-\lambda t}$$

If the initial concentration (C_0) and the present concentration (C) are known, then the length of time that the ^{36}Cl has resided in the subsurface groundwater system can be calculated. Chlorine-36 dating method is capable of dating groundwaters with an age range of 46,000 to 1,000,000 years if we assume that 10% and 90%, respectively, of the original ^{36}Cl atoms are disintegrated.

Main advantages of the method are: (i) the long half-life of ^{36}Cl makes this method suitable for dating very old groundwaters, and (ii) measured ^{36}Cl values may be used for other hydrologic applications, in addition to their use in dating.

Disadvantages of ^{36}Cl method are: (i) too many sources for ^{36}Cl and initial value problem, (ii) underground sources and sinks of ^{36}Cl, limiting to suggest limiting the dating range of ^{36}Cl to 500,000 to 1,000,000 years only (Phillips 2000), (iii) applicable only to limited regions, because this method is suited only for deep regional aquifers, (iv) not applicable to saline aquifers with chloride concentration of more than 150 mg/L, and (v) high cost of analysis and sample preparation, and inadequate availability of laboratories worldwide.

4.4 Applications of Groundwater Age Data

Groundwater age data can be used to evaluate the renewability of groundwater reservoirs, to constrain the parameters of groundwater flow and transport models, to study groundwater flow paths and vertical and horizontal flow velocities, to identify paleo-climate conditions (in combination with isotopes), to estimate groundwater recharge, to determine fracture and matrix properties and water velocities in fractured rock environments, to help study the trend of groundwater pollution, to identify past seawater level fluctuation, to manage groundwater-driven dryland salinity, to map susceptibility of groundwater systems to contamination, and to be used in many more hydrological applications such as mixing, groundwater–surface water interaction, and seawater intrusion.

4.4.1 Replenishment of Groundwater Reservoirs

The most important and the unique application of groundwater age concept is renewability or replenishment of the groundwater resources. Groundwater age is, as yet, the only sound and concrete piece of scientific evidence to show that groundwater resources are recharged by modern precipitation, or else, the extracted groundwaters were accumulated in the aquifers by slow infiltration processes that happened a very long time ago. This application is more highlighted in the arid zones where due to the scarcity and periodicity of the rainfall, the question of recharge (if any) often remains open (Payne 1988). An important fraction of young water within an extracted water sample is an indication of an actively renewable reservoir; the opposite, i.e., a considerable amount of old water in the sample, depicts a poorly recharging reservoir and/or significant internal mixing processes.

Information about the age of groundwater is required if one is to confidently define the sustainability of groundwater resources of any particular well field. Estimates of renewable groundwater resources and an understanding of related hydrological processes are critically dependent upon determining the presence and age of modern groundwater.

4.4.2 Prevention of Over-Exploitation and Contamination of Aquifers

Increase in the population density often leads to an exponential increase in the demand on the aquifer. Once residences or industries are established, it is very difficult to limit their water supply. Over-development can eventually lead to limited supply, with the greatest effects being to those districts farthest from the aquifers recharge zone (supply source). By measuring the age of the water at certain time intervals within a district's well field (say once every five years), it would be possible to identify over-exploitation before it happens.

If the groundwater, being extracted, increases in age with time (becomes older and older), it means that a higher proportion of water is drawn from slow-moving storage. In contrast, if the age of groundwater being withdrawn decreases with time (becomes younger and younger), it means that a higher proportion of extracted water is derived from active present recharge. This shows that either the pumping rate has increased or the source water has changed (i.e., river recharge instead of rainfall recharge). This condition though does not imply groundwater mining, but it may not be a good sign in terms of contamination because eventually surface contaminants (if present) dissolved in very young waters (which may be contaminated) will reach the well field. Hence, regular dating of the groundwater from well fields can provide a mechanism to monitor, understand, and control exploitation and contamination of the aquifer.

4.4.3 Estimation of Rate of Groundwater Recharge

This particular usage of groundwater age data is perhaps the most widely applied of all. Figure 4.12 illustrates in a simple way the approach to calculate the recharge rate to a groundwater system by age data. The procedure is to have either:

1. A minimum of two ages along the vertical line at the point of interest, i.e., age data must be obtained from a piezometer nest, which comprises at least two piezometers opened to the aquifer at different depths, or

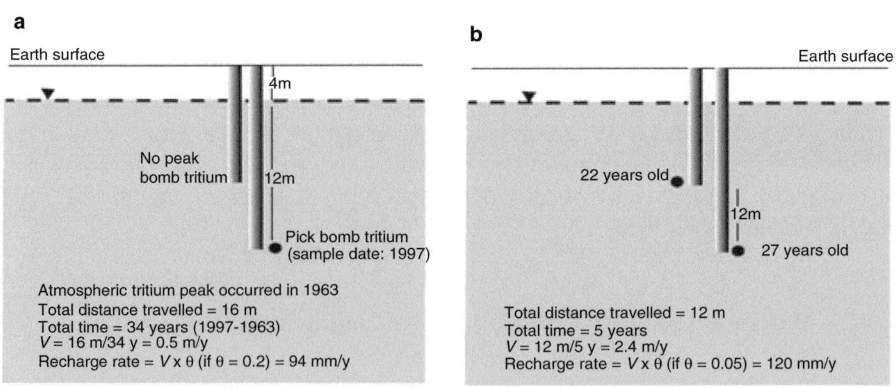

Fig. 4.12 Determination of groundwater recharge rate: (**a**) by locating bomb peak tritium; (**b**) by measuring groundwater ages at two points along a vertical profile such as at a piezometer nest

2. The vertical position of the bomb peak tritium in the aquifer. It should be noted that the groundwater flow should consist of only one vertical component with negligible horizontal movement.

The second approach may not be particularly accurate because of the difference between the flow rates in the saturated and unsaturated zones.

4.4.4 Estimation of Groundwater Velocity

The velocity of groundwater flow can be calculated if we measure the age of groundwater at two separate points along a particular horizontal flow line. The ages should be measured at nearly the same depth and on the same flow line in order to avoid the effect of three-dimensional flows (Fig. 4.13).

The important point is that groundwater flow rates for aquifers can be gained from artificial (applied) tracer experiments as well, but age data offer the only realistic alternative if time scales of years or decades have to be taken into account (Zoellmann et al. 2001). Having obtained groundwater velocity, we can also back-calculate the hydraulic conductivity of the aquifer if we have an estimation of the effective porosity of the aquifer through $V = KI/\theta$ (the assumption is that the hydraulic gradient is easily obtainable).

4.4.5 Identification of Groundwater Flow Paths

Groundwater flow paths in both vertical and horizontal directions can be determined by having ages that increases along the inferred flow lines. Accurate information

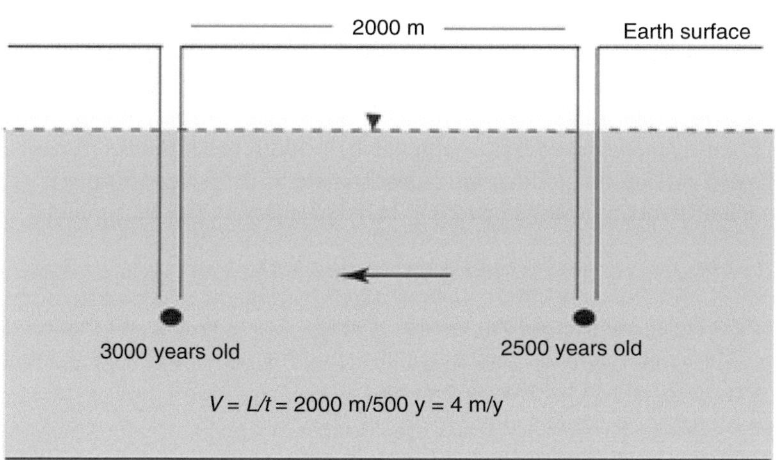

Fig. 4.13 Measurement of groundwater flow rate using groundwater age determination in two different boreholes

about the groundwater flow path is needed in many water resources projects such as in construction of dams (determining different routes that water stored in the dam lake/reservoir may escape), movement of plumes, mixing between different quality groundwaters, and study of surface water–groundwater interaction. Establishing flow directions in various parts of a deep, long regional aquifer is another example for use of groundwater age data.

4.4.6 Confirming the Parameters of Groundwater Flow and Transport Models

Groundwater flow models are increasingly employed as a powerful tool to help manage the groundwater resources. These models usually require an extensive volume of data and parameters, the quality of which is the key to the success of the modelling exercise. Estimation of some of these parameters such as hydraulic conductivity values, specific yields, and aquifer geometrics is always prone to some degree of error. Groundwater age data are the more precise data that can be incorporated into the models to complement the existing data, to eliminate some uncertainties, to serve as model independent calibration data for groundwater transport models (Dörr et al. 1992), and to verify flow models especially those that predict the travel time of source water to wells.

4.4.7 Identification of Mixing between Various End Members

The ages of those groundwaters that are mixtures of various components are helpful to calculate the mixing ratios and to identify the origin of each end member. For this purpose, it is necessary to apply a minimum of two dating methods from different ranges; for instance, an old dating method and a young dating method (it is also possible to identify mixing by having the ratios of the three CFC-11, CFC-12 and CFC-13, which are all for dating young groundwaters). The ages obtained from different dating methods are cross-plotted. On the plots, the best-case scenario would be to have a mixing line joining the oldest sample to the youngest sample.

In such a situation, a simple two end-member equation can be applied.

$$As = Ay(x) + A_0(1 - x)$$

where As is the age of sample in question, Ay is the age of youngest sample, A_0 is the age of oldest sample, x is the percentage of young fraction in the sample, and $(1 - x)$ is the percentage of old fraction in the sample.

Such mixing calculations allow hydrogeologists to study surface water–groundwater interaction and to investigate diffuse flow and conduit flow in fractured rocks aquifer.

4.4.8 Study of the Pre-Holocene Climate

Evidence from groundwater systems may be used to help interpret the timing and nature of climatic changes during the Pleistocene era (Metcalfe et al. 1998). Groundwaters that are old or very old can be studied to identify the type of climate during which they entered the subsurface environment. To achieve this objective, first we have to know the age of groundwater. Then we have to study those parameters of groundwater that are climate indicators such as oxygen-18 isotope values. Combination of these two pieces of evidence will lead to an understanding of the pre-Holocene climate.

4.4.9 Evaluating the Pollution of Groundwater

The isotopic tools for groundwater dating are important because shallow, young groundwaters provide drinking water supply in many parts of the world and also are the most vulnerable to contamination from anthropogenic activities.

Most contaminants, especially human-induced ones, have entered aquifers recently, perhaps not earlier than 100 years ago. Further, groundwaters that were polluted 100 years ago or earlier have so far had enough time to be purified through natural processes because of their long contact with the subsurface environment. Based on these assumptions, the following conclusions can be drawn:

(i) If a water sample is older than 100 years, it should be pollution free.
(ii) For old groundwaters, the contamination risk is low.
(iii) For young groundwaters, the contamination risk is high.
(iv) Along the age line, the concentration of the contaminants should decrease, i.e., older groundwater should show less contamination (concentration of the contaminants should be less).
(v) For a set of groundwater samples, if a contaminated groundwater sample is dated as young, we can be sure that the dating exercise was most probably undertaken correctly and the age data obtained can be used for other purposes.
(vi) If there is a meaningful negative correlation between the age of groundwater and the concentration of the contaminants (the older the age, the lower the concentration of the contaminants), it would then be possible to predict the extent and timing of contamination plume, i.e., concentration reaches to what level and at what time.
(vii) If there is a positive correlation between the age of groundwater and the concentration of contaminants (the older the age, the higher the concentration of the contaminants), it would then be logical to conclude that contaminants are gradually degraded over time.

4.4.10 Estimating the Travel Time of Groundwater Plume to the Points of Interest

By having the age of groundwater, we can:

(i) Predict the time required for a present-day polluting source (a leaking fuel station, for instance) to inflict damage on the water quality at the location of interest.
(ii) Predict the timing of the effect of land-use changes.
(iii) Determine the likely sources of contaminants whose initial applications occurred during a specific time period. The age of the groundwater will help to determine if the sources of contamination can be traced to recent events, or if other unknown sources of contamination exist that would warrant further investigation.
(iv) Estimate the time span required for self-purification of a polluted aquifer after removal of the pollutant.
(v) Determine whether there is enough time for natural purification of surface waters recharging aquifers (e.g., river water) whose quality should improve during the course of travel underground to make them suitable for domestic usages.

4.4.11 Mapping of Vulnerable Shallow Aquifers

Construction of vulnerability map for the aquifers is a relatively modern exercise. These maps can be used by resource protection agencies to focus prevention programmes on areas of the greatest concern and to help prevent contamination of groundwater resources by identifying areas that are at greater risk of pollution. One of the earliest methods to evaluate vulnerability of the aquifers is DRASTIC, which considers seven factors: *d*epth to water, net *r*echarge, *a*quifer media, *s*oil media, *t*opography, *i*mpact of vadose zone media, and hydraulic *c*onductivity of the aquifer (Aller et al. 1985). Generally, groundwaters with a high percentage of young water should be rated as highly susceptible, while those with a high fraction of old water should be regarded as insusceptible or less susceptible.

4.4.12 Assessment of Radioactive Waste Disposal Facilities

Among all of the investigations concerning the safety and feasibility of storing radioactive wastes underground, a high degree of certainty is needed to assure that the contaminants will not leach from the wastes. The likely maximum rates of future groundwater movements can be obtained from knowledge of the past history of flow events in the area under consideration. Aquifer pumping tests, applied tracer tests, and water-level measurement studies all lack the capability to measure aquifer hydraulic properties, *and associated likely changes*, over very large temporal and spatial scales. For this reason, groundwater age data are of prime importance and can provide invaluable information on the hydraulic properties of the waste disposal facilities.

4.4.13 Site Specific Applications

Some of the applications of groundwater age data have been undertaken on a site-specific basis. These include the evaluation of atmospheric acid deposition (Robertson et al. 1989; Busenberg and Plummer 1996), estimating rates of geochemical and geo-microbial processes in the unsaturated zone (Plummer et al. 1990; Chapelle et al. 1987), estimating longitudinal dispersivitiy coefficient (Solomon et al. 1993), determining hydraulic conductivity and transmissivity of the aquifers (Phillips et al. 1989; Hanshaw and Back 1974), finding the origin of methane in gas hydrate deposits (Fehn et al. 2003), identifying groundwater inflow into artificial lakes (Weise et al. 2001), evaluating ecosystem health (Chesnaux et al. 2005), quantitatively describing groundwater flow and hydrogeology (Boronina et al. 2005), and estimating volume of aquifer storage and location of recharge and extracting information on the rates of geochemical and microbiological processes in aquifers.

5 Application of Isotopes in Groundwater Management

Groundwater assessment and management has great significance for a country like India. It is becoming more critical with the growth of population and rapid industrialization. The issue of groundwater management involves reliable assessment of available water, scope for augmentation, distribution, reuse/recycle, pollution, and its protection from depletion and degradation.

Isotopes are being used extensively for studying the soil moisture variation and its movement; recharge through unsaturated zone; origin, age, occurrence and distribution of groundwater in a region recharge mechanism, determination of groundwater flow direction and velocity; interconnections and interaction between aquifers; and identification of recharge areas and sources. Isotopes can also be applied to study surface water and groundwater interaction; tracing sources of pollutants including sea water intrusion and salinization mechanism.

Isotopic methods are normally used in conjunction with conventional hydrological, hydrogeological and geochemical or water-chemical techniques, to provide additional and valuable information for solving hydrological problems. In recent years, a number of complex groundwater problems have been resolved successfully using isotopic methods.

5.1 Groundwater Recharge

Groundwater recharge can be determined by various methods, such as, from water balance computations, from pumping over very long periods, from mass balance of different artificial or environmental tracers, or from hydraulic interpretations of the soil moisture movement below the active roots in the unsaturated zone. The hydraulically based methods, though highly developed in theory and practice, are hampered by the complex relationship between hydraulic conductivity and hydraulic gradient in the unsaturated zone. Tracer techniques have the advantage that old soil water can be differentiated from relatively fresh water. A rather recent method of estimating groundwater recharge is to use isotopes as tracers. The methodology of using artificial as well as environmental isotope tracers for estimating percolation rate and groundwater recharge is discussed here.

5.1.1 Artificial Radioactive Isotope Tracers (Tritium Tagging) for Estimating Groundwater Recharge

Artificial radioactive isotopes (produced in laboratory or reactor) can be used as water tracers. The artificial tracers have the advantage over environmental tracers that they are injected in a controlled way and that the concentrations are high enough to be easily detected. The disadvantage is that it is non-natural, which means it may be environmental hazardous and that experiments can only be made at specific points and at specific times.

The most commonly used artificial tracer isotope for groundwater studies is tritium (^{3}H) as HTO, which is applied below the root zone or in the groundwater depending on the purpose of the study.

Artificial radioactive isotope tracer method, also known as, *Tritium tagging technique* was developed by Zimmermann et al. (1967a, b), Blume et al. (1967) and Munnich (1968a, b) with the assumption that the movement of soil moisture in an unsaturated zone is similar to piston type flow. Any water applied to the ground surface, either from precipitation or from irrigation, will infiltrate and percolate by pushing equal amount of water beneath it further down. The amount of moisture content of the last layer in the unsaturated zone is added to the groundwater as recharge. Blume et al. (1967) suggested that the tracer should be introduced/injected below the ground surface (preferably below the root zone), otherwise it may be lost by evaporation or evapotranspiration.

Soil water moves along a range of different pathways. Local field heterogeneity should bring about a considerable dispersion of pollutants or of a tracer. However, field experiments of Zimmermann et al. (1967a and 1967b) and Blume et al. (1967), showed "piston flow" type behaviour of soil moisture in nearly homogeneous soils; infiltrating water simply pushes the old water downward. This means that the soil moisture profile may change shape, but no newly percolated water bypasses water that has previously percolated below the root zone. In the above-mentioned

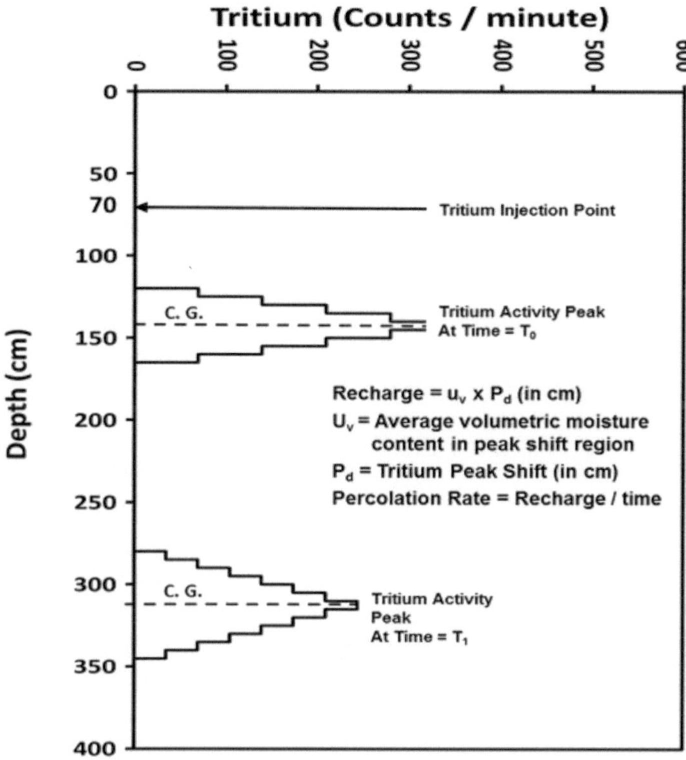

Fig. 4.14 Shifting of injected tritium with respect to the movement of infiltrated water

experiments, the broadening of the peak of the injected tracer was of the same magnitude as expected by molecular diffusion only. In addition, experiments conducted in the alluvial tracers of north India by Bahadur et al. (1977) showed broadening of the tracer peak comparable to the spread by molecular diffusion. It seems that if the flow is slow, the lateral mixing, mostly from molecular diffusion, in rather homogeneous soils between moisture packets having different flow velocities is quite effective, thus indirectly counteracting vertical dispersion.

In the course of water infiltration, an injected or environmental tracer is carried along with the soil water. The position of the peak tracer concentration can be monitored in the soil profile. From the temporal displacement of the tracer, percolation rate or moisture flux, and the groundwater recharge can be estimated, provided the measurements are taken below the root zone so that all water movements are directed downwards. Zimmermann et al. (1967a) used deuterium and tritium as tracers and called the method of tracing the peak as "tracer tagging" technique. The principle of how percolation rates are determined from tracer monitoring is illustrated in Fig. 4.14.

Shortly after the tracer injection, the peak concentration, which is at depth z_1, moves down and is found at depth z_2 after a certain time. Provided no vertical mixing takes place during the downward movement, the mean moisture flux, q at the lower depth z_2 over the time period, Δt, between the two observations is:

$$q = \theta_v(z_2 - z_1)/\Delta t$$

where θ_v is the average volumetric moisture content between the two depths at the time of the first observation after reduction of eventual residual moisture content, that is, interstitial water or water that adheres to the soil particles. This residual moisture content is generally negligible, except for very fine soils.

If the experiment is for a full season or full year, the moisture flux corresponds to the recharge over a season or year, although the particular soil water particles do not reach the groundwater during the particular year when the observations are made.

Case Study: Recharge to Groundwater in Bundelkhand Region of U.P., India Using Tritium Tagging Technique

Bundelkhand region in India faces acute water deficiency due to higher losses of rain and surface waters. Keeping in view the prevailing conditions in Bundelkhand region, it was necessary to estimate the correct value of recharge to groundwater due to monsoon rains, which is the main source (Kumar and Nachiappan 1995).

Tritium was injected at 25 sites before the start of monsoon rains. Soil samples were collected from the injected sites in the month of November and recharge percentages were determined. Since sampling was carried out in November, the water input for the irrigation was also taken into account while determining the percentage of recharge. Estimated recharge to groundwater ranged from 6 to 34 cm. This variation may be due to different types of soil, topography, hydrogeology, groundwater level conditions, cropping pattern, rainfall pattern, evapo-transpiration and several other local factors which are very difficult to account for.

Based on the estimated data and rainfall for the 25 sites, an empirical relation was developed (Fig. 4.15) between rainfall-recharge process that fairly satisfy the variation of recharge values. The equations developed distribute the data in two groups A and B given by the equations:

$$\text{Group A} \quad \text{Rg} = 29.316 \ln(P) - 111.259 \quad (r = 0.83) \tag{4.1}$$

$$\text{Group B} \quad \text{Rg} = 12.861 \ln(P) - 48.757 \quad (r = 0.85) \tag{4.2}$$

where Rg is recharge to groundwater in cm and P is rainfall/precipitation in cm.

5.1.2 Environmental Tritium Technique

The radioactive isotope of hydrogen, i.e. tritium (^3H), released from thermonuclear explosions in the atmosphere made possible a way of estimating groundwater recharge. The cosmogenically produced tritium, which is produced and assimilated

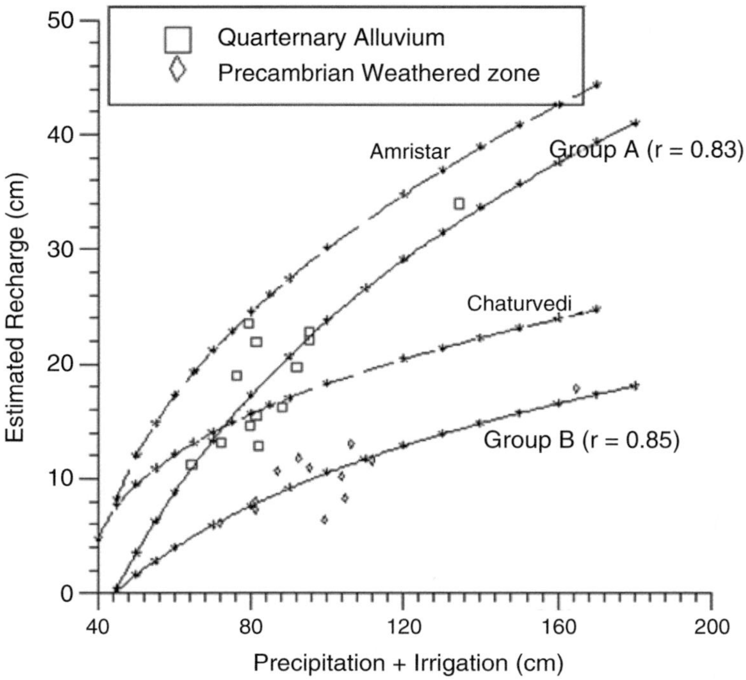

Fig. 4.15 Correlation of estimated recharge (cm) with precipitation

in the atmospheric water vapour, is brought down to earth's surface by precipitation. Before 1952, the tritium concentration in precipitation was low, but after thermonuclear testing began in the atmosphere in 1952, tritium concentrations in precipitation suddenly increased and reached a record-high concentration in 1963–64 in the northern hemisphere. In India, the peak concentration of bomb derived tritium in 1963 was more than 1000 TU. The fact that water originating from precipitation, which has fallen before 1952, has lower tritium concentration than water contributed by later precipitation has been used for tracing groundwater.

Assuming that more recent infiltrating water pushes previously infiltrated water, the bomb tritium of the infiltrated precipitation of a particular year can be found in a soil profile. The tritium concentrations in the soil profile may be moderated due to dispersion and molecular diffusion. Among others, Munnich et al. (1967), Sukhija and Shah (1976) in India, and Andersen and Sevel (1974) in Europe have used bomb-released tritium for the evaluation of groundwater recharge in Europe and India, respectively. An example of tritium concentration in soil profile is given in Fig. 4.16.

In this method, it is assumed that the amount of water from the soil surface to the soil depth, where the 1963–64 tritium peak is located, is the measure of recharge from that time until the time of investigation. In the other method, the tritium concentration of the water lost as evaporation of surface runoff as well as of the

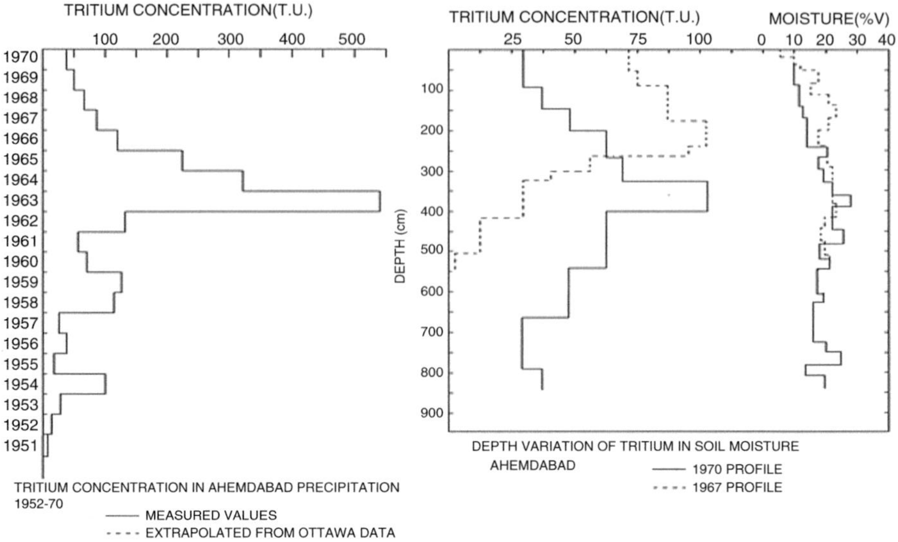

Fig. 4.16 Tritium concentration (without decay correction) in precipitation at Ahmedabad (1952–1970). (*Source:* Sukhija and Shah 1976)

water percolating below the root zone is, at any time, assumed to equal the concentration of the precipitation. The accumulated percolation, R, which will contribute to groundwater recharge, is simply given by,

$$R = P(M_s/M_p)$$

where 'P' is accumulated rainfall since the beginning of the bomb tests, M_p is the total amount (per unit area) of tritium in the precipitation, and M_s is the amount of tritium found in the soil above the depth where the tritium concentration is at pre-1952 level. The method can be adjusted to be applied from the peak concentration time instead of from 1952, that is, from 1964.

5.1.3 Environmental Stable Isotopes Technique

The stable isotopes ^{18}O and D (^2H) in precipitation are being used for a long time as potential tracers for natural waters yet they have been little exploited for measuring percolation. The flux of HDO and $H_2{}^{18}O$ from an open water body to the atmosphere is relatively low as compared to the flux of the lighter $H_2{}^{16}O$ because of the lower vapour pressure of the former species, which causes fractionation in evaporation and condensation processes. In cold climate, seasonal stable isotopic composition of precipitation is rather well reflected in soil moisture, whereas in semi-arid climate, the isotope picture of soils is rather complex due to strong fractionation caused by high evaporation rates from the soil.

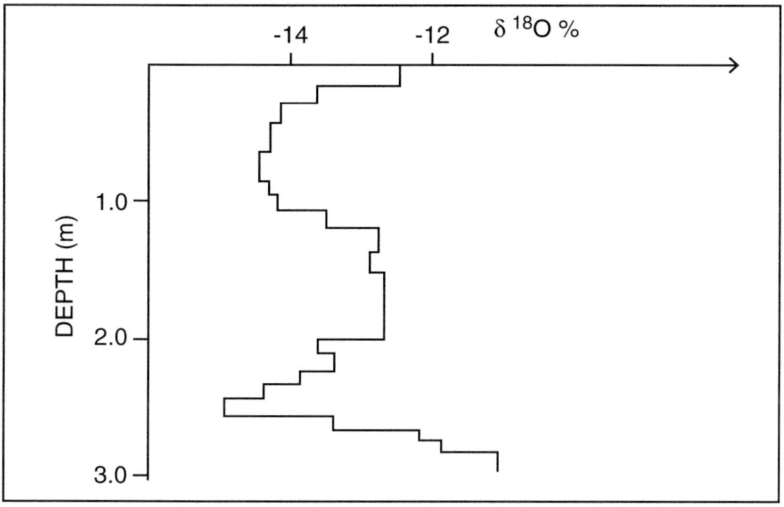

Fig. 4.17 Oxygen-18 profiles in soil moisture patterns indicating infiltrated water in different years

Soil water infiltrated during two periods with a time span of a year can be identified within a soil profile on the basis of stable isotopic composition of either deuterium (^{2}H or D) or oxygen-18 (Fig. 4.17). Therefore, the total amount of percolated water during the year can be estimated simply by totalling the amount of soil water between the two depths where the soil water from the two periods is found. Reduction should be made for eventual residual water.

Mathematically expressed, the annual percolation, R, which will later contribute to groundwater recharge, is

$$R = \int_{z_1}^{z_2} \theta_v \, dt$$

where z_1 and z_2 are the identified depths and θ_v is the volumetric soil moisture content after reduction of eventual residual water. The method cannot be applied in areas where the groundwater level is shallow during some periods of the year.

The method was first applied to estimate the recharge to groundwater in sand dunes with depth fluctuations in deuterium in soil moisture profiles (Thoma et al. 1979). The seasonal variations of $\delta^{18}O$ in precipitation have been traced in the soil moisture and estimates of groundwater recharge and rates of moisture movement were estimated for Swedish glacio-fluvial deposits and moraine formations (Saxena 1984). Now, this technique is used wherever the measurement of stable isotopes is possible.

Application of both the environmental isotope methods (radioactive as well as stable isotope method) is restricted to sites where the percolating water of peak tritium concentration has not yet reached the groundwater table. In fact, the environmental tritium method was relatively more useful only until mid-1970s. Most of bomb tritium has by now mixed with the groundwater and the soil profiles have more or less a constant concentration of tritium.

5.2 *Groundwater Flow Velocity and Direction*

Single well and multi-well dilution techniques are employed using artificial radioactive isotopes to determine the groundwater velocity and direction of flow. In fact, single well dilution method, also called point dilution method can be used to determine both, the velocity and direction of flow using some special probe but multi-well method can be used to determine filtration velocity, transit velocity and direction of flow without using any special probe.

5.2.1 Single Borehole Dilution Technique

Single borehole dilution technique can be employed to obtain direct measurement of filtration velocity in a water bearing formation under natural or induced hydraulic gradient. The filtration velocity interpreted in conjunction with other parameters can provide valuable information about an aquifer.

The dilution rate of the solution, which is homogeneously dispersed in a volume V in the borehole, is related to the horizontal water flow velocity V_f by the equation:

$$V_{\mathrm{f}} = (\pi d/4\varphi)\ \ln\ (Co/C)$$

where 'd' is the diameter of bore hole, Co is the initial concentration of the tracer at time $T = 0$ and C is tracer concentration at time $T = t$ and φ is the distortion factor which compensate the effect deformation caused by the presence of the bore hole (distortion of flow lines takes place).

The value of φ depends on borehole geometry and permeability in the area of borehole (aquifer, filter tube and filter gravel used) and can be determined theoretically or through model experiments. However, for small borehole, having diameter of 2", its value can be taken as 1, while for large diameter borehole, its value may be taken as 2 or even more. Other factors which affect the dilution rate of the tracer are vertical currents, artificial mixing of the tracer, molecular diffusion etc.

Actual filtration velocity in a borehole can be determined by using the empirical relation,

$$V_{\mathrm{f}} = Vo\{1 - \exp\ (-hD/L)\}$$

where Vo is the actual filtration velocity averaged over the filtration velocities of different currents, h is the height of water column in bore hole and D is the distance of the borehole from the central line of canal or river. L is length of one side curved surface of the canal or river. This relation is not valid in the case of natural groundwater flow velocity measurements. The diffusion velocity in case of tritium has been observed very low, i.e., of the order of 0.5 cm per day by many investigators.

5.2.2 Multi-well Dilution Technique

In this technique, the radioisotope is injected in one borehole, called the main borehole, and either a probe is placed or water samples are collected in/from another borehole constructed at some distance. In fact, if the natural groundwater velocity is to be observed, boreholes are made in a circle keeping the main borehole at the centre. By knowing the travel time of radioisotope from main borehole to the borehole under observation, and distance of borehole, the velocity of transit is determined. The multiplication of specific yield of the aquifer with transit velocity provides the estimate of filtration velocity or groundwater flow velocity. Similarly, if the flow velocity is observed in all the bore holes made around the main borehole, the direction of flow is the direction of borehole in which the maximum velocity is observed.

5.3 Origin of Groundwater

All groundwater of economic interest originates as precipitation. Thus the amount of recharge to a groundwater system, along with the storage capacity of the aquifer, determines the maximum available resources for exploitation. Under favourable conditions (that is, aquifers where the barrier boundaries and the inputs and outputs are well defined) it is possible to construct a simple conceptual model of the system to obtain a water balance. Nevertheless, in many cases, where recharge and water flow are rather complex, more information of the actual process is desirable. Knowledge of the recharge process is also important for preventing deterioration of water quality by salinization and pollution. The environmental isotopic methods provide a valuable approach to understand these complex phenomena as well as to test the validity of the alternative hypothesis.

The application of the environmental isotope methods to understand the origin of groundwater with respect to its recharge is based on the spatial and temporal variability of the isotopic contents of water. The spatial variability can be grouped in four different topics:

Altitude Effect Groundwater recharge from high altitude either directly or by rivers draining high altitude catchment basins can be distinguished from recharge originating from low-altitude precipitation due to altitude effect in precipitation. This effect is most useful in regimes having orographic precipitation, where there is a regular relationship between land-surface altitude and condensation temperature of precipitation (isotopic composition of heavier isotopes depletes in precipitation with increasing altitude). When saturated air moves upward it cools, which causes condensation. Consequently heat is released which counteracts cooling. The resulting change in temperature with altitude is called adiabatic lapse rate. The wet adiabatic lapse rate varies with altitude, but a value of 0.6 °C/100 m is reasonable. For ^{18}O, the temperature dependence during wet adiabatic cooling is about 0.5‰/°C. The observed range of variation of $\delta^{18}O$ per 100 m is between −0.16 and −0.7‰ with an average value of −0.25‰, while for $\delta^{2}D$, the variation per 100 m is between −1.0 and −4.0‰ with an average value of −2.0‰.

Latitude Effect The stable isotope content of precipitation shows a marked dependence on latitude (depletes with increasing latitude); thus, groundwater replenished by rivers spanning significant latitude zones can be distinguished from local recharge.

Deuterium Excess Another geographic effect on the stable isotope content that can be applied effectively is the shift in the d-excess (defined as δD-$\delta^{18}O$), relating to the source of atmospheric moisture. It is noticed that in the eastern Mediterranean and similar marine belts of rapid evaporation, the resulting precipitation has a d-excess of 22; quite different from the general northern hemisphere precipitation d-excess of 10. This difference can be used in the border areas of climatic zones, where precipitation on the coastal mountains can be of Mediterranean or oceanic origin.

Seasonal Recharge In regions, where temperature and precipitation are distinctly seasonal and groundwater flow occurs in crystalline rock or karst limestone, the stable isotope content of groundwater may indicate the seasonal dependency of recharge and in some instances even the influence of specific large storm events. The stable isotope content can also be used to determine the ratio of seasonal recharge and base-flow of springs having such a composite discharge regime.

Case Study: Identification of Recharge Zones and Recharge Sources in Districts Haridwar and Saharanpur

National Institute of Hydrology carried out a study to identify recharge zones and recharge sources in the Saharanpur district of Uttar Pradesh in the year 1998 (NIH 1999; NIH 2000). For the study, about 130 groundwater samples were collected, from all the geohydrological regions viz., Bhabhar, Tarai and plains, from shallow (upto 50 m), intermediate (50–100 m), and deeper aquifers (>100 m) during the pre-monsoon season of 1998. A few samples for rainfall were also collected. The samples were analyzed for stable isotope and tritium dating. The stable isotopic index for rainfall at various sites in the study area was estimated using IAEA data for rainfall available at New Delhi and using the altitude effect of 0.31‰ per 100 m increase in altitude. The shallow aquifer samples were found to be depleted with respect to rainfall composition while deeper aquifers were found depleted. Using tritium and stable isotopic data, the recharge zones were mapped and the flow velocity for groundwater flow in the deeper aquifer in the eastern Saharanpur district (now Haridwar District) were computed (Fig. 4.18). It was found that the recharge zone of the deeper aquifers is located in the Siwalik (TU~16). Tritium content in

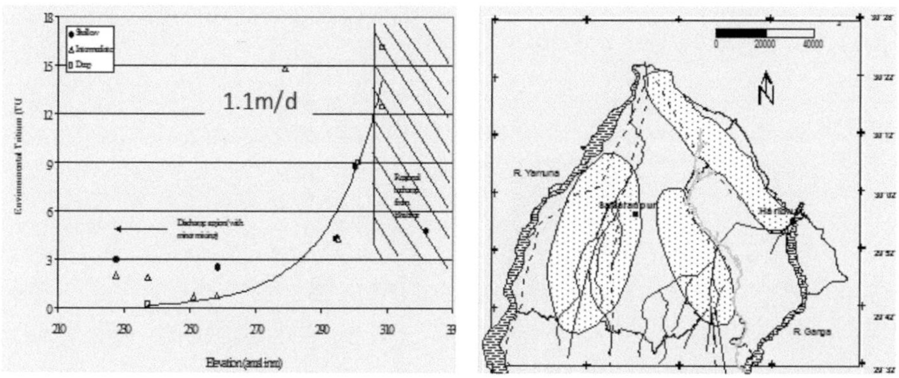

(A) (B)

Fig. 4.18 (**a**) Estimation of groundwater velocity from variation in environmental tritium content in Solani-Ganga Interfluve, and (**b**) Recharge zones identified on the basis of tritium content in Districts Haridwar and Saharanpur

groundwater at a site at Laksar town (in the groundwater flow direction) was measured to be ~1 TU, indicating groundwater velocity to be 1.1 m/d for deeper aquifer.

Major conclusions drawn from the isotopic investigations are: (i) the major recharge zones are located in Bhabhar-Siwalik area and extend over an altitude range of 500–1000 m above the mean sea level, (ii) a local recharge zone exists in the south of Roorkee towards Purkazi town, (ii) shallow aquifers in the Solani-Yamuna interfluve are recharged locally and those in Solani-Ganga interfluve are recharged mainly from Bhabhar-Siwalik region, (iii) the Upper Ganga Canal and Eastern Yamuna Canal are effective in recharging the shallow aquifer over a width of few kilometres, (iv) the deeper aquifers are semi-confined to confined type and are not recharged vertically, and (v) mean groundwater velocity in the Solani-Ganga interfluve is about 1.1 m/d.

The recharge zones identified on the basis of isotopic analysis are given in Tables 4.5 and 4.6.

5.4 Recharge Zones and Sources to Aquifers/Springs

5.4.1 Recharge Zones of Deeper Aquifers

Until the end of twentieth century, groundwater was mostly abstracted from the shallow aquifers and was believed to be safe, free from pathogenic bacteria and from suspended matter. However, with the increase in population and urbanization, and due to technological advancement, the stress on deeper aquifer has increased, mostly because the shallow aquifers have either drying-up or been contaminated. In fact, the deeper aquifers are not only catering the present need of fresh water but also these

Table 4.5 Isotope characterization of shallow aquifers

Area type	Site name & altitude (msl)	$\delta^{18}O$ (‰) Local rainfall	$\delta^{18}O$ (‰) Groundwater	TU	Interfluve type	Possible recharge sources
Tarai-Plains away from rivers and canal network	Chhutmalpur (295 m)	−6.8	−5.8	Intermediate aquifer TU = 17.8)	Solani-Yamuna	100% local precipitation recharge Enrichment due to evaporation
Plains, close to canal	Roorkee (265 m)	−6.7	−7.6	–		Rainfall - 73% and canal seepage 27%
Plain region away from rivers and canal system	Gagalhedi (280 m)	−6.7	−5.3	7.2		100% local precipitation recharge
	Nakud (263 m)	−6.7	−5.5	11.7		100% local precipitation recharge
Plains, near river Yamuna	Toda (247 m)	−6.6	−6.7	–		Rainfall - 96% and Canal seepage 4%
Plains, close to River Ganga	Chandpuri Kalan (228 m)	−6.5	−7.8	3.0	Solani-Ganga	87% from middle Siwalik (~700 m) and 13% from local precipitation

Table 4.6 Isotope characterization of deeper aquifers in Solani-Yamuna interfluve

Region	Site name and altitude (msl)	$\delta^{18}O$ Local rainfall	$\delta^{18}O$ Groundwater	TU	Recharge source
Siwalik foot hills	Nagal Kothari (417 m)	−7.1	−9.0	11.1	100% precipitation recharge at upper Siwalik (~1000 m msl)
	Timli (550 m)	−7.7	−8.0	10.4	100% recharge from middle/ upper Siwalik (~700 m)
Tarai-Plain interfringe	Chhutmalpur (295 m)	−6.8	−7.3	1.8	100% recharge at lower Siwalik (~450 m)
Plains near Upper Ganga Canal	Roorkee (265 m)	−6.7	−8.8	5.0	14% recharge from shallow aquifer ($\delta^{18}O = -7.6$) + 86% from upper Siwalik (~1000 m)
	Purkazi (232 m)	−6.5	−7.1	10.7	83% local precipitation ($\delta^{18}O = -6.5$) +17% canal ($\delta^{18}O = -10$)
Plains, near the river Yamuna	Nakud (263)	−6.7	−7.3	ND	100% precipitation recharge from lower Siwalik (450 m)

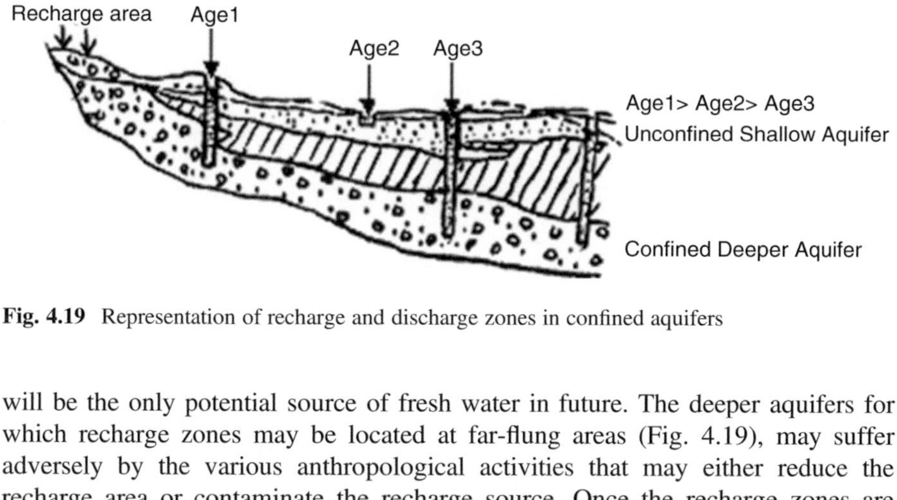

Fig. 4.19 Representation of recharge and discharge zones in confined aquifers

will be the only potential source of fresh water in future. The deeper aquifers for which recharge zones may be located at far-flung areas (Fig. 4.19), may suffer adversely by the various anthropological activities that may either reduce the recharge area or contaminate the recharge source. Once the recharge zones are identified, these can be protected from the anthropogenic activities.

Environmental isotopes like ^{3}H, ^{14}C, ^{2}H and ^{18}O are used to identify the recharge zones and recharge sources of aquifers and springs. Geohydrological details like groundwater level conditions, geological cross sections etc., and water quality data like major and minor ion chemistry, physico-chemical parameters etc., are used as supporting tools. Groundwater samples are collected from different aquifers for the measurement of ^{3}H, ^{14}C, ^{2}H and ^{18}O. The dating of groundwater using ^{3}H and ^{14}C provide information of recharge zones, groundwater flow velocity and flow pattern while the δD and $\delta^{18}O$ analysis help in understanding the contribution of different recharge sources and also to pinpoint the most important recharge source.

Case Study: Groundwater Recharge in Bist Doab, Northwest India

Comparing the groundwater and precipitation isotope signatures, Lapworth et al. (2015) established the dominance of local modern meteoric sources in both shallow and deep regional groundwater recharge in Bist Doab of Northwest India (Fig. 4.20). In few shallow groundwater samples and one canal sample, the isotopic signatures were found to deviate from the regional meteoric water line, indicating significant evaporative enrichment prior to recharge.

The study further indicates that even at a depth of 160 m, groundwater isotope signatures are consistent with local modern rainfall sources (Fig 4.20a and c) and together with the Noble Gas Recharge Temperature (NGT) and $\delta^{18}O$ results from this study suggest that palaeo-water (e.g., Wieser et al. 2011), if present in this region, is deeper than 160 m below ground level (bgl).

Further, the analysis of modern tracers (CFC-12 and CFC-11) has indicated that majority of deep boreholes have a significant component of modern recharge reaching depths of at least 150 m. The presence of significant quantities of tracer in deep boreholes has been considered to be due to the vertical migration of groundwater in the subsurface. This is supported by sedimentary logs from this region which show that while low K horizons are prevalent within the sedimentary sequence (0–150 mbgl), they have limited lateral continuity (Bowen 1985; Singh

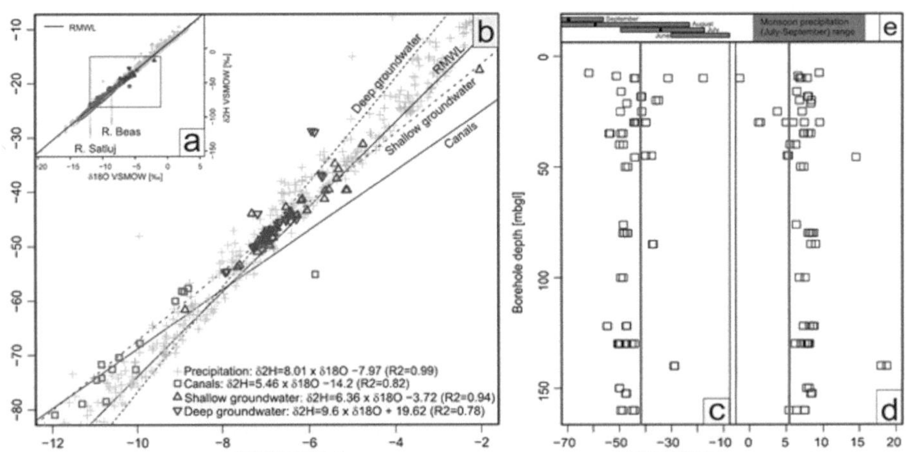

Fig. 4.20 (**a**) Cross plot of $\delta^{18}O$ versus $\delta^{2}H$ for groundwater, canals and precipitation within the study area. (**b**) Cross plot of $\delta^{18}O$ versus $\delta^{2}H$ for data extent for groundwater and canal points. (**c**) Depth variation in $\delta^{2}H$ for groundwater and solid line shows amount-weighted $\delta^{2}H$ for precipitation. (**d**) Depth variation in groundwater d-Excess and solid line shows amount-weighted $\delta^{2}H$ and d-Excess for precipitation. (**e**) median and interquartile ranges for local precipitation data, and monsoon (June–September) precipitation ranges for d-Excess. Note *x* axis on same scale as for Fig. 4.20c and d

et al. 2015), and analogous sedimentary settings in India (Kumar et al. 2007; Samadder et al. 2011; Sinha et al. 2013).

5.4.2 Recharge Zones of Springs

In order to identify the recharge zones of springs, the water samples of springs are collected along with precipitation samples from different altitude. The δD and $\delta^{18}O$ of samples are analysed along with environmental tritium of spring water. The altitude effect in the study area is established. The δD vs $\delta^{18}O$ plot is used to determine the isotopic value of spring water. If the sample data fall on the evaporation line, the line is extended back to get the real isotopic value of spring water if it would have not been subjected to evaporation. After determining the isotopic value of spring water (either δD or $\delta^{18}O$), the altitude of the recharge area is determined using the altitude effect equation.

Case Study: Identification of Recharge Zones of Springs at Gaucher, Uttarakhand

A study was conducted jointly by Isotope Applications Division, Bhabha Atomic Research Centre, Mumbai and Himalayan Environmental Studies and Conservation Organization, Dehradun to identify the recharge areas at Gauchar area of Chamoli, Uttarakhand (Shivanna et al. 2008).

Stable isotope data of precipitation collected from three different heights of the valleys, viz. 1180, 990 and 800 m above mean sea level (amsl) during September 2004 show that $\delta^{18}O$ varied from −10.0 to −8.2‰ and δ D varied from −69.2 to −56.3‰. The isotopic composition of spring water varied from −7.7 to −7.0‰ for $\delta^{18}O$ and from −57.9 to −50.3‰ for δD.

For estimating the altitude effect, the stable isotopic compositions of the precipitation samples were plotted against their corresponding altitudes. The altitude effect was calculated as the inverse of the slope of the best-fit line. It was found to be −0.55‰ for $\delta^{18}O$ and −3.8‰ for δD per 100 m rise in altitude (Fig. 4.21). Generally, the discharge rates of springs ranged from 0.7 to 120 l/min during the

Fig. 4.21 $\delta^{18}O$ vs altitude of the rainwater samples from Gaucher area. Dotted line indicates the recharge altitudes of the low-altitude springs. (*Source:* Shivanna et al. 2008)

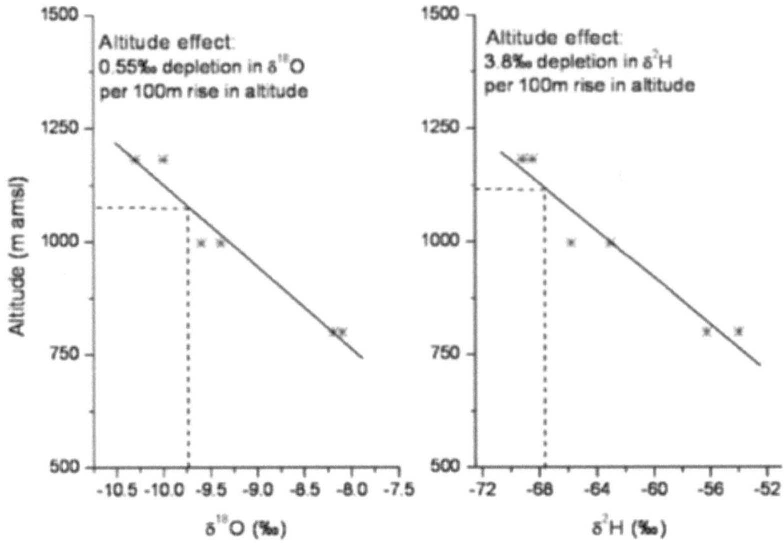

Fig. 4.22 (Right) Subsurface dykes; (Left) New spring that appeared subsequently

monsoon season, which reduced to 0.2–99 l/min within a month after the monsoon. The discharge measurements indicated that all the springs responded instantaneously to the precipitation events.

During summer, most of the high-altitude springs dry up, whereas low-altitude springs continue to discharge with low rates. This shows that low-altitude spring discharges were a mixture of groundwater and precipitation, whereas only precipitation contributed to the high-altitude springs. The tritium content of the low-altitude springs was about 9.5 TU, whereas it was about 11.5 TU for high-altitude springs. This slight variation in tritium content can be attributed to relatively longer residence time of groundwater in low-altitude springs compared to that of high altitude springs. The topography and geology of the area suggests that the presence of thin, weathered, top soil cover underlain by hard and compact quartzite, phyllite and granite with a steep gradient does not allow the rainwater to percolate down, and hence the spring waters were mostly derived from seepage waters. Based on geology, geomorphology, hydrochemistry and isotope information of the study area, the recharge zones to the springs were located at altitudes of 1270, 1330 and 1020 m amsl respectively. At these identified altitudes, water-conservation and artificial recharge structures like subsurface dykes (5 nos.) in valley-1, check bunds (2 nos) in valley-2 and a few trenches in valley-3 were constructed for rainwater harvesting (Fig. 4.22). Monthly spring discharge measurements were carried out during December 2004–April 2005 before the construction of the structures and also during the same period in the following year, after the construction. The cumulative discharge rate of ten springs increased from 375 to 708 l/min during the post-monsoon period. It was found that the discharge rates had not only increased considerably, but also were sustainable even during the dry period. In addition to the existing springs, two new springs also appeared close to the subsurface dykes. The cumulative discharge rate of these new springs was about 67 l/min. The significant increase in the spring discharge rates, their longer duration and formation of new springs can be attributed to proper identification of recharge areas and implementation of artificial recharge structures.

5.4.3 Recharge Sources/Surface Water and Groundwater Interaction

In regions where direct and rapid infiltration of rain occurs, the composition of the groundwater will have isotope ratio as that of the precipitation or slightly enriched due to evaporation effect during the process of infiltration as when evaporation precedes, the heavier isotope gets enriched in water phase. The analyses of stable isotopic composition together with that of groundwater dating using radioisotopes gives a reliable 'fingerprint' to identify each group of water of different sources. Water infiltrated at higher altitude and is transmitted over long distances will show depleted δ^{18}O and low tritium compared to water infiltrated from local precipitation.

One of the important uses of isotopic characterisation of waters is to study the surface water and groundwater interaction. It is based on the fact that the surface water, particularly rivers originating at the higher altitude, normally has different stable isotopic composition than that of groundwater recharged by infiltration of local precipitation where the surface-groundwater relation is under investigation.

In the areas near to bank of the river or lake, there are two possible sources of recharge to groundwater, viz. infiltration of local precipitation and infiltration of river water. In such conditions, the accuracy of the estimate of the proportion of infiltrated river water depends upon the accuracy of the estimates of stable isotopic indices (most common value of the source water) of these two potential sources of recharge and the difference between these indices. An estimate of the isotopic indices of the surface water body is made on the basis of isotopic values measured at different times and especially at different stages/discharge to ascertain whether there is any significant variations in stable isotopic composition with stage discharge. If variations are evident, then the mean value weighted for various stages/discharges is used. The preferable approach is to sample groundwater close to the river where piezometer indicates river water as the source of recharge. The estimation of the index for groundwater generated by infiltration of local precipitation is based on measurements of groundwater away from the influence of the river. If the errors in estimates of the indices of the two potential sources of recharge are not greater than the analytical error, then the accuracy in the estimate of the proportion is better than 10%. In practice the limitations of the method are not in the method itself, but in the availability of meaningful samples.

In most of the cases, the stable isotopes, ^{18}O and D (^{2}H) are utilized for determining the contribution of groundwater to the surface water or vice-versa. If R_1 and R_2 are the isotopic composition of the groundwater and the surface water body, respectively and m_1 and m_2 are the fractions of groundwater and surface water, respectively in the admixture, while R_{am} is the isotopic composition of the admixture, then the isotopic balance and mass balance equations can be written as:

$$m_1R_1 + m_2R_2 = R_{am} \text{ and } m_1 + m_2 = 1$$

From the above two equations, we have

$$m_2 = (R_{am} - R_1)/(R_2 - R_1)$$

Therefore, by knowing the value of R_1, R_2 and R_{am}, the fraction of surface water mixed with groundwater can be evaluated.

Case Study: Surface Water–Groundwater Interaction at Palla, Delhi, Case Study of the Yamuna River at Delhi

To understand the interaction of surface water and groundwater at Palla village in Delhi, National Institute of Hydrology conducted a detailed study during 2007–2010 (Kumar et al. 2012). The study was based on the fact that the River Yamuna originates at higher elevation and normally has a different stable isotopic composition than that of groundwater being recharged by infiltration of local precipitation. In case of a mountainous river, the river transports water, which has generally been originated from precipitation falling at higher elevations than the area where the surface-groundwater relation is under investigation. The difference in isotopic composition of these waters is due to altitude effect. The isotopic composition for ^{18}O in precipitation changes between −0.2 and −0.3‰ per 100 m with altitude. Thus, the stable isotopic composition of the river water is more depleted than that of groundwater derived from infiltration of local precipitation. This distinct difference helps in identifying the contribution of one to the other. The studies carried out by NIH, Roorkee and few others have revealed that the River Yamuna has stable isotopic signatures ($\delta^{18}O$) in the range of −8 to −9‰ while the groundwater in Delhi region varies between 6 and 7‰ where recharge due to precipitation dominates. Therefore, stable isotopes of hydrogen and oxygen have been used to determine the contribution of groundwater to river or vice versa at the selected locations in the study area.

Similarity of $\delta^{18}O$ and $\delta^{2}H$ values between stream and aquifer may indicate interconnectivity, whereas isolated aquifers may contain waters with different $\delta^{18}O$ and $\delta^{2}H$ values. Stable isotope systematics of waters around surface water bodies may be used to trace movement of seepage into/from nearby groundwater systems.

Groundwater samples were collected at an interval of seven days (every Sunday) from piezometers on Delhi side and at an interval of 14 days (every alternate Sunday) from piezometers on UP side. While collecting the samples water levels were also monitored in the piezometers (Fig. 4.23).

Based on the $\delta^{18}O$ variation in groundwater, the river water and the ranney well, the component of river water in the water supply wells was computed (Fig. 4.24). The percentage of water from the floodplain in various months is given in Table 4.7.

The isotopic analysis of the groundwater, river water and pumped wells indicates that percentage of river water in pumped water vary from season to season and from

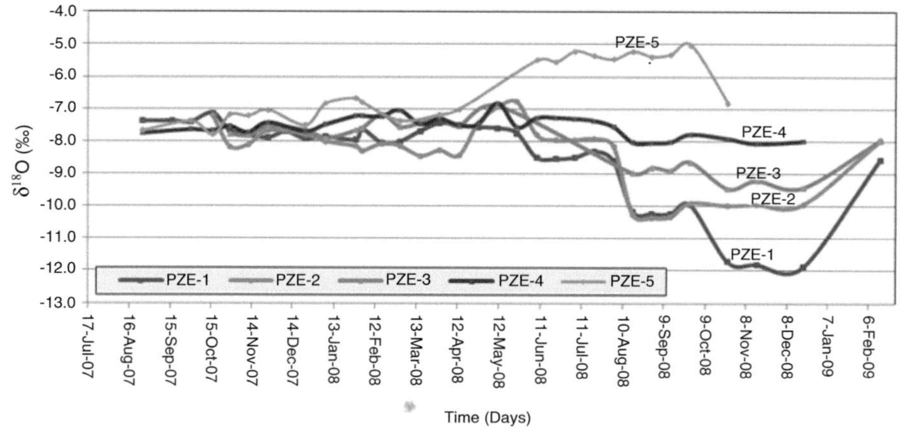

Fig. 4.23 Variation of water levels in piezometers and River Yamuna at Palla in Delhi

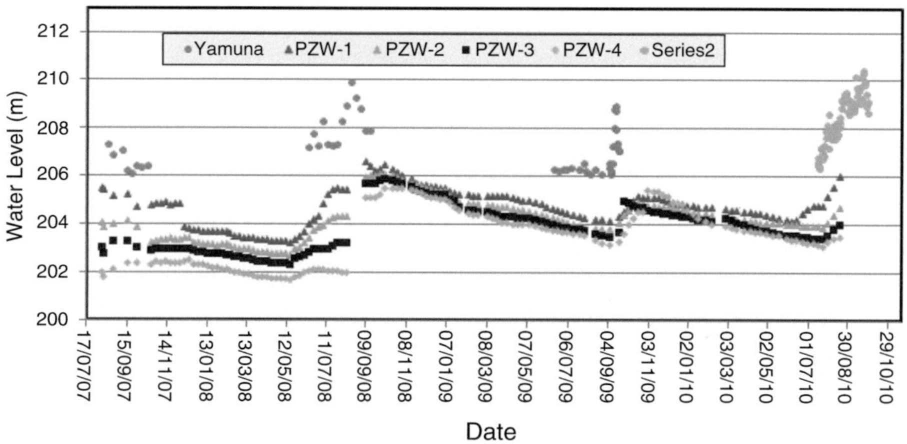

Fig. 4.24 Variation of $\delta^{18}O$ with time in the floodplain of a section across River Yamuna at Delhi

Table 4.7 Component of river water in groundwater pumped from flood plains

Months	$\delta^{18}O$ in ranney well water (‰)	$\delta^{18}O$ in ground water (‰)	Flood plain water in pumped water (%)
Aug-07	−7.9	−6.8	41
Oct-07	−7.9	−6.9	43
Dec-07	−7.8	−7.1	33
Feb-08	−7.1	−6.9	9
Apr-08	−7.4	−6.8	24
Jun-08	−7.6	−6.9	30
Aug-08	−8.0	−7.4	32

year to year. Isotopic analysis of the water samples indicates that River Yamuna recharges groundwater on the Delhi side (Palla sector) more than the UP side during monsoon season (July to October). The recharge on the UP side takes place only during high floods, but on the Delhi side, the recharge occurs even during low floods also. During the years when there is no flood or the floods are of short duration, the contribution of water from the river Yamuna and its floodplain in the pumped water goes upto 50% during monsoon season, which gradually reduces and reaches to the tune of 20% during non-monsoon. During the year of major flooding, the contribution from the River Yamuna and its floodplain to the pumped water becomes 100% during monsoon season, which goes on reducing but continues for longer duration depending upon the availability of recharged water in the floodplain.

5.5 *Effectiveness of Artificial Recharge Measures*

Management of depleting groundwater table in the urban areas and in semiarid/arid regions has drawn the attention of water resources managers. This situation is also common in areas where surface water bodies such as rivers, canals and natural or artificial lakes/reservoirs do not exist. In order to mitigate the increasing shortage of groundwater, artificial recharge of groundwater by making earthen bunds, through injection wells or roof top rainwater harvesting programmes have been given priority by many organizations and individuals. However, the effectiveness of these programmes has not been assessed at the desired scale as it is difficult using conventional techniques. Isotope techniques have the potential to assess the effectiveness of these programmes using environmental isotopes.

The effectiveness of artificial recharge measures can be studied with the use of environmental isotopes (either δD or $\delta^{18}O$ and ^3H, if required) provided the artificially recharged water has different isotopic composition than the natural recharge. However, if the same water is used for artificial recharge through prolonged infiltration by constructing earthen bunds, the isotopic composition of ponded water is changed due to evaporative enrichment. The isotopic indices (δD or $\delta^{18}O$) of precipitation, groundwater (without artificial recharge component), and surface water being used for artificial recharge are determined and then the isotopic composition of the groundwater (from the study area) are determined at different time intervals (minimum monthly frequency). The use of two-component model can reveal the percentage of mixing of artificially recharged water at different time. If the groundwater samples are collected from different locations around the site/s of artificial recharge, then the effect and extent of recharge (%) can be determined in different directions. ^3H values of groundwater can confirm the recent recharge due to artificial measures.

For example, the contribution of rainfall (m_p) or channels (m_{ch}) in groundwater can be determined using the following relation based on $\delta^{18}O$ values of end members.

$$m_p = \left(\delta^{18}O_{gw} - \delta^{18}O_p\right)/\left(\delta^{18}O_{ch} - \delta^{18}O_p\right)$$
$$\text{or} \quad m_{ch} = \left(\delta^{18}O_{gw} - \delta^{18}O_{ch}\right)/\left(\delta^{18}O_p - \delta^{18}O_{ch}\right)$$

^3H values of groundwater can confirm the recent recharge due to artificial measures.

Case Study: Ozar Watershed Study in District Nasik, Maharashtra

Groundwater, precipitation and Hantur Canal water samples from the Ozar watershed were analyzed for $\delta^{18}O$ and ^3H (Kumar et al. 2009). Tritium in groundwater ranged between 14 and16 TU. These values were comparable to the Hantur canal (~12.6 TU), the water of which is used for artificial recharge through earthen channels, and much different from the rainfall in the area (^3H ~5 to 6 TU). Therefore, it is inferred that the groundwater is more dominated by recharge from canal waters through earthen channels and rainfall recharge component is comparatively very less. Moreover, the data reflect that the groundwater is young with negligible aquifer storage.

The contribution of rainfall or channels in groundwater is determined using the following relation based on $\delta^{18}O$ values of end members.

$$m_p = \left(\delta^{18}O_{gw} - \delta^{18}O_p\right)/\left(\delta^{18}O_{ch} - \delta^{18}O_p\right)$$
$$\text{or} \quad m_{ch} = \left(\delta^{8}O_{gw} - \delta^{18}O_{ch}\right)/\left(\delta^{18}O_p - \delta^{18}O_{ch}\right)$$

where m_p and m_{ch} are the contributions of precipitation and channel water to groundwater respectively while $\delta^{18}O_{gw}$, $\delta^{18}O_p$ and $\delta^{18}O_{ch}$ are the corresponding oxygen-18 values of groundwater, precipitation and channel water. Figure 4.25 clearly indicates the percent contribution of rainfall in recharging groundwater in Ozar watershed. The percent artificial recharge to groundwater through earthen channels can be estimated by subtracting percent of rainfall recharge from 100. A straight line relation between the amount of rainfall and $\delta^{18}O_p$ values also enables to determine the percent contribution of rainfall to groundwater.

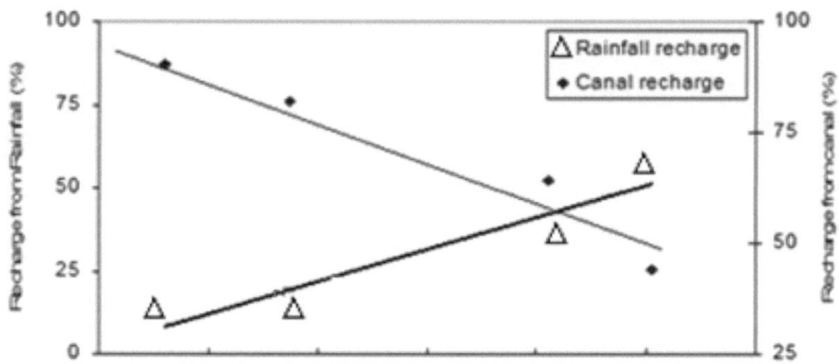

Fig. 4.25 Rainfall recharge to groundwater and the $\delta^{18}O$ of precipitation

5.6 Groundwater Salinization

The build-up of salinity in groundwater and soil represents a major wastage of natural resources in arid and semi-arid zones. The salinity of groundwater may occur due to several processes, but, two of them are of major importance, that is (i) salts leached by percolating water; these salts may be evaporative deposits, aeolian transported salts (usually of marine origin) or products of weathering of surface rocks and soils, and (ii) infiltration or mixing of saline surface water, seawater, brines and connate water with groundwater.

Conventionally, ionic ratios, such as, Na^{+}/Cl, Ca^{+2}/Mg^{+2}, SO_4^{2}/Cl etc. are used to study the source of salinity. However, after transport of the salts by percolating water to the water table and after primary mixing with saline waters, the chemistry of the water may undergo further secondary changes which may make it difficult to identify the mechanism of salinization precisely. The problem of identifying the mechanism of salinization becomes more acute in irrigated areas in arid zones. Leaching is usually local and the soluble salts may not be transported very far from the area of generation/accumulation. Thus, low rainfall and high evaporation rates, which are characteristic of arid zones, tend to concentrate salts in the groundwater. Also, surface drainage is poorly developed in arid regions and has no outlet to permanent stream. These further result in accumulation of salts in these regions.

The use of environmental isotopes, D and ^{18}O, allow distinction of leaching from evaporation. Leaching causes no isotopic change, whereas evaporation leads to isotopic enrichment. Further, the S-34 ($\delta^{34}S/\delta^{32}S$) isotope provides very clear picture about the salinity of marine origin. The isotopic approach is particularly useful in coastal areas where all salinity ultimately originates from marine environment so that little chemical distinction exists amongst salinity originating for example, from sea spray or from direct sea water encroachment from surface lagoons or connate sea water. The stable isotopes along with dating techniques of groundwater in the study of salinization mechanism have been extensively used abroad but comparatively less in India.

In studies dealing with seawater intrusion, the significant difference between sea water and freshwater, particularly for stable isotopes of hydrogen and oxygen, provides a direct means of identifying and studying dynamics of sea water intrusion (pathways, mixing ratios). Furthermore, isotopic evolution during different groundwater salinization processes (i.e. mineral dissolution, leaching of salt formations, or mixing with saline formation waters) exhibit different characteristics, to enable process identification through observations to be made on stable isotope concentrations. Particularly, the use of environmental isotopes, D and ^{18}O, allow distinction of leaching from evaporation. As stated earlier, leaching causes no isotopic change, whereas evaporation leads to isotopic enrichment. In addition, the stable isotopes of sulphur -34 ($\delta^{34}S/\delta^{32}S$) isotope provides very clear picture about the salinity of marine origin.

Radioactive isotopes of ^{3}H (tritium) and ^{14}C (radiocarbon), with their known input concentrations into the hydrological cycle (both natural and anthropogenic origin) also provide a label for different water bodies enabling tracing of sea water intrusion processes. Their natural production in the atmosphere due to interaction of cosmic radiation with the constituents of air is rather steady state. However, large amounts of these isotopes were also released in to the atmosphere by the nuclear weapon tests carried out during the period 1953 to 1963. The unique radioactive decay property of these isotopes, particularly of radiocarbon with a steady-state input concentration, also facilitates time-domain estimation to be made of the physical parameters related to circulation dynamics of groundwater.

Freshwater derived from precipitation is always isotopically different (contains less species of heavier isotopes) than sea water due to isotopic fractionation processes occurring during evaporation and condensation. In addition, the evolution of stable isotope concentrations during different natural processes, and particularly the resulting relationships between ^{18}O and ^{2}H concentrations, provide an effective tool for many hydrological applications, such as assessment of the "genesis" (origin) of water, particularly in groundwater systems; for the processes involved in replenishment (process tracing), for estimating mixing proportions of different sources or component flows (component-tracing); and studying hydraulic relationships between groundwater and surface waters or between different aquifer units in a given groundwater system. Thus, stable isotopes provide an effective label for sea water and freshwater to enable tracing of sea water intrusion, as well as identifying processes that may be responsible for water salinization.

During the processes of leaching salt formations or mineral dissolution, the stable isotope content of the water is not affected while the salinity of water increases. This is a unique feature which will enable identification of such processes based on isotopic and chemical data. The stable isotopes ^{18}O and ^{2}H are the most conservative tracers during their transport in hydrological systems and their relationship with salinity changes is univocal.

The identification of sea water intrusion could also be made through the use of tritium and/or radiocarbon concentrations. However, since the concentration of these isotopes also is removed from the system through radioactive decay and changes that may be induced by complex geochemical reactions in the case of radiocarbon, they may present difficulties for this purpose. This is why these radioactive isotopes are more often used for estimating the travel times (transit time) of component flows and to distinguish the present and palaeo-origin of salinity i.e., high salinity with high ages of groundwater indicates either salinity of palaeo-origin or due to very slow process of dispersion and diffusion of marine dissolve salts.

Case Study: Sea Water-Groundwater Interaction in the Coastal Zone of Krishna Delta

The groundwater-sea water inter-connection in the Krishna river delta, India was studied by Nachiappan et al. (2003). To establish the interconnection, groundwater samples (from depths: <30 m, 30–60 m and >60 m) and surface water samples from River Krishna, its distributaries, sea (Bay of Bengal) water and Prakasam Reservoir

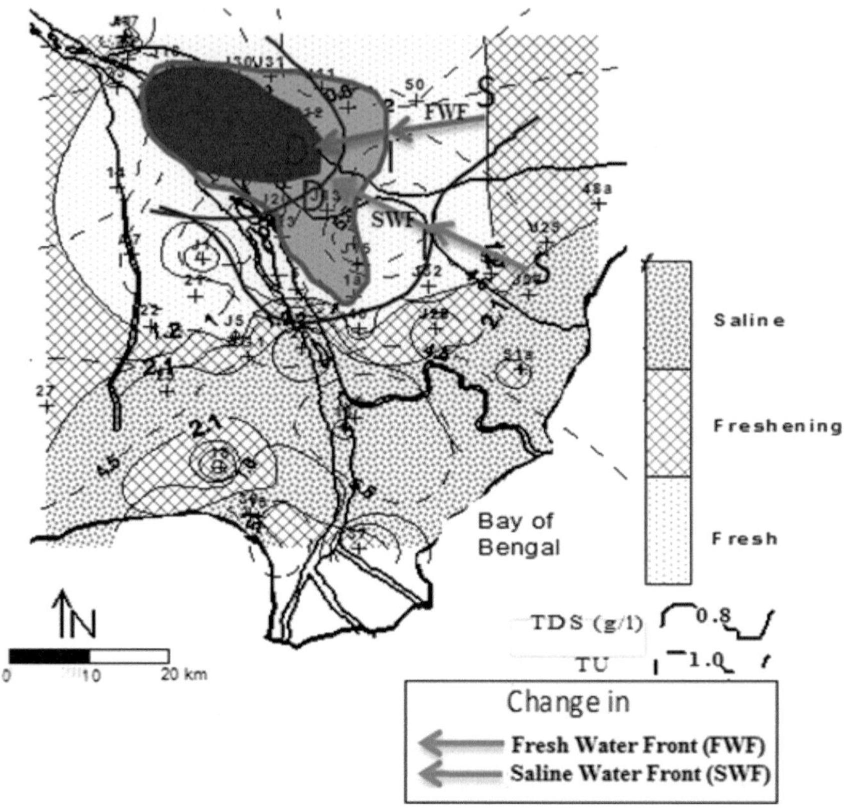

Fig. 4.26 Change in groundwater saline and fresh water interface with depth (S – Shallow, I – Intermediate, D – Deep) in Krishna delta

were collected during 1999–2002. The samples were analyzed for EC, ^{3}H and stable isotopes. A few samples were also analyzed for ^{14}C. Isotopic and chemical data indicated that the present day salinity in shallow, intermediate and deep aquifers extended over a distance up to 27 km, 32 km and 50 km from the present day seacoast respectively. Groundwater age increases towards the coast. The old groundwater in the area is interpreted to be of palaeo-marine origin. The fresh groundwater recharge taking place from Prakasam Barrage, along the canal tract (particularly in the area near Kaza in eastern delta) and along few paleochannels (on which Krishna canals are laid) is mixing with old groundwater and thereby diluting the pre-existed salinity (Fig. 4.26). The mixing was found to decrease at deeper levels. The results were corroborated with reference to the palaeo-coast line, dates on fossil shells, peat/wood and calcrete materials, and chemical data.

5.7 *Assessment of Groundwater Potential and Sustainability in Fractured Hard Rocks*

In order to address the problem of realistic assessment of groundwater potential and its sustainability in fractured hard rocks, it is vital to study the recharge processes and mechanism of groundwater flow, where heterogeneities and discontinuities play a dominant role. Wide variations in chloride, $\delta^{18}O$ and ^{14}C concentrations of the groundwaters observed in space and time could only reflect the heterogeneous hydrogeological setting in the fractured granites of Hyderabad (India). Isotopic and environmental chloride variations of the groundwater system put forth two broad types of groundwaters involving various recharge processes and flow mechanisms in the granitic hard rock aquifers found in Andhra Pradesh. Relatively high ^{14}C ages (1300 to ~6000 yr. B.P.), $\delta^{18}O$ content (−3.2 to −1.5‰) and chloride concentration (<100 mg/l) are the signatures that identified one broad set of groundwaters resulting from recharge through weathered zone and subsequent movement through extensive sheet joints. The second set of groundwaters possessed an age range Modern to ~1000 yr. B.P., chloride in the range 100 to ~350 mg/L and $\delta^{18}O$ from −3.2 to +1.7‰. The $\delta^{18}O$ enrichment and chloride concentration further help in the segregation of the groundwater into different sub-sets characterized by different recharge processes and sources. Based on these processes and mechanisms, a conceptual hydrogeologic model can be evolved to understand the fracture network ant its linkage with recharge sources and their contribution.

5.8 *Groundwater Flow in Fractured Rocks*

Radon (^{222}Rn) concentrations in unpurged bores can be used as a qualitative indicator of groundwater flow rate. Radon is produced from decay of uranium and thorium minerals in the aquifer and has a half-life of 3.8 days. High concentrations of radon in streams have been used as a quantitative indicator of groundwater discharge. In a similar way, high concentrations of radon in the borehole should indicate active groundwater inflow. If there is zero flow from the aquifer to the borehole, we would expect radon concentrations in the borehole to be zero due to radiogenic decay. However if flow through the well is faster than radon can decay then we would expect significant concentrations of radon in the well. If we assume that the concentrations of ^{222}Rn are uniform in the aquifer over the length of the borehole, then radon concentrations can be related to groundwater flow rate.

A large number of tracer tests have often been used to characterise fractured flow at small scale field sites. However, little effort has been made to use environmental tracers to characterise fracture flow. Environmental tracer techniques have advantages over hydraulic methods, as they can integrate over temporal and spatial scales. Hydraulic methods provide only a 'snap shot' of the present day flow regime. However, where possible, both environmental tracers and hydraulic data should be used together to help develop or constrain conceptual and numerical models.

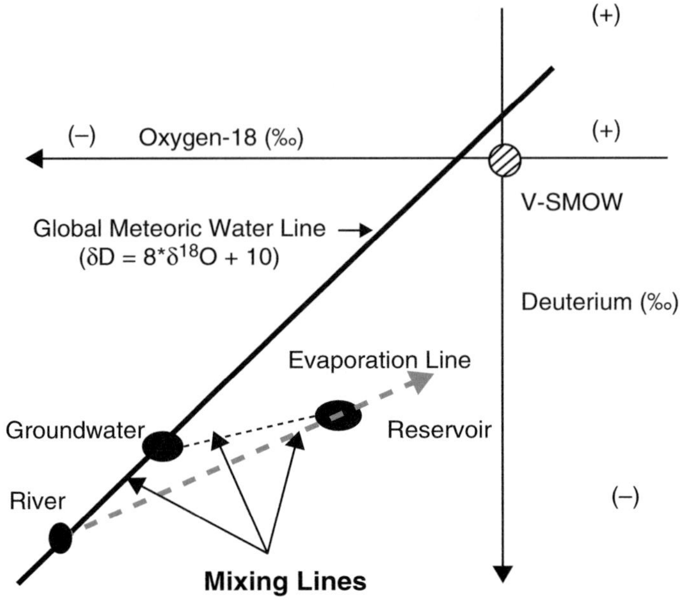

Fig. 4.27 δD and $\delta^{18}O$ relationship of different types of water near a reservoir

5.9 Identifying Leakages from Dams and Reservoirs

Water can escape from reservoirs either as leakage or seepage. It frequently appears below the dam foundation as well as at the abutments through concentrated pathways and emerges downstream forming springs. Concentrated leakage can be a serious problem for these big engineering constructions. In this context the question is asked whether the given leakage (spring) carries water from the reservoir behind the dam or whether it is simply a manifestation of the local hydrological system.

The fact that a surface water body subject to evaporation is changing its isotopic composition in a characteristic way (Fig. 4.27) makes it possible to distinguish between groundwater recharged under typical conditions and waters with an "evaporation history". This distinction becomes obvious when plotting the δD and $\delta^{18}O$ data. The evaporated waters plot on the right-hand side of the global (local) meteoric water line. Significantly, lower or higher ^{3}H values in the leakages, as compared to the characteristic reservoir value, also suggest a different origin of the water. Similarly, $^{13}\delta$ of the dissolved inorganic carbon of reservoir water that is in exchange with atmospheric CO_2 and biological activity is usually significantly higher than that of river inflow and local groundwater. In addition, the chemical composition of the water may provide an additional set of parameters for comparison. Environmental isotopes, both stable and radioactive, can be helpful in identifying the origin of these new occurrences of water near the given reservoir.

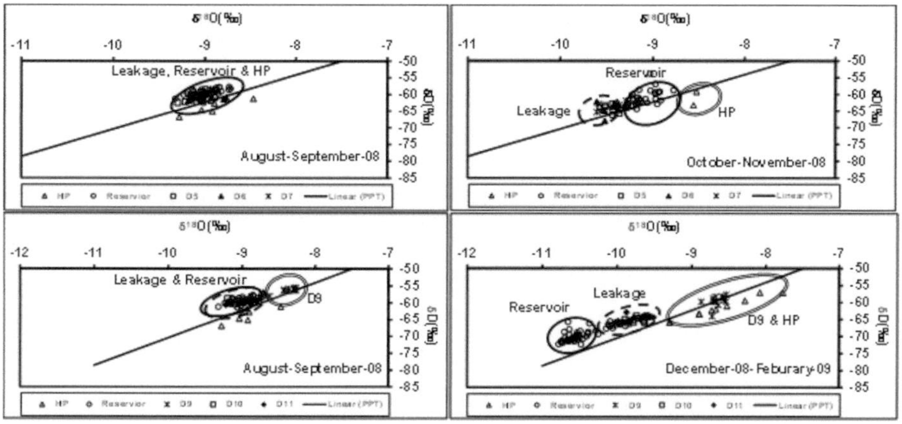

Fig. 4.28 Seasonal variation of δD vs $\delta^{18}O$ in groundwater, reservoir surface water and seepage water in inspection galleries Tehri Dam

Case Study: Leakage in Tehri Dam

The isotopic technique was successfully applied to identify the seepage in the inspection galleries of Tehri Dam (Rai et al. 2012). Using the discharge and isotopic composition, two separate fracture zones, responsible for seepage were identified. The reservoir water was the major component in leakage water occurring through seepage galleries.

In order to identify the source of leakage, the δD and $\delta^{18}O$ relationship of reservoir and leakage water was studied for the monsoon and post monsoon months (Fig. 4.28).

The seasonal variation in isotopic composition indicated that during monsoon months (August–September), when the reservoir level is above 770 m, isotopic values (δD and $\delta^{18}O$) of reservoir and leakage water fall at the same place. This clearly indicates that the source of seepage water in the drainage galleries was reservoir during monsoon months. However, a different isotopic composition during summer months indicated that when the reservoir level falls below 770 m, the seepage water was groundwater. During the winter months, segregation of isotopic values (δD and $\delta^{18}O$) of the reservoir and the seepage sites indicated that isotopic values (δD and $\delta^{18}O$) of reservoir and leakage water falls under different groups. This was attributed to less contribution from reservoir to leakage/seepage source due to fall of reservoir level after the month of October. During the summer months, isotopic composition became entirely different and discharge was minimal. Thus, the δD and $\delta^{18}O$ relationship of reservoir and seepage water confirmed the connectivity of seepage water in drainage galleries D5, D6, D7, D10 and D11 water with the reservoir when the reservoir level is above 770 m. The isotopic values at other locations remained in another group, indicating seepage from groundwater only.

Case Study: Seepage during Tunneling for Salal Hydropower Project

During construction of Tail Race Tunnel for the Salal Hydropower Project in Jammu several seepages were encountered. The tunnelling was through a dolomite

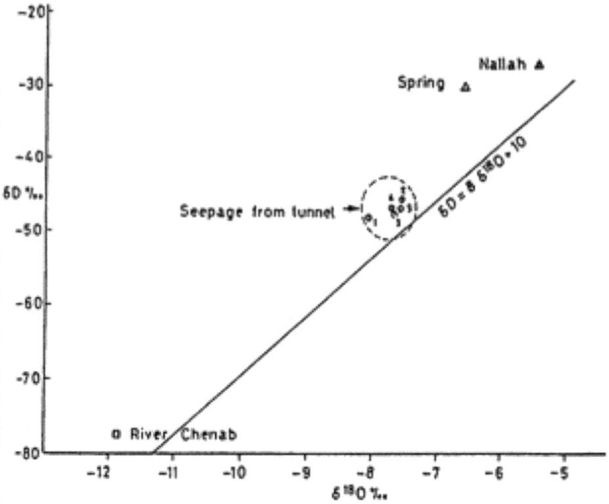

Fig. 4.29 Isotopic composition of sample waters collected in Salal H.P. Project

rock which is highly jointed and often crumbly and sheared. The total length of the tunnel was 2.4 km long and 11 m in diameter. The purpose of the tunnel was to carry the tail waters from the power house and to put them back into the River Chenab. Samples for the analysis were collected from neaby springs, nallahs, River Chenab and the seeping water.

The isotopic results (Fig. 4.29) clearly indicated that the seepage water that has isotopic composition ($\delta^{18}O$ –7.7‰) all along the tunnel is different from isotopic composition of River Chenab (~ −12‰). ^{3}H and EC profile of seepage waters showed a good correlation along the length of the tunnel and was different from the water of River Chenab. Using Piston Flow Model, the age of seepage water was estimated to be 10–15 years. It was concluded that the seepage water is old precipitation water stored or percolating in the fractured dolomite.

In addition to the applications mentioned above, isotope technique has been successfully used to solve problems in arid regions. This includes: isotope studies along the buried river course (considered to belong to the legendary Saraswati) near Jaisalmer; canal (IGNP) groundwater interaction and groundwater salinity canal and return flow in area in Stage I of the command area; groundwater recharge studies in Barmer and Bikaner districts of Rajasthan etc.

5.10 *Soil Salinisation*

The origin of sulphate—important to differentiate between seawater arid evaporite salt dissolution—can, in some cases, be ascertained with the help of $\delta^{34}S$ and $\delta^{18}O$ values of dissolved sulphate. This is not a common tool, but it is used under special circumstances.

Table 4.8 Role and applications of chemical elements and isotopes in salinity problems

Geochemical/ isotopic tool	Role in evaluating salinity
$\delta^{18}O$, $\delta^{2}H$	Essential indicators with Cl of evaporative enrichment and to quantify evaporation rates, in shallow groundwater environments. Diagnostic indicators of marine and palaeomarine waters.
^{14}C, $\delta^{13}C$	Additional indicator for modern seawater and for dating of saline waters. Half-life 5730 years. Understanding of carbon geochemistry is essential to interpretation.
$^{37}Cl/^{35}Cl$	Fractionation in some part of the hydrological cycle, mainly in saline/ hypersaline environments may allow fingerprinting.
$\delta^{87}Sr$	Additional indicator of source of groundwater salinity especially in carbonate environments.
$\delta^{11}B$	Additional indicator of salinity source.
$\delta^{34}S$	Indicator of evolution of seawater sulphate undergoing diagenesis. Characterisation of evaporite and other SO_4 sources of saline waters.
^{36}Cl	Half-life 3.01×10^5 years. Thermonuclear production; use as tracer of Cl cycling in shallow groundwater and for recharge estimation. Potential value for dating over long time spans and also for study of long term recharge processes. However, in situ production must be known.
Cl:	Master variable: inert tracer in nearly all geochemical processes; use in recharge estimation and to provide record of recharge history.
Br/Cl	To determine geochemical source of Cl.
Mg/Ca	Diagnostic ratio for (modern) sea water.
Sr, I, etc.	Diagenetic reactions release incompatible trace elements and may provide diagnostic indicators of palaeomarine and other palaeowaters.
Nutrients (NO_3, K, PO_4)	Nitrate accumulation may accompany Cl in aerobic arid environments. Nutrient elements characteristic of irrigation returns.
Organics	Indicator species (e.g. fatty acids) to characterize marine waters of different age. Pesticides etc. diagnostic of irrigation sources of salinity.

A distinct difference in boron isotopes between seawater and terrestrial water is emphasized by ^{11}B values for seawater and groundwater. The isotopic composition of boron in groundwater can be used to quantify seawater intrusion and identify intrusion types, e.g. seawater or brine intrusions with different chemical and isotopic characteristics, by using the relation of $\delta^{11}B$ and chloride concentration.

The fundamental relationships between $\delta^{18}O$ and $\delta^{2}H$ and between $\delta^{18}O$ and salinity may be used to identify different salinization pathways. $\delta^{37}Cl$ and $\delta^{11}B$ can be used to identify the source of salinity in coastal areas. Saline waters are often old waters and advances using long lived radioisotopes (^{39}Ar and ^{81}Kr for example) and gas accumulations (^{4}He) are becoming popular. Advances in the measurement (AMS techniques), use and understanding of ^{14}C, especially using organic carbon, have also provided new impetus to dating approaches. In the real world, therefore, where salinity problems appear to be of a mixed origin, there is now a need to apply a multiple isotope and geochemical approach to understand the salinisation of groundwater systems (Table 4.8).

6 Conclusion

Isotopes have wide applications in groundwater investigations and provide vital information for the better management of water resources. Isotopes also help in understanding various hydrological processes. Isotope techniques using "environmental isotopes" are commonly used in the developed countries by meteorologists, hydrologists and hydrogeologists in the study of water. The use of these techniques is also increasing in our country, but still it requires momentum and training of the field persons in this subject. Study of the isotopes of oxygen and hydrogen in water or of elements contained in dissolved salts which have the same behaviour as water, enable exact recording of phenomena affecting the occurrence and movement of water in all its forms.

These isotopes, both stable and radioactive, occur in the environment in low and varying concentrations with respect to the most abundant isotope of the same element. Environmental isotopes are natural or man-made. In either case, their distribution in the environment is governed by natural processes. Although small, the variations in concentrations of environmental isotopes are measured with high accuracy and provide valuable information on hydrological systems. Processes in the hydrological cycle and interactions between the hydrosphere and atmosphere are responsible for isotopic variations observed in natural waters.

Isotopic methods are normally used in conjunction with conventional hydrological, hydrogeological and geochemical or water-chemical techniques, so as to provide additional and valuable information for solving hydrological problems. In recent years, in hundreds of difficult cases, isotopic methods have provided definite, satisfactory results.

References

Aller, L., Bennett, T., Lehr, J.H. and Petty, R.J. (1985). DRASTIC – A standardized system for evaluating ground water pollution potential using hydrogeologic settings, US EPA. Robert S. Kerr Environmental Research Laboratory, Office of Research and Development, EPA/600/2-85/018, 163 p.

Amberger, A. and Schmidt, H.L. (1987). Natürliche isotopengehalte von Nitrate als Indikatoren für dessen Herkunft. *Geochim Cosmochim Acta*, **51:** 2699–2705.

Andersen, L.J. and Sevel, T. (1974). Six years' environmental tritium profiles in the unsaturated and saturated zones, Gronhoj, Denmark. Proc. IAEA Symp. Isotope Techniques in Groundwater Hydrology, 9–13 March, Vienna, 3–20.

Bahadur, J., Saxena, R.K. and Mookerjee, P. (1977). Soil moisture movement and groundwater recharge by tritium tracer tagging technique. *Proc Ind Acad Sc*, **Ser. A 85:** 462–471.

Bethke, CM, Johnson TM (2002). Ground water age. *Ground Water*, **40(4):** 337–339.

Bigeleisen, J. (1965). Chemistry of isotopes. *Science*, **147(3657):** 463–471.

Blume, H.P., Zimmerman, U. and Munnich, K.O. (1967). Tritium tagging of soil moisture: The water balance of forest soils. *In:* Proc. Symp. Isotope and Radiation Techniques in Soil Physics and Irrigation Studies (Istanbul): 315–332. IAEA, Vienna.

Boronina, A., Renard, P., Balderer, W. and Stichler, W. (2005). Application of tritium in precipitation and in groundwater of the Kouris catchment (Cyprus) for description of the regional groundwater flow. *Applied Geochemistry*, **20:** 1292–1308.

Böttcher, J., Strebel, O., Voerkelius, S. and Schmidt, H.L. (1990). Using isotope fractionation of nitrate-nitrogen and nitrate-oxygen for evaluation of microbial denitrification in a sandy aquifer. *J. Hydrol.*, **114:** 413–424.

Bowen, R. (1985). Hydrogeology of the Bist Doab and adjacent areas, Punjab, India. *Nord. Hydrol*, **16(1):** 33–44.

Bu, X. and Warner, M.J. (1995). Solubility of chlorofluorocarbon 113 in water and sea-water. *Deep-sea Research*, **42:** 1151–1161.

Busenberg, E. and Plummer, L.N. (1996). Concentrations of chlorofluorocarbons and other gases in ground water at Mirror Lake, New Hampshire. *In:* Morganwalp, D.W. and Aronson, D.A. (eds). USGS Toxic substances hydrology program. Proc. of the Technical Meeting, Colorado Springs, Colorado, September 20–24, 1993: USGS WRIR 94-4014, 151–158.

Chapelle, F.H., Zelibor, J.L., Grimes, D.J. and Knobel, L.L. (1987). Bacteria in deep coastal plain sediments of Maryland: A possible source of CO_2 to ground water. *Water Resour. Res.*, **23:** 1625–1632.

Chesnaux, R., Molson, J.W. and Chapuis, R.P. (2005). An analytical solution for ground-water transit time through unconfined aquifers. *Ground Water*, **43:** 511–517.

Clark, I. and Fritz, P. (1997). Environmental Isotopes in Hydrogeology. Lewis Publishers, New York.

Cook, P.G. and Herczeg, A.L. (1998). Groundwater chemical methods for recharge studies. Part 2 of the Basics of Recharge and Discharge (Ed. L. Zhang). CSIRO Publishing, CSIRO Australia.

Cook, P.G. and Solomon, D.K. (1995). Transport of atmospheric trace gases to the water table: Implications for groundwater with chlorofluorocarbons and dating krypton 85. *Water Resources Research*, **31:** doi: https://doi.org/10.1029/94WR02232.

Cornaton, F. and Perrochet, P. (2005). Groundwater age, life expectancy and transittime distributions in advective-dispersive systems; 1. Generalized Reservoir Theory. *Advances in Water Resources* **29(9):** 1267–1291.

Craig, H. (1961). Isotopic Variations in Meteoric Waters. *Science*, **133:** 1702–1703. https://doi.org/10.1126/science.133.3465.1702.

Dansgaard, W. (1964). Stable isotopes in precipitation. *Tellus*, **16:** 436–468.

Dörr, H., Schlosser, P., Stute, M. and Sonntag, C. (1992). Tritium and ^{3}He measurements as Calibration data for groundwater transport models. *In:* Progress in Hydrogeo-chemistry. Mattess et al. (eds.). Springer, Berlin, 461–466.

Emiliani, C. (1966). Isotopic Paleotemperature. *Science*, **154(3751):** 851–857.

Epstein, S., Buchsbaum, R., Lowenstam, H. and Urey, H.C. (1953). Revised carbonate-water isotopic temperature scale. *Bull. GSA*, **64:** 1315–1326.

Etcheverry, D. and Perrochet, P. (2000). Direct simulations of groundwater transit-time distributions using the reservoir theory. *Adv. Water Res.*, **8:** 200–208, doi:https://doi.org/10.1007/s100400050006.

Fehn, U., Snyder, G.T., Matsumoto, R., Muramatsu, Y. and Tomaru, H. (2003). Iodine dating of pore waters associated with gas hydrates in the Nankai area, Japan. *Geology*, **31:** 521–524.

Forstel, H. (1982). $^{18}O/^{16}O$-ratio of water in plants and in their environment. *In:* Stable Isotopes. H-L Schmidt, H. Forstel and K. Heizinger (Eds). Pp. 503–509. Elsevier Scientific Publishing Co, Amsterdam.

Freeze, R.A. and Cherry, J.A. (1979). Groundwater. Prentice Hall, Englewood Cliffs, p. 604.

Gleick, P.H. (ed.) (1993). Water in Crisis: A Guide to the World's Fresh Water Resources. Oxford University Press, New York.

Gonfiantini, R., W. Stichler, and K. Rozanski (1995), Standards and intercomparison materials distributed by the International Atomic Energy Agency for stable isotope measurements, IAEA-TECDOC-825, Int. At. Energy Agency, Vienna.

Goode, D.J. (1996). Direct Simulation of groundwater age. *Water Resources Research*, **32(2):** 289–296.

Hanshaw, B.B. and Back, W. (1974). Determination of regional hydraulic conductivity through ^{14}C dating of groundwater. *Memoirs of the Intl. Association of Hydrogeologists*, **10:** 195–196.

Harvey, C.F. and Gorelick, S.M. (1995). Temporal moment-generating equations: Modeling transport and mass transfer in heterogeneous aquifers. *Water Resources Research*, **31:** doi: https://doi.org/10.1029/95WR01231. ISSN: 0043-1397.

Krouse, H.R. (1980). Sulphur isotopes in our environment. *In:* Fritz, P. and Fontes, J. Ch. (Eds). Isotope Geochemistry, Vol. 1. The Terrestrial Environment. Elsevier, Amsterdam. 435–471.

Kumar, B. and Nachiappan, Rm. P. (1995). A mathematical approach based on tritium tagging technique to evaluate recharge to groundwater due to monsoon rains. Tracer Technologies for Hydological System. IAHS Publ. No. 229.

Kumar, R., Suresh, N., Sangode, S.J. and Kumaravel, V. (2007). Evolution of the Quaternary alluvial fan system in the Himalayan foreland basin: Implications for tectonic and climatic decoupling. *Quat. Int.*, **159(1):** 6–20.

Kumar, B., Rao, M.S., Navada, S.V., Verma, S.K. and Shrivastava, S. (2009). Evaluation of effectiveness of artificial recharge measures in parts of Maharashtra using environmental isotopes. *Current Science*, **97(9):** 1321–1330.

Kumar, Sudhir, Mishra, K. and Kumar, B. (2012). Surface and ground water interaction at selected locations along the river Yamuna, NCT Delhi. Proc. of National Symposium on "Water Resources Management in Changing Environment (WARMICE-2012)", NIH Roorkee, Feb. 8–9, 2012.

Lapworth, D.J., MacDonald, A.M., Krishan, G., Rao, M.S., Gooddy, D.C. and Darling, W.G. (2015). Groundwater recharge and age-depth profiles of intensively exploited groundwater resources in northwest India. *Geophys. Res. Lett.*, **42:** 7554–7562, doi:https://doi.org/10.1002/2015GL065798.

Létolle, R. (1980). Nitrogen-15 in the natural environment. *In:* Handbook of Environmental Isotope Geochemistry. 1. The Terrestrial Environment. P. Fritz and J.C. Fontes (eds). Elsevier, Amsterdam, The Netherlands. 407–433.

Mazor, E., and Nativ, R. (1992) Hydraulic calculation of groundwater flow velocity andage: Examination of the basic premises. J. Hydrol., 138:211–222.

McCrea, J.M. (1950). On the isotopic chemistry of carbonates and a paleotemperature scale. *J. Chem. Phys.*, **18:** 849–857.

Metcalfe, R., Hooker, P.J., Darling, W.G. and Milodovski, A.E. (1998). Dating Quaternary groundwater flow events: A review of available methods and their application. *In:* Parnell, J. (ed.), Dating and Duration of Fluid Flow and Fluid-Rock Interaction. Geological Society, London. Special publications, No. 144: 233–260.

Modica, E., Buxton, H.T. and Plummer, L.N. (1998). Evaluating the source and residence times of groundwater seepage to streams, New Jersey Coastal Plain. *WRR*, **34:** 2797–2810.

Munnich, K.O. (1968a). Moisture movement measured by isotope tagging. *In:* Guide book on nuclear techniques in hydrology. IAEA, Vienna, 112–117.

Munnich, K.O. (1968b). Use of nuclear techniques for the determination of groundwater recharge rates. *In:* Guide book on nuclear techniques in hydrology. IAEA, Vienna, 191–197.

Munnich, K.O., Roether, W. and Thilo, L. (1967). Dating of groundwater with tritium and C-14. Proc. IAEA Symp. Isotopes in Hydrology, 14–18 Nov. 1966, Veinna. pp. 305–319.

Nachiappan, Rm. P., Rao, M.S., Kumar, B., Navada, S.V. and Satyanarayana, P. (2003). Chemical and isotopic techniques for the development of groundwater management strategies in a coastal aquifer, Krishna River Delta, South India. International Symposium on Isotope Hydrology and Integrated Water Resources Management organized by IAEA, Vienna, Austria, 19–23 May, 2003.

NIH (1999). Study of soil moisture movement and recharge to groundwater due to monsoon rains and irrigation using tritium tagging technique in Hardwar district. National Institute of Hydrology, Roorkee. Report No. TR/BR-14/98-99.

NIH (2000). Study of soil moisture movement and recharge to ground water due to monsoon rains and irrigation using tritium-tagging technique in Saharanpur district. National Institute of Hydrology, Roorkee. Report No. CS/AR-23/1999-2000.

Nier, A.O. (1950). A Redetermination of the Relative Abundances of the Isotopes of Carbon, Nitrogen, Oxygen, Argon and Potassium. *Phys. Rev.*, **77:** 789.

Ozima, M. and Podosek, F. (2001). Noble Gas Geochemistry. Second edition. Cambridge University Press, Cambridge. 286 p.

Payne, B.R. (1988). The status of isotope hydrology today. *J Hydrol*, **100:** 207–237.

Phillips, F.M. (2000). Chlorine-36. *In:* Cook, P. and Herczeg, A. (eds). Environmental Tracers in Subsurface Hydrology. Kluwer Academic Publishers, Boston. 299–348.

Phillips, F.M., Tansey, M.K., Peeters, L.A., Cheng, S. and Long, A. (1989). An isotopic investigation of groundwater in the central San Juan basin, New Mexico: ^{14}C dating as a basis for numerical flow modeling. *Water Resources Research*, **25:** doi: https://doi.org/10.1029/89WR01275. ISSN: 0043-1397.

Plummer, L.N., Busby, J.F., Lee, R.W. and Hanshaw, B.B. (1990). Geochemical modelling of the Madison aquifer in parts of Montana, Wyoming and South Dakota. *Water Resources Research*, **26:** 1981–2014.

Rai, S.P., Kumar, Sudhir, Kumar, Bhishm and Rawat, Y.S. (2012). Identification of Source of Leakage in Drainage Gallery of Tehri Dam using Isotopic Techniques. *Hydrology J*, **35(3&4):** 63–75.

Robertson, W.D., Cherry, J.A. and Schiff, S.L. (1989). Atmospheric sulfur deposition 1950–1985 inferred from sulfate in groundwater. *Water Resources Research*, **25:** doi: https://doi.org/10.1029/89WR00133. ISSN: 0043-1397.

Rozanski, K., Araguás-Araguás, L. and Gonfiantini, R. (1993). Isotopic patterns in modern global precipitation. American Geophysical Union, Washington DC, **78:** 7–6. doi: https://doi.org/10.1029/GM078p0001.

Samadder, R.K., Kumar, S. and Gupta, R.P. (2011). Paleochannels and their potential for artificial groundwater recharge in the western Gangaplains. *J Hydrol*, **400(1):** 154–164.

Saxena, R.K. (1984). Seasonal variations of oxygen-18 in soil moisture and estimation of recharge in esker and moraine formations. *Nordic Hydrology*, **15:** 235–242.

Shivanna, K., Tirumalesh, K., Noble, J., Joseph, T.B., Singh, G., Joshi, A.P. and Khati, V.S. (2008). Isotope techniques to identify recharge areas of springs for rainwater harvesting in the mountainous region of Gaucher area, Chamoli District, Uttarakhand. *Current Science*, **94(8):** 1003–1011.

Solomon, D.K., Schiff, S.L., Poreda, R.J. and Clarke, W.B. (1993). A validation of the $^{3}H/^{3}He$ method for determining groundwater recharge. *WRR*, **29:** 2951–2962.

Singh, A., Gupta, S., Sinha, R., Carter, A., Thomsen, K.J., Mark, D.F., Buylaert, J.P., Mason, P.J., Murray, A.S., Jain, M. and Debajyoti, P. (2015). Large-scale avulsion of the late Quaternary Sutlej river in the NW Indo-Gangetic foreland basin. *In:* EGU General Assembly Conference Abstracts, **17:** 6661.

Sinha, R., Yadav, G.S., Gupta, S., Singh, A. and Lahiri, S.K. (2013). Geo-electric resistivity evidence for subsurface palaeochannel systems adjacent to Harappan sites in northwest India. *Quat. Int.*, **308:** 66–75.

Sukhija, B.S. and Shah, C.R. (1976). Conformity of groundwater recharge rate by tritium method and mathematical modelling. *J Hydrol*, **30:** 167–178.

Thoma, G., Esser, N., Sonntag, C., Weiss, W. and Rudolph, J. (1979). New technique of in-situ soil moisture sampling for environmental isotope analysis applied at Pilat sand dune near Bordeaux. Proc. Symp. Isotope Hydrology, 19–23 June, 1978, Neuherberg, IAEA, Vienna. pp. 753–768.

Thompson, G.M. and Hayes, J.M. (1979). Trichlorofluoromethane in groundwater – A possible tracer and indicator of groundwater age. *Water Resour. Res.*, **15:** 546–554.

Urey, H.C., Brickwedde, F.G. and Murphy, G.M. (1932). A hydrogen isotope of mass 2 and its concentration. *Physical Review*, **40:** 1–15.

Varni, M. and Carrera, J. (1998). Simulation of groundwater age distributions. *Water Resources Research*, **34:** doi: https://doi.org/10.1029/98WR02536. ISSN: 0043-1397.

Warner, M.J. and Weiss, R.F. (1985). Solubilities of chlorofluorocarbons 11 and 12 in water and seawater. *Deep Sea Research*, **32:** 1485–1497.

Weise, S.M., Stichler, W. and Bertleff, B. (2001). Groundwater inflows into an excavated artificial lake (gravel pit) indicated by apparent $^3H/^3He$ ages. *In:* Seiler, K.P. and Wohnlich, S. (eds.) Proc. XXXI IAH Congress "New Approaches Characterizing Groundwater Flow", Swets and Zeitlinger Lisse, Munich. 225–228.

Wieser, M., Aeschbach-Hertig, W., Schneider, T., Deshpande, R.D. and Gupta, S.K. (2011). A temperature and monsoon record derived from environmental tracers in the groundwater of northwest India. *In:* Proceedings of IAEA International Symposium on Isotopes in Hydrology.

Wilson, R. and Mackay, D.M. (1993). The use of sulfur hexafluoride as a conservative tracer in saturated sandy media. *Ground Water*, **31(5):** 719–724.

Yurtsever, Y. and Gat, J.R. (1981). Atmospheric waters. *In:* Gat, J.R. and Gonfiantini, R. (eds). Stable Isotope Hydrology: Deuterium and oxygen-18 in the water cycle. IAEA Tech Rep Ser 210: 103–142.

Zimmermann, U., Ehhalt, D. and Munnich, K.O. (1967a). Soil water movement and evapotranspiration; changes in the isotopic composition of the water. *In:* Isotopes in Hydrology, IAEA, Vienna. pp. 567–584.

Zimmermann, U., Munnich, K.O. and Roether, W. (1967b). Downward movement of soil moisture traced by mean of hydrogen isotope. *In:* Geophysical Monograph No. 11. Isotope Techniques in the Hydrological Cycle. Stout, G.E. (ed.). Amer. Geophys. Union. Washington. pp. 28–36.

Zoellmann, K., Kinzelbach, W. and Fulda, C. (2001). Environmental tracer transport (^3H and SF_6) in the saturated and unsaturated zones and its use in nitrate pollution management. *J Hydrol*, **240:** 187–205.

Chapter 5
Geostatistics in Groundwater Modelling

Bhabesh C. Sarkar

1 Introduction

Geostatistics is a collection of statistical techniques for the analysis of spatial data. Geostatistics has been described by several authors (Matheron 1971; David 1977, 1986; Isaak and Srivastava 1989; Kitanidis 1997). In recent years, these tools have developed from research topics into basic techniques in the design and, such as mining, geology and hydrology, among others. The aim of this chapter is to present application of geostatistical tools in groundwater modelling and mapping. A typical spatial data set, such as groundwater levels, monthly precipitations, or transmissivities, is composed of scattered readings in space, denoted by $z(x)$, where x represents the measurement location. Having such information, geostatistics provides many techniques to solve a variety of hydrogeological resources problems, such as: (i) Estimation of z at an unmeasured location: interpolation and mapping of z; (ii) Estimation of one variable based on measurements of other variables: co-estimation of piezometric head and transmissivity; (iii) Estimation of the gradient of z at an arbitrary site: estimation of groundwater flow velocity based on observed heads; (iv) Estimation of the integral of Z over a defined block: estimation of contamination volume based on point measurements; and (v) Design of sampling and monitoring networks, such as groundwater quality monitoring. Many of the groundwater related variables are spatial functions presenting complex variations that cannot be effectively described by simple deterministic functions, such as polynomials. Such phenomena are subject of geostatistics that are named as regionalized variables. Annual point precipitation is an example of a regionalized variable. Transmissivity also displays spatial variations due to complex processes governing the transport, deposition and compression of materials in sedimentary deposits.

B. C. Sarkar (✉)
Department of Applied Geology, Indian Institute of Technology (ISM), Dhanbad, India
e-mail: bhabeshsarkar2005@gmail.com

P. K. Sikdar (ed.), *Groundwater Development and Management*,
https://doi.org/10.1007/978-3-319-75115-3_5

Another example of a regionalized variable is the concentration of a chemical compound in groundwater that varies in both space and time.

Geostatistics goes beyond the interpolation problem by considering the studied phenomenon at unsampled locations as a set of correlated random variables. Let $Z(x)$ be the value of the variable of interest at a certain location x. This value is unknown (e.g. temperature, rainfall, piezometric level, geological facies, etc.). Although there exists a value at location x that could be measured, geostatistics considers this value as random since it was not measured, or has not been measured yet. However, the randomness of $Z(x)$ is not complete, but defined by a cumulative distribution function (CDF). Typically, if the value of Z is known at locations close to x (or in the neighbourhood of x) one can constrain the CDF of $Z(x)$ by this neighbourhood. If a high spatial continuity is assumed, $Z(x)$ can only have values similar to the ones found in the neighbourhood. Conversely, in the absence of spatial continuity $Z(x)$ can take any value. The spatial continuity of the random variables is described by a model of spatial continuity that can be either a parametric function in the case of variogram-based geostatistics, or have a non-parametric form when using other methods such as multiple-point simulation or pseudo-genetic techniques. By applying a single spatial model on an entire domain, one makes the assumption that Z is a stationary process. It means that the same statistical properties are applicable on the entire domain.

The variations of these processes can be so complicated that estimating their values are difficult, even if measurements from nearby locations are available. Geostatistics recognizes these difficulties and provides statistical tools for: (i) calculating the most accurate (according to well defined criteria) predictions, based on measurements and other relevant information, (ii) quantifying the accuracy of these predictions, and (iii) selecting the parameters to be measured, and where and when to measure them, if there is an opportunity to collect more data. Considering that spatial data represents only an incomplete picture of the natural phenomenon of interest, it is logical to use statistical techniques to process such information. Geostatistics has adopted the procedure and some of the most practical and yet powerful applicable tools of probability theory (Rouhani 1986).

2 Statistical Modelling

The possible outcome of a random selection of a sample is expressed by its probability distribution that may or may not be known. In the case of a *discrete distribution*, which can only assume integer values, the distribution would associate to each possible value X, a probability $P(X)$. The individual value of $P(X)$ will be positive and the sum of all possible $P(X)$ will be equal to 1. The function $f(x)$ is a mathematical model that provides the probability that the random variable X would take on any specified value x, i.e. $f(x) = P(X = x)$. This function, $f(x)$ is called the probability distribution of the random variable X and describes how the probability values are distributed over the possible values, x of a random variable X. In the case of a *continuous distribution*, to each possible value x, a density of probability $f(x)$ is associated so that probability of a value lying between x and $x + dx$ is $f(x)\,dx$, where dx is infinitesimal. This serves as a mathematical model for describing the uncertainty of an outcome for a continuous variable. The probability of x lying between lower limit, (a) and upper limit, (b) is expressed as:

$$\text{Prob}(a \leq X \leq b) = \int_a^b f(x)dx$$

The individual probability density value will be positive and the sum of all such values extending from $-\propto$ to $+\propto$ will be 1. The probability of X being smaller than or equal to a given value x is called the cumulative probability distribution function $F(x)$:

$$\text{Prob}(X \leq x) \int_{-a}^{b} f(x)dx = F(x); F(-\propto) = 0; \text{and}\, F(+\propto) = 1$$

The following holds true for the cumulative distribution function, $F(x)$:

(i) $0 \leq F(x) \leq 1$ for all x;
(ii) $F(x)$ is non-decreasing.

The usual practice to determine the characteristics of an aquifer is to collect drill hole samples, analyse the properties of those samples and infer the characteristics of the aquifer from the properties. *If one uses classical statistics to represent the properties of sample values*, an assumption is made that the values are realisations of '*a random variable*'. The relative positions of the samples are ignored and it is assumed that all sample values in aquifer have an equal probability of being selected. The fact that two samples taken close to each other is more likely to have similar values than if taken far apart is also not taken into consideration.

Sample spacing remains wide in the initial stages of groundwater exploration that provide broad knowledge of an aquifer. It is in this early stage of exploration, quality of the aquifer is examined by estimating mean (average) value, '*m*' of the aquifer. For this purpose, '*n*' samples of same support (size, shape and orientation) are taken at points X_i. The sample values are used to estimate '*m*' of the population mean, μ and the confidence limits of the mean. The estimator for this purpose would vary according to the probability distribution of sample values. In classical statistical analysis, since it is assumed that all sample values are independent (i.e. random), the location X_i of the sample is ignored. The parameters estimated from a classical statistical model refer to variables such as thickness, permeability, porosity, etc. Theoretical models of probability distributions which are commonly encountered in aquifers to represent sample value frequency distribution are either Normal (Gaussian) or Lognormal. Various other distributions are known but the assumption of either normality or lognormality can be made for most aquifers and the use of more complex distributions is not justified.

2.1 The Normal Distribution Theory

This distribution is characterised by a symmetrical bell-shape and its probability density function (p.d.f.), $f(X)$ is expressed (Davis 1986) as:

$$\text{p.d.f.}, f(X) = \left[\left(1/S\sqrt{2\pi}\right)\right]\exp\left[-\left((1/2)\left(X_i - \bar{X}\right)^2\right)/S^2\right] \text{ for } -\infty \leq X \leq \infty$$

where $\bar{X}$ is the sample mean which is an estimate of the population mean μ, and S is the sample standard deviation, an estimate of the population standard deviation σ. The distribution can be standardised by expressing $\left[\left(X_i - \bar{X}\right)/S\right]$ equal to Z:

$$f(Z) = \left[1/\sqrt{2\pi}\right]\exp\left[-\tfrac{1}{2}Z^2\right]$$

This standard normal distribution has a zero mean and unit standard deviation, i.e. $N(0,1)$. The cumulative probability density function (c.d.f.), F(X) of a normal distribution has the expression:

$$\text{c.d.f.}, F(X) = \left[1/\sqrt{2\pi}\right]\int_{-\alpha}^{x} \exp\left[-\left((1/2)\left(X_i - \bar{X}\right)^2\right)/S^2\right]dx$$

2.2 Fitting a Normal Distribution

To check the assumption of normality, or in other words, to fit a normal distribution to an experimental histogram, a convenient graphical method known as the probability-paper method can be used. Cumulative frequency distribution of the values are calculated and plotted in an arithmetic-probability paper against the upper limits of the class values. From the definition of arithmetic-probability scale, the cumulative distribution of a normally distributed variable will plot as straight line on arithmetic-probability paper. If the points obtained by this approach can be considered or closely approximated as distributed along a straight line, the assumption of normality can be accepted, and the theory of normal distribution to estimate the mean, variance and confidence limits of mean can then be applied.

Other methods to test the fit of a normal distribution include: (i) measures of degree of skewness and kurtosis, and (ii) χ^2 (Chi-squared) goodness of fit test. For a normal variate, the degree of skewness is zero and that of kurtosis is 3, and the calculated value of χ^2 must be less than or equal to the table value of χ^2 at 'α' level of significance and 'f' degrees of freedom.

2.3 Estimation of Mean, Variance and Confidence Limits

The sample mean and sample variance for a normal distribution are estimated as follows:

$$\text{Sample mean}, \bar{X} = [1/n]\sum_{i=1}^{n} X_i$$

$$\text{Sample variance}, S^2 = [1/(n-1)]\sum_{i=1}^{n}(X_i - \bar{X})^2$$

where $S = \sqrt{S^2}$ which is an estimate of the population standard deviation. The mean value, 'm' of the aquifer is estimated by:

$$m = \bar{X}; \text{with variance}, V = S^2/n$$

If m_p be confidence limits of the true mean 'm' such that the probability of 'm' being less than m_p is p, then $m_{1\text{-}p}$ is the confidence limit such that the probability that 'm' is larger than $m_{1\text{-}p}$ is $1 - p$. The probability that 'm' falls between m_p and $m_{1\text{-}p}$ is $1 - 2p$ confidence limits of the mean. The following equations can be used to calculate m_p and $m_{1\text{-}p}$ for the mean value, 'm' of an aquifer:

$$\text{Lower limit}, m_p = m - t_{1-p}(S/\sqrt{n}); \text{and}$$

$$\text{Upper limit}, m_{1\text{-}p} = m + t_{1-p}(S/\sqrt{n})$$

where $t_{1\text{-}p}$ is the value of student's t-variate for $f = n - 1$ degrees of freedom, such that the probability that 't' is smaller than '$t_{1\text{-}p}$' is $1 - \text{p}$.

2.4 Measures of Skewness, Kurtosis and Chi-squared goodness of Fit

Degrees of skewness and kurtosis of a sample distribution are given by the equations:

$$\text{Skewness}, Sk = [1/(n-1)]\sum_{i=1}^{n}(X_i - \bar{X})^3/S^3$$

$$\text{Kurtosis}, Ku = [1/(n-1)]\sum_{i=1}^{n}(X_i - \bar{X})^4/S^4$$

Once the optimum solution for '*m*' has been determined, it is desirable to check for the goodness of fit of a normal distribution to the sample distribution. Chi-squared $(\chi)^2$ test provides a robust technique for the fit. The test statistics is given by:

$$\chi^2 \text{Calculated} = \sum_{i=1}^{n} (O_i - E_i)^2 / E_i$$

where O_i = observed frequency in group *i* and E_i = expected frequency in group *i*.

2.5 The Lognormal Distribution Theory

In many aquifers, where the distribution of the sample values is asymmetrical, either positively or negatively skewed, it has been observed that this skewed distribution can be represented either by a 2-parameter or a 3-parameter lognormal distribution. If $\log_e(X_i)$ has a normal distribution, we call it a 2-parameter lognormal distribution, and if $\log_e(X_i + C)$ has a normal distribution, we call it a 3-parameter lognormal distribution (where *C* is the additive constant). The value of the additive constant, *C* is:

(i) Positive for a positively skewed distribution, i.e. a distribution showing an excess of low values with tail towards high values; and
(ii) Negative for a negatively skewed distribution, i.e. a distribution showing an excess of high values with tail towards low values.

The p.d.f. of a lognormal distribution is given by the expression:

$$f(X) = \left[1/\left(x\beta\sqrt{2\pi}\right)\right]\exp\left[-1/2\left(\frac{(\ln x - \alpha)}{\beta}\right)^2\right]$$

where α = logarithmic mean, i.e. log mean and β^2 = logarithmic variance, i.e. log variance.

The probability distribution of a 3-parameter lognormal variate, X_i is defined by:

- the additive constant, *C*;
- the logarithmic mean of $(X_i + C)$
- the logarithmic variance of $(X_i + C)$.

2.5.1 Fitting a Lognormal Distribution

For '*n*' samples with values X_i ($i = 1, 2, ..., n$), the cumulative frequency distribution of a 2-parameter lognormal variate plots as a straight line on logarithmic probability paper. If the variate is 3-parameter lognormal, the cumulative curve shows either an excess of low values for positively skewed distribution and or an excess of high values for negatively skewed distribution. In such cases, plot of (X_i + C) will be a straight line on logarithmic probability paper conforming to a lognormal distribution.

2.5.2 Estimation of Additive Constant (C)

If a large number of samples are available, the cumulative distribution may be plotted on a log-probability paper. Different values of 'C' can then be tried until the plot of (X_i + C) is reasonably assumed to be a straight line. Alternatively, the value of 'C' can be estimated using the following approximation:

$$C = \frac{M_e^2 - F_1 F_2}{F_1 + F_2 - 2M_e}$$

where M_e is the sample value corresponding to 50% cumulative frequency (i.e. the median of the observed distribution) and F_1 and F_2 are sample values corresponding to 'p' and '$1 - p$' percent cumulative frequencies respectively. In theory, any value of 'p' can be used but a value between 5% and 20% gives best results.

2.5.3 Estimation of Logarithmic Mean and Logarithmic Variance

$$\text{Let, } y_i = \log_e(X_i + C)$$

$$\log_e \text{ mean, } \alpha \text{ or } \bar{Y} = [1/n] \sum_{i=1}^{n} y_i$$

$$\log_e \text{ variance, } \beta^2 \text{ or } v(y) = [1/(n-1)] \sum_{i=1}^{n} (y_i - \bar{y})^2$$

2.5.4 Estimation of Average for a Deposit

$$\begin{aligned} m^* &= e^{\bar{y}+v(y)} \\ &= e^{(\alpha+(\beta^2))/2} \\ &= e^{\alpha}.e^{(\beta^2/2)} \\ \text{Average value,} \quad m &= (m^* - c) \\ \text{Variance,} \quad S^2 &= m^2[\exp(v) - 1] \end{aligned}$$

2.5.5 Estimation of Central 90% Confidence Limits

The lower and upper limits for the estimation of Central 90% confidence interval of the mean of a lognormal population can be obtained by using factors $\psi_{0.05(\mathrm{v,n})}$ and $\psi_{0.95(\mathrm{v,n})}$:

$$\text{Lower limit} = \left(\psi_{0.05(\mathrm{v,n})} m^*\right) - C; \text{ and}$$
$$\text{Upper limit} = \left(\psi_{0.95(\mathrm{v,n})} m^*\right) - C.$$

3 Geostatistical Modelling

Classical statistics produce an error of estimation stated by confidence limits but ignores the spatial relations within a set of sample values. These limitations point to the need for an estimation technique that is capable of producing estimates with minimum variance. Such estimates are achieved with the use of geostatistics based on the *'Theory of Regionalised Variables'*, i.e. a variable that is related to its position in space and has a constant support.

The underlying assumption of geostatistics is that the values of samples located near or inside a block of ground are most closely related to the value of the block. This assumption holds true if a relation exists among the sample values as a function of distance and orientation. The function that measures the spatial variability among the sample values, is known as the semi-variogram function, $\gamma(h)$. Comparisons are made between each sample of a data set with the remaining ones at a constantly increasing distance, known as the lag interval.

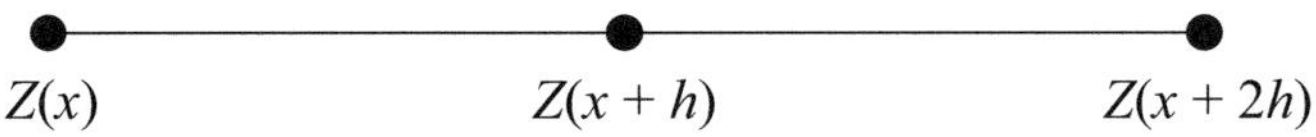

Thus, a semi-variogram function numerically quantifies the spatial correlation of aquifer parameters (e.g. thickness, bedrock elevation, porosity, permeability etc.). If $Z(x_i)$ be the value of a sample taken at position x_i and $Z(x_i + h)$ be the value at 'h' distance away from x_i position, the mathematical formulation of a semi-variogram function, $\gamma(h)$ is given by the expression:

$$\gamma(h) = (1/2N) \sum_{i=1}^{N} (Z(x_i) - Z(x_i + h))^2$$

where N is number of sample value pairs, $Z(X_i)$ is the value of Regionalized Variable at location X_i and $Z(X_i + h)$ is the value of Regionalized Variable at a distance 'h' away from X_i.

Spatial variance changes from arrangement to arrangement. The function $2\gamma(h)$ is called the variogram function. It is the semi-variogram function $\gamma(h)$ that is used rather than variogram function $2\gamma(h)$ because the relation between semi-variogram and covariogram (i.e. plot of covariance between $Z(x_i)$ and $Z(x_i + h)$ with constantly increasing values of 'h') is straight forward:

$$2\gamma(h) = E\left[2\gamma(h)^*\right]$$

where E is the Expected Value which is the probability weighted sum of all possible occurrences of regionalized variable; and $2\gamma(h)^*$ is the experimental variogram function based on sample values;

$$\begin{aligned}
\text{or,}\quad & 2\gamma(h) \\
&= E\left[2\gamma(h)^*\right] = E\left[(Z(x_i) - Z(x_i + h))^2\right] \\
&= E\left[(Z(x_i) - m + m - Z(x_i + h))^2\right] \text{ where } m \text{ is the sample mean} \\
&= E\left[((Z(x_i) - m) - (Z(x_i + h) - m))^2\right] \\
&= E\left[(Z(x_i) - m)^2 + (Z(x_i + h) - m)^2 - 2(Z(x_i) - m)(Z(x_i + h) - m)\right] \\
&= E\left[(Z(x_i) - m)^2\right] \\
&\quad + E\left[(Z(x_i + h) - m)^2 - 2E[(Z(x_i) - m)(Z(x_i + h) - m)\right] = 2 \text{ variance} \\
&\quad - 2 \text{ covariance } (h)
\end{aligned}$$

Hence the fundamental relation: $\gamma(h) = \sigma^2 - CV(h)$.

Graphical representation of a semi-varigram is given in Fig. 5.1.

An experimental semi-variogram permits the interpretation of several characteristics of the aquifer as follows:

(i) *The Continuity (C):* The continuity is reflected by the rate of growth of $\gamma(h)$ for constantly increasing values of 'h'.
(ii) *The Nugget Effect (C_o):* This is the name given to the semi-variogram value, $\gamma(h)$ at $h \to 0$. It expresses the local homogeneity (or lack thereof) of aquifer. The nugget effect represents an inherent variability of a data set which could be due to both the spatial distribution of the values together with any error encountered in sampling.

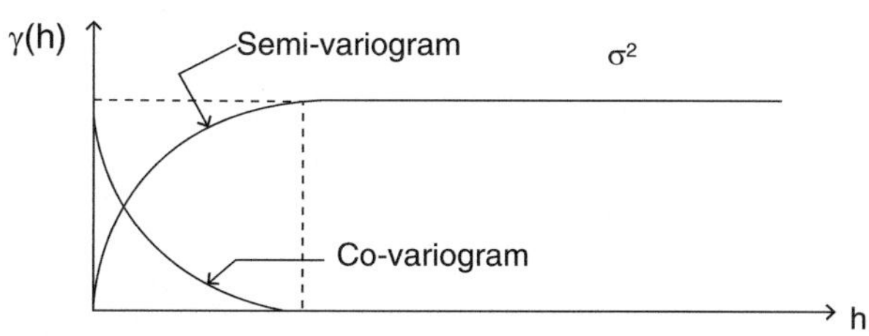

Fig. 5.1 Relation between semi-variogram and co-variogram

(iii) *The Sill Variance* ($C_o + C$)*:* The value where a semi-variogram function $\gamma(h)$ plateaus is called the sill variance. For all practical purposes, the sill variance is equal to the statistical variance of all sample values used to compute an experimental semi-variogram.
(iv) *The Range (a):* The distance at which a semi-variogram levels off at its plateau value is called the range (or zone) of influence of semi-variogram. This replaces the conventional geological concept of an area of influence. Beyond this distance of separation, values of sample pairs do not correlate with one another and become independent of each other.
(v) *The Directional Anisotropy:* This denotes whether or not the aquifer has greater continuity in a particular direction compared to other directions. This characteristic is analysed by comparing the respective ranges of influences semi-variograms computed along different directions. Where the semi-variograms in different directions are very similar, it is said to be isotropic.

In practice, since sampling grids are rarely uniform, semi-variograms are computed with a tolerance on distance (i.e., $h \pm dh$) and a tolerance on direction (i.e. $\alpha \pm d\alpha$) to accommodate sample pairs not falling on the grid. The tolerances on distance and direction should be kept as low as possible in order to avoid any directional overlapping.

4 Semi-Variogram Models

There are several mathematical models of semi-variogram. However, three most commonly encountered models in aquifer modelling (Fig. 5.2) are:

4.1 Spherical Model

This model is encountered most commonly in aquifer where sample values become independent once a given distance of influence (i.e. the Range) '*a*' is reached. The equations are given by:

$$\begin{array}{ll}
\gamma(h) = Co + C\left[3/2(h/a) - ½\left(h^3/a^3\right)\right] & \forall h < a; \\
\gamma(h) = Co + C & \forall h \geq a; \\
\gamma(h) = Co & \forall h \text{ tends to } 0; \\
\gamma(h) = 0 & \forall h = 0.
\end{array}$$

This model is common in most aquifers and said to describe transition phenomena as it is the one which occurs when one has geostatistical spatial structures independent of each other beyond the range but, within it, sample values are highly correlated.

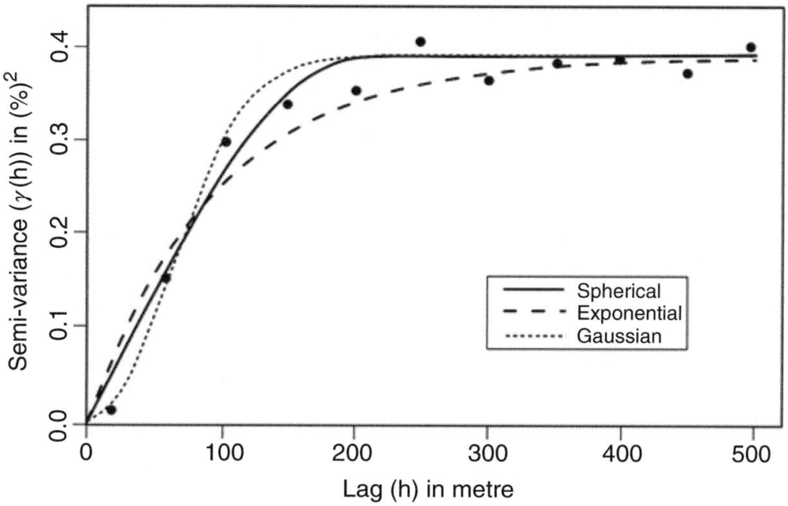

Fig. 5.2 Common semi-variogram models

4.2 Exponential Model

This model is not encountered too often in aquifers since its infinite range is associated with a too continuous process. The equation is: $\gamma(h) = C\ [1 - e^{-h/a}]$. The slope of the tangent at the origin is *C*/*a*. For practical purposes, the range can be taken as 3*a*. The tangent at the origin intersects the sill at a point where '*h*' equals '*a*'.

4.3 Gaussian Model

This model is characterised by two parameters *C* and *a*. The curve is parabolic near the origin and the tangent at the origin is horizontal, which indicates low variability for short distances. Excellent continuity is observed which is rarely found in geological environments. Practical range is $\sqrt{3a}$. The equation is: $\gamma(h) = C\left[1 - e^{\left(-h^2/a^2\right)}\right]$.

5 Practice of Semi-VARIOGRAM Modelling

The behaviour at the origin for both nugget effect and slope plays a crucial role in fitting of a model to an experimental semi-variogram. While the slope can be assessed from the first three or four semi-variogram values, the nugget effect can be estimated by extrapolating back to the $\gamma(h)$ axis. The choice of nugget effect is extremely important since it has a very marked effect on kriging weights and in turn on kriging variance. Three methods for semi-variogram model fitting include:

5.1 Hand Fit Method

The sill ($Co + C$) is set at the value where experimental semi-variogram stabilizes. In theory, this should coincide with the statistical variance. Estimate of nugget effect is achieved by joining the first three or four semi-variogram values and projecting this line to the $\gamma(h)$ axis. By projecting the same line until it intercepts the sill provides 2/3rd the range. Using the estimates of *Co*, *C* and '*a*', calculate a few points and examine if the model curve fits the experimental semi-variogram. Although this method is straight forward, and simple to practice, there is an element of subjectivity involved in the estimation of model parameters.

5.2 Non-linear Least Squares Fit Method

Like any curve fitting technique, this method uses the principle of polynomial fit by least squares to fit a model with sum of the deviations squared of the estimated values from the real values being minimum. Unfortunately, polynomials obtained by least squares do not guarantee the positive definite function (otherwise semi-variance could turn out to be negative).

5.3 Point Kriging Cross-Validation Method

Point kriging cross-validation (PKCV) is a technique referred to as a procedure for checking the validity of a mathematical model fitted to an experimental semi-variogram that controls the kriging estimation (Davis and Borgman 1979).

The principle underlying the technique is as follows:

> '......... a sample point is chosen in turn on the sample grid that has a real value. The real value is temporarily deleted from the data set and the sample value is kriged using the neighbouring sample values confined within its radius of search. The error between the estimated value and the real value is calculated. The kriging process is then repeated for rest of the known data points'. A crude semi-variogram model is initially fitted by visual inspection to the experimental semi-variogram. Estimates of the initial sets of semi-variogram parameters (viz., *Co*, *C* and '*a*') are made from the initial model and cross-validated through point kriging empirically. The error statistics such as mean error, mean variance of errors and mean kriging variance are then computed. The model parametes are varied and adjusted until: (i) a ratio of mean variance of the errors (estimation variance) to mean kriging variance approximating to unity (in practice, a value of 1 ± 0.05 has been

observed to be the acceptable limits); (ii) a mean difference between sample values and estimated values close to zero; and (iii) an adequate graphical fit to the experimental semi-variogram are achieved. For a good estimate, most of the individual errors should also be close to zero. A model approximated or fitted by this approach eliminates subjectivity.

6 Geostatistical Estimation – Kriging

Kriging is an optimal spatial interpolation technique. In general terms, a kriging system calculates an estimated value, G^* of a real value, G by using a linear combination of weights, a_i of the selected surrounding 'n' values such that:

$$G^* = \sum_{i=1}^{n} a_i g_i, \text{ where } \sum_{i=1}^{n} a_i = 1 \text{ and } g_i \text{ are the sample values.}$$

If G* is the estimate of a block average grade G by applying straight average method, i.e.

$$G^* = 1/\mathrm{n} \sum_{i=1}^{n} g_i$$

then equal weight is given to all the sample values, and the error of estimation of G is:

$$\sigma_E^2(S \text{ to } V) = E\left[(G^* - G)^2\right] = -\overline{\gamma}(S, S) - \overline{\gamma}(v, v) + 2\overline{\gamma}(S, V)$$

In many cases, however, we know that to assign equal weight to all selected surrounding samples may not provide the best possible estimate. Consider the case of a block valued by a centre sample and a corner sample as configured below:

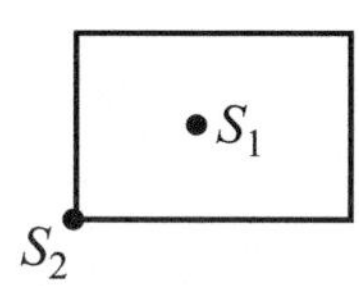

Clearly, the centre sample should be given a greater weight than the corner sample. Say, we give weight a_1 to S_1 and a_2 to S_2. The new grade estimate would be:

$$G^* = a_1 g_1 + a_2 g_2$$

The weights of selected surrounding sample values are so chosen that:

- G^* is an unbiased estimate of G, i.e. $E\,[(G^* - G)] = 0$; and
- Variance of estimation of G by G^*, i.e. $E\,[(G^* - G)^2]$ is minimum.

By definition, Kriging is known as Best (because of minimum estimation variance) Linear (because of weighted arithmetic average) Unbiased (since the weights sum to unity) Estimator – BLUE.

7 Practice of Kriging

Once the model semi-variogram parameters characterizing all information about the expected sample variability are defined, the subsequent step involves estimation of grid cell values together with their associated variances through kriging. At this stage, a geological domain is considered within an aquifer which is further divided into smaller grids equalling the size of a geocellular block. Decision on the choice of a geocellular block size is generally influenced by several factors such as measurement points, hydrogeological framework, precision of measurement data, desired use of grid cell, and capability of manipulating a huge number of grid cells. The arrays of geocellular blocks are kriged producing kriged estimate and kriging variance for each of them and an overall average. The following input parameters are found to be adequate for geocellular block kriging:

- a minimum of four measurement points (because of the necessity to define a surface) and a maximum of 16 measurement points (because of reasonable computational time and cost) with at least one measurement point in each quadrant to krig a geocellular block; and
- the radius of search for measurement points around a geocellular block centre to be within the semi-variogram range of influence.

The individual values are averaged to produce a mean kriged estimate and a mean kriging variance in order to provide global estimates. The 95% geostatistical confidence limits are calculated as:

$$m \pm 1.96\sqrt{\sigma_k^2},$$

where m = mean kriged estimate and σ_{k}^2 = mean kriging variance.

8 Applications

Geostatistics can be used in a variety of groundwater modelling studies, such as: (i) mapping of spatial variables; (ii) simulation of hydrogeological fields; (iii) co-estimation of hydrological fields using physical relationships, such as co-mapping of piezometric head and transmissivity using groundwater flow equations; (iv) sampling and monitoring designs; and (v) groundwater resource management under uncertainty. One of the earliest applications of geostatistics in groundwater was in the area of mapping spatial variables, such as transmissivity maps, piezometric surfaces, and precipitation fields. Kitanidis (1997) in his book has dealt with geostatistics and applications to hydrogeology. In fact, the power of the methods described becomes most useful when utilizing measurements of different types, combining these with deterministic flow and transport models, and incorporating geological information to achieve the best characterization possible.

Geostatistical mapping also yields the accuracy map that indicates the areas of high and low precision. Simulation of hydrogeological fields is another application of geostatistics. Simulation usually means the generation of spatial data, such that their mean and their covariance are the same as the original data. There are various useful applications for simulated data. For example, by generating different spatial rainfall patterns, one can determine the statistical distribution of runoff. Co-estimation allows the user to utilize the information in one variable in the estimation of another. In some instances, a variable that is sampled at a lower cost can be used to improve the accuracy of another variable which is costly to measure. If there are known physical relationships between variables, they can be used to further improve our estimation. The estimation variance is a measure for the accuracy of estimated fields. This measure can help us to design sampling activities based on the maximization of gained information. In some instances, such as groundwater quality monitoring, the estimated magnitude of the variable of interest is as important as its accuracy. So the sampling may be designed not only for improving the precision of the estimated field, but also for targeting those areas which exhibit critical estimated values. Water resources management problems usually include many variables that exhibit uncertainty. Ignoring the stochastic nature of these problems may yield non-optimal solution. Geostatistics provides the framework to quantify these uncertainties and incorporate them in our decisions.

9 A Case Study

A study has been aimed at modelling of spatial phenomena of groundwater distribution during pre-monsoon and post-monsoon periods in respect of the year 2014 using geostatistical techniques with reference to rainwater harvested groundwater level inside the IIT(ISM) campus. The groundwater level data of 44 recharge bore wells located within the ISM campus area were collected on a monthly basis during the year 2014. Pre-monsoon (May to June), monsoon (July to September) and post-monsoon (October to December) measurements of groundwater level data of the recharge bore wells have been utilized for spatial modelling of the groundwater fluctuation employing the theory and applications of 'Regionalised Variable'. The modelling study reveals the spatial variability of the fluctuation and estimates the rise in the groundwater level employing Ordinary Kriging.

Statistical analyses of groundwater level data for these periods were carried out to compute the distribution parameters. Geostatistical methods utilize an understanding of the inter-relations of measurement (sample) values and provide a basis for quantifying the geological concepts of (i) an inherent variability; (ii) a change in the continuity of inter-dependence of measurement (sample) values according to the spatial variability; and (iii) a range of influence of the inter-dependence of measurement (sample) values. Based on these quantifications, geostatistics produces an estimated map with minimum variance, and provides an error of estimation both on a local and a global scale. The underlying assumption of geostatistics is that the values of samples located nearby are most closely related to one other than the distant ones. This assumption holds true if a relation exists among the sample values as a function of distance and orientation. The function that measures the spatial variability among the sample values, is known as the semi-variogram function, $\gamma(h)$. Comparisons are made between each sample of a data set with the remaining ones at a constantly increasing distance, known as the lag interval.

Geostatistical analysis was initiated with computation of experimental semi-variogram and fitting appropriate mathematical model to it that characterizes the spatial variability of the groundwater level. A semi-variogram model exhibit various spatial characteristics, viz. nugget effect (C_0), continuity (C), sill ($C_0 + C$), range of influence (a) and directional anisotropy. Semi-variogram constitute the major tool in geostatistics to express the spatial dependence among neighbouring values measured in pairs. Most commonly used mathematical models of semi-variogram include spherical, exponential, gaussian, and pure nugget effect. The behaviour at the origin for both nugget effect and slope plays a crucial role in fitting of a model to an experimental semi-variogram. While the slope can be assessed from the first three or four semi-variogram values, the nugget effect can be estimated by extrapolating back to the γ (h) axis. The choice of nugget effect is extremely important since it has a very marked effect on kriging weights and in turn on kriging variance. The appropriateness and rationality of a semi-variogram model fit was carried out employing point kriging cross-validation technique.

3D omni-directional experimental semi-variograms for pre- and post-monsoon groundwater levels and that of the fluctuations have been carried out using GEXSYS software (Sarkar 1988). The spatial variability analyses revealed experimental semi-variograms with moderately low nugget effect and increasing tendency of semi-variogram values with constantly increasing distances levelling off at respective range of influences. Cross-validated models as obtained employing point kriging cross-validation technique for Pre-monsoon, Post-monsoon, and Fluctuation are given in Figs 5.3, 5.4 and 5.5 and semi-variogram model parameters obtained through point kriging cross-validation are given in Table 5.1.

Prior to the grid cell kriging, the grid size of the study area was decided by taking into account the various parameters i.e. area, fluctuation of groundwater and the best fitted grid cell which can cover the maximum extent near to the boundary of the ISM. A grid cell size of 25 m × 25 m dimensions was selected on the basis of appropriate fitting of the cells in the periphery of the boundaries. Having delineated the cells of the dimension of 25 m × 25 m ordinary kriging was performed cell by cell to provide kriged estimate and kriged standard deviation. The plots show the spatial distribution of groundwater levels generated in the study area. The plots of pre-monsoon (Fig. 5.6) and post-monsoon (Fig. 5.7) groundwater levels display the spatial

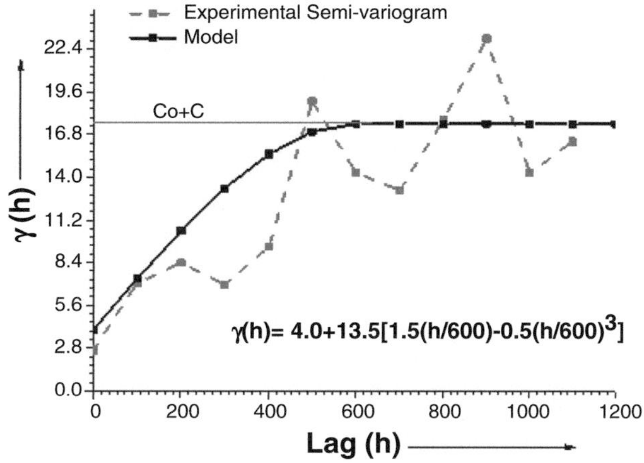

Fig. 5.3 Semi-variogram model of pre-monsoon period

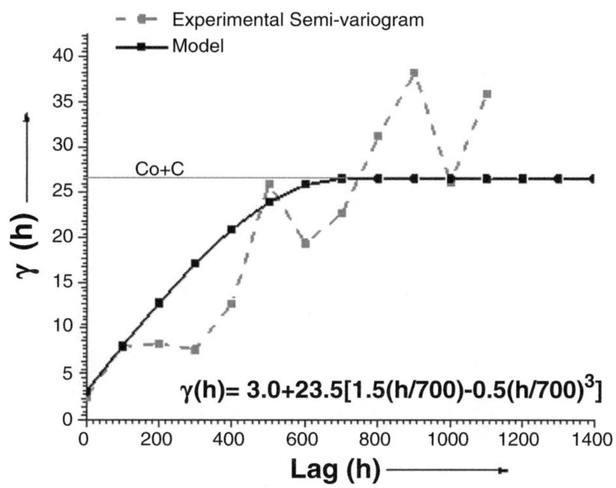

Fig. 5.4 Semi-variogram model of post-monsoon period

distribution maps of kriged estimate and kriged standard deviation of groundwater levels along with that of the fluctuation (Fig. 5.8) in the study area. Groundwater flow maps of pre- and post-monsoon periods have been developed that provide the direction of flow of groundwater (Fig. 5.9).

Statistical analyses of pre-monsoon and post-monsoon groundwater levels provided a negatively skewed characteristic while that in respect of the fluctuations between pre- and post-monsoon provided a positively skewed characteristic. Estimated mean and standard deviation values corresponding to each of these periods and that of the fluctuation are (240.55 m; 4.08 m), (242.44 m; 3.40 m) and (1.09 m; 1.44 m) respectively.

Spatial variability analyses of pre-monsoon and post-monsoon groundwater levels and that of the fluctuation between pre- and post-monsoon periods revealed a spherical function fit. Pre-monsoon period exhibited a nugget effect of 3.0 m^2, a

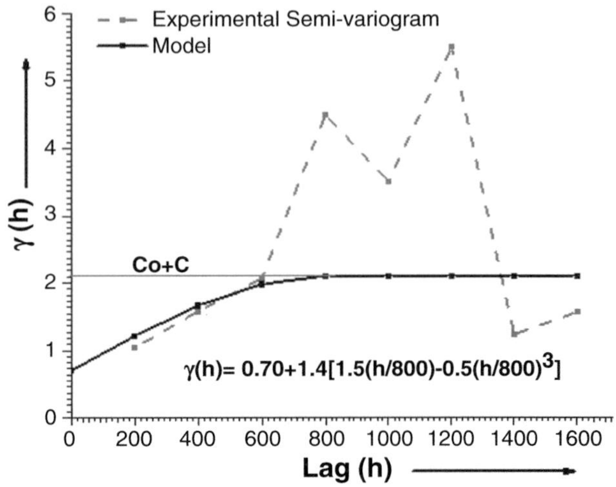

Fig. 5.5 Semi-variogram model of fluctuation

Table 5.1 Semi-variogram model parameters obtained through Point Kriging Cross-validation

Semi-variogram parameters	Pre-monsoon period	Post-monsoon period	Fluctuation
Nugget effect, C_0	3.0	4.0	0.7
Continuity, C	23.5	13.5	1.40
Sill, $C_0 + C$	26.5	17.5	2.1
Range (a)	700	600	800
Ratio of KV:EV	0.98	1.04	1.03

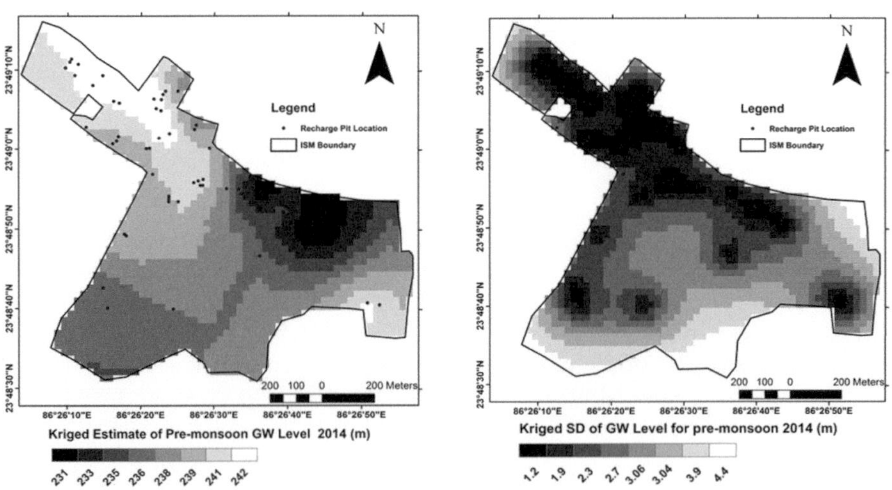

Fig. 5.6 Spatial distribution of kriged estimate and kriged standard deviation of groundwater levels in respect of pre-monsoon period

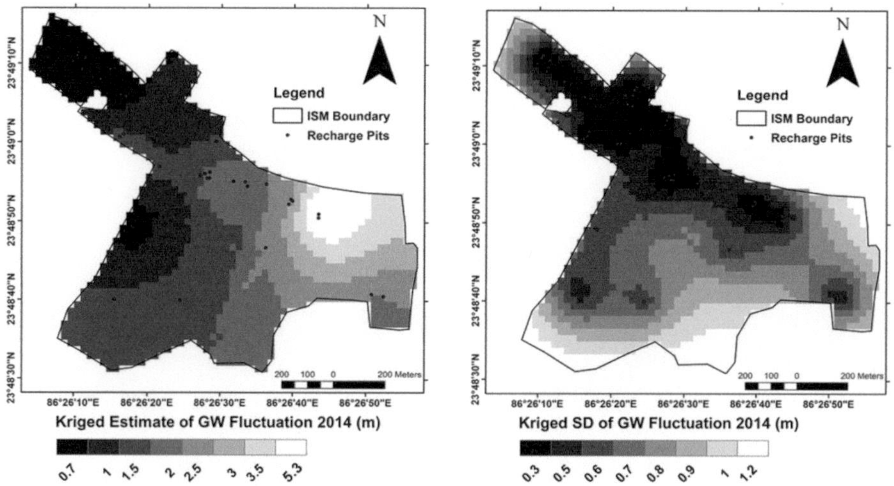

Fig. 5.7 Spatial distribution of kriged estimate and kriged standard deviation of groundwater levels in respect of post-monsoon period

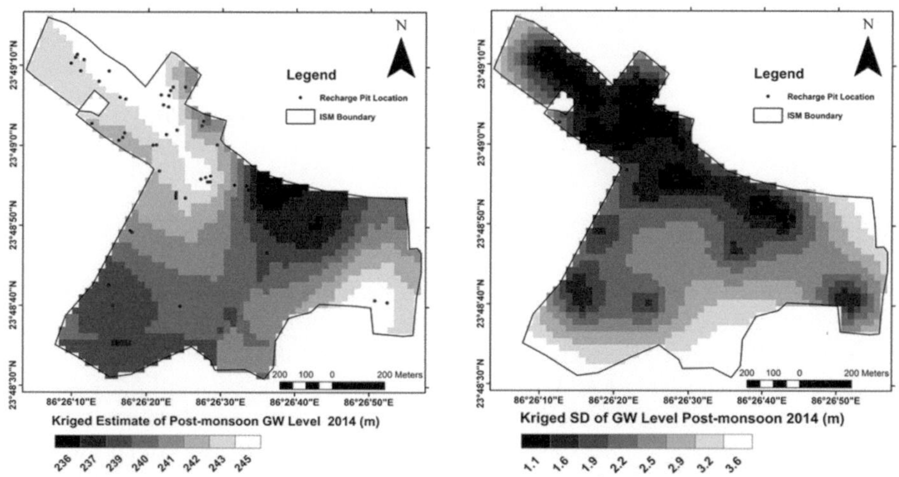

Fig. 5.8 Spatial distribution of kriged estimate and kriged standard deviation of groundwater levels in respect of fluctuation

continuity of 23.5 m^2 and a range of 700 m; post-monsoon period exhibited a nugget effect of 4.2 m^2, a continuity of 13.5 m^2 and a range of 600 m; fluctuation between pre- and post-monsoon periods displayed a nugget effect of 0.7 m^2, a continuity of 1.40 m^2 and a range of 800 m. Model fitting exercise has been carried out employing point kriging cross-validation technique yielding a ratio of estimation variance to

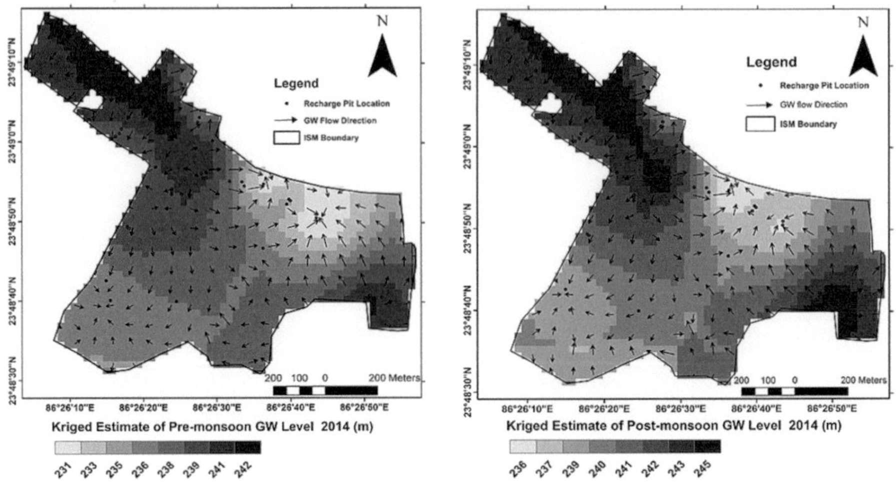

Fig. 5.9 Groundwater flow maps of pre- and post-monsoon periods

kriging variance as 0.98, 1.04 and 1.03 respectively with adequate graphical fit to experimental semi-variograms. Following are the geostatistical model equations of groundwater levels:

$$\text{Pre-monsoon} \quad : \gamma(h) = 3.0 + 23.5\left[1.5(h/700) - 0.5(h/700)^3\right]$$

$$\text{Post-monsoon} \quad : \gamma(h) = 4.0 + 13.5\left[1.5(h/600) - 0.5(h/600)^3\right]$$

$$\text{Fluctuation} \quad : \gamma(h) = 0.7 + 1.40\left[1.5(h/800) - 0.5(h/800)^3\right]$$

Geostatistical estimation was initiated with gridding the IIT (ISM) campus area into cells 25 m × 25 m with each cell defined in space in terms of northing and easting. Block kriging has been carried out for each of these cells which provided kriged estimate and associated kriged standard deviation in respect of pre-monsoon and post-monsoon groundwater levels and that of the fluctuation between pre- and post-monsoon periods. Assessment of the goodness of fit (R^2) of kriged estimates was carried out through a correlation plot of the measured (true) values versus the kriged estimated values of groundwater level and assuming the best fit straight line as the regression model. The R values of pre- and post-monsoon and that of fluctuation are 0.94, 0.88 and 0.88 respectively and have been found to be significant through t-test of significance on R. The calculated values of 't' on R are 18.40, 12.80 and 11.90 as compared to the critical value of 1.67, thereby indicating a significant correlation.

Spatial distribution maps of kriged estimate values in respect of pre-monsoon and post-monsoon groundwater levels exhibit a distinct high in the north-western zone and in the south-eastern periphery of the campus associated with geomorphic high. The groundwater levels gradually decrease towards east-central periphery and towards south-western zone with a distinct low observed in the east-central periphery. The kriged standard deviation maps exhibit a relatively high error in the north-western, south-western, southern and eastern peripheries of the campus reflecting a reduced reliability of kriged estimate due to presence of few recharge bore wells (sample location) while in rest of the part, the error reduces towards the areas with high density of recharge bore wells as prominently observed in the north-western zone trending towards east-central part and also in the south-eastern part and north-western part. As regards the spatial distribution map of fluctuation is concerned, it is observed that the kriged estimated fluctuation is maximum in the east-central periphery owing to the presence of geomorphic depression. There is a gradual decrease in the fluctuation towards the north-western side and also towards the western side of the campus. The associated kriged standard deviation has the same distribution pattern as with that of the pre- and post-monsoon kriged standard deviation maps for the same reason of presence of few recharge bore wells. The fluctuation map at the north-western side has the least rise of 0.70 m in groundwater level which gradually increases in the eastern side to 5.3 m. It may be stated that areas showing higher density of recharge bore wells are associated with lower kriged standard deviation which gradually increases towards areas with lesser density of recharge bore wells as evident from the spatial distribution maps of kriged standard deviation. Groundwater flow maps exhibit a similar pattern, i.e. the direction of flow of groundwater is from northwest, west, southwest and southeast sides to the east owing to the geomorphic low.

Attempt made to estimate the replenishable groundwater resource within the campus area using the norms of Groundwater Resource Estimation Committee (GEC) 1997 led to calculation of total replenishable volume of water or dynamic groundwater resource using mean kriged rise of groundwater level of 2.77 m as:

$$\text{Volume of water recharge} = (\text{Area} \times \text{fluctuation} \times \text{specific yield}) + \text{Draft}$$

where specific yield considered for hard rock as per GEC, 1997 is 0.03. Hence, volume of water recharge = $(60 \times 56 \times 25 \times 25 \times 2.77 \times 0.03\ \text{m}^3)$ + Draft. The first term of the equation 174,510 $\text{m}^3 \times 1000 =$ 174,510,000 litres and the Draft, which is related to the consumption of groundwater for the year 2014 is $19{,}54{,}000 \times 12 \times 30 = 703{,}440{,}000.00$ litres. Dynamic groundwater resource thus calculated is 877,950,000 litres.

Spatial variability phenomena of groundwater level within the campus of IIT (ISM) Dhanbad for pre-monsoon and post-monsoon periods and that of the fluctuation between pre- and post-monsoon have been analysed and modelled. The spatial variability analyses revealed experimental semi-variograms with moderately low nugget effect and increasing tendency of semi-variogram values with constantly increasing distances and levelling off at respective range of influences. Point Kriging Cross-validation technique has been used for fitting a mathematical model to experimental semi-variograms. This is followed by construction of block grid cells of 25 m $\times$ 25 m for which kriged estimate and kriging standard deviation values have been arrived at employing ordinary kriging to estimate the rise in the rainwater harvested groundwater level during the year 2014. The modelling study led to generation of kriged estimate and kriged standard deviation spatial distribution maps in respect of pre-monsoon, post-monsoon and the fluctuation for the year 2014. The study revealed a mean rise of 2.77 m in the groundwater level owing to the rainwater harvesting. The rise in the groundwater level during the study period has led to an estimate of groundwater resource to 174,510,000 litres as compared to the consumption of 703,440,000 litres. The study estimated that about 80% of total volume of groundwater available is consumed and thereby maintaining a balance of about 20%. This figure of groundwater resource balance is expected to improve over the years with continued monitoring study of the fluctuating trend of the groundwater level with implementation of rainwater harvesting and artificial recharge in the campus. Similar rainwater harvesting study can be of use in other areas for assessing spatial and temporal phenomena leading to the usefulness of geostatistical modelling for sustainable development and management of groundwater resource.

10 Conclusion

Geostatistics is a branch of statistics focusing on spatial or spatio-temporal datasets. Developed originally to predict probability distributions of ore grades for mining operations, it is currently applied in diverse disciplines including petroleum geology, hydrogeology, hydrology, meteorology, oceanography, geochemistry, geography, forestry, environmental control, landscape ecology, soil science and agriculture (especially, in precision farming). Geostatistics is intimately related to interpolation methods, but extends far beyond simple interpolation problems in aquifer modelling. Geostatistical techniques rely on statistical model that is based on random function (or random variable) theory to model the uncertainty associated with spatial estimation. Empirical semi-variogram is used in geostatistics as a first estimate needed for spatial interpolation by kriging. Kriging is a group of geostatistical techniques to interpolate the value of a random field (e.g., the elevation of the bedrock as a function of the spatial location) at an unobserved location from observations of its value at nearby locations.

References

David, M. (1977). Geostatistical Ore Reserve Estimation. Elsevier, The Netherlands. 364 p.

Davis, J.C. (1986). Statistics and Data Analysis in Geology. Wiley, New York. 646 p.

Davis, B.M. and Borgman, L.E. (1979). A test of hypothesis concerning a proposed model for the underlying variogram. Proc. 16th APCOM, pp 163-181.

Isaak, M. and Srivastava, R.M. (1989). An Introduction to Applied Geostatistics. Oxford University Press, New York. 561 p.

Kitanidis, P.K. (1997). Introduction to Geostatistics: Applications in Hydrogeology. Cambridge University Press. 249 p.

Matheron, G. (1971). The Theory of Regionalized Variables and its Application. Paris School of Mines, Cah. Cent. Morphologie Math., 5. Fontainebleau.

Rouhani, S. (1986). Comparative Study of Ground Water Mapping Techniques. *J Ground Water*, **24(2):** 207-216.

Sarkar, B.C. (1988). An Integrated System for Geology-controlled Geostatistical Evaluation. Ph.D. thesis. Imperial College of Science and Technology, London. 218 p.

Chapter 6
Application of Remote Sensing and Geographical Information System in Groundwater Study

Surajit Chakraborty

1 Introduction

The availability of water, both in quantity and quality is the prime factor in deciding the sustainable growth of mankind and also for the development of towns and cities. With the advance in space technology, now it is possible to employ remote sensing techniques to estimate surface and subsurface water over large areas. The need for remote sensing based groundwater survey is necessary to define all possible features connected with localization of groundwater. Satellite data provide quick and beneficial baseline information about numerous factors that directly or indirectly control the occurrence and movement of groundwater such as geomorphology, soil types, slope, land use/land cover, drainage patterns, lineaments, etc. (Jha et al. 2007). These features are extracted from the appropriate satellite data products and integrated with the thematic details obtained from topographic sheets of the desired scale. The use of conventional techniques (e.g., geophysical, geostatistical, numerical modelling, etc.) for groundwater management, is often limited by the lack of adequate data. Frequent and long-term monitoring of groundwater and vadose zone systems by using all these conventional methods is expensive, laborious, time-consuming and destructive (Jha and Chowdary 2007). Therefore, innovative technologies, such as remote sensing (RS) and Geographic Information Systems (GIS), have an immense role to play (Jha and Chowdary 2007). Airborne and spaceborne imagery have also become better and increasingly detailed. The recent development of viewers like Google Earth, Bhuvan ISRO Portal etc. with very high resolution has allowed users to explore 2D/3D representation of the surface of the Earth more accurately. The application package 'WARIS' (Web enabled Water Resources Information System)

S. Chakraborty (✉)
Department of Environment Management, Indian Institute of Social Welfare and Business Management, Kolkata, India
e-mail: surajitchak@rediffmail.com

P. K. Sikdar (ed.), *Groundwater Development and Management*,
https://doi.org/10.1007/978-3-319-75115-3_6

developed by ISRO provides a comprehensive, credible and contextual view of India's water resources data along with allied natural resources data and information. It will allow users to search, access, visualize, understand, analyze and look into context and study spatial patterns.

The traditional use of remotely sensed images and their interpretations lies in qualitative characterization of hydrogeological mapping units (Meijerink 2000). For about three decades, the application of remote sensing technology in groundwater resources evaluation has been practised. Remote sensing is without doubt the backbone of hydrogeological reconnaissance in areas of the world where the coverage of detailed geological maps and field data are insufficient (Hoffmann and Sander 2007). These data, in conjunction with ground truth information, provide details on geology, geomorphology, structural pattern and recharge conditions, which ultimately define the groundwater regime. The groundwater prospect/potential maps can show the range in groundwater yield at different depths, besides indicating probable sites for recharging aquifers. GIS provides a means for introducing information and knowledge from other data sources into the decision-making process and help in handling and management of large and complex data bases. RS combined with GIS facilitates better data analysis and interpretation.

Even though remote sensing has proved to be a useful tool in providing data for GIS to study various environmental aspects including groundwater, a number of factors must be considered before using this data in GIS. Among these factors are those summarised by Baban and Luke (2000), which include the spatial resolution, the spectral resolution and the temporal resolution. In summary, satellite data can provide reliable up-to-date information on land cover, land use, geology, vegetation, geomorphology etc. which will provide an improved understanding of the hydrogeological system.

This chapter tries to outline how RS and GIS technologies are used for groundwater studies and the various approaches related to groundwater potential assessment in geological (consolidated and unconsolidated) formations using RS and GIS.

2 Development of Groundwater Mapping Using RS

Hydrogeologists were late to use satellite data for an obvious reason; groundwater lies in the subsurface. Also, the radar and radiometer based satellite has very few centimetres penetration capability into the ground. In spite of this apparent obstacle, RS holds tremendous potential for regional groundwater flow studies (Becker 2006). The remote-sensing data, with its ability for synoptic view, repetitive coverage with calibrated sensors to detect changes, and observations at different resolutions, provide a better alternative for natural resources management as compared to traditional methods (Lillesand et al. 2004; Chandra and Ghosh 2007; Ranganath et al. 2007; Mukherjee 2008). Remotely sensed indicators of groundwater may provide important data where conventional alternatives are not available. An example of measured data includes groundwater heads, changes in groundwater storage, heat signatures, and subsidence data (Becker 2006). Ground-based RS (geophysics) is usually more expensive than space and airborne RS but is still more accurate and cheaper than invasive methods (drilling) (Meijerink 2007). These indicators include vegetation, surface water, water discharging to the surface carrying heat energy and runoff (Becker 2006). Satellite technology is reviewed in Table 6.1 with respect to its ability to measure groundwater potential, storage and fluxes. Continuous advancement in satellite sensor technology over the years has led to the availability of metre and sub-metre ground resolution.

2.1 *Some Benefits of Using Satellite Data in Groundwater Mapping*

The utility of remotely sensed data as an efficient tool for groundwater exploration and mapping in different lithological, structural and geomorphological features of various rock types has been well established through a number of studies worldwide (Krishnamurthy et al. 1996, 2000; Ai et al. 1998; Saraf and Choudhury 1998; Murthy 2000; Dhiman and Keshari 2002; Shahid et al. 2000; Sikdar et al. 2004; Chandra et al. 2006; Sultan et al. 2007; Vijith 2007; Sharma and Thakur 2007; Dhakate et al. 2008; Madrucci et al. 2008; Ganapuram et al. 2009; Suja Rose and Krishnan 2009; Dar et al. 2010; Pothiraj and Rajagopalan 2012; Waikar and Nilawar 2014; Ramamoorthy and Rammohan 2015) which are as follows:

1. For extracting the drainage network information from multisensor, multisatellite images such as Landsat-3, Return Beam Vidicon (RBV), Landsat-5 Thematic Mapper (TM) with SPOT Panchromatic Linear Array (PLA) stereo pairs as well as topographic sheets showed that satellite images can be used effectively.
2. The TanDEM-X (TerraSAR-X add-on for digital elevation measurement) mission was launched by the German Space Agency in June 2010 is expected to provide a globally standardized and accurate three-dimensional (3D) digital elevation model of Earth with a measuring point density of 12 m to obtain the depth to groundwater level from the digital elevation model which is more advance than shuttle radar topography mission (SRTM) data set.
3. For lineament extraction Landsat TM combining with SPOT-MS and SIR (Subsurface Interface Radar) images were used extensively because reflection on Landsat TM imagery shows as straight to curvilinear topographic breaks. At present, the HRSI (High Resolution Satellite Image) clearly depicts minor faults and lineaments, and provides highly accurate spatial information products such as orthoimages, DEM, linear feature extraction and terrain modelling.

Table 6.1 Selected list of active space-based sensors that report data of potential use for investigations of groundwater resource

Sensor	Launch year	Ground resolution (m)	Precipitation	Surface temp	Soil moisture	Water storage	Snow water	Land cover	Topography
AMSR-E	2002	5400 – 56,000	X	X	X		X		
ASTER	1999	15,30,90		X				X	X
AVHRR	1991–2003	1100	X	X	X		X		
GRACE	2002	300,000				X			
ENVISAT- RA	2002	1000		X					X
Landsat-7	1999	30,60		X				X	
MODIS	1999	250, 500, 1000		X				X	
Orb View-2	1997	1100		X				X	
Orb View-3	2003	1,4		X				X	
RADARSAT-1	1995	10,30,100			X				
SRTM	2000	30,90			X				X
IKONOS-2	2000	1,4						X	
Quick Bird-2	2009	0.6,2.4						X	
KOMPSAT-2	2006	1,4						X	
EROS-B1	2006	0.7						X	
Cartosat-2	2016	1			X			X	X
GeoEye-1	2008	0.4,1.65						X	
WorldView-2	2009	0.52, 2.08						X	
Pleiades-HR 1 & 2	2011, 2012	0.7, 2.8						X	X
KOMPSAT-3	2012	0.7						X	X

4. The recent data from the GRACE (Gravity Recovery and Climate Experiment) Satellite are available to assess changes in terrestrial water over larger areas. GRACE is a twin-satellite mission used to identify mass changes due to variations in water storage, assist in determination of groundwater depletion and residual basin-scale estimates of evaporation or validation of hydrological models. GRACE data are expressed in mm equivalent to water.
5. Microwave or radar images have several applications in hydrogeological studies. Radar altimetry for detecting lake levels is important for the groundwater heads and flow studies around lakes. Moreover, Digital Elevation Models (DEMs) and precise measurement of land subsidence can be done by radar data.
6. Meteosat thermal, Landsat optical and Modis normalized vegetation index (NDVI) data along with GIS have been extensively used by various scientists all over the world to identify land-surface features that control location of groundwater recharge and discharge areas in a large data-scarce, semi-arid basin.

3 Development of Groundwater Mapping Using GIS

The growth and use of GIS in groundwater investigations is increasing tremendously. It is used for groundwater potential (Krishnamurthy et al. 1996) and vulnerability assessment (Laurent et al. 1998), groundwater modelling (Watkins et al. 1996; Pinder 2002) and management (Patra et al. 2016). In regional scale, it requires to handle large volume of geo-referenced (spatial) and attribute (aspatial) data. GIS is an ideal problem solving environment where RS data and interpretations can be merged with discrete and continuous data from various primary and secondary sources (Burrough 1986; Manap et al. 2013; Rahmati et al. 2014; Deng et al. 2016).

GIS uses both operational and analysis tools for generating thematic maps. There are several commercial GIS packages available in the industry, namely Arcinfo, Integrated Land and Water Information System (ILWIS) and Earth Resource Data Analysis System (ERDAS), Geomatica, ENVI, TNTmips, QGIS etc. developed by various software vendors.

A GIS approach comprises three distinct phases: (1) data acquisition, (2) data processing, and (3) data analysis. The data acquisition phase includes establishing control of the data quality, which consists of positional accuracy, and reliability of observations. There are several ways of digitizing map data for inserting in a GIS. The data can be directly digitized from the map using a digitizing table or it can be digitized by tracing the outline of required classes on a transparent overlay in image processing software. A general approach to prepare different thematic maps for the development of a GIS database for hydrogeological mapping is shown in Fig. 6.1.

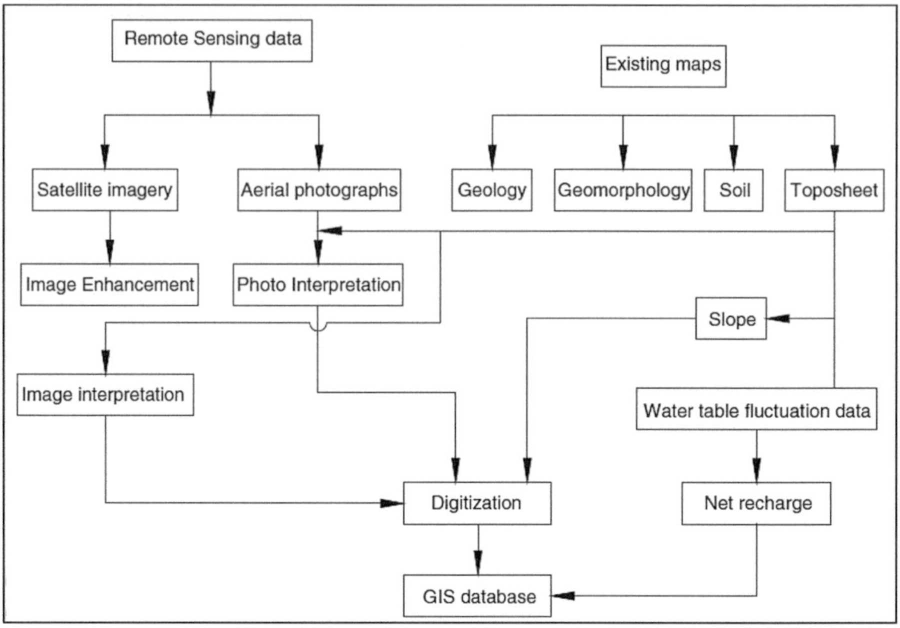

Fig. 6.1 Preparation of thematic map

4 Benefits of Using GIS

Some benefits of using GIS in groundwater mapping are as follows:

1. Teeuw (1995) proposed an integrated approach of RS and GIS techniques to improve the site selection for borehole drilling.
2. Nagarajan and Singh (2009) and Chowdhury et al. (2009) highlighted the importance of coupling RS and GIS in groundwater potential assessment studies.
3. Gosain and Sandhya (2004) used GIS as a pre-processor to the SWAT (Soil and Water Assessment Tool) model for assessment of existing and anticipated water uses and water shortages in chronically drought prone area. The output from the model was combined with water availability, sediment yield, demographic and socio-economic data in the GIS environment for prioritization of sub-watersheds for water-resources planning. Using GIS capabilities the possible locations of water-harvesting structures were identified for future development.
4. Elewa and Qaddah (2011) integrated Enhanced Thematic Mapper Plus (ETM+) images, GIS, a watershed modelling system (WMS) and weighted spatial probability modelling (WSPM) to identify the groundwater potential areas.
5. Dawoud et al. (2005) developed a GIS-based model to simulate the water resources in the Western Nile Delta. GIS was used to manage the spatially distributed input parameters such as the time invariant spatial data and outputs of the model.

Table 6.2 MCDA mapping studies relevant to groundwater recharge applications

Study	Country of application	MCDA
Rahman et al. (2012)	Portugal	AHP, WLC and OWA
Chowdhury et al. (2010)	West Bengal	AHP
Sargaonkar et al. (2011)	India	AHP
Srivastava and Bhattacharya (2006)	India	AHP
Machiwal et al. (2011)	India (Rajasthan)	AHP
Anane et al. (2008)	Tunisia	AHP and Boolean

6. Lubczynski and Gurwin (2005) applied GIS cross-overlay procedure to merge different component maps and weight information relating to controls on recharge processes. The mapping of relative recharge rates has been applied to studies in aquifer vulnerability modelling, i.e. using the DRASTIC approach.
7. Hajkowicz and Higgins (2008) suggested that multi-criteria decision analysis (MCDA) is a very efficient tool for groundwater assessment, and one of the most widely used MCDA types is the Analytic Hierarchy Process (AHP) method. It is implemented within GIS, which defines weights for criteria in many environmental and groundwater management problems. Table 6.2 shows some of the studies that used AHP as a MCDA method and Weighted Linear Combination (WLC) and Ordered Weighted Averaging (OWA) techniques to select site for groundwater recharge.

The AHP approach was used by Chowdhury et al. (2010), and Srivastava and Bhattacharya (2006) for deriving criteria weights, whilst Rahman et al. (2012) used WLC and OWA techniques to select the optimum sites for groundwater recharge, Machiwal et al. (2011) and Anane et al. (2008) used an AHP approach to support decision making, incorporating AHP to identify site for groundwater recharge.

5 Preparation of Groundwater Potential Maps Using RS and GIS

For the purpose of systematic and controlled development and planning of groundwater resources, groundwater mapping is one of the major tool used by many scientists. These maps are extensively used by engineers, planners and decision makers in order to govern the planning and development of groundwater resources and their optimum utilization as formulated in the National Water Policy by the Ministry of Water Resources, Government of India (NWP 2012). Jaiswal et al. (2003) proposed a schematic plan for exploitation of groundwater by integrating a number of thematic maps (Fig. 6.2) in a GIS environment for depicting village-wise groundwater prospect zones. Thematic maps such as topographical, geological,

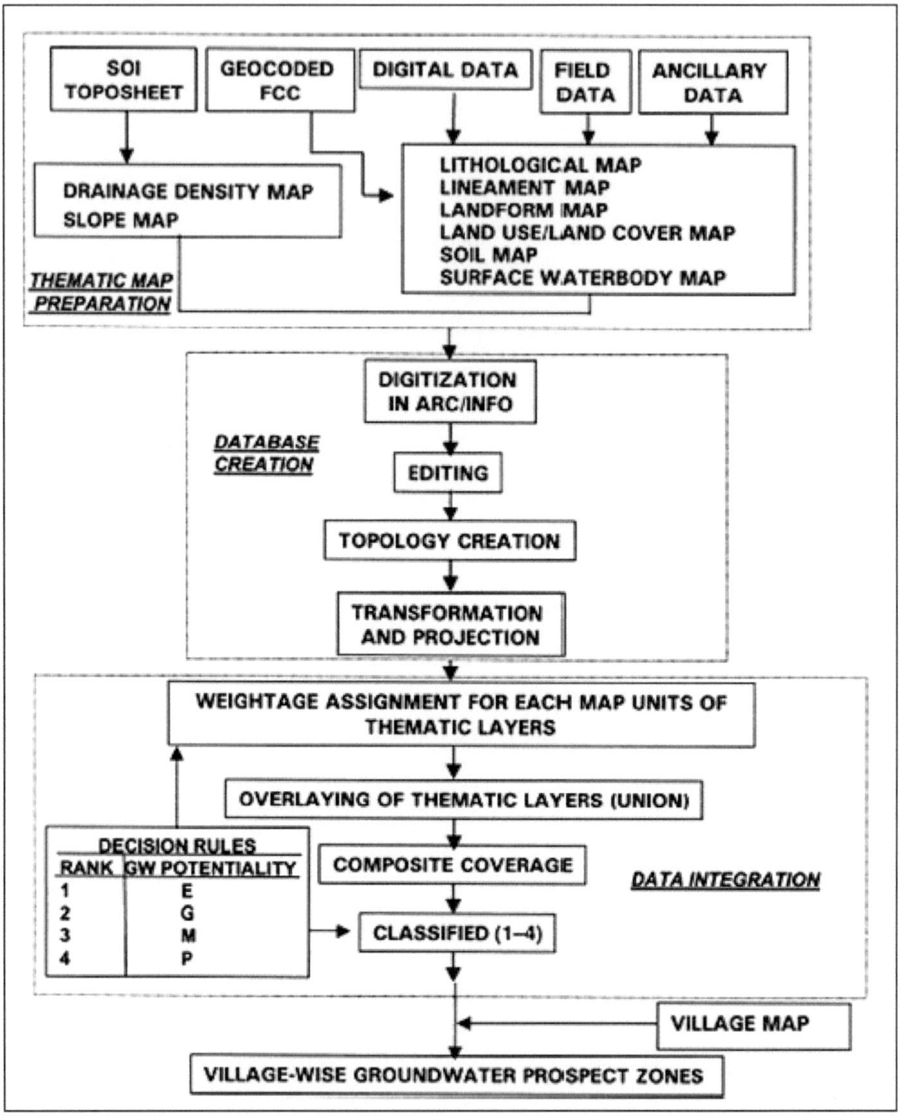

Fig. 6.2 Approach for delineating groundwater prospect zone (Adapted from Jaiswal et al. 2003).

hydrographical, hydrological, hydrochemical, drainage density and lineament density were prepared using the Indian Remote Sensing (IRS) data. The feature-based categorizations of thematic maps and the exact weightage were assigned to each of the thematic features based on their relative merit and demerit with respect to groundwater occurrence.

These maps were based on determining the most important contributing parameters that control groundwater storage. These parameters include the following:

1. Slope, which controls the runoff of water or remains on the ground surface for time long enough to infiltrate (Sl),
2. Stream network, which influences the distribution of runoff and groundwater recharge (Dr),
3. Lineament, which enhance significantly the permeability by inducing secondary porosity and hence vertical water percolation to recharge the aquifers (Lin),
4. Lithology or rock type, which determines the soil and exposed rocks infiltration capabilities and govern the flow and storage of water (Geo), and
5. Topographical map layers (top).

These input layers were combined together mathematically by the equations (6.1) and (6.2) using the Raster Calculate Module in a GIS model to produce the final groundwater prospective zones of the investigated area. Each thematic map has been given a weight value depending on its strength of influence/contribution with respect to the groundwater storing.

$$\mathrm{GWP} = \sum Wi \times CVi \tag{6.1}$$

where the GWP = groundwater potential, Wi = map weight and CVi = capability value (weight of inter-map class).

$$\mathrm{GWP} = \sum (\mathrm{Dr}, \mathrm{Lin}, \mathrm{SI}, \mathrm{Geo}, \mathrm{top}) \tag{6.2}$$

where Dr = stream network class, Lin = lineaments density class, Sl = slope class Geo = lithology class and top = topography/elevation class.

6 Applications of RS and GIS in Consolidated and Unconsolidated Formations for Groundwater Studies

Groundwater is extracted from either consolidated rock formations or unconsolidated loose sediments. The occurrence of groundwater in a geological formation and the scope for its exploitation primarily depends on its porosity and permeability (Kesavulu 1993). A detailed methodology regarding the groundwater exploration in India has been summarized by the National Remote Sensing Agency (1993, 2008) and RGNDWM Project Team (2009). The steps includes:

1. Preparation of drainage map showing surface-water bodies, drainage channels, paleochannels (buried as well as abandoned) and flood plains.
2. Preparation of lineament map showing faults, fractures, dykes and lineament intersections.

3. Preparation of landuse/landcover map from RS data by incorporating observations on vegetation, landforms and topography.
4. Collection of soil, geology, water level and precipitation data and preparation of respective thematic maps.
5. Assigning suitable weights for each layer depending on their contribution to groundwater potential.
6. Integration in GIS environment.

Various groundwater exploration studies were successfully attempted by combining RS and GIS in the semi-arid, arid and hard rock geological formations of the world (Murthy 2000; Shahid et al. 2000; Krishnamurthy et al. 2000; Majumdar and Pal 2005; Kamla et al. 2006; Galonos and Rokos 2006; Vijith 2007; Sultan et al. 2007; Dhakate et al. 2008; Agarwal and Kachhwaha 2009; Kushwaha et al. 2010; Ganapuram et al. 2009; Dar et al. 2010; Khodaei and Nassery 2011; Madani and Niyazi 2015).

The studies by these authors suggest that the most important themes necessary for groundwater assessment in consolidated formations are lineament, geomorphology, slope, drainage density and land use/land cover. The major themes necessary for groundwater assessment in unconsolidated formations are lithology, geomorphology, slope, drainage density, soil type and land use/land cover. Further, themes such as rainfall intensity, surface-water bodies, topographic elevation and net recharge have also been considered in various studies and found to have improved the accuracy. The following salient features of the landscape have been used for assessing groundwater condition from RS data (Table 6.3).

Various models such as logical conditioning overlay analysis, weighted-index method, groundwater-potential index and multi-criteria analysis, overlay analysis, water-balance techniques, and correlation with NDVI values all in combination with remote sensing and GIS were attempted to arrive at the promising zones of groundwater potential in unconsolidated formations. A comparative study between these methods was not possible as the methods are site specific.

7 Case Studies

7.1 Assessment of Groundwater Potential in Hard Rock Area through RS and GIS

The Raniganj area of West Bengal has a long history of coal mining starting from 1744. This has resulted in major change in land use pattern and high groundwater abstraction leading to drinking water crisis especially during the pre-monsoon period.

Sikdar et al. (2004) investigated landuse/landcover change over the period of 26 years (1972–1998), geology and geomorphic set up and also groundwater potential zoning for future development using RS and GIS in Raniganj area, West Bengal. The study indicated that land covered with vegetation and settlement has

Table 6.3 Salient features of the landscape used for assessing groundwater condition from remote-sensing data

Features	Classification	Groundwater potential
Topography		The local and regional relief setting gives an idea about the general direction of groundwater flow and its influence on groundwater recharge and discharge
Slope	Low slope (0–5°)	High groundwater potential
	Medium slope (>5–20°)	Moderate groundwater potential
	High slope (>20°)	Poor groundwater potential
Vegetation	Phreatophytes	Shallow groundwater under unconfined conditions
	Xerophytes	Deep groundwater under unconfined conditions
	Halophytes	Shallow brackish or saline groundwater under unconfined conditions
Geologic landform	(a) Modern alluvial terraces, alluvial plains, floodplains and glacial moraines	(a) Favourable sites for groundwater storage
	(b) Sand dunes	(b) Presence of underlying sandy glacio-fluvial sediments indicating the presence of groundwater
	(c) Thick weathered rocks	(c) Moderate groundwater potential
	(d) Rocks with fractures/fissures	(d) Very good or excellent potential for groundwater
	(e) Rocks without fractures/fissures	(e) Unfavourable sites for groundwater occurrence
	(f) Hillocks, mounds and residual hills	(f) Unfavourable sites for groundwater existence
Lakes and streams	Oxbow lakes and old river channels	Favourable sites for groundwater extraction
	Perennial rivers and small perennial and intermittent lakes	High to moderate potential of groundwater
Drainage density	High	Unfavourable for groundwater availability
	Moderate	Moderate ground water potential
	Less/no	High groundwater potential
Drainage pattern	Gives an idea about the joints and faults in the bedrock which, in turn, indicates the presence or absence of groundwater	
Drainage texture	Coarse	Favourable site for groundwater storage
	Medium	Moderate groundwater occurrence
	Fine	Unfavourable site for groundwater storage

(continued)

Table 6.3 (continued)

Features	Classification	Groundwater potential
Lineaments	In pediplain or valley fill	Groundwater occurrence
	In denudational hills, residual hills, in areas of high-drainage density, high-slope areas, clay zones	Less significant for groundwater availability
Spring type	Depression springs, contact springs, and artesian springs.	Presence of potential aquifer
	Moist depressions, seeps and marshy environments	Presence of shallow groundwater under unconfined conditions

Adapted from: Jasmin and Mallikarjuna (2011)

decreased at the expense of mining activity. For delineating groundwater potential zone overlay analysis using multi-criteria such as drainage, texture, geomorphology, lithology, land use and steepness of slope were utilized to understand the potentiality of groundwater for future development.

The following data sources and methods were used to delineate groundwater potential zones for Raniganj area.

7.1.1 Data

- Survey of India Toposheet (73 M/2)
- The corresponding satellite imagery IRS-ID LISS-III geocoded FCC of 1998 on 1:50,000 scale, and geological map of the Geological Survey of India.

7.1.2 Methodology

The methodology used consists of following steps:

- Visual interpretations of satellite imagery to delineate the geology, drainage, geomorphologic units, lineament and land use/land cover.
- Field verification of interpreted units.
- Preparation of various thematic maps using GIS (ILWIS 3.3 Academic version).
- Preparation of slope map using Digital Elevation Model (DEM).
- Preparation of landuse/landcover change map over the period of 26 years.
- Preparation of groundwater potential zone map using overlay analysis in GIS platform.

7.1.3 Preparation of Thematic Maps

All the thematic maps were prepared in 1:50,000 scale with a spatial resolution of 0.1 km^2 using GIS package ILWIS to find out the potential groundwater bearing zone (Figs. 6.3 and 6.4, Table 6.4).

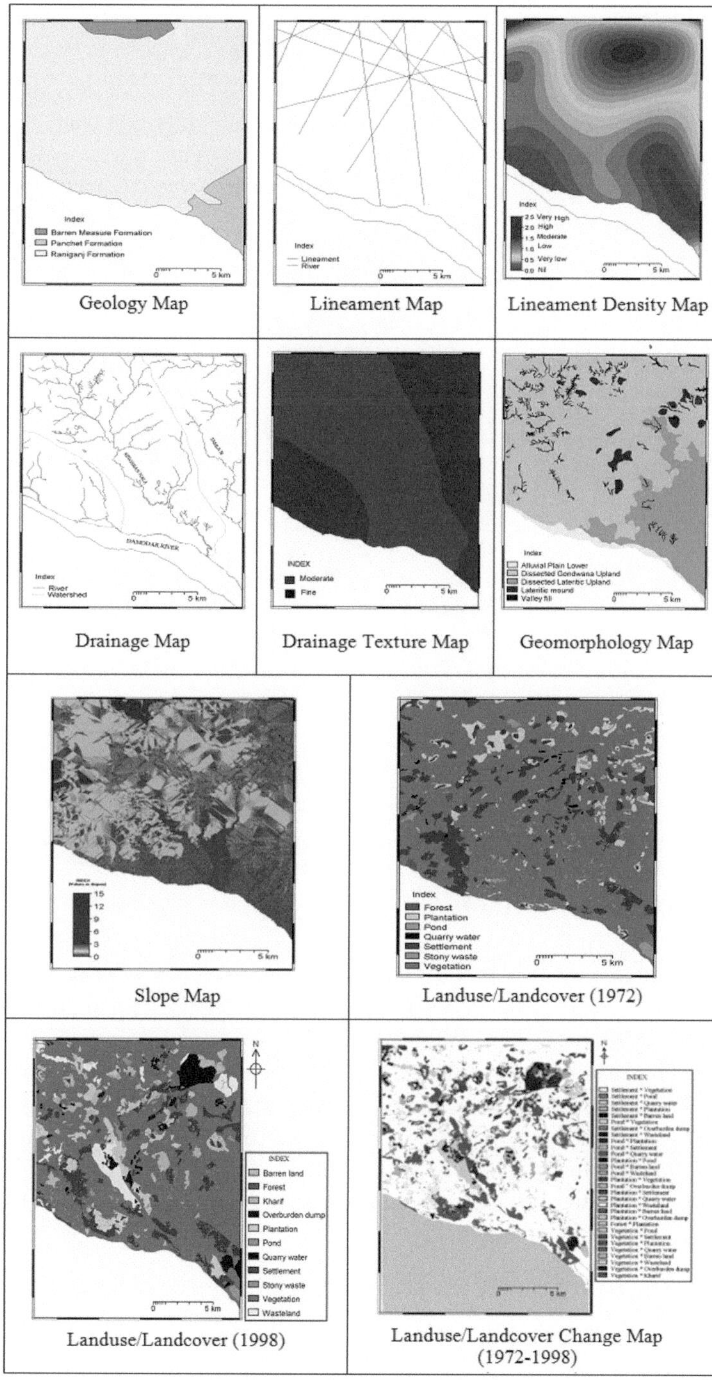

Fig. 6.3 Thematic maps of various layers (Adapted from Sikdar et al. 2004)

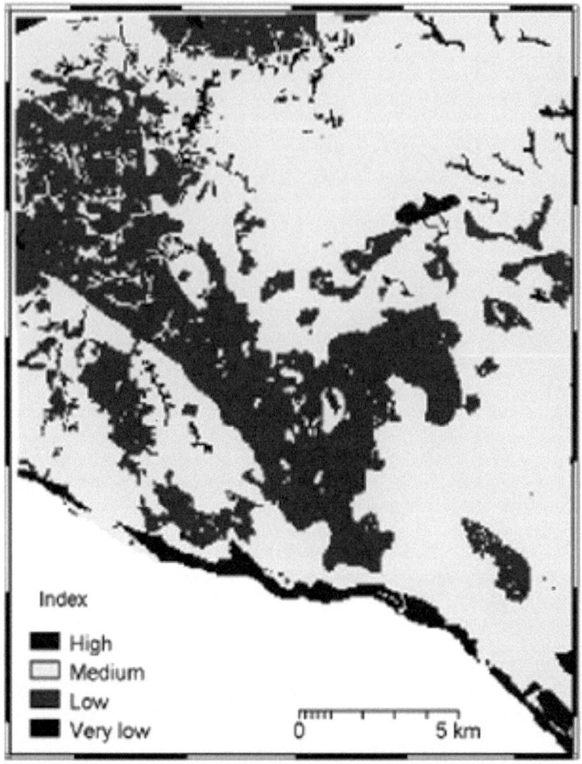

Fig. 6.4 Groundwater potential map of Raniganj area (Adapted from Sikdar et al. 2004)

7.2 Assessment of Groundwater Potential in Soft Rock Area through RS and GIS

Shahid et al. (2000) applied GIS to demarcate the groundwater potential zone in a soft rock area in Midnapur District, West Bengal, India which falls under the Gangetic West Bengal region. He used seven hydrogeologic themes on the basis of its direct control on the groundwater which include lithology (L), geomorphology (G), soil (S), net recharge (R), drainage density (D), slope (S) and surface water bodies (W). Each theme was assigned a value from 1 to 7. Each feature of an individual theme was ranked in the 1–10 scale in the ascending order of hydrogeologic significance. The Ground Water Potential Index (GWPI) for an integrated layer was calculated using GIS as

$$\mathrm{GWPI} = (\mathrm{LwLr} + \mathrm{GwGr} + \mathrm{SwSr} + \mathrm{RwRr} + \mathrm{DwDr} + \mathrm{EwEr} + \mathrm{WwWr})/\Sigma w$$

where the index 'w' represents the weight of a theme and 'r' the rank of a feature in the theme. GWPI is a dimensionless quantity that helps in indexing the probable groundwater potential zones in an area. All the thematic maps were prepared in the

Table 6.4 Area covered by different groundwater potential zone

Category	Area covered (km^2)	Percentage of the total area	Yield range (m^3/hr)	Depth/ thickness of aquifer	Groundwater structures feasible
High	14.70	5.2	50–100	Unconfined 15–25 m	Dug wells fitted with low power pumps and tubewell fitted with hand pump and submersible pump
Medium	184.33	64.8	25–50	Confined 45–60	Dug well, dug-cum-bored well and tube well fitted with hand pump
Low	85.38	29.9	<25	Unconfined <15 m	Dug well, dug-cum-bored well
Very low	0.19	0.1	<25	Unconfined <15 m	Generally groundwater structures will not be successful. Dugwell, dug-cum-bored well may be constructed. Surface water should be harnessed and rooftop rainwater harvesting schemes may be adopted
Total	285.10	100			

Adapted from Sikdar et al. (2004)

1:50000 scale with a spatial resolution of 0.1 km^2 using the GIS package ARC/INFO. The evolved GIS-based model of the study area was found to be in strong agreement with available borehole and pumping test data. The field verification of this model also established the efficacy of the GIS in demarcating the potential groundwater areas in soft rock terrain.

8 Conclusion

This chapter highlights the capability of RS with suitable sensors to generate informations on the spatial and temporal domain and significance of GIS technology in managing large and complex databases for groundwater assessment and management studies. This chapter also focuses on the integration of RS and GIS based analysis and models in finding groundwater potential areas in the form of thematic maps in consolidated and unconsolidated formations. These maps are used by engineers, planners and decision makers to allocate, develop and manage groundwater resources. Satellite data derived geological and hydro-geomorphologic features, preparation of corresponding thematic maps, assigning appropriate weights and integration in a sophisticated GIS platform assist in prospecting the groundwater resources to plan recharging of aquifer, constructing water harvesting structures and drinking water sources. The use of advance high resolution satellite and aerial data in the future will remarkably augment the remote sensing services extending it to infrastructure, planning and management.

References

Agarwal, A.K. and Kachhwaha, T.S. (2009). Water resources management plan in hilly terrain using large scale IKONOS satellite data. Map World Forum, Hyderabad, India, February 2009.

Ai, Nakagawa, Ochi, S. and Shibaski, R. (1998). An application of RS and GIS for hydrological modeling in continental scale. Proceedings of the 19th Asian Conference on Remote Sensing, Manila, Philippines, 16–20 November 1998.

Anane, M., Kallali, H., Jellali, S. and Ouessar, M. (2008). Ranking suitable sites for soil aquifer treatment in Jerba Island (Tunisia) using Remote Sensing, GIS and AHP-Multicriteria Decision Analysis. *International Journal of Water*, **4(1/2):** 121–135.

Baban, S. and Luke, C. (2000). Mapping agricultural land use using retrospective ground referenced data, satellite sensor imagery and GIS. *International Journal of Remote Sensing*, **21(8):** 1757–1762.

Becker, M.W. (2006). Potential for satellite remote sensing of ground water. *Ground Water*, **44(2):** 306–318.

Burrough, P.A. (1986). Principles of geographical information systems for land resources assessment. Oxford University Press, Oxford, London.

Chandra, A.M. and Ghosh, S.K. (2007). Remote sensing and geographical information system. Narosa, New Delhi.

Chandra, S., Rao, V.A., Krishnamurthy, N.S., Dutta, S. and Ahmed, S. (2006). Integrated studies for characterization of lineaments used to locate groundwater potential zones in a hard rock region of Karnataka, India. *Hydrogeol J*, **14:** 1042–1051.

Chowdhury, A., Jha, M.K. and Chowdary, V. (2010). Delineation of groundwater recharge zones and identification of artificial recharge sites in West Medinipur district, West Bengal, using RS, GIS and MCDM techniques. *Env Erth Sc*, **59(6):** 1209–1222.

Chowdhury, A., Jha, M.K., Chowdary, V.M. and Mal, B.C. (2009). Integrated remote sensing and GIS-based approach for assessing groundwater potential in West Mednipur district, West Bengal, India. *Int J Remote Sens*, **30:** 231–250.

Dar, A., Sankar, K. and Dar, M.A. (2010). Remote sensing technology and geographic information system modeling: An integrated approach towards the mapping of groundwater potential zones in Hardrock terrain, Mamundiyar basin. *J Hydrol*, **394(3–4):** 285–295.

Dawoud, M.A., Darwish, M.M. and EL-Kady, M.M. (2005). GIS-based groundwater management model for Western Nile Delta. *Water Resour Manage*, **19:** 585–604.

Deng Feifan, Deng Zhengdong, Lv Dan, Wang Daqing, Duan Huajie and Xing Zheli (2016). Application of remote sensing and GIS analysis in groundwater potential estimation in west Liaoning Province, China. *J Engg Res*, **4(3):** 1–17.

Dhakate, R., Singh, V.S., Negi, C., Chandra, S. and Rao, V.A. (2008). Geomorphological and geophysical approach for locating favourable groundwater zones in granitic terrain, Andhra Pradesh, India. *J Environ Manag*, **88:** 1373–1383.

Dhiman, S.D. and Keshari, A.K. (2002). GIS based correlation between groundwater quality parameters and geological units. Proceedings of the Map India Conference, New Delhi.

Elewa, H.H. and Qaddah, A.A. (2011). Groundwater potentiality mapping in the Sinai Peninsula, Egypt, using remote sensing and GIS-watershed-based modeling. *Hydrogeo J*, **19(3):** 613–628.

Galonos, I. and Rokos, D. (2006). A statistical approach in investigating the hydrogeological significance of remotely sensed lineaments in the crystalline mountainous terrain of island of Naxos, Greece. *Hydrogeol J*, **14:** 1569–1581.

Ganapuram, S., Kumar, G.T.V., Krishna, I.V.M., Kahya, E. and Demirel, M.C. (2009). Mapping of groundwater potential zones in the Musi basin using remote sensing data and GIS. *Adv Eng Softw*, **40:** 506–518.

Gosain, A.K. and Sandhya, R. (2004). GIS-based technologies for watershed management. *Curr Sci*, **87:** 948–953.

Hajkowicz, S. and Higgins, A. (2008). A comparison of multiple criteria analysis techniques for water resource management. *Eur J Oper Res*, **184(1):** 255–265.

Hoffmann, J. and Sander, P. (2007). Remote sensing and GIS in hydrogeology. *Hydrol J*, **15:** 1.

Jaiswal, R.K., Mukherjee, S., Krishnamurthy, J. and Saxena, R. (2003). Role of remote sensing and GIS techniques for generation of groundwater prospect zones towards rural development – An approach. *Int J Rem Sens*, **24(5):** 993–1008.

Jasmin Ismail and Mallikarjuna, P. (2011). Review: Satellite-based remote sensing and geographic information systems and their application in the assessment of groundwater potential, with particular reference to India. *Hydrogeol J*, **19:** 729–740.

Jha, M.K. and Chowdary, V.M. (2007). Challenges of using remote sensing and GIS in developing nations. *Hydrogeol J*, **15:** 197–200.

Jha, M.K., Chowdhury, A., Chowdary, V. and Peiffer, S. (2007). Groundwater management and development by integrated remote sensing and geographic information systems: Prospects and constraints. *Water Resou Manag*, **21(2):** 427–467.

Kamla, R., Pramod, C.S. and Hrushikesh, S. (2006). Targeting groundwater in tribal dominated Bonai area of drought prone Sundargarh district, Orissa, India: A combined geophysical and RS approach. *J Hum Ecol*, **20:** 109–115.

Kesavulu, C. (1993). Text book of Engineering Geology. Macmillan, Delhi.

Khodaei, K. and Nassery, H.R. (2011). Groundwater exploration using remote sensing and geographic information systems in a semi-arid area (southwest of Urmieh, Northwest of Iran). *Arab. J. Geosci.*, **6:** 1229–1240, doi: https://doi.org/10.1007/s12517-011-0414-4.

Krishnamurthy, J., Kumar, Venkates N., Jayaraman, V. and Manivel, M. (1996). An approach to demarcate ground water potential zones through remote sensing and a Geographic Information System. *Int J Remote Sens*, **17(10):** 1867–1884.

Krishnamurthy, J., Mani, A.N., Jayaram, V. and Manivel, M. (2000). Groundwater resources development in hard rock terrain: An approach using remote sensing and GIS techniques. *Int. Jour. Appld Earth Obser Geoinformatics*, **2:** 204–215.

Kushwaha, S.P.S., Mukhopadhyay, S., Prasad, V.H. and Kumar, S. (2010). Sustainable development planning in Pathri Rao sub-watershed using geospatial techniques. *Curr Sci*, **98:** 1479–1486.

Laurent, R., Anker, W. and Graillot, D. (1998). Spatial modeling with geographic information system for determination of water resources vulnerability application to an area in Massif Central (France). *J Am Water Resour Assoc*, **34(I):** 123–134.

Lillesand, T.M., Kiefer, R.W. and Chipman, J.W. (2004). Remote sensing and image interpretation. 5th ed. John Wiley and Sons.

Lubczynski, M.W. and Gurwin, J. (2005). Integration of various data sources for transient groundwater modelling with spatio-temporally variable fluxes: Sardon study case, Spain. *J Hydrol*, **306:** 71–96.

Machiwal, D., Jha, M.K. and Mal, B.C. (2011). Assessment of groundwater potential in a semi-arid region of India using remote sensing, GIS and MCDM techniques. *Water Resou Manag*, **25(5):** 1359–1386.

Madani, A. and Niyazi, B. (2015). Groundwater potential mapping using remote sensing techniques and weights of evidence GIS model: A case study from Wadi Yalamlam basin, Makkah Province, Western Saudi Arabia. *Environ Earth Sci*, **74(6):** 5129–5142.

Madrucci, V., Taioli, F. and César de Araújo, C. (2008). Groundwater favourability map using GIS multicriteria data analysis on crystalline terrain, São Paulo State, Brazil. *J Hydrol*, **357(3–4):** 153–173.

Majumdar, R.K. and Pal, S.K. (2005). Geoelectric and borehole lithology studies for groundwater investigation in alluvial aquifer of Munger district, Bihar. *J Geol Soc*, **66:** 463–474.

Manap, M.A., Sulaiman, W.N.A., Ramli, M.F., Pradhan, B. and Surip, N. (2013). A knowledge-driven GIS modeling technique for groundwater potential mapping at the Upper Langat Basin, Malaysia. *Arab J Geosc*, **6(5):** 1621–1637.

Meijerink, A.M.J. (2000). Groundwater. *In:* Remote sensing in hydrology and water management. Springer Berlin Heidelberg.

Meijerink, A.M.J. (2007). Remote sensing applications to groundwater. IHP-VI, series on groundwater no. 16, UNESCO.

Mukherjee, S. (2008). Role of satellite sensors in groundwater exploration. *Sensors*, **8:** 2006–2016.

Murthy, K.S.R. (2000). Groundwater potential in a semi-arid region of Andhra Pradesh: A GIS approach. *Int J Remote Sens*, **21:** 1867–1884.

Nagarajan, M. and Singh, S. (2009). Assessment of groundwater potential zones using GIS techniques. *J Ind Soc Remote Sens*, **37:** 69–77.

National Remote Sensing Agency (1993). Integrated Mission for sustainable Development. Technical guidelines. National Remote Sensing Agency, Hyderabad, India. 134 p.

National Remote Sensing Agency (2008). Groundwater prospects mapping using remote sensing and geographic information system. Rajiv Gandhi National Drinking Water Mission, Project manual. National Remote Sensing Agency, Hyderabad, India. 256 p.

NWP (National Water Policy of India) (2012). Available at http://mowr.gov.in/writereaddata/linkimages/NWP2012Eng6495132651.pdf (last accessed on March 10, 2017).

Patra, H.P., Adhikari, S.K. and Kunar, S. (2016). Groundwater Prospecting and Management. Springer Hydrogeology, XXI. 291 p.

Pinder, G.F. (2002). Groundwater modeling using geographical information system. Wiley, New York.

Pothiraj, P. and Rajagopalan, B. (2012). A GIS and remote sensing based evaluation of groundwater potential zones in a hard rock terrain of Vaigai sub-basin, India. *Arab J Geosci*, DOI: https://doi.org/10.1007/s12517-011-0512-3.

Rahman, M.A., Rusteberg, B., Gogu, R., Lobo Ferreira, J. and Sauter, M. (2012). A new spatial multi-criteria decision support tool for site selection for implementation of managed aquifer recharge. *J Env Manag*, **99:** 61–75.

Rahmati, O., Samani, A.N., Mahdavi, M., Pourghasemi, H.R. and Zeiniv, H. (2014). Groundwater potential mapping at Kurdistan region of Iran using analytic hierarchy process and GIS. *Arab J Geosc*, 1–13.

Ramamoorthy, P. and Rammohan, V. (2015). Assessment of Groundwater potential zone using remote sensing and GIS in Varahanadhi watershed, Tamilnadu, India. *International Journal for Research in Applied Science & Engineering Technology* (IJRASET), **3(V):** 695–702.

Ranganath, R.N., Jayaraman, V. and Roy, P.S. (2007). Remote sensing applications: An overview. *Curr Sci*, **93:** 1747–1766.

RGNDWM Project Team (2009). Satellite remote sensing applications. Bull NNRMS, **33(B):** 7–10.

Saraf, A.K. and Choudhury, P.R. (1998). Integrated RS and GIS for groundwater exploration and identification of artificial recharge sites. *Int J Remote Sens*, **19:** 1825–1841.

Sargaonkar, A.P., Rathi, B. and Baile, A. (2011). Identifying potential sites for artificial groundwater recharge in sub-watershed of River Kanhan, India. *Environmental Earth Sciences*, **62(5):** 1099–1108.

Shahid, S., Nath, S.K. and Roy, J. (2000). Groundwater potential modelling in a soft rock area using GIS. *Int J Remote Sens*, **21:** 1919–1924.

Sharma, A.K. and Thakur, P.K. (2007). Quantitative assessment of sustainability of proposed watershed development plans for Kharod watershed, western India. *J Ind Soc Remote Sens*, **35:** 231–241.

Sikdar, P.K., Chakraborty, S., Adhya, E. and Paul, P.K. (2004). Land use/land cover changes and groundwater potential zoning in and around Raniganj coal mining area, Bardhaman district, West Bengal: A GIS and RS approach. *J Spat Hydrol*, **4:** 1–24.

Srivastava, P.K. and Bhattacharya, A.K. (2006). Groundwater assessment through an integrated approach using remote sensing, GIS and resistivity techniques: A case study from a hard rock terrain. *Int J Remote Sens*, **27(20):** 4599–4620.

Suja Rose, R.S. and Krishnan, N. (2009). Spatial analysis of groundwater potential using remote sensing and GIS in the Kanyakumari and Nambiyar basins, India. *J Indian Soc. Remote Sens*, **37:** 681–692.

Sultan, M., Wagdy, A., Manocha, N., Sauck, W., Abdel Gelil, K., Youssef, A.F., Backer, R., Milewski, A., El Alfy, Z. and Jones, C. (2007). An integrated approach for identifying aquifers in transcurrent fault systems: The Najd shear system of the Arabian Nubian shield. *J Hydrol*, **349:** 475–488.

Teeuw, R.M. (1995). Groundwater exploration using RS and a low cost GIS. *Hydrogeol J*, **3:** 21–30.

Vijith, H. (2007). Groundwater potential in the hard rock terrain of Western Ghats: A case study from Kottayam district, Kerala using Resources at (IRS-P6) data and GIS techniques. *J Indian Soc. Remote Sens*, **35:** 163–171.

Waikar, M.L. and Nilawar, P.A. (2014). Identification of Groundwater Potential Zone using Remote Sensing and GIS Technique. *Int J Innov Res Sc, Eng and Techn*, **3(5):** 12163–12174.

Watkins, D.W., Mckinncy, D.C., Maidment, D.R. and Lin, M.D. (1996). Use of geographic information system in ground water flow modeling. *J Water Resour Plann Manag*, **122:** 88–96.

Chapter 7
Numerical Groundwater Modelling

Pradip K. Sikdar

1 Introduction

A model is a representation of a physical system to predict its behaviour over time. Models may be conceptual, physical or mathematical. A conceptual model is a representation of a system using general rules and concepts. A physical model is a physical copy of the system in a reduced scale. A mathematical model is a representation of a system using mathematical concepts and language. Mathematical models may be analytical or numerical. Numerical models are those that use numerical time-stepping procedure to understand the models behaviour over time, whereas analytical models have a closed form solution which can be expressed as a mathematical analytic function. Analytical models do not require much data, but their application is limited to obtain solutions for simple problems. But numerical models are useful for handling more complicated problems.

Groundwater modelling is a way to replicate the behaviour of an aquifer system by defining the essential features of the aquifer in some controlled physical or mathematical manner (Bear et al. 1992). It is a tool for hydrogeologists and civil engineers for understanding, developing, managing and protecting groundwater systems. Groundwater models represent the simplified version of a complex aquifer system and therefore are not perfect, but provide insight to the behaviour of a groundwater system with a reasonable degree of confidence and have been proved to be useful tools for solving many groundwater problems.

Groundwater modelling requires systematic understanding of the groundwater system to predict possible impacts of proposed future development. Groundwater modellers seek to obtain the best representation of the subsurface hydrogeological

P. K. Sikdar (✉)
Department of Environment Management, Indian Institute of Social Welfare and Business Management, Kolkata, India
e-mail: p_sikdar@hotmail.com

P. K. Sikdar (ed.), *Groundwater Development and Management*,
https://doi.org/10.1007/978-3-319-75115-3_7

conditions by field investigation and collecting data on water level, thickness of the aquifer, recharge and discharge rates and aquifer parameters such as porosity, hydraulic conductivity storage coefficient, etc. These data along with other factors like boundary conditions, initial condition and time and space discretisation are then incorporated into the model to obtain quantitative data and qualitative information in a predictive framework (Ghosh and Sharma 2006). There are various methods for groundwater modelling. The most commonly used numerical modelling methods are the "finite difference" method and the "finite element" method. Each method has its advantages and disadvantages.

This chapter describes the purpose, objectives and steps of groundwater modelling, the governing equations for groundwater flow and contaminant transport, and the data required for modelling. An outline of two case studies on groundwater modelling is also provided.

2 Purpose of Groundwater Modelling

The purpose of groundwater modelling is to simulate groundwater flow and transport of pollutants in groundwater. Flow models deal with quantities, that is, to find out the solution for $h = f(x, y, z, t)$ and transport models deal with quality of groundwater, that is, to find out the solution for $c = f(x, y, z, t)$.

Flow models are used for interpretation of observed heads, understanding the response of a groundwater system to changes in natural and anthropogenic recharge and discharge, prediction of drawdown, estimation of water balances and delineation of catchment areas of wells. Transport models are used for interpretation of concentration data, mass balance of contaminants, prediction of pollutant plumes, design of pump and treat management, planning of monitoring strategy and risk assessment in case of waste disposal.

3 Process of Modelling

The first step in groundwater modelling is to be clear about the objective of the modelling exercise which is followed by data collection and interpretation of the collected data to develop the conceptual hydrogeological model. This is followed by the selection of model type and design. Then the model is run for estimation of error by changing the various model parameters which is the process of calibration. The model with the minimum error is selected. After selection of the model, sensitivity analysis is carried out. The model is the run for various scenarios based on the objectives of the modelling exercise. The stepwise methodology of groundwater modelling is shown in Fig. 7.1.

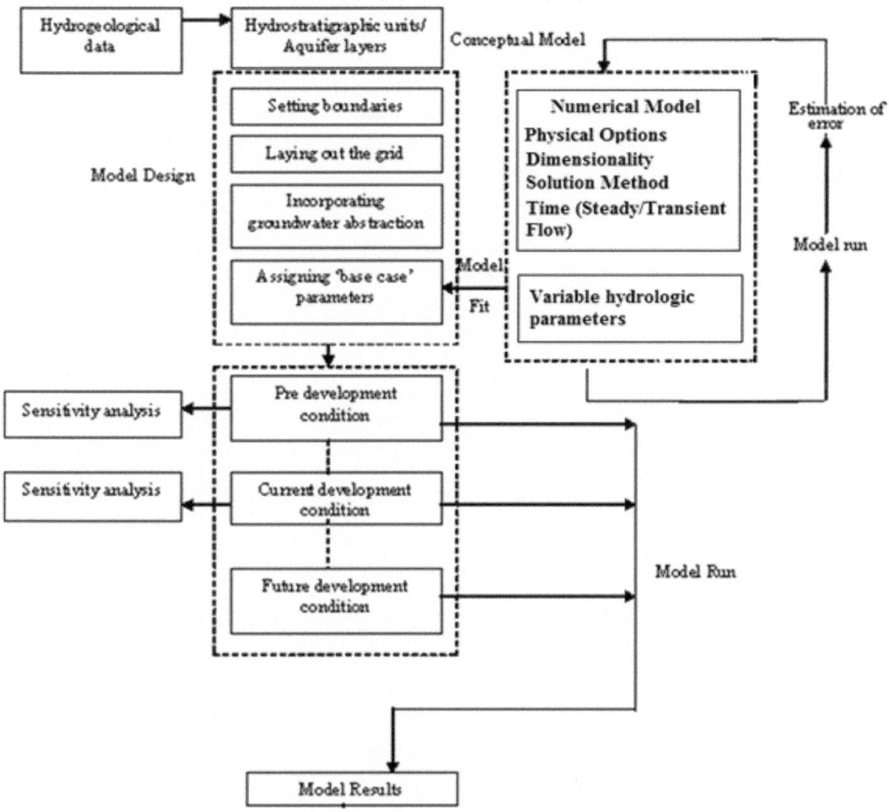

Fig. 7.1 Flow chart showing the process of groundwater flow modelling

3.1 *Objectives of Modelling*

Groundwater modelling is essential to manage groundwater resource in a sustainable way and to support management decisions regarding groundwater development programme with respect to quality and quantity. Groundwater models can be used as (i) interpretive tool for analysing groundwater flow patterns and contaminant transport, (ii) predictive tool for predicting change in head or pollutant concentration in response to stress in the future, (iii) generic tool for analysing recharge, discharge and storage of groundwater, and different scenarios of groundwater development or remediation schemes, and (iv) visualising tool for communicating key messages to the public and decision-makers.

Therefore, the objectives of a groundwater modelling exercise can be, but not limited to, as follows:

- Predicting groundwater flow and head in time and space due to urbanisation.
- Investigating the effect of groundwater abstraction by a well, group of wells or well field on the flow pattern and predicting the resultant drawdown.
- Predicting migration of pollutant by change in depth of pumping.
- Understanding the most likely pathway of contaminants if there is a leak in a proposed underground repository for radioactive waste or from waste water discharge on land or from solid waste dumping ground.
- Modelling sea water intrusion in coastal aquifers.
- Analyzing different management programmes on the groundwater system, quantitatively and qualitatively.

3.2 *Governing Groundwater Flow and Contaminant Transport Equations*

A general form of the groundwater flow in three dimensions in a porous media with the assumption that the density and viscosity of the fluid remain constant can be expressed by combining Darcy's Law with the Laplace Continuity Equation by the following partial differential equation:

$$\frac{\partial}{\partial x}\left(K_x\frac{\partial h}{\partial x}\right)+\frac{\partial}{\partial y}\left(K_y\frac{\partial h}{\partial y}\right)+\frac{\partial}{\partial z}\left(K_z\frac{\partial h}{\partial z}\right)=S_s\frac{\partial h}{\partial t}\pm W \tag{7.1}$$

where K_x, K_y and K_z are the hydraulic conductivity along, x, y and z coordinates ($\mathrm{L\,T^{-1}}$) h is the piezometric head (L); S_s is the specific storage of porous medium ($\mathrm{L^{-1}}$); t is the time and W is the volumetric flux per unit volume and represents sources and/or sinks of water ($\mathrm{T^{-1}}$).

Detailed derivation of the above equation are given in Bear (1979), Freeze and Cherry (1979), Rushton and Redshaw (1979) and many others.

The generalized solute transport equation for a 3-dimensional advection-dispersion-adsorption-decay equation for contaminant transport in porous media is given by (Bear 1972):

$$\begin{aligned}&\frac{\partial C}{\partial t}+\frac{l}{R}\left\{v_x\frac{\partial C}{\partial x}+v_y\frac{\partial C}{\partial y}+v_z\frac{\partial C}{\partial x}-D_x\frac{\partial^2 C}{\partial x^2}-D_y\frac{\partial^2 C}{\partial y^2}-D_z\frac{\partial^2 C}{\partial z^2}\right\}\\&+\lambda C+\sum_{k=1}^{N}\pm R_k=0\end{aligned} \tag{7.2}$$

where C is the concentration of contaminant, ($\mathrm{M\,L^{-3}}$); R is the retardation factor = $(1 + \rho_b K_d/n)$; ρ_b is the bulk density of solid material ($\mathrm{M\,L^{-3}}$); K_d is a distribution factor, (L^3 M^{-1}); n is the porosity; v_x, v_y and v_z are the velocity vectors along x, y and z directions, respectively (L T^{-1}); D_x, D_y and D_z are the dispersion coefficient tensors along x, y and z directions, respectively (L^2 T^{-1}); λ is the decay rate coefficient, (T^{-1}); and $\sum_{k=1}^{N} R_k$ is the source or sink of the contaminant, (M L^{-3} T^{-1}).

Detailed derivation of the above equation are given in Reddell and Sunada (1970), Bear (1972), Bredehoeft and Pinder (1973), Konikow and Grove (1977) and Javandel (1984).

The transport equation is linked to the flow equation through the relationship:

$$v_{ii} = \frac{K_{ii}}{n}\frac{\partial h}{\partial x_i} \tag{7.3}$$

where K_{ii} is the principal component of the hydraulic conductivity tensor, [L T^{-1}]; and h is the hydraulic head, [L].

3.3 *Numerical Modelling Techniques*

Numerical modelling employs approximate methods to solve the partial differential equation describing the flow in porous medium. These techniques do not emphasize on obtaining an exact solution of the partial differential equation but on obtaining a reasonably approximate solution. Finite difference and finite element methods are presently used extensively in numerical groundwater modelling. For limitation of space only the finite difference method will be discussed in this chapter.

3.3.1 Finite Difference

Finite difference method (FDM) has been widely used in groundwater studies since the early 1960s. The method was studied by Newton, Gauss, Bessel and Laplace (Pinder and Gray 1977).

The Finite Difference Method is a computational procedure based on dividing the model area into a grid with discrete set of points called nodes (Fig. 7.2). Various hydrogeological parameters are assigned to each of these nodes and then the flows associated within a single zone of the aquifer are analyzed (Trescott and Larson 1976). In order to solve a finite difference equation, one has to start with the initial distribution of heads and then compute the heads at later time instants. This is an iterative process using fast converging iterative algorithms to solve the set of algebraic equations obtained through discretization of equation (7.1) (Trescott and Larson 1976). The solutions obtained are only an approximate one and not an exact one.

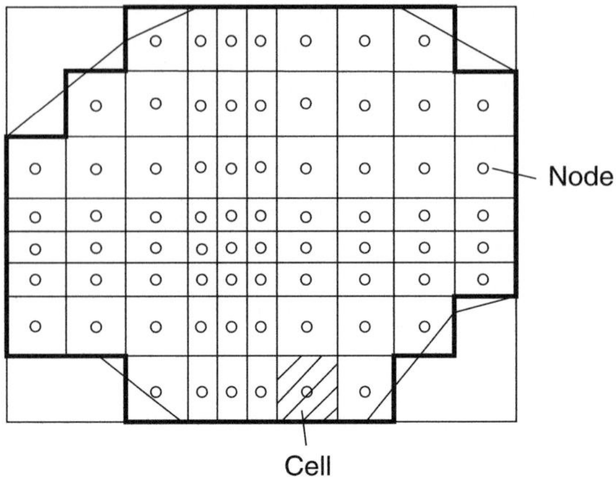

Fig. 7.2 Finite difference cells and nodes

The advantages of finite difference method are that it is fairly simple to understand and to program, easy to implement, the numerical properties are well known and produces reasonably good results. The finite difference grid overlain over the aquifer improves the accuracy in the calculation of flow rate and direction and hence is an advantage in particle tracking within the aquifer domain. However, this method has some disadvantages. The main disadvantage is that the irregular model boundaries cannot be approximated with this method. In addition, the distribution of grids, their size, and size difference highly affects the accuracy and computation effort. Output accuracy of the finite difference method is not good in the case of solute transport modelling. Mass balance is not guaranteed if conductivity or grid spacing varies (Cirpka 1999).

The most widely used finite difference based groundwater model is MODFLOW (Harbaugh and McDonald 1996).

3.4 Field Data and Maps

All the parameters in equations (7.1) and (7.2) should be known at maximum points in time and space to conceptualize the groundwater flow and mass transport processes and to prepare a model of it. The parameters are:

1. Topograhic data
2. Sub-surface lithological data
3. Boundary conditions
4. Hydraulic conductivity in horizontal and vertical directions of the aquifer.
5. Storage coefficient of the aquifer
6. Input stress (Recharge rate)
7. Output stress (Discharge rate – Natural and Anthropogenic)
8. Water level elevation in space and time

9. Dynamic or effective porosity
10. Groundwater concentration in space and time of different ions
11. Dispersivity or mixing length
12. Evapo-transpiration
13. Top and bottom of screens in the wells

Based on these data, various maps, geological sections and tables should be prepared. They are

1. Surface topographic contour map
2. Geological map, panel/fence diagrams showing the aereal and vertical extent and boundaries of the system
3. Contour maps of the top and bottom of aquifers and aquitards
4. Water table/piezometric surface elevation contour maps
5. Maps showing distribution of various aquifer parameters such as hydraulic conductivity and storage coefficient
6. Soil map
7. Landuse/Landcover map showing rivers, lakes, wetlands, other surface water bodies, springs, forests, agricultural land, urban areas etc.
8. Rainfall and evapo-transpiration maps
9. Map showing distribution of the net recharge of the aquifers
10. Maps showing locations of groundwater pumping

Detailed assessment of all the data could include statistical analysis together with an analysis of errors that can be used in later uncertainty analysis.

3.5 Conceptual Model

Conceptual model is a pictorial representation of the groundwater flow system and provides the basis for model designing and communicates the physical processes that control groundwater occurrence and movement in the study area. The model is generated using feature objects such as, points, arcs and polygons of the data and maps discussed above. These feature objects are stored in the modelling system and later transferred to the model grids. The conceptual model defines the dimensions of numerical model, how the grid is to be designed and oriented. Conceptual models are, therefore, independent of the horizontal (spatial) and vertical (temporal) discretization of the numerical model. The data are automatically assigned to the grids when the conceptual model is converted to the numerical model. Since the conceptual model is independent of the grid resolution of the numerical model, a new numerical model can be generated in a short time by changing the conceptual model only. This enables the modeller to evaluate many alternative scenarios in short time, resulting in a more precise and efficient modelling process. Another advantage of storing attributes with feature objects is that it reduces some of the instability that is inherent in finite difference models such as MODFLOW especially in applying the boundary conditions to the grid cells and thereby produces a model that more accurately represents the field condition.

There are three steps to construct a conceptual model. They are:

1. Define hydrostratigraphic units based on available geologic and hydrogeologic information.
2. Prepare a water budget by quantifying all sources of flow into the system (precipitation, snow melt, flow across aquifer boundaries, recharge from surface water bodies and wells) and out of the system (baseflow, evapo-transpiration, spring flow and groundwater pumping).
3. Define the flow system using recharge, head, baseflow, evapo-transpiration and geochemical data.

3.6 Model Design

3.6.1 Boundary Conditions

Boundary conditions are essential to get a unique solution of the Laplace Equation. In groundwater flow problems, boundary conditions are not only mathematical constraint, they also represent the sources and sinks within the system (Reilly and Harbaugh 2004). Selection of boundary conditions is critical to the development of an accurate model (Franke et al. 1987). Since groundwater flowpaths may span a regional scale, the model area should be a hydrologically closed system with major hydrologic boundaries on all sides.

Boundary conditions (Fig. 7.3) can be classified into five types (Baalousha 2008):

1. *Prescribed head/Specified head/Fixed head/Constant head boundary*: In this type of boundary condition, head and/or concentration remain constant throughout the simulation period. It can be expressed in a mathematical form as: $h\ (x, y, z, t) =$ constant. Constant head boundaries assume that the head is constant over time.

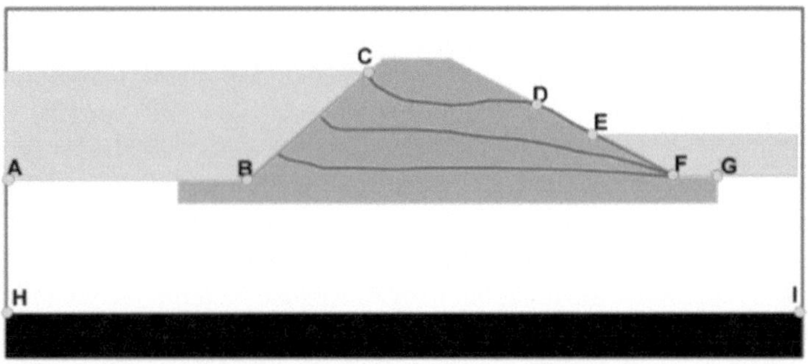

Fig. 7.3 Different types of boundaries. (*Source:* Baalousha 2008)

Lines ABC and EFG in Fig. 7.3 are examples of constant head boundaries, where part of the aquifer occurs underneath a reservoir. Typical prescribed head boundaries are: (i) elevation of a surface water body (e.g. stream, canal, river, lake, ocean) well-connected to groundwater, (ii) sub-surface measurements of head where the system is thought to be at steady state or a known transient, (iii) approximate head values at distant locations, and (iv) water table elevation (for modelling deeper units).

2. *Specified flux boundary*: It is represented by source (sink) of fluid, that is, wells or drains where pumping rates are known or estimated or when flow across the boundary can be specified in time and location. An example of a specified flux boundary is recharge across the water table in a phreatic aquifer. Line CD in Fig. 7.3 is a specified flux boundary. A special case of a specified flux boundary is a no-flow boundary where zero flux is specified. In this type of boundary condition discharge (Q) is kept constant over time. This occurs at a line normal to streamline (i.e. normal to flow direction). This case normally occurs where impermeable media exist. Line HI in Fig. 7.3 represents a no-flow boundary. A water divide can be used as a no-flow boundary but with caution, as position of the water divide may move with time as a result of stresses on the aquifer.
3. *Head dependent flux boundary or General Head Boundary (GHB)*: In this type of boundary the flux is calculated for variable head and assigned accordingly. This type of boundary condition is called a mixed boundary condition as it relates the boundary flows to the boundary heads. The simulation of river-aquifer interaction, aquifer-drain seepage, and general head-stream are called head dependent flux boundaries. A semi-confined aquifer, where the head depends on the flux through the semi-confining layer, is an example of this type of boundary. This can be represented by lines ABC and EFG in Fig. 7.3.
4. *Free-surface boundary*: The water table and the fresh-saline water interface in a coastal aquifer are examples of free-surface boundaries. Line CD in Fig. 7.3 represents a free-surface boundary. Pressure head at free-surface boundary is always zero and the total head equals elevation head.
5. *Seepage face boundary*: This occurs at the boundary between saturated flow and the atmosphere. The face of a landfill dam, as shown by line DE in Fig. 7.3 is an example of a seepage face boundary.

3.6.2 Grid Design or Discretization

An important feature of numerical modelling is the representation of the real world by discrete volumes of material (Reilly and Harbaugh 2004). These volumes of material are called cells in the finite difference method. Cells are square or rectangular blocks. Each cell has a node which is a discrete point where the head, concentration, hydraulic properties and stresses are computed. The size of the cells determines the variation of the hydraulic properties and stresses of the model area. A grid is a network of cells and nodes (Fig. 7.2).

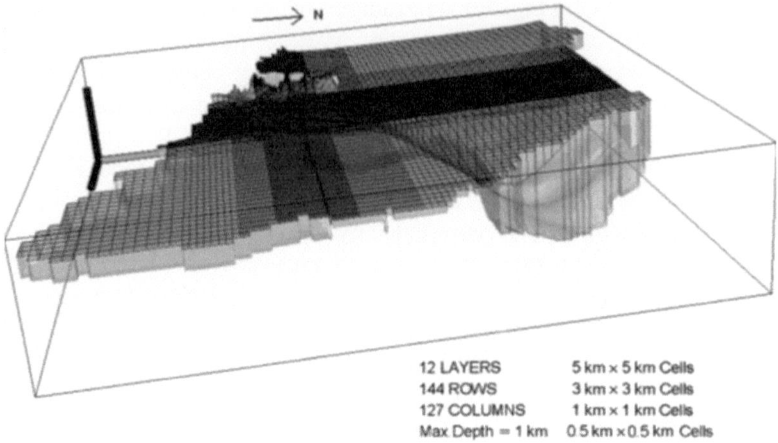

Fig. 7.4 Discretization of model area of North Bengal Plain. (*Source:* Sikdar and Chakrabarty 2017)

The grid should be covering the entire area to be modelled. The finite difference grid is designed by manipulating rows, columns and layers of cells. Cells oriented in the x–direction are called a row and those oriented in the y-direction are called a column. A horizontal 2-dimensional network of cells is called a layer. Cells are designated using the row and column coordinates, with the origin in the upper left corner of the mesh. The upper left cell is designated as: row 1 and column 1. The upper layer is layer 1 and layers increase in number downwards. It is essential to orient the grid to minimize the number of cells that fall outside the boundaries of the modelled area. These cells are called inactive cells. The inactive cells are formed in case of finite difference model only because of the discretization of the irregular model boundary into rectangular or square grids. For best results the grid should have finer discretization (that is smaller cell size) in the study area compared to that of the surrounding model area (Fig. 7.4). Fine discretization allows detailed representation of the groundwater flow in the study area, while coarse discretization in the larger area allows efficient modelling of the entire regional system.

Further, for transient models, time discretization is represented by discrete increments of time called time steps. The size of the time steps also has an impact on the accuracy of a model. It is desirable to use small time steps (Δt) for stable numerical modelling. But it may be difficult to find an optimal time step to ensure stable solution. An approximate time step can be estimated based on transmissivity (T), storage coefficient (S) and grid spacing (a) as per the following equation (Thangarajan 2004):

$$\Delta t = Sa^2/4T$$

This time step is called critical time step and it can be increased or decreased during the simulation process.

3.6.3 Specifying Properties of Cells

Each cell in the grid must be assigned some hydraulic properties such as hydraulic conductivity, storage coefficient, head etc. These properties are measured at limited locations in the field and hence are not known at every cell in the grid. To overcome this difficulty the simplest way is to give a uniform value to a group of cells. Various geostatistical interpolation methods such as kriging, distance-squared interpolation and bi-variate interpolation methods can also be used to assign data to each cell.

Specifying storage properties of the groundwater system (specific storage for confined storage and specific yield for unconfined storage) has some complexities. In a transient system, to simulate changes in storage, unconfined storage or specific yield value should be specified only in all the cells of the uppermost active boundary. In the cells of the lower layers confined storage or specific storage should be assigned.

3.6.4 Model Calibration

Model calibration is the most important and difficult task in the modelling process. The objective of model calibration is to minimize the difference between the simulated data and the measured field data. The data which can be used for calibration are heads and flow rates. A large difference indicates an error either in conceptualization of the system or in the database. Model parameters that can be adjusted during calibration are hydraulic conductivity in horizontal and vertical directions (K_h and K_v), storage coefficient (S) and input/output stresses. Calibration is done through trial-and-error method. In this method, parameter values are initially assigned to each node of a cell and then the values are adjusted in sequential model runs to arrive at a better match between the simulated head/flow data and the measured field head/flow data. This trial-and-error method of calibration is an arduous one but very effective.

There are several quantitative measures of calibration. Some of these are: sum of squared residuals, residual mean, residual standard deviation, the absolute mean error, and the root mean squared error (Anderson and Woessner 1992). The areal distribution of residuals (differences between measured and simulated values) also is important to determine whether some areas of the model are biased either too high or too low.

3.6.5 Sensitivity Analysis

Sensitivity analysis is important to evaluate the impact of uncertainty in values of important hydrogeologic parameters including bulk anisotropy, boundary conditions and vertical discretization on the modelled heads and flow paths. Sensitivity analysis indicates which parameter or parameters have greater influence on the output.

The common method of sensitivity analysis is to manually estimate the rate of change in model output as a result of change in a certain parameter. This requires several model runs. The changes to the model output for all of the parameter changes may be displayed in tables or graphs for evaluation.

Automatic sensitivity analysis can also be done. In this method user intervention is not required. The automatic parameter adjustment algorithm uses parameter sensitivity to compute the parameter values that cause the model to best match observed heads and flows.

3.7 Prediction Runs

Prediction runs comprise those model simulations that provide the outputs to address the questions defined in the modelling objectives, which will involve changes to stresses and/or boundary conditions with respect to the calibration model.

The most important task in predictive simulation is the time length for which predictive simulation is required. For long-term prediction of groundwater responses to various assumed levels of extraction steady state models are particularly useful. If transient predictive scenarios are used it is essential to consider the model simulation time period and an appropriate time discretization for the calculations. When the predictive model duration substantially exceeds the period of transient calibration the uncertainty associated with the prediction increases. Therefore, it is suggested that the duration of predictive model runs should be less than five times the duration of the calibration (Barnett et al. 2012). But in cases such as mine dewatering scenarios the model needs to be run for the duration of the mining operations. The predictive analysis is followed by an analysis of the implications of the uncertainty associated with the modelling outputs.

3.8 Particle Tracking

Particle tracking is the determination of the path a non-reactive particle will take through a three-dimensional groundwater flow system in steady state. The particle tracking code is MODPATH (Pollock 1994) and is used in conjunction with MODFLOW. The determination of the paths of water in the flow system aids in conceptualizing and quantifying the sources of water in a modelled system. It also helps to determine advective transport of chemicals.

In particle tracking the user designate the location of a set of discrete particles of equal mass. The particles are then tracked through time, assuming they are transported by advection using the flow field computed by MODFLOW. Particles can be tracked either forward in time or backward in time. Particle tracking analyses are particularly useful for delineating capture zones or areas of influence for wells.

The disadvantages of particle tracking are: (i) do not directly account for retardation, dispersion or diffusion, and (ii) ignores the vadose zone.

4 Uncertainties in Model Prediction

The predictive results of groundwater modelling often deviate from the true values, which is attributed to the uncertainty of groundwater numerical simulation. This uncertainty may have adverse effect on management decision and policy. The sources of uncertainty are conceptual model, model parameters and observation data. The field hydrogeological conditions are often simplified in the conceptual model. Again, there may be uncertainties in assigning boundary conditions, aquifer disposition, recharge and discharge quantities etc. Therefore, uncertainties start from the conceptual model stage which is the foundation of the numerical model. Next, there are uncertainties in assigning the spatially variable field parameters such as hydraulic conductivity, specific yield, specific storage, etc. in the model. Thirdly, the field observation data such as head data have inherent error associated around their measurement. Therefore, all numerical models have uncertainty, in spite of much effort and money spent in the field. Uncertainty in model predictions for large model area are less compared to that of small model area because characterization methods are well-suited to discern bulk properties, and field observations directly reflect bulk system properties.

There are three approaches for estimating uncertainty (Barnett et al. 2012). They are:

(i) Quick uncertainty estimates: linear methods
(ii) Encompassing uncertainty estimates: non-linear methods
(iii) Other methods: ensemble, global and heuristic uncertainty estimation

Detailed description of guidelines and software tools that are available for groundwater uncertainty analysis are given by Doherty et al. (2010).

(i) Quick Uncertainty Estimates: Linear Methods The simplest and quickest method of linear uncertainty analysis is the Jacobian matrix. It is a matrix that relates the sensitivity of changes in model parameters to changes in model outputs. Model outputs are those for which field measurements are available. The uncertainty is considered linear because it assumes that the sensitivity calculated by the parameters specified and encapsulated in the Jacobian matrix applies for all possible values that the parameters might attain.

(ii) Encompassing Uncertainty Estimates: Non-linear Methods Non-linear methods for calculating uncertainty are attractive because the linearity restriction does not apply, but are computationally intensive. Amongst the non-linear methods, Monte Carlo analysis is widely used in uncertainty estimates. In this method, a probability distribution type is identified or assumed for each uncertain characteristic in the model. The model is then run repeatedly for pre-determined number of times, each

time with the value for each uncertain characteristic being randomly selected from the probability distribution. The result is a distribution of values for the model prediction. This distribution can be regarded as the prediction error, representing the combined effects of all uncertainties in characterizing the model (Thangarajan 2004).

(iii) Other Methods: Ensemble, Global and Heuristic Uncertainty Estimation Beven and Binley (1992) proposed the concept of equifinality in the uncertainty analysis of hydrological model and developed a generalized likelihood uncertainty estimation (GLUE) method based on regionalized sensitivity analysis. The key idea of GLUE method is getting rid of the framework of global optimal parameter solution. GLUE techniques evaluate the family of possible outcomes between equifinal models and assess how good the associated representations of uncertainty are. Bayesian model averaging (BMA) (Hoeting et al. 1999) provides an optimal way to combine the predictions of several competing models and to assess their joint predictive uncertainty. However, it tends to be computationally demanding and relies heavily on prior information about model parameters. In these methods the objective is to accept all the models that are 'behavioural' (i.e. that reproduce historic observations within some tolerance and contain features consistent with the conceptual model), rank them; and reject the ones that are not behavioural.

5 Case Studies

5.1 Impacts on Groundwater Recharge Areas of Megacity Pumping: Analysis of Potential Contamination of Kolkata, India, Water Supply (Sahu et al. 2013)

This work highlights that heavy pumping of groundwater beneath the megacity of Kolkata, India has altered the hydrological system and shifted the recharge area locations closer to the city centre where there are areas of known contamination. Starting from a conceptual model of the groundwater system of the southwest Bengal Basin, an anisotropic, homogeneous, steady-state groundwater flow model of the area was developed to evaluate the spatial distribution of piezometric heads and recharge areas under groundwater pre-development, current-development and possible future-development scenarios. The three-dimensional groundwater code MODFLOW (McDonald et al. 2000) was used to construct the model with the pre-processor ModflowGUI (Winston 2000) based on ArgusONE^{TM} commercial software. Groundwater flowpaths and travel times to wells were estimated by backward particle tracking from pumping areas of East Kolkata Wetlands and the city solid-waste dumping ground to recharge locations and also by forward tracking of particles to discharge locations under the three development scenarios.

Model calibration was done through trial-and-error method by adjusting the parameter values, i.e., stress, hydraulic conductivity and anisotropy, to arrive at a good fit by calculating the RMS error between the simulated head data and the measured field head data.

In order to evaluate the impact of uncertainty in the values of important hydrogeological parameters/conditions, including anisotropy, hydraulic conductivity and lateral boundary, on the modelled head and flow paths, sensitivity analyses were carried out for both pre-development and current-development conditions. The sensitivity of the model was tested by varying the parameter values/conditions from those of the base case for pre-development and current-development conditions and measuring the horizontal path length (hpl), i.e. the lateral distance from recharge to discharge, and travel time.

The study highlighted that (i) pumping has likely induced recharge from areas of known contamination that were initially areas of groundwater discharge, and (ii) the long-term development of the city's aquifer as a source of clean water for the residents may be in danger.

5.2 *Numerical Modelling of Groundwater Flow to Understand the Impacts of Pumping on Arsenic Migration in the Aquifer of North Bengal Plain (Sikdar and Chakraborty 2017)*

In this paper, numerical simulations of regional-scale groundwater flow of North Bengal Plain have been carried out to understand whether in the near future arsenic (As) will migrate to urban area of English Bazar block, Malda district, West Bengal, where large water table declines have taken place, from the adjacent As-polluted aquifer.

The groundwater flow model was run in steady state condition using 'base case' parameter values with a prescribed head layer everywhere at the elevation of the land surface. Sensitivity analysis was done to evaluate the impact of uncertainty in values of important hydrogeologic parameters.

Simulation indicates that current pumping which has resulted in a drawdown of >10 m in the urban area has significantly changed the groundwater flowpaths from pre-development condition. The drawdown due to pumping in the urban area extends to the rural areas and wells experience interference effect. The recharge areas of the urban area wells are As-free and hence the wells pump out water which contains <10 μg/L As. With increase of pumping rate the predicted flow field may spread over a wider area and hence may draw As-polluted water. At the present pumping rate, the wells of the urban area may remain uncontaminated for the foreseeable future, considering only pure advection of water. In the adjacent rural areas irrigation pumping is predominant, the groundwater is As-rich and the flow paths are near-vertical. If As-rich groundwater is applied for irrigation in the agricultural field, it will reach the well depths within 10 years and will continue to occur in the pumping wells. Therefore, flushing of As from the shallow aquifer may not take place and there will be recycling of As between the surface and the shallow pumping zone. Modelling also indicated that placing all the pumping wells at depths below 100 m may not provide As-free water permanently.

6 Conclusions

Numerical models are an excellent tool for understanding groundwater systems if limitations and sources of error are considered and accounted for. To begin a modelling process, firstly, the conceptual model needs to be built with adequate geological, hydrogeological and geophysical data such that it closely represents the actual groundwater system. In the next step the numerical model should be built by assigning boundary conditions, designing appropriate grids, and specifying hydrological properties to each of the cells of the grid. The model should then be calibrated and the error calculated. The model is now ready to be used to simulate impacts of human activities on the groundwater flow systems, to formulate alternative development scenarios, and to communicate the results to the public and decision makers.

It should be remembered that a model is not a reality and that a good model includes important features of reality. A good modeller should not overstretch a model and should explore the uncertainty of the predictions. It is recommended that model predictions should not be used to take management decisions until it is validated.

References

Anderson, M.P. and Woessner, W.W. (1992). Applied Groundwater Modelling –Simulation of Flow and Advective Transport. Academic Press, Inc., San Diego, Calif. 381 p.

Baalousha, H. (2008). Fundamental of Groundwater Modelling. *In:* Groundwater: Modelling, Management (Eds L.F. Konig and J.L. Weiss). Nova Science Publishers, Inc.

Barnett, B., Townley, L.R., Post, V., Evans, R.E., Hunt, R.J., Peeters, L., Richardson, S., Werner, A.D., Knapton, A. and Boronkay, A. (2012). Australian groundwater modelling guidelines. National Water Commission, Australia. 191 p.

Bear, J. (1972). Dynamics of fluids in porous media. Elsevier, New York, Amesterdam. 764 p.

Bear, J. (1979). Hydraulics of Groundwater. McGraw Hill, New York. 567 p.

Bear, J., Bejin, M.S. and Ross, R.R. (1992), Fundamentals of groundwater modelling, EPA-Groundwater Issue. US-EPA, Solid Waste and Emergency Response. Report No. EPA/540/S-92/005, April 1992.

Beven, K. and Binley, A. (1992). The future of distributed models: Model calibration and uncertainty prediction. *Hydrol Processes*, **6:** 279-298.

Bredehoeft, J.D. and Pinder, G.F. (1973). Mass transport in flowing groundwater. *Water Resou Research*, **9(1):** 194-210.

Cirpka, O. (1999). Numerical methods of groundwater flow and transport. Technical report. Stanford University, Department of Civil and Environmental Engineering.

Doherty, J.E., Hunt, R.J. and Tonkin, M.J. (2010). Approaches to highly parameterized inversion: A guide to using PEST for model-parameter and predictive-uncertainty analysis. US Geological Survey Scientific Investigations Report 2010–5211.

Franke, O.L., Reilly, T.E. and Bennett, G.D. (1987). Definition of boundary and initial conditions in the analysis of saturated groundwater flow systems – An introduction: Techniques of Water Resources Investigations of the United States Geological Survey, Book 3.

Freeze, R.A. and Cherry, J.A. (1979). Groundwater. Prentice Hall, Englewood Cliffs, N.J. 604 p.

Ghosh, N.C. and Sharma, K.D. (eds) (2006). Groundwater Modelling and Management. Capital Publishing Company, New Delhi, 634 p.

Harbaugh, A. and McDonald, M. (1996). User's documentation for MODFLOW-96: An update to the U.S. Geological Survey modular finite-difference groundwater flow model. U.S. Geological Survey Open-File Report 96-485.

Hoeting, J.A., Madigan, D., Raftery, A.E. and Volinsky, C.T. (1999). Bayesian model averaging: A tutorial. *Stat Sci*, **14(4):** 382-417.

Javandel, I. (1984). Methods of evaluating vertical groundwater movement. Technical Report of Lawrence Berkley Laboratory, University of California, Berkley. 134 p.

Konikow, L.F. and Grove, D.B. (1977). Derivation of equations describing solute transport in groundwater. U.S. Geol. Survey Water Resources Investigations, 77-19.

McDonald, M.G. et al. (2000). MODFLOW-2000, the US Geological Survey modular groundwater model: User guide to modularization concepts and the ground water flow process. US Geological Survey, Reston, VA. US Geological Survey Open-File Report 00-92.

Pinder, G. and Gray, W. (1977). Finite element simulation in surface and subsurface hydrology. Academic Press Inc. 295 p.

Pollock, D.W. (1994). User's guide for MODPATH/MODPATH-PLOT, Version 3: A particle tracking post-processing package for MODFLOW, the US Geological Survey finite-difference groundwater flow model. US Geological Survey, Reston, VA. USGS Open-File Report.

Rushton, K.R. and Redshaw, S.C. (1979). Seepage and Groundwater Flow. John Wiley, New York. 330 p.

Reddell, D.L. and Sunada, D.K. (1970). Numerical simulation of dispersion in groundwater aquifers. Colorado State Univ., Hydrology Paper 41.

Reilly, T. and Harbaugh, A. (2004). Guidelines for evaluating Groundwater Flow Models. Scientific Investigations Report 2004-5038. U.S. Department of Interior, U.S. Geological Survey.

Sahu, P., Holly, A., Michael, H.A., Voss, C.I. and Sikdar, P.K. (2013). Impacts on groundwater recharge areas of megacity pumping: Analysis of potential contamination of Kolkata, India, water supply. *Hydrolog Sc J*, **58(6):** 1340-1360.

Sikdar, P.K. and Chakraborty, S. (2017). Numerical Modelling of Groundwater Flow to Understand the Impacts of Pumping on Arsenic Migration in the Aquifer of North Bengal Plain. *J. Earth Syst. Sci.*, **126:** 29. DOI https://doi.org/10.1007/s12040-017-0799-x.

Thangarajan, M. (2004). Regional Groundwater Modelling. Capital Publishing Company, New Delhi. 340 p.

Trescott, P.C. and Larson, S.P. (1976). Finite Difference Model for Aquifer Simulation in Two-Dimensions with Results of Numerical Experiments. U.S. Geological Survey Techniques at Water-Resources Investigations, Book 7.

Winston, R.B. (2000). Graphical user interface for MODFLOW, Version 4. US Geological Survey, Reston, VA. US Geological Survey Open-File Report 00-315.

Chapter 8
Development and Management of Coastal Aquifer System Through Seawater Intrusion Modelling

Subrata Halder

1 Introduction

Sea is a dynamic entity which is often partly enclosed by land and coastal aquifers occupy these land areas in the vicinity of the sea. Coastal areas represent zones where land and sea meet and comprises variety of complex environments including deltas, estuaries, bays, marshes, dunes and beaches. Coastal aquifers have boundaries in contact with seawater and are always under dynamic equilibrium with it. Withdrawal of fresh ground water from these aquifers may result in inequilibrium resulting in intrusion of saline water in coastal aquifers. In many aquifers, the natural occurrence and movement of seawater i.e. salt water has been changed by the development of groundwater resources for human uses. Groundwater development has lowered water levels and caused salt water to intrude into many of the most productive coastal aquifers. To meet up the need for fresh groundwater supply to support coastal populations and economic prosperity is a great challenge for hydrologists, water resources developers and decision makers. A flexible mathematical model capable of predicting salt water concentration in the aquifer over a wide range of field conditions may be an important tool in groundwater management programme of coastal aquifers.

The coastal aquifers in different parts of the world form important sources of water supply. Some of the coastal areas are heavily urbanized leading to a greater demand for groundwater. Usually groundwater gradient is seaward and so a large amount of fresh groundwater is discharged into sea as (1) submarine springs, and (2) leakage through the aquifers and semi-confining layers exposed in the sea bottom (Singhal and Gupta 1999). Coastal aquifers have site-specific characteristics and

S. Halder (✉)
State Water Investigation Directorate, Department of Water Resources Investigation & Development, Government of West Bengal, Kolkata, West Bengal, India
e-mail: shalder_2007@yahoo.co.in

P. K. Sikdar (ed.), *Groundwater Development and Management*,
https://doi.org/10.1007/978-3-319-75115-3_8

differ from other aquifers in terms of characteristic parameters. Coastal hydrogeology has large amount of uncertainty due to two facts: (i) the dynamism of sea and hydrological principle of water flow and movement has linkages, which remains unanswered fully, and (ii) the hydro-geological science involved in such evaluation is dependent on the actual field conditions encountered and existing parametric variations, due to which it is difficult and cumbersome for precise evaluation.

In addition, coastal aquifers within the zone of influence of mean sea level (MSL) are threatened by an accelerated rise in global MSL. This rise in global MSL, 50 cm for the coming century as the present best estimate, could even more jeopardize vulnerable coastal aquifers than they are threatened today. Subsequently, the salinisation of coastal aquifers will accelerate. This could mean a reduction of fresh groundwater resources. The present capacity of the discharge systems in several coastal low-lands may be insufficient to cope with the excess of seepage water, especially in those coastal areas which are below MSL. This seepage will probably have a higher salinity than at present (Essink 2001).

2 Seawater Intrusion

Seawater intrusion is a natural process that occurs in virtually all coastal aquifers and limited to coastal areas only. It consists of salt water (approximately 35,000 mg/L TDS, which includes about 19,000 mg/L of chloride) inflow from sea towards freshwater aquifers and flowing inland. This inflow behaviour is caused by the fact that seawater has a higher density (due to carrying more solutes) than freshwater. An imaginary line called interface is developed all along which earmarks the salt-fresh water interaction. The first physical formulation of seawater intrusion was made by W.B. Ghyben in 1888 and A. Herzberg in 1901 which is known as Ghyben-Herzberg formulation. The Ghyben-Herzberg formulation states that for every metre of fresh water in an unconfined aquifer above sea level, there will be 40 m of fresh water in the aquifer below sea level. Salinity (higher TDS concentration), temperature, rock heterogeneities, solute transport (density dependent flow) and aquifer characteristics namely hydraulic conductivity (K) and transmissivity (T) are some key parameters that affect the performance of coastal aquifers and seawater intrusion phenomenon.

In some areas, coastal hydro-geologic condition is represented by an individual confined, unconfined or island aquifer system, whereas in other cases the hydrogeologic setting may be that of a multi-layer aquifer system. In either situation, the aquifer system has a sea front so that there is a direct contact between continental freshwater and marine saltwater. Besides differences in viscosity between these two fluids, there exists density changes as well, which depends mainly on undisturbed conditions, a seaward hydraulic gradient exist in the aquifer with freshwater discharging into sea. The heavier saltwater flows in from the sea and a wedge-shaped body of salt water develop beneath the lighter freshwater, with the freshwater thickness decreasing from the wedge towards the sea. The fresh-salt water interface

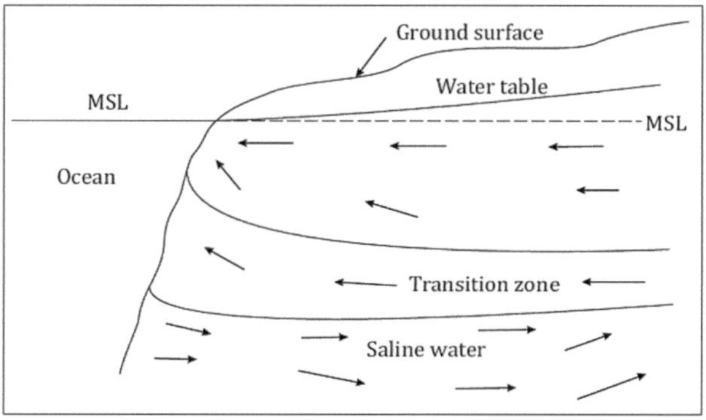

Fig. 8.1 Vertical cross-section showing flow patterns of fresh-salt water in an unconfined coastal aquifer. (*Source:* http://iwmi.dhigroup.com/solute_transport/saltwaterintrusion.html. Accessed 12 January 2017)

is stationary under steady state condition with its shape and position determined by the freshwater head and gradient (Fig. 8.1).

3 Coastal Groundwater Development Issues

Normally, saline water bodies owe their origin to entrapped sea water (connate water), seawater ingress, and leachates from navigation canals constructed along the coast, leachates from salt pans etc. In general, the following situations are encountered in coastal areas having multi-layered aquifers:

(i) Fresh water overlying saline water (in unconfined coastal aquifer)
(ii) Saline water aquifer (unconfined) overlying fresh water aquifer (confined)
(iii) Alternating sequence of fresh water and saline water aquifers (multi-layered)

The position of the fresh-salt water interface in coastal aquifer is essentially determined by the rate of outflow to sea. Any increase in the groundwater development would lead to reduction of the outflow rate and hence landwards and upwards movement of the interface. As a result the interface may encroach upon the screens of a few wells leading to their failure.

Thus in order to ensure a sustainable groundwater development pattern in a coastal aquifer, groundwater development may need to be planned both at regional scale and at the localized scale. Regional scale problems pertain to large areas in which the interface moves gradually upwards and in inland direction. The large scale movement of the interface is caused by a large scale modification of the outflow rate. Recognizing that the outflow rate in a given hydro-geologic scenario is determined by the recharge and pumpage pattern, planning at the regional scale can be conceptualized as designing such a regional pumping and recharge pattern, which keeps the interface between the freshwater and salt water at a pre-assigned large enough depth. Planning at the localized scale will imply the design of wells and pumping schedule (duration of pumping spell and the necessary rest period between two consecutive spells) to ensure salt-free fresh water (Sharma 2006).

4 Regional Mathematical Models

A geo-hydrologic system can be described by its hydraulic properties and associated boundary conditions. The steps involved in mathematical modelling are (Karanth 1987):

1. Examination of physical problems and understanding the physical behaviour of the system in relation to cause and effect to formulate a conceptual model of how the system operates.
2. Translation of physical problem into mathematical terms, making appropriate simplifying assumptions and developing governing equations.
3. Calibration and verification for reliability, and application of the mathematical results in terms of physical problems that require solution.

Seawater Intrusion Modeling (SIM) can be applied for the coastal aquifers with the two main objectives:

(i) Variable density flow (between fresh water and seawater) modelling in sub-surface, and
(ii) Optimization of pumping rates and well placement in coastal aquifers for freshwater extraction and controlled seawater intrusion.

The goal of the first objective is to find out the unique situation of fresh groundwater supply that coastal aquifers have and the second objective is to maximize freshwater extraction and to minimize the distance between wells and coastline as site specific criteria. Thus the main aim of the modelling approach is to recommend a viable techno-economic management plan for sustainable development of the groundwater resources from coastal aquifers without jeopardizing the coastal ecology and environment.

To combat and mitigate the problem of seawater intrusion at regional scale, various alternative strategies may be employed as follows:

- Modification in pumping and recharge patterns (including re-location of wells)
- Construction of physical barriers
- Construction of injection barriers and/or extraction barriers
- Tapping alternative aquifers etc.

In order to gauge the feasibility of a particular strategy and its relative effectiveness, a prior knowledge of immediate and long-term response of the coastal aquifers to the stress is vital. This can be accomplished by mathematical modelling of the coastal aquifer system. A mathematical model of a coastal aquifer essentially involves a coupled solution of differential equations governing flow and mass transport. In practice some assumptions may be made for simplifying the mathematics without sacrificing the predictive capacity of the model (Sharma 2006).

The mathematical analysis of the variable density salt water intrusion problem at regional scale may involve several simplifying assumptions depending upon whether the freshwater and saltwater are taken as miscible or immiscible fluids.

Therefore, there are two distinct approaches to model coastal aquifer system namely, (i) Sharp Interface Approach (SIA) (considering freshwater and seawater as immiscible fluids separated by an abrupt interface and width of fresh-seawater transition zone is small relative to thickness of the aquifer), and (ii) Miscible Transport Approach (MTA) (considering freshwater and seawater as miscible fluids and fresh-seawater transition zone is wide, which accounts for the effects of hydrodynamic dispersion) (Rastogi 2007).

5 Mathematical Description of Modelling

Numerical models based on Sharp Interface Approach account only for advective transport of salt water while those based on Miscible Transport Approach account for both advection and hydro-dynamic dispersion and explicitly represent the disperse interface.

5.1 *Sharp Interface Model*

The Sharp Interface Model requires simultaneous solution of freshwater and saltwater flow equations coupled by the boundary conditions that specific discharge and pressure must be equal on either side of sharp interface.

The equation of flow in freshwater (f) region is (Bear 1979):

$$\begin{aligned} &\partial/\partial x\left(K_{xf}\partial h_f/\partial x\right) + \partial/\partial y\left(K_{yf}\partial h_f/\partial y\right) \\ &+\partial/\partial z\left(K_{zf}\partial h_f/\partial z\right) + q_f = S_{sf}\partial h_f/\partial t \end{aligned} \tag{8.1}$$

The equation of flow in salt water (s) region is (Bear 1979):

$$\begin{aligned} &\partial/\partial x(K_{xs}\partial h_s/\partial x) + \partial/\partial y\left(K_{ys}\partial h_s/\partial y\right) \\ &+\partial/\partial z(K_{zs}\partial h_s/\partial z) + q_s = S_{ss}\partial h_s/\partial t \end{aligned} \tag{8.2}$$

where h = Hydraulic head, K_x, K_y, K_z = Hydraulic conductivity in x, y and z directions, S_s = Specific storage and Q_s = Source/sink volume per unit volume of porous medium (positive for inflow).

The solution of equations (8.1) and (8.2) yields the spatial distribution of the freshwater head (h_f) and salt water head (h_s) at steady state. The corresponding position and the shape of the interface are then obtained using the famous Ghyben-Herzbag equation with the following assumptions:

Assumption 1: To reduce the dimensionality of the problem, seawater intrusion may be simulated in x-y (horizontal) plane invoking Dupit-Forcheimmer assumption i.e. equipotential lines are in vertical direction or flow is in horizontal direction.

The two-fluid approach, which accounts for the flow of both freshwater and salt water, may be further simplified by adopting the one-fluid approach taking saltwater to be stationary.

Assumption 2: Saltwater is taken to be stationary, thus making it possible to solve for freshwater flow only.

However, since the one fluid approach neglects the influence of salt water flow on the freshwater head distribution, it is suitable for reproducing the long-term responses only. The two-fluid approach is more appropriate for investigation of short-term responses.

Assumption 3: One fluid approach assumes the coastal aquifer as unconfined (although this theory is valid for confined aquifers), homogeneous and isotropic.

Assumption 4: The interface is sharp (freshwater and seawater is considered as immiscible at the interface).

Assumption 5: Static equilibrium and hydrostatic pressure distribution conditions prevail in the freshwater region with stationary salt water, which states that

$$z = -\left[\rho_f/\left(\rho_s - \rho_f\right)\right]h_f$$

For typical sea water conditions, let ρ_s = 1.025 gm/cm^3 and ρ_f = 1.025 gm/cm^3 so that

$$\begin{aligned} z &= -[1.00/(1.025 - 1.00)]h_f \\ z &= -40h_f \end{aligned} \tag{8.3}$$

where z = Elevation of interface (i.e. depth of fresh-salt water interface below mean sea level (msl) at any location), ρ_s, ρ_f = Mass density of salt water and freshwater and h_f = Hydraulic head of freshwater (i.e. elevation of water table above mean sea level (msl) at the same location).

This leads to a very popular thumb-rule (Ghyben 1889; Herzberg 1901 principle) which states that the depth of the salt-freshwater interface (below mean sea level) at any location is 40 times the height of the water table above it (Fig. 8.2).

The erroneous result in equation (8.3) is that the thickness of freshwater zone is represented as zero at the shore, where the elevation of water table is zero. This is because the Ghyben-Herzberg principle relates the head at the water table to the position of the interface. In 1940, Hubbert improved upon Ghyben-Herzberg principle by formulating as

$$\begin{aligned} z &= \left[\gamma_s/\left(\gamma_s - \gamma_f\right)\right]h_s - \left[\gamma_f/\left(\gamma_s - \gamma_f\right)\right]h_f \\ &= \left[\rho_s/\left(\rho_s - \rho_f\right)\right]h_s - \left[\rho_f/\left(\rho_s - \rho_f\right)\right]h_f \end{aligned} \tag{8.4}$$

where z = Elevation of interface, γ_s (= $\rho_s g$), γ_f (= $\rho_f g$) = Specific weights of salt water and freshwater respectively, ρ_s, ρ_f = Mass density of salt water and freshwater respectively, g = Acceleration due to gravity and h_s, h_f = Hydraulic head of saltwater and freshwater respectively.

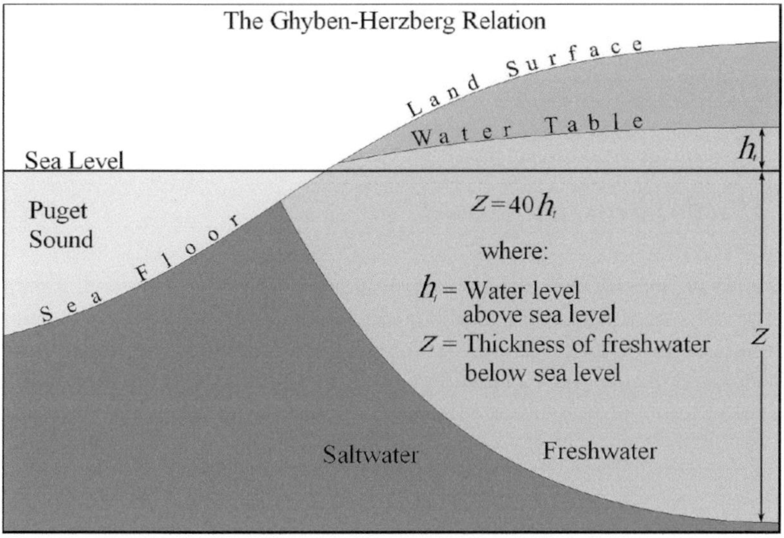

Fig. 8.2 Vertical cross-section showing flow patterns of fresh-salt water in an unconfined coastal aquifer. (*Source:* Island County Water Resources Management Plan 2005)

Hubbert's equation relates the freshwater head and salt water head at the point on the interface to its elevation. In 1959, Glover expresses the position of interface accounting the movement and discharge of freshwater from coastal aquifer under steady flow conditions as

$$z^2 - (2Q/\Delta\gamma K)x_0 - \left(Q^2/(\Delta\gamma)^2K^2\right) = 0 \tag{8.5}$$

The width gap (x_0) through which freshwater discharges into the sea as

$$x_0 = Q/2\Delta\gamma K \tag{8.6}$$

where Q = Freshwater flow per unit length of shore, K = Hydraulic conductivity, x = Distance from the shore, z = Depth from mean sea level and $\Delta\gamma = (\gamma_s - \gamma_f)/\gamma_f$

These analytical solutions are easier to program and use, and are very convenient for arriving at a quick initial estimate of the interface profile; but its use in field studies is restricted since the analytical solutions have been derived for homogeneous and isotropic aquifers. In field, complex boundary conditions exist and natural recharge and pumping are variable in space and time. To simulate such scenarios the governing partial differential equations need to be solved numerically. Most of these solutions are based on Finite Difference Method (FDM) and Finite Element Method (FEM) and pertain to transient cases. Though Sharp Interface Model cannot provide details concerning nature of the salt-fresh water transition zone but it represents the overall flow dynamics of a coastal aquifer system and can reproduce the general response of the interface to the applied hydraulic stresses. Therefore, Sharp Interface

Models are very useful for understanding the overall behaviour of the system before applying Miscible Transport Approach for examining the smaller scale effect (Sharma 2006).

5.2 Miscible Transport Model

Miscible Transport Model accounts for both advective and dispersive transport of saltwater. This model requires simultaneous solution of the coupled groundwater flow advective-dispersive equations accounting for dependence of density upon the salt concentration. For the variable density in time and space, the flow equation is expressed in terms of pressure as (Bear 1979):

$$\begin{aligned}&\partial/\partial x(k_x\gamma/\mu\,\partial p/\partial x)+\partial/\partial y\left(k_y\gamma/\mu\,\partial p/\partial y\right)\\&+\partial/\partial z[k_z\gamma/\mu\,(\partial p/\partial z)+\gamma]+q\gamma'=S_s\partial p/\partial t\\&+\theta(\partial r/\partial c)(\partial c/\partial t)\end{aligned}\tag{8.7}$$

where p = Fluid pressure, k_x, k_y, k_z = Intrinsic permeability in x, y and z direction, γ (= ρg) = Specific weight of fluid, μ = Dynamic viscosity of fluid, γ' = Specific weight of source/sink fluid, S_s = Specific storage coefficient, θ = Porosity, c = Solute concentration (mass of solute/volume of solvent), t = Time and q = Source/sink volume flux per unit volume of porous medium (+ve for inflow).

The advective-dispersive equation describing the transport of dissolved salt (assuming no absorption and no chemical reactions with solid matrix) (Bear 1979):

$$\begin{aligned}&\partial/\partial x\left[\theta\left(D_{xx}\partial c/\partial x+D_{xy}\partial c/\partial y+D_{xz}\partial c/\partial z\right]\right.\\&+\partial/\partial y\left[\theta\left(D_{yx}\partial c/\partial x+D_{yy}\partial c/\partial y+D_{yz}\partial c/\partial z\right]\right.\\&+\partial/\partial z\left[\theta\left(D_{zx}\partial c/\partial x+D_{zy}\partial c/\partial y+D_{zz}\partial c/\partial z\right]\right.\\&-v_x\partial c/\partial x-v_y\partial c/\partial y-v_z\partial c/\partial z+q(c'-c)=\theta(\partial c/\partial t)\end{aligned}\tag{8.8}$$

where D_{xx}, D_{xy}, $D_{xz}\ldots D_{zz}$ = Coefficients of hydrodynamic dispersion, $v_x = -\,k_x/\mu\ \partial p/\partial x$, $v_y = -\,k_y/\mu\ \partial p/\partial y$, $v_z = -\,k_z/\mu\ (\partial p/\partial z + \gamma)$ [Darcy's velocities in x, y and z directions] and c' = Solute concentrations in source/sink fluid.

The ability of the Miscible Transport model is to simulate the flow and miscible transport phenomena and also, position of the disperse interface. Moreover, through a Miscible Transport model, the amount of contaminant present in water arriving at supply wells through upconing of the interface can be estimated. Therefore, Miscible Transport approach is useful for designing wells and operation schedules in coastal aquifers.

6 Modelling Methodology

As regards with coastal areas, aquifer boundaries and heterogeneous hydrologic and structural properties of aquifers are extremely important from evaluation point of view. Assessment of geology of coastal aquifers considering geologic parameters becomes complex for scientific evolution. In the twentieth century, the higher computing power allowed the use of numerical methods, usually Finite Difference Method (FDM) or Finite Element Method (FEM) for salt-water intrusion studies as they are very easy to use and versatile.

Seawater Intrusion Modelling (SIM) has three steps to approach to the solutions: (1) specify Initial Conditions (IC) and Boundary Conditions (BC) for both the concentrations and fluid flow, (2) employ Sharp Interface Model (using governing PDEs for fresh water and seawater flows coupled by the BC that the specific discharge and pressures must be equal on either side of the sharp interface) and (3) Miscible Transport Model (using coupled groundwater flow advection-dispersion governing partial differential equations (PDE) with variable density in time and space).

A flexible mathematical model capable of predicting saltwater concentration in the aquifer over a wide range of field conditions constitutes an important tool in groundwater management programme of coastal aquifers. A wide variety of models are available for modelling of groundwater with variable densities as in the case of saltwater intrusion. A complete model to describe the saltwater intrusion should be three dimensional, transient and account for varied densities and dispersion. The recent development in modelling is the coupling of the model with a Geographical Information System (GIS) for the input data and presentation of output model (Kumar 2006). A computer program for groundwater modelling numerically solves a system (matrix) of algebraic equations. This matrix represents an approximation of the mathematical model formulated by Partial Differential Equations (PDE) of groundwater flow. Finite Difference Method (FDM) and Finite Element Method (FEM) are two common ways in which this approximation is made (Kresic 1997). The basic idea of these methods is to replace derivatives at a point by ratios of the changes in appropriate variables over a small but finite interval (Remson et al. 1971). The computations from these methods must be convergent and stable to give a result that has some sense close to the solution of the original problem.

SEAWAT, a computer program developed by the U.S. Geological Survey for simulation of three dimensional variable-density transient groundwater flows coupled with multi-species solute and heat transport can be used for the purpose of Seawater Intrusion Modelling (SIM). SEAWAT combines two widely used computer programs, a flow code (MODFLOW 2000) and a solute-transport code (MT3DMS). SEAWAT derives the governing equation for variable density groundwater flow in terms of equivalent freshwater head (Visual MODFLOW 2006). The equivalent freshwater head (h_f) at a point is defined as $h_f = (\rho_s/\rho_f)\ h_s + ((\rho_s - \rho_f)/\rho_f)z$ and the governing equation of groundwater flow in terms of fresh water head used in SEAWAT is (Guo and Langevin 2002):

$$\partial/\partial\alpha(\rho_s K_{f\alpha}[\partial h_f/\partial\alpha + ((\rho_s - \rho_f)/\rho_f)\partial z/\partial\alpha]) \\ +\partial/\partial\beta(\rho_s K_{f\beta}[\partial h_f/\partial\beta + ((\rho_s - \rho_f)/\rho_f)\partial z/\partial\beta]) \\ +\partial/\partial y(\rho_s K_{f\gamma}[\partial h_f/\partial\gamma + ((\rho_s - \rho_f)/\rho_f)\partial z/\partial\gamma]) = \rho_s S_{sf}\partial h_f/\partial t + \theta(\partial\rho_s/\partial c) \\ \times(\partial c/\partial t) - \widetilde{\rho}_s q_s \tag{8.9}$$

where α, β = Principal directions of permeability parallel to the bedding, γ = Directions of permeability normal to the bedding, $\widetilde{\rho}_s$ =Mass density of solute entering from a source or leaving through a sink, q_s = Source or Sink volume solute flux per unit volume of porous medium (positive for inflow), θ = Porosity and c = Solute concentration (mass of solute or volume of solvent).

Verification of a density dependent groundwater modelling code is necessary to check the validity of the formulation before it can be applied to real problem (Simpson and Clement 2001). The SEAWAT code was tested by simulating mainly four benchmark problems involving variable density groundwater flow:

1. **Box problem:** The purpose of Box problem was to verify that fluid velocities were properly calculated by SEAWAT. The original problem studied by Elder (1967) concerned a closed rectangular box modelled in cross section. The flow was initiated by a temperature gradient across the box and thermally induced gradients caused a complex pattern of fingering of denser water mixing in through the box (Simpson and Clement 2001). The initial conditions within the box consisted of a layer of freshwater overlying saltwater – a stable configuration for fluid density. There were two different cases of the box problem. For the first case, SEAWAT was tested by simulating groundwater flow within a two dimensional, vertical cross-sectional model with no flow boundaries on all sides. Longitudinal dispersivity values were set to a length similar in size to the length of a model cell, and the diffusion coefficient and transverse dispersivity value was set as zero. When the model was run with steady state conditions, the interface between fresh water and salt water remained in the same layer of the model. For the second case, horizontal flow was induced by specifying different, but hydrostatic constant heads on the left and right sides of box.
2. **Henry problem:** Henry problem was associated with groundwater flowing towards seawater boundary. The Henry problem was designed with (i) a constant flux of fresh groundwater with zero concentration applied to left side boundary of the box, (ii) a constant head boundary applied to the right side boundary of the box to represent seawater hydrostatic conditions, and (iii) zero flow applied to the upper and lower boundaries of the box.
3. **Elder problem:** Elder problem was used to verify variable-density groundwater codes. The Elder problem was designed with (i) a constant concentration boundary specified for a portion of the upper boundary and salt from the constant concentration boundary diffuses into the model domain and initiates complex vortices that redistribute the salt mass throughout the model, (ii) a constant concentration boundary with a value of zero specified for the lowest layer of the model, and (iii) two outlet cells with constant head values of zero specified for

the upper left and right boundaries. These constant head cells allow salt to diffuse into the model by providing an outlet for the fluid and salt mass.

4. **Hydrocoin problem:** The purpose of **HYDRO**logic **CO**de **IN**tercomparison (HYDROCOIN) was to evaluate the accuracy of selected groundwater modelling codes. The HYDROCOIN problem was designed with (i) a sloping pressure boundary imposed across the top of box that was surrounded on the sides and bottom by no flow conditions, and (ii) along the base of the middle part of the model, a constant concentration condition applied to represent the top of a salt dome. As groundwater flows along the bottom boundary and over salt dome, salt disperses into the system and collects in the lower right corner of the model domain (Guo and Langevin 2002).

Alternatively, **SUTRA**, a 3D groundwater **S**aturated-**U**nsaturated **TRA**nsport model developed by Voss (1984) can be used for Seawater Intrusion Modelling (SIM). It simulates the groundwater flow and solute transport in saturated aquifer and unsaturated vadose zone (Thangarajan 2004). SUTRA solves two partial differential equations – one for fluid flow and another for either concentration or temperature (Anderson et al. 1992).

The required databases to develop a conceptual groundwater flow model are as follows:

- Hydro-meteorological data (rainfall, evapotranspiration, maximum and minimum temperature, wind velocity etc.).
- Lithology of subsoil.
- Hydro-geological parameters (hydraulic conductivity, transmissivity, storage coefficient, porosity, leakage factor etc.).
- Groundwater level data at fixed locations in different time (phreatic and peizometric level).
- Salt concentrations (Total Dissolved Solids (TDS), chloride concentrations, Sodium concentrations etc.).
- Groundwater pumpage (location, rate and duration of individual pumping well).
- Natural recharge to groundwater (land covered area, rate of infiltration, coefficient of surface runoff etc.).
- Artificial recharge to groundwater (location, rate and duration of individual recharge structure).
- Relative sea level rise (present and predicted in future).

7 Utility of Model

Mathematical models provide a practical tool with which to understand the nature and severity of the salinity problem in a coastal aquifer and aid in arriving at feasible solutions. Thus it becomes possible for policy managers to frame a clear set of objectives for planning the future groundwater development of a coastal aquifer

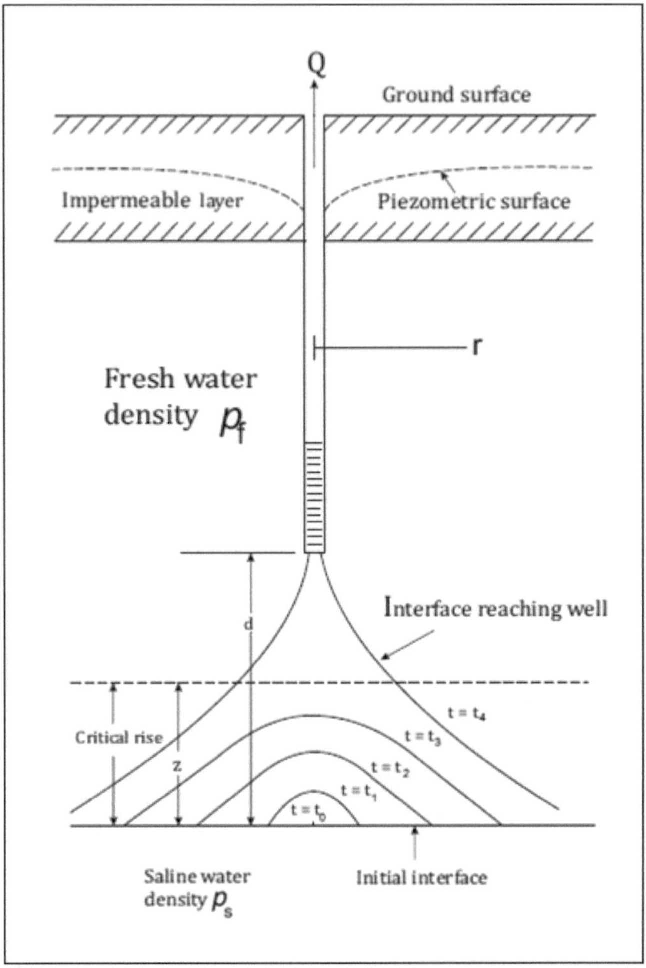

Fig. 8.3 Upcoming of saline groundwater under a pumping well (Essink 2001)

without any negative consequences, but at the same time meeting the demand of water supply. The possible outcomes of the model may be the following:

- Seawater intrusion in coastal aquifer may occur as a localized and regionalized problem.
- Current groundwater pumpage does or does not produce significant movement of interface.
- Saltwater does or does not enter pumping well as a result of upconing of interface within short or long period (Fig. 8.3).
- Extent of lateral migration of interface may be predicted for the later few years.

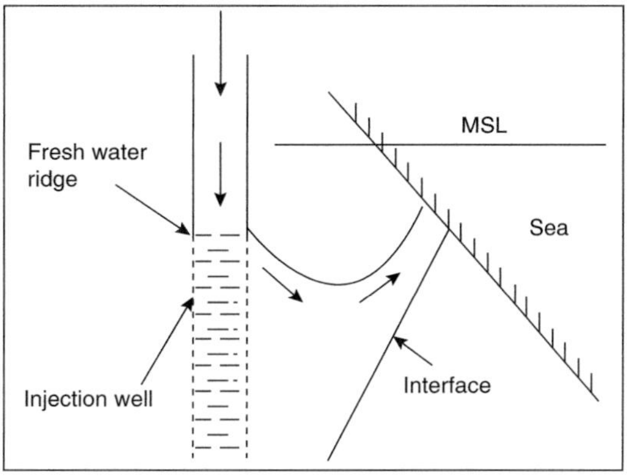

Fig. 8.4 Fresh water ridge along the sea coast formed by Injection well (Rastogi 2007)

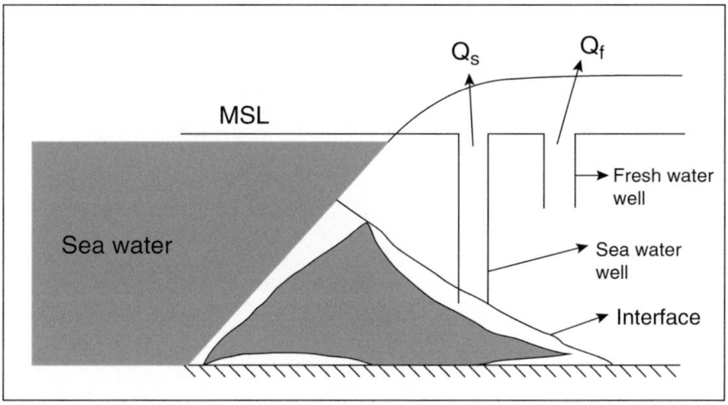

Fig. 8.5 Simultaneous withdrawals of sea water and fresh water to prevent seawater intrusion (Rastogi 2007)

Mathematical models can be used effectively to apply appropriate strategies to control seawater intrusion in coastal aquifers, which are as follows:

- Development of pumping trough adjacent to coastline by means of lines of pumping wells parallel to the coastline to create an external barrier for intruding saltwater and protect pumping wells further inland.
- Development of a pressure ridge adjacent to coastline by means of lines of recharge wells parallel to the coastline to create an injection barrier to prevent saltwater intrusion from sea into unaffected portion of the aquifer (Fig. 8.4).
- Increasing natural recharge by proper land use and land tillage practices in coastal areas.
- Lowering of salt-freshwater interface using coupled well pumping of both seawater and freshwater (Fig. 8.5).
- Checking upconing of salt-freshwater interface by dual pumping technique (Fig. 8.6).

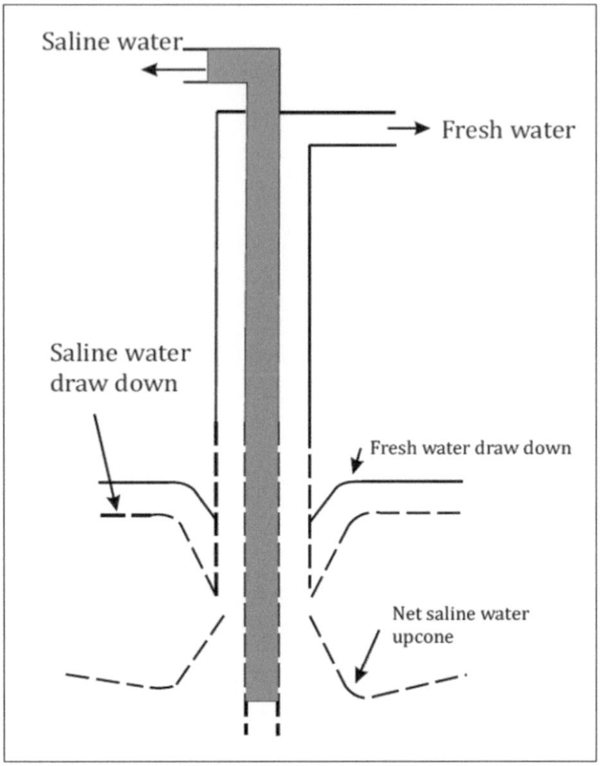

Fig. 8.6 A coupled well system to prevent salt water intrusion (Rastogi 2007)

- Construction of impermeable or semi-pervious subsurface physical barriers e.g. slurry of walls or sheet piles to prevent inflow of seawater.
- Opting for tapping alternative groundwater sources from multi-layered coastal aquifer system and then blending saltwater with fresh water to reduce its salinity up to the desired level
- Reduction of pumping rates in order not to exceed the sustainable freshwater yield.
- Modifying pumping schedule to control salt-freshwater interface upconing problem.
- Relocating pumping wells to reduce the intensity of pumping in a pocket of coastal area or relocation of production wells to areas that are less susceptible to intrusion.
- Coordinating for conjunctive use of groundwater and surface water to overcome excessive abstraction of groundwater from coastal aquifers.
- Treating saline water up to its tolerable limit using cost-effective treatment unit for use of it in drinking and irrigation purposes.

Implementation of any above alternative solutions should be based on evolution from the model. However, during decision making process the economic, environmental, social, technical, and political and other such issues of coastal area underground development should be given adequate considerations.

8 Case Studies

Elango and Sivakumar (2008) carried out a study in the coastal aquifer located south of the city of Chennai, India with the objective of developing a numerical model for this area in order to understand the behaviour of the system with the changes in hydrological stresses. The finite difference computer code MODFLOW with Groundwater Modelling System (GMS) was used to simulate the groundwater flow in this study. Simulation of groundwater flow by three-dimensional mathematical model indicated that increase in 10% pumping rate will lead to lowering of the groundwater head by 0.5–1.5 m and this would result in seawater intrusion in coastal regions of this aquifer system. The study indicated that the total abstraction from this aquifer had to be restricted to 4.25 mgd to prevent seawater intrusion.

Sarsak (2011) did his research on numerical simulation of seawater intrusion in response to climate change impacts in North Gaza coastal aquifer using SEAWAT. Various scenarios were simulated to study the impacts of climate change into seawater intrusion in the study area due to sea level rise, recharge and pumping rates variability. The results showed that the in-land movement for seawater intrusion for the reference scenario which reflected the continuation of the current situation was about 4200 m with a rate of 65 m/yr. The most critical extent of salinity was found in scenario with recharge –30% which causes inland intrusion movement of about 4500 m with a rate of 80 m/yr while the inland intrusion movement due to increasing pumping rates in scenario with pumping +30% was about 4300 m with a rate of 70 m/yr. The best results for the inland intrusion found in management scenario to solve the high salinity problems and water deficit in Gaza aquifer, was about 2900 m with a rate of 35 m/yr. As a result, seawater intrusion in the study area was very sensitive to recharge decrease as compared to pumping rates increase. As such, the most critical impact on seawater intrusion for the study area was recharge variability due to climate change.

9 Conclusions

Seawater Intrusion Model (SIM) gives a detailed dynamic behaviour of the saline-fresh groundwater interface and the extent of seawater ingress from shoreline in different time period. Thus a transient state density-dependent groundwater flow model should be developed to simulate regional flow patterns and general TDS concentration distributions in the coastal region, and to predict potential seawater intrusion. Also, optimized pumping rate and ideal location of pumping well can be obtained from this model, which is very useful to regulate groundwater abstraction from coastal aquifers restricting up-coning salt-freshwater interface, simultaneously lessening the seawater intrusion. The outcome of modelling can be used for prediction purpose by updating its input boundary conditions and hydrologic stresses.

As the water supply demand increases in coastal areas, groundwater contamination due to saltwater intrusion and coastal aquifer management will become more important. Modelling of groundwater flow with variable density is commonly required as a Decision Support System (DSS) for groundwater resource management in coastal areas, especially towards groundwater irrigated agricultural practices. Seawater Intrusion Modelling (SIM) evaluation outcomes may be linked with socio-economic parameters for effective groundwater management in coastal regions.

References

Anderson, M.P., Ward, D.S., Lappala, E.G. and Prickett, T.A. (1992). Computer models for subsurface water. *In:* Maidment, D.R. (ed.), Handbook of Hydrology. McGraw Hill, Inc., New York.

Bear, J. (1979). Hydraulics of groundwater. McGraw Hill, New York.

Elango, L. and Sivakumar, C. (2008). Regional simulation of a groundwater flow in coastal aquifer, Tamil Nadu, India. *In:* S. Ahmed, R. Jayakumar and A. Salih (es), Groundwater Dynamics in Hard Rock Aquifers. Springer, The Netherlands.

Elder, J.W. (1967). Transient convection in a porous medium. *J Fluid Mech*, **27:** 609-623.

Essink, G.H.P.O. (2001). Improving fresh groundwater supply: Problems and solutions. *Ocean & Coastal Management*, Elsevier, **44:** 429-449.

Ghyben (Baden Ghyben), W.; 1889. Nota in verband met de voorgenomen Putboring Naby Amsterdam:The Hague, Tydyschrift van het Koninkhyk Institute van Ingenieurs, p. 21

Guo, W. and Langevin, C.D. (2002). User's guide to SEAWAT: A computer programme for simulation of three-dimensional variable-density groundwater flow. US Geological Survey, Florida.

Herzberg, B.; 1901. Die Wasserversorgung einiger Nordsubader, Munch, J. *Gasbelenchtung und Wasserversorgung*, V. **44**, p. 815–819, 842–844.

Island County Water Resources Management Plan (2005). Seawater intrusion topic paper. http://www.islandcountyeh.org/Uploads/FKCEditor/file/TopicPaper%20SWI.pdf. Accessed 24 February 2017.

Karanth, K.R. (1987). Groundwater assessment development and management. Tata Mc-Graw Hill Publishing Co Ltd, India.

Kumar, C.P. (2006). Management of groundwater in salt water ingress coastal aquifers. *In:* Ghosh, N.C. and Sharma, K.D. (eds), Groundwater Modelling and Management. Capital Publishing Company, New Delhi.

Kresic, N. (1997). Quantitative solutions in hydrology and groundwater modeling. Lewis Publishers, New York.

Rastogi, A.K. (2007). Numerical groundwater hydrology. Penram International Publishing (India).

Remson, I., Hornberger, G.M. and Molz, F.J. (1971). Numerical methods in subsurface hydrology with an introduction to the finite element method. John Wiley and Sons.

Sharma, A. (2006). Freshwater-saltwater interaction in coastal aquifers. *In:* Ghosh, N.C., Sharma, K.D. (eds), Groundwater Modelling and Management. Capital Publishing Company, New Delhi.

Simpson, M.J. and Clement, T.P. (2001). Density dependent groundwater flow modeling: An evaluation of common benchmark problems. *In:* UNESCO-IHP sponsored international training workshop on modelling in hydrogeology (MODHYD). Allied Publishers Ltd, India.

Singhal, B.B.S. and Gupta, R.P. (1999). Applied hydrogeology of fractured rocks. Kluwer Academic Publishers, United Kingdom.

Sarsak, R.F. (2011). Numerical simulation of seawater intrusion in response to climate change impacts in North Gaza coastal aquifer using SEAWAT. Post Graduate Thesis, An-Najah National University, Nablus, Palestine.

Thangarajan, M. (2004). Regional groundwater modeling. Capital Publishing Company, New Delhi.

Visual MODFLOW (2006). User's Manual v4.2. Waterloo Hydrologic Inc., Canada.

Voss, C.I. (1984). SUTRA: A finite-element Simulation model for saturated-unsaturated fluid density-dependent groundwater flow with energy transport or chemically-reactive single species solute transport. U.S. Geological Survey Water-Resources Investigations Report 84-4369, 409 p. htpp://water.usgs.gov/nrp/gwsoftware/sutra.html.

Chapter 9
Spring Protection and Management: Context, History and Examples of Spring Management in India

Jared Buono

1 Introduction

Springs support hundreds of millions of people across India, providing safe, reliable water for domestic, agricultural and other purposes. But springs are in crisis as ecological degradation, climate change and increasing demand for water put pressure on these sensitive and limited natural resources. Improved management of springs is required to ensure water security for people in mountain regions, as well as downstream communities in the plains. In response to the crisis, a new approach to spring management is being adopted in many parts of the country. It combines the science of hydrogeology with local knowledge and participation to map the catchment area – the springshed – for targeted treatments such as restoration or protection. Results have been generally positive in terms of increased spring discharge and higher water quality. Local governance has also been seen to improve because decentralization of science and revival of traditional knowledge empowers local institutions. The springshed management approach has been scaled with public investment in several locations. This chapter describes this approach and highlights experiences in implementation from around the country. It also makes the case for the importance of springs and the need for protection, giving both context and history to the current state of spring management. Knowledge gaps and recommendations are discussed at the close of the chapter.

J. Buono (✉)
Eco-Hydrologist and Independent Consultant, Ruby, NY, USA
e-mail: jared.buono@gmail.com

P. K. Sikdar (ed.), *Groundwater Development and Management*,
https://doi.org/10.1007/978-3-319-75115-3_9

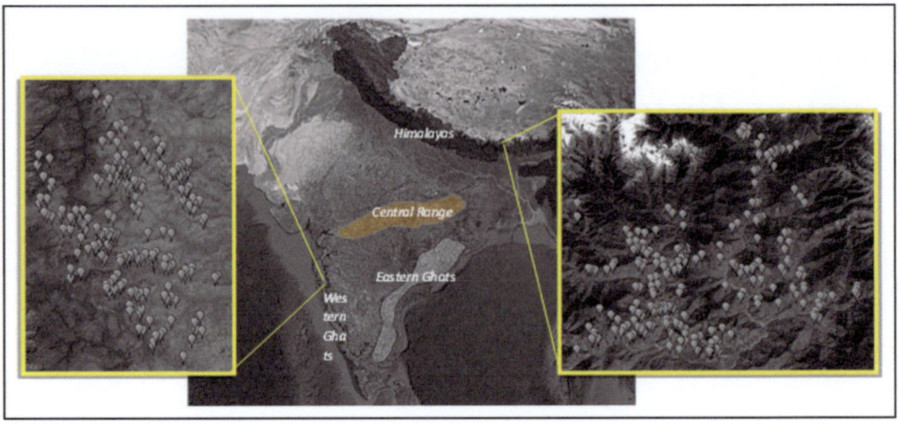

Fig. 9.1 Major mountain ranges and example locations showing typical spring density and spatial distribution in the Western Ghats on the left inset and the Himalayas on the right inset. Blue pins indicate the location of one or more springs at a given location – estimates from ground surveys range from 2 to 5 springs per square kilometre. Similar densities are found across most mountain regions leading to a national estimate of upwards of five million springs with 200 million people directly dependent. Data shown here are from Buono et al. (2016) on the left inset and www.sikkimsprings.org on the right inset

1.1 Springs in Context

Occurring where groundwater intersects the surface, springs have been used by communities across India for generations. From the Himalayas to the Niligris, springs support people and ecosystems providing a source of safe and reliable drinking water, as well as water for irrigation, livestock, culture, biodiversity and industry. In the Western Ghats, 120 million people rely directly on the mountain range's freshwater ecosystems for food, livelihoods and energy (Molur et al. 2011). Vital components of biodiversity and ecosystems, springs also impact the well-being of people outside of mountain ranges; in the Western Ghats downstream communities constitute almost 400 million people. Most of the rivers in central and south India are spring-derived including the Krishna, Godavari, and Cauvery. The Krishna, Govdari and Cauvery Rivers all originate as springs.

A reliance on springs can be found across the country's mountain ranges. In Uttarakhand, 90% of drinking water supply is spring based. In Meghalaya, it is estimated by the government that all villages in the state depend on one or more springs for drinking, irrigation or other purpose (Shabong and Swer 2015).

In total, it is estimated that there are as many as five million springs in India with 200 million people directly dependent on them (Fig. 9.1).

1.2 Springs Under Threat

Despite the widespread nature and importance of springs, they are in crisis. Spring discharge is declining due to groundwater pumping under increased demand, changing land use patterns, ecological degradation and a changing climate. A survey in Sikkim found that water production has declined in half of all springs in the state – a dangerous sign that aquifers are depleting in a state almost entirely dependent on springs for drinking water (Tambe and Arrawatia 2012).

Similar effects are being observed in mountain ranges across India. In the Western Ghats, communities are struggling to meet demand as once perennial springs are drying out in the hot season. Perennial springs that were places of worship are now dry as upstream wells compete for drinking and irrigation water (Buono et al. 2016). Exacerbating the problem, water quality also appears to be deteriorating under changing land use and improper sanitation. Springs, traditionally used as a village drinking water source, are being abandoned due to the proximity and contamination from latrines and dumping grounds.

Any reduction in water supply or water quality will have increasingly profound impacts not only on village drinking water, but also on livelihoods, health, biodiversity, agriculture, tourism, power generation, and industry – in mountain regions and the plains they support.

1.3 Institutional Response to Declining Springs

In response to the crisis, communities have been seeking solutions from government and NGO partners. Unfortunately, there is a lack of specialized knowledge about springs, and groundwater in general, across public and civil society. This has been hampering efforts to find sustainable water sources. In many instances, springs are being destroyed by desperate, misguided attempts to maximize water supply.

To date, the institutional response to a declining spring is largely to add or replace pipe and tank distribution systems with the goal of improving efficiency in the capture, transfer and storage of water. This usually means replacing the existing spring box with a larger, more efficient tank, as well as replacing any pipes and building for more water storage capacity structures such as tanks, ponds or open wells. While these improvements are often justified, and can lead to increases in water supply through conservation, they cannot increase spring discharge and often do little to improve water quality. In other words, they do not often address the problem or underlying cause of spring decline in terms of degradation of the catchment, aquifer and spring outlet. Without a greater understanding of where spring water comes from, local engineers are left with no choice but to try additional improvements – many of which are harmful.

For example, wells are being dug on or adjacent to spring sites. The idea being that a larger hole in the ground will allow more groundwater to discharge. But

Fig. 9.2 A spring converted to a spring-cum-well in the Western Ghats. The spring outlet is on the left behind the stone wall, and can be seen emerging into the well. This type of conversion from a groundwater to surface water source can lead to contamination of the water. The colour of the water is green which indicates algal growth. It is recommended that a springbox be built at the outlet so that groundwater can be harvested prior to any potential surface contamination, with the overflow collected in the open well for other uses. Also, as this is a contact spring type, there is no increase in water yield due to excavation of the well

digging below the level of the outlet in a contact-spring (defined as a spring that emerges near the contact of two geologic formations due to differences in transmissivity) will not result in increased spring discharge. And this converts a relatively safe groundwater source to an open water source prone to contamination (see Fig. 9.2). Many springs have also been blasted, excavated with machines or subjected to lateral bore wells to increase yields. Unfortunately, many of these methods end up destroying the springs. Springs are sensitive systems where excavation can lead to drying of the springs. And manipulation of the spring outlet leads to loss of habitat and biodiversity. These approaches ignore the fact that the spring's behaviour is governed by the aquifer and its catchment area.

1.4 An Emerging Model for Sustainable Management

There are successful models for sustainably managing springs and a new approach is gaining ground across the country. About a decade ago, NGOs in the Himalayas, as well as the Government of Sikkim, realized that sustainable spring management required moving beyond supply-side improvements to focus on partnering with communities to map and protect spring source areas (Kulkarni et al. 2015). Organizations with specialized knowledge in hydrogeology, such as the Advanced Centre for Water Resources Development and Management (ACWADAM), began providing technical and capacity-building support, particularly in hydrogeology and aquifer management. An approach evolved that shifted focus from the traditional ridge-to-valley model to a valley-to-valley model seeking to pinpoint spring recharge areas for targeted restoration and protection.

Results from these efforts have demonstrated significant increases in discharge and water quality. Increases in spring discharge by up to fivefold and fecal coliform reduced in drinking water supplies after social fencing. The approach also allows the decentralization of science and revival of traditional knowledge – hydrogeologic mapping is taken up by communities and local institutions fostering collective and participatory action.

The Central Himalayan Action and Research Group (CHIRAG) and People Science Institute (PSI) now have well established spring protection programmes in Uttarakhand and Himachal Pradesh. In other parts of the country over the last few years, similar initiatives were independently developed in the Eastern and Western Ghats. The Government of Sikkim and Meghalaya have initiated state-wide programmes to survey, monitor, protect and rejuvenate springs at scale. Similar spring protection programmes are now being emulated in at least ten states. Below is a description of the general approach of these programmes with examples and lessons from around the country.

2 Springshed Management – A Valley-to-Valley Approach

Sustainable management of springs means moving beyond supply-side, infrastructure improvements, to look at the larger picture of the water source and the community. One must recognize that springs are part of a freshwater ecosystem that includes the forests above, rivers below, and all the people, plants and animals in between.

A spring itself is comprised of multiple components including a catchment, aquifer and an outlet – each being distinct, but connected, part of the system that requires specific management considerations. The components of a spring system can be distantly separated with a spring outlet and its catchment occurring in different watersheds. Therefore, it is essential to employ hydrogeology to map the entire system, the springshed. This allows one to trace back, from the spring outlet, to the aquifer and to source of the water at the catchment and thus enables managers to restore and/or protect areas that have been degraded. Unlike traditional watershed ridge-to-valley approach, this requires a conceptual shift to valley-to-valley perspective (see Fig. 9.3).

It is also important to note that there are a variety of types of springs. There are karst type springs that form when naturally-acidic rainfall dissolves limestone to form caves and macropores capable of discharging large amounts of water. There are

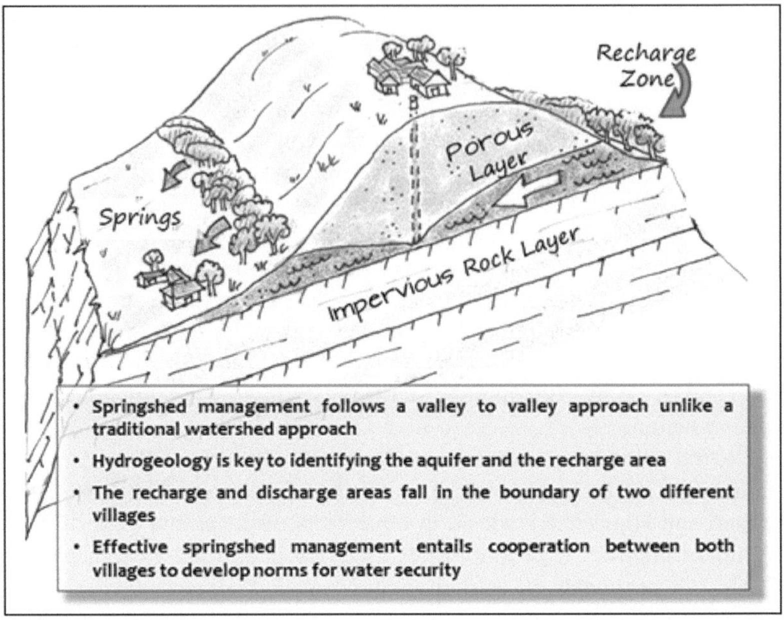

Fig. 9.3 A simplified hydrogeologic model of a spring showing that recharge occurs outside the watershed in a different village than where the springs emerge. This is typical of spring situations around the country

depression type springs that form where a break in the slope of the surface intersects groundwater. Other springs form along faults where water bearing rocks are displaced, or along cracks and fissures of a fracture zone. And there are contact springs which form at the boundary between different rock types – such as the one in Fig. 9.3 (Mahamuni and Kulkarni 2012). These classifications are largely based on the structural mechanism causing the spring to emerge, but they can also tell us about the amount, timing and duration of water discharge, the aquifer and how to map and manage a spring. For example, finding many springs along a contour line of a mountain may indicate a contact type spring. This information can tell you what rock formation constitutes the aquifer, and where to protect the recharge area.

Springsheds are often also separated administratively with an aquifer spanning multiple villages. One community accessing a spring may not have rights or permission to the catchment area. Hence, community participation is key to sustainable management. The science of hydrogeology needs to be open-sourced to show the interconnectedness of the water resource between stakeholders. Local knowledge needs to be leveraged and capacity built for applying scientific management in the long term. And local institutions need to be made a part of the process to inspire cooperation between communities sharing the common resource.

In addition, scientific decision making requires data and information, thus the importance of spring monitoring. This represents another opportunity for community engagement as participatory data collection and citizen science is employed.

The springshed management project cycle generally includes the following steps:

1. Work with community members to identify spring issues using traditional community-based methods; initiate participatory monitoring and begin training para-professionals.
2. Conduct hydrogeologic investigation to map the springshed, involving community in all aspects to demystify the science and incorporate local knowledge.
3. With a scientific understanding of the springshed, define collective management objectives; identify target areas for restoration, protection and management in community action plan.
4. Community-led implementation of interventions; continue participatory monitoring.
5. Evaluate impact; revisit objectives if needed; transfer monitoring and management to local institutions.

2.1 Mapping a Springshed

One of the most important aspects of spring management is hydrogeologic survey and mapping of the springshed. This is also one of the most challenging aspects because there is a general lack of understanding and expertise in hydrogeology across stakeholders. Yet, it has been shown that these skills can be learned and employed by almost anyone (see Section para-professionals below).

The field mapping process generally starts with a ground survey of the spring and surrounding area. Rock type and structures are mapped, including strike and dip of formations and features, and a geologic cross section is developed. Infrastructure, socioeconomic and other environmental data are collected. Hydrologic monitoring of discharge and water quality is initiated. Google earth satellite imagery, topo maps or other sources are used to delineate the local watersheds and the field survey information is placed on the map. This provides the first draft of a conceptual model of the spring and provides information about the aquifer and the rough location of the catchment and recharge areas.

A more specific description of exactly how a springshed is mapped is warranted because of the dearth of available resources on this topic. The North-Western Ghats mountain range, along the Deccan Plateau, represents a good case study for the mapping process. Rocks in this area are mainly horizontal beds of basalt. Each bed, or layer, represents a different period of volcanic eruptions resulting in extensive lava flows many metres thick. These rock layers vary in their ability to store and transmit water due to differences in cooling time, rock chemistry, etc. (Kulkarni et al. 2000). The result is a layer cake landscape characterized by broad flat plateau mountains and wide valleys. Monsoon rains absorbed by the more porous basalt

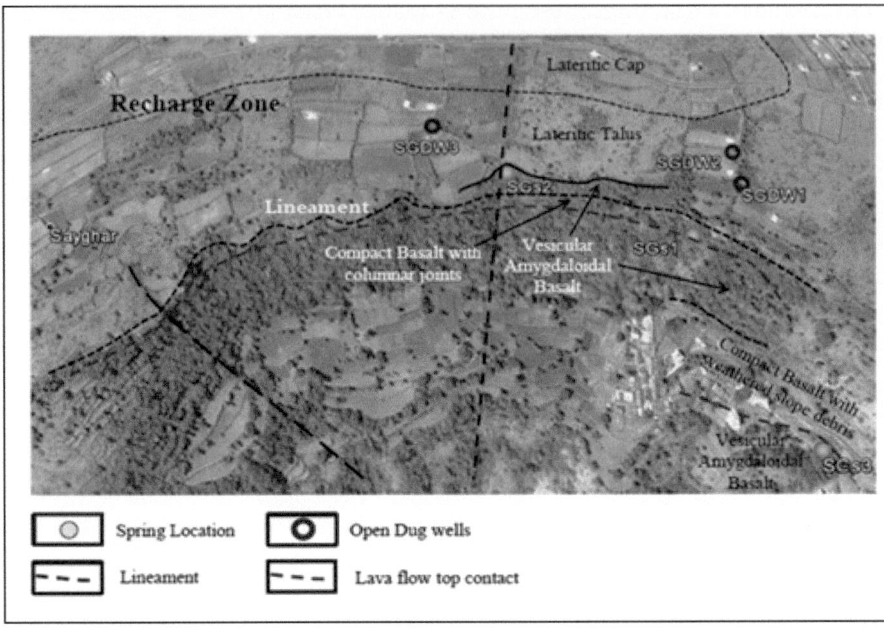

Fig. 9.4 A satellite image showing geologic field observations and compiled on Google Earth. Important features such as contact between rock formations, fracture systems, and the location of wells and springs are included. To the trained eye, these allow a landscape-scale understanding of the hydrogeology and help inform management decisions

layers percolates down until encountering an impervious layer – the water then moves laterally where it may emerge on the side of a mountain as a spring. There are also occasional lineaments, or landscape-scale fractures in the rocks that can transmit water between basalt layers. With this basic geologic understanding, it is possible to map springsheds in this area by surveying the rock type and structural features.

For example, field surveys generally begin at the spring outlet, describing visible features. Transects are walked uphill, as well as across the hillslope, to map contacts between rock formations and presence of any obvious fractures or local joints in the basalt. Satellite imagery from Google can be used to see landscape-scale features not visible from the ground. Other information is incorporated such as landscape position of springs. For example, springs found adjacent to each other on the contour tends to mean a contact spring system at the intersection of two different basalt layers, while springs found adjacent to each other up-and-down a valley likely indicate a fracture-type spring system. These observations are marked on satellite images and or toposheets to give a landscape-scale understanding of the hydrogeology (see Fig. 9.4). Toposheets can then be used to make vertical profiles showing the landscape in cross-section.

While there are much more rigorous and precise means to identify and map spring sources, such as dye or radio isotopes, this approach is sufficient for most applications when applied by a trained and experienced field team.

2.2 *Para-Professionals and Capacity Building in Hydrogeology*

The process of hydrogeologic mapping is quite specialized and requires a fundamental understanding of basic geology and hydrology. For example, one must identify the general rock types in the area, at least relative to their ability to store and transmit water. In addition, the feature and structures of the rock need to be mapped, relative to the topography and location of the spring. These are complex concepts, particularly for those without geosciences training, but it has been shown that a basic understanding can be conveyed after a few short training sessions to local community members.

This is, in fact, an essential component of the participatory process. Panchayat Raj members, Self Help Groups, and individual community members help locate springs that were traditionally used by their villages and are encouraged to map geology and monitor hydrology. The survey and discovery process is open to all and the hydrogeology is shared. This 'demystification' of the science provides a powerful tool for stakeholder engagement. This, along with community based monitoring of discharge and water quality are used to transfer long-term ownership of the project to the community. Communities are trained in measuring discharge, rainfall and sometimes even water quality parameters such as fecal coliform (using H_2S field testers or similar).

In many communities, para-professional hydrogeologists have emerged from these exercises. They are using their newfound skills to help adjacent communities better manage their water, and even training government or NGO staff. Farmers, ladies' groups and last-mile workers have all taken up this work. Several organizations now provide specialized training in spring mapping and management including ACWADAM in Pune and PSI in Uttarakhand.

2.3 *Community Action Plans*

Once the springshed has been delineated, management goals are established and a plan is developed. Action plans may include restoration and protection of the springshed, or even augmentation of groundwater recharge. For example, if the community is facing dry-season water shortages, the goal would be increased discharge. If the springshed was found to be affected by intense grazing, tree cutting and annual burning, then the plan would be to reduce the human pressures in the springshed. Secondarily, restoration best practices may be used to restore ecological function – i.e. plantation of native species to increase soil cover. In addition, structures such as chaals (traditional ponds dug for rainwater harvesting) or staggered trenches may be used to increase natural rates of infiltration. It is important to keep in mind all the components of the springshed to inform management. For example, the amount, quality and seasonality of spring discharge are all functions of the type and size of aquifer and its catchment area. Degradation of the catchment area tends to reduce infiltration of rainfall in turn reducing the spring discharge. This can even lead to a perennial spring becoming seasonal. Degradation of the outlet of the spring tends to impact biodiversity, habitat and water quality.

2.4 Monitoring

Spring monitoring is essential for making science-based decisions. Collecting data not only allows one to track potential changes in the spring over time, and to assess the impacts of the management programme, but it can also reveal hydrogeologic information important to managing the spring. Monitoring also represents an opportunity to engage the local community through participatory data collection and even citizen science.

The basic data that should be collected includes spring discharge, rainfall and water quality parameters. Discharge is generally collected via time-volumetric sampling and reported in litres per minute. It is collected at least once a month, at roughly the same hour and day within the month (e.g. at 11 am on the 1st of every month). After 12 consecutive months, this provides a rough annual hydrograph and is the basis for estimating total annual discharge, and for planning water consumption during low flows, etc. Ideally, however, more frequent measurements of discharge would be conducted. For example, the transition between seasonal low flows and peak flows, generally occurring at the onset of the monsoon, should be sampled at more frequent interval such as on a weekly basis. The reason being is that the rising limb of the hydrograph can tell you about the aquifer governing the spring. If the hydrograph is 'peaky' and responsive to rainfall events the aquifer may contain macropores such that there is low storativity and short residence; the latter of which may affect water quality where short residence times have been correlated with lower dissolved solids (Naik et al. 2002). In contrast, a relatively flat annual hydrograph that shows no response to rainfall events may indicate a larger, regional aquifer with a potentially greater residence time. Multiple years of monthly discharge data should be collected to ascertain larger processes such as impacts of climate, etc.

In terms of participatory data collection, measuring discharge can be a challenge for most farmers in that it involves conversion between units (i.e. going from the seconds it took to fill a 5-litre vessel into the standard of litres per minute can be confusing). That said, many communities have taken up monthly, and even weekly monitoring of discharge.

Rainfall should be collected near to the spring catchment, on an event basis. Simple, non-recording rain gauges are adequate. Rainfall is generally summed by month and compared to discharge, even though some information may be lost in terms of rainfall-runoff response. Rainfall is an excellent parameter to incorporate into participatory data collection schemes as the concept is straight forward, does not require conversions, and is fun for most people to do. It is recommended that several inexpensive rain gauges be spread across individual within a springshed whenever possible so that human error will be reduced during averaging.

Water quality should be monitored whenever discharge is recorded; however, this may depend on the management goals of the project and the parameters under consideration. The constituents for baseline monitoring include pH, temperature, total dissolved solids (TDS), and potentially salinity and/or electrical conductivity (EC) and fecal coliform presence/absence. These can be readily measured with relatively inexpensive field kits and instrumentation, and hence should be collected frequently. Others, such as arsenic, fluoride and nitrogen that may require laboratory testing may not be feasible monthly.

In terms of participatory collection, the baseline parameters measured by field instruments are generally a good way for people to engage and take ownership of spring water quality. Everyone feels like an engineer when using an electronic pH meter. And simple testing of fecal coliform presence, when conducted using an H_2S bottle for example, is useful because positive samples turn black and enforce a feeling of disgust.

3 Efficacy and Examples

3.1 Impacts on Water Quantity and Quality

Evidence suggests that it is possible to restore and even augment spring discharge. In Sikkim, the government took up 50 pilot springs starting in 2008 under its Dhara Vikas programme. Communities developed Village Water Security Plans that focused on protection and restoration of springsheds. This involved restoration or placement of traditional chaals (rainwater ponds) in the upper springshed to increase infiltration and recharge. The results were positive with some springs yielding increased discharge of up to fivefold (Norbu 2015). While it is likely that some springs cannot be augmented in terms of recharge, due to natural characteristics of their aquifers, the results in Sikkim were strong enough that this has now been taken up at the state level.

In Uttarakhand and Himachal Pradesh, NGOs such as CHIRAG and People Science Institute (PSI) have well established spring programmes yielding excellent results. A typical example of impact on spring discharge is from a PSI project in Himachal Pradesh. Multiple springs were monitored and treated over the course of three monsoon seasons providing paired-springshed and before-and-after evaluations. After mapping the hydrogeology and identifying catchment areas, an average of seven hectares was treated for each spring. To increase spring discharge, the community implemented plantation and infiltration basin measures. Increases in discharge can be seen in the treated watershed for all seasons of monitoring post-intervention (see Fig. 9.5). Even more compelling is that discharge remained

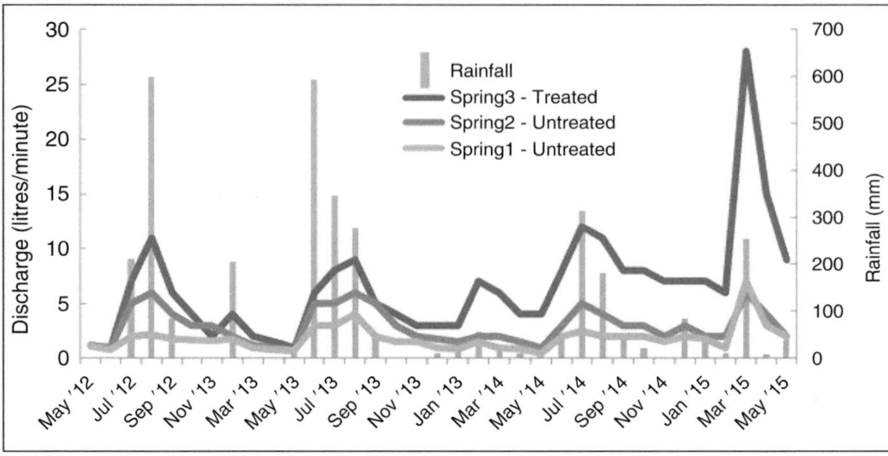

Fig. 9.5 Hydrographs and rainfall over three seasons showing the impact of springshed management on water discharge. Increases in total, peak and lean-season discharges can be seen on the treated spring

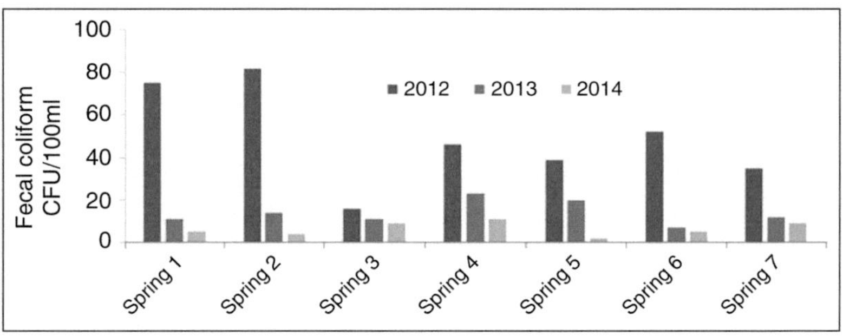

Fig. 9.6 Reduction in fecal coliform after social fencing and other interventions in a springshed. (From a project conducted by PSI)

relatively high despite significantly lower rainfall in 2012 as compared to 2010 (Bisht and Mahamuni 2015). This suggests that drought mitigation and climate adaptation may be possible for springs.

In addition to increasing spring water discharge, springshed management can also improve water quality. In the same project by PSI above, water quality was determined to be an issue for the drinking water in several springs. The solution was social fencing to reduce open defecation in the springshed to reduce microbial load. Figure 9.6 shows that for the 2 years post-intervention, fecal coliform significantly declined across seven treated springs based on testing of the water supply (Sharma and Kumar 2015).

3.2 *Efforts at Scale*

The Sikkim government began a state-wide springshed management programme in 2008. After piloting 50 springs with significant results, they began scaling up and have since mapped over 1000 springs and have initiated Village Water Security Plans. Over the last 5 years, it is estimated that over 900 million litres of water have been recharged annually (Norbu 2015). This work involved a one-time investment of one paisa per litre. A key element in this success was the partnerships the government built between departments and with other stakeholders such as NGOs, academic institutions and donors, for knowledge transfer and training in hydrogeology.

More recently, other state governments are showing interest in this approach. Meghalaya has trained hundreds of government field staff and volunteers. Thousands of springs have been inventoried and there are plans to protect springs in nearly all villages in the state – the first such activity at this scale anywhere in the nation (Shabong and Swer 2015).

In the Eastern Ghats, Visakha Jilla Nava Nirmana Samithi (Siva Kumar 2015) has demonstrated that gravity-fed spring water supply and village-maintained filtration systems can provide safe and sustainable drinking water – a programme now being replicated in hundreds of villages by the Andhra Pradesh government (Siva Kumar 2015).

3.3 *Challenges*

There have also been challenges. Springs used by one village are often recharged in another village or occur on Forest Department land – complicating management measures. There is also a lack of knowledge on groundwater and springs in particular across all stakeholders. There are not enough trainers to meet the current demand for springs-related planning and capacity building – something that the Springs Initiative is addressing through individual efforts, such as ACWADAM's 2 week training on groundwater, and PSI's 2-week training on Participatory Ground Water Management, and through group efforts such as development of a pan-India spring management curriculum.

3.4 *Knowledge Gaps and Research*

There are several knowledge gaps that need to be examined to improve spring management and policy. The total number of springs, and how many people rely on them, remains largely unknown. And the downward trend in water quality and quantity experienced by communities around the country may be part of a larger, long-term trend, or not; because there is no database of spring conditions we simply do not know for certain. Below are three areas of research that need to be explored.

3.4.1 Spring Density

No one really knows how many springs are there and exactly how many people they support because there have been no large-scale surveys at the national or mountain range level. The numbers presented here, of five million springs supporting roughly 200 million people, are based on an extrapolation from local spring density surveys and efforts. It is typical to find between two and five springs per square kilometre in many parts of India's various mountain ranges. Multiplied by the spatial extent of the mountain ranges, this provides an estimate of up to five million springs within the country. These numbers are based on the assembled knowledge and experience of the author and the many other organizations and government agencies who have been working to protect springs for the past decade. While this is a gross estimate, it provides the first rough calculation of spring density and the total number of springs in the country. However, more rigorous and comprehensive surveys would be required to derive spring density numbers with any degree of precision.

3.4.2 Population Dependent on Springs

The number of people dependent on springs is similarly estimated based on the spatial extent of the mountain ranges. Governments in mountain states have been exploring this in more detail as interest in springs grows. For example, approximately 90% of drinking water sources in Uttarakhand are spring based. In Meghalaya, close to 100% of drinking water sources come from springs. Therefore, it can be said that nearly the entire population of those states is directly dependent on springs. And this is just for drinking water related uses, if other uses are considered, such as agriculture, industry, culture and tourism, then the number of dependent people increases further. Many more people are indirectly dependent on springs, particularly outside of mountain ranges. In the Western Ghats for example, 120 million people are directly dependent on freshwater ecosystems for food, livelihoods, energy and other services, while up to 400 million people indirectly depend on those same freshwater ecosystems downstream. All the rivers in South and Central India originate as springs in the Western and Eastern Ghats mountain ranges. When springs are degraded, what happens to base flows, ecological flows of river systems? What happens to biodiversity, food webs and fisheries? And the communities that depend on them?

4 Conclusions

While the results above are encouraging, springs remain off-radar for the vast majority of stakeholders, from regulatory agencies, to engineers, research institutions, gram panchayats, and last mile workers. Greater awareness and understanding of springs will improve management of aquifers and groundwater in general. Springs also represent an opportunity. They are the nexus between groundwater and surface water, a growing area of research. They also improve monitoring due to ease of access. And in many places, infrastructure such as tanks and pipes is already in place due to historic use – something that facilitates monitoring. Finally, springs are the proverbial canary in the coal mine for the mountain aquifers that feed rivers and support hundreds of millions of people downstream. Springs should therefore be a national priority in terms of research, policy and implementation. At the very least, improved methods for managing springs should be disseminated far and wide.

References

Bisht, B.K. and Mahamuni, K. (2015). CHIRAG's Spring Protection Programme. PowerPoint presented at 3rd Annual Meeting of the Springs Initiative, Bhimtal, Uttarakhand, May 15, 2015.

Buono, J., Thomas, R., Kulkarni, H., Mahamuni, K., Ambrale, D., More, A., Karandikar, M., Ghate, K. and Kulkarni, K. (2016). Ecohydrologic Description of Springsheds in the North-western Ghats, Pune, Maharashtra. J Ecolog Soc, 28**:** 2015–2016.

Kulkarni, H., Shah, M. and Vijay Shankar, P.S. (2015). Shaping the Contours of Groundwater Governance in India. *J Hydro*, Part A, 4**:** 172–192.

Kulkarni, H., Deolankar, S.B., Lalwani, A., Joseph, B. and Pawar, S. (2000). Hydrogeological Framework of the Deccan Basalt Groundwater Systems, West-Central India, 8(4)**:** 368–378.

Mahamuni, K. and Kulkarni, H. (2012). Groundwater Resources and Spring Hydrogeology in South Sikkim, with Special Reference to Climate Change. *In:* Arrawatia, M.L. and Tambe, S. (eds), Climate Change in Sikkim: Patterns, Impacts and Initiatives. Government of Sikkim, Gangtok, Sikkim.

Molur, S., Smith, K.G., Daniel, B.A. and Darwall, W.R.T. (Compilers) (2011). The status and distribution of freshwater biodiversity in the Western Ghats, India. IUCN, Cambridge, UK and Gland, Switzerland and Zoo Outreach Organisation, Coimbatore, India.

Naik, P.K., Awasthi, A.K. and Mohan, P.C. (2002). Springs in a Headwater Basin in the Deccan Trap Country of the Western Ghats, India. Hydrogeol J, 10**:** 553–565.

Norbu, P. (2015). Dara Vikas, Sikkim. PowerPoint presented at 3rd Annual Meeting of the Springs Initiative, Bhimtal, Uttarakhand, May 15, 2015.

Shabong, L. and Swer, A.A. (2015). Meghalaya's Spring Initiative under IBDLP. PowerPoint presented at 3rd Annual Meeting of the Springs Initiative, Bhimtal, Uttarakhand, May 15, 2015.

Sharma, S. and Kumar, A. (2015). People Science Institute's Spring Protection Programme. PowerPoint presented at 3rd Annual Meeting of the Springs Initiative, Bhimtal, Uttarakhand, May 15, 2015.

Tambe, S. and Arrawatia, M.L. (2012). Climate Change Initiatives. *In:* Arrawatia, M.L. and S. Tambe (eds), Climate Change in Sikkim: Patterns, Impacts and Initiatives. Government of Sikkim, Gantok, Sikkim.

Siva Kumar, A. (2015). Gravity Fed Water Supply Systems at VJNNS. PowerPoint presented at 3rd Annual Meeting of the Springs Initiative, Bhimtal, Uttarakhand, May 15, 2015.

Chapter 10
Water Well Drilling, Well Construction and Well Development

Ashis Chakraborty

1 Introduction

Special ability of exploration and exploitation of resources available in nature is the key to the advancement of mankind. Though some of the resources are readily available in the surroundings, some of the resources are hidden; hence they need some devices and means to reach them.

Groundwater resource available beneath earth surface is excavated by mechanically powered equipment referred as Drilling Rig. Huge numbers of tube wells are drilled every year for irrigation, drinking water and also to meet industrial requirements. The type of well to be constructed depends on the (a) geological formation of the area, (b) intended use of the well, and (c) available financial resources.

2 Activities

Wells are drilled either for exploration or for exploitation. The object of exploratory well is to collect information of the hydro-geology of the underground aquifer or aquifer system. Apart from information gathered about geological layers encountered during exploratory drilling, the physical properties of aquifer are commonly assessed by conducting various tests.

Exploration of groundwater consists mainly of three steps: (a) site selection, (b) test drilling, and (c) bore hole logging.

A. Chakraborty (✉)
Central Ground Water Board, Faridabad, India
e-mail: chakrabortyashis54@gmail.com

P. K. Sikdar (ed.), *Groundwater Development and Management*,
https://doi.org/10.1007/978-3-319-75115-3_10

2.1 Site Selection

It is made by various hydro-geological and geo-physical studies followed by synthesis of data obtained from those studies.

Hydro-geological studies are mainly based on historical data of the area, hydro-geological information obtained from any government, non-government or local agencies, study of topography of the area etc. In some cases, if required, bail test or slug test are conducted. Bail test involves the instantaneous removal of measured volume of water from a borehole/dugwell followed by time measurement of water level recovery. A slugtest involves the injection of a measured volume of water into a borehole/dug well followed by time monitoring of the water level decay.

Geo-physical studies involve measurement of basic physical parameters to infer sub-surface condition. Many geo-physical techniques developed for oil exploration have also been adopted in groundwater exploration. Most common techniques (Dellur 1998) are:

(i) Electrical Resistivity: This technique is applied to determine depth to layer boundaries, fresh/saline water interface, and availability of water in fractures.
(ii) Seismic Refraction: This technique is used to determine depth to water table, number and depth of layer boundaries and bedrock.
(iii) Gravity: This technique is used to determine relative depth to bedrock and thickness of alluvium within an area

Thermal survey is useful to map high permeability zones for well location. This method is also helpful to identify area of shallow concealed bedrock and fault barriers.

Any one of the above technique or combination of more than one technique can be used to obtain adequate sub-surface information as a guiding factor for selection of site.

Synthesis of data is an important factor where soft skill is required. In this process, all information generated from various surveys and all results obtained from different tests are compiled. Now it is the responsibility of the concern scientist/engineer to derive a conceptual hydro-geological model that best fits the observed data.

From static water level measurement and historical well records, an idea can be drawn about the number and type of aquifers. From static water level and depth of various boundaries, aquifer thickness can be estimated. The investigator scientist/engineer must have some confidence on his predictions. Confidence level builds up with experience and number of success as well as failure studies.

2.2 Test Drilling

It is the method to obtain most authentic sub-surface information. There is a basic difference between test well drilling and production well drilling. The objective of test well drilling is basically to obtain sub-surface information to determine casing policy and design a production well. On the basis of the promising result, the test wells are often converted to a production well adopting necessary production well completion method.

Cable Tool drilling method is the best method to obtain most authentic formation samples. But this is very slow method as well as cost intensive. The test bore hole is usually a small diameter hole to reduce cost, but at the same time the hole-diameter must be large enough to conduct geo-physical logging wherever necessary.

In case of consolidated hard rock formations, the rock itself acts as a conduit and retains the bore hole. Only the top soil is cased by suitable pipe to prevent collapsing. In unconsolidated alluvium formation surface casing is used up to certain depth depending upon the looseness of the formation. Below that depth hydrostatic pressure of the drilling fluid filled in the bore hole maintains the bore hole and prevents from collapsing. As per Pascal's law hydrostatic pressure in the bore hole is directly proportional to the depth (Eq. 10.1).

$$P = h\rho g \tag{10.1}$$

where $P =$ hydrostatic pressure at particular depth, $h =$ depth, $\rho =$ density of drilling fluid, $g =$ acceleration due to gravity.

In case of unconsolidated boulder formation cable tool drilling is adopted and entire bore hole is cased to prevent the bore hole from collapsing. In such cases normal electrical logging is not possible and gamma-ray logging is recommended for such cased wells.

2.3 Logging

It is a very common word in drilling industry which means measurement of any parameter in respect of depth. In any test well, mainly three types of loggings are carried out to gather maximum information of the aquifer. They are: (i) litho-logical logging, (ii) time logging, and (iii) geo-physical logging.

(i) **Litho-logical logging:** We have already discussed about sampling of bore hole cuttings while test drilling. Tabulation of physical properties like colour, grain size, formation material of sample cuttings chronologically in respect of depth gives an ideal lithological log. From lithological log characteristics of aquifer is predicted.

(ii) **Time logging:** This is basically the measurement of rate of penetration in respect of depth at the time of test drilling. In usual practice time consumed for drilling every 3 m is monitored. In case any change in geology is observed during drilling, the frequency of measurement can be increased. From time log, depth-wise relative compactness of the formation can be judged. It is commonly found that availability of formation water is inversely proportional to its compactness. Alternately loose formation can contribute more water. So, time log is an important input towards first hand estimation of water availability.

(iii) Geo-physical loggings are sub-surface studies after completion of test well. Common geo-physical loggings in groundwater exploration are (a) electrical logging and (b) gamma-ray logging.

(a) Electrical logging: This logging refers to record of apparent resistivity of underground formations and spontaneous potential generated in bore holes. Both are plotted in respect of depth below ground surface. Electrical logging equipment can be a simple hand operated arrangement or it can be a truck mounted power driven equipment. Usually beyond 300 m depth, cable weight becomes so high that some mechanical device is required. Interpretation of electrical logging data is influenced by diameter of the bore hole, porosity of formation, amount of dissolved mud in the bore hole and mainly by the chemical quality of ground water. Resistivity of the formation inversely varies with the dissolved solid content in the formation water, for example, water with 600 mg/L dissolved solids will show half the resistivity than the water with 300 mg/L dissolved solids. A typical electrical logging arrangement is shown in Fig. 10.1.

(b) Gamma-Ray logging: Gamma-Ray logging is based on measurement of natural radiation of gamma-rays from certain radioactive elements that occur in sub-surface formations. Gamma-ray log is a diagram showing relative emission of gamma-rays (Fig. 10.1). The rays are measured in count per second and plotted against depth. The resulting curve is similar in appearance to the resistivity curve of an electric log. But probe and detecting devices are different in gamma-ray logging and suitable counter is used as down the hole sensing unit. At ground the counter is calibrated to convert electric pulses per second received from the probe into voltage. The voltage is continuously recorded on a film as the probe is pulled up the hole. Unlike electrical logging, gamma-ray logging is not influenced by the chemical quality of water. So gamma-ray logging can identify thickness of different formations. Gamma-ray logging can be conducted in both cased as well as uncased wells.

3 Drilling Methods

Groundwater is in use since pre-historic time. Springs were the easily available source around which habitations developed. Subsequently dug wells were drilled in river-bed to excavate groundwater. Bucket and pulley arrangement to lift water from dug wells was developed in China. It is also believed that percussion drilling originated in China. In modern era it is recorded that a well was drilled in Paris in 1842. The first oil well was drilled in USA in 1859. Now a days drilling technology has developed to a large extent and it is possible to drill through any geological formation and up to any depth within the outer crust of the earth surface.

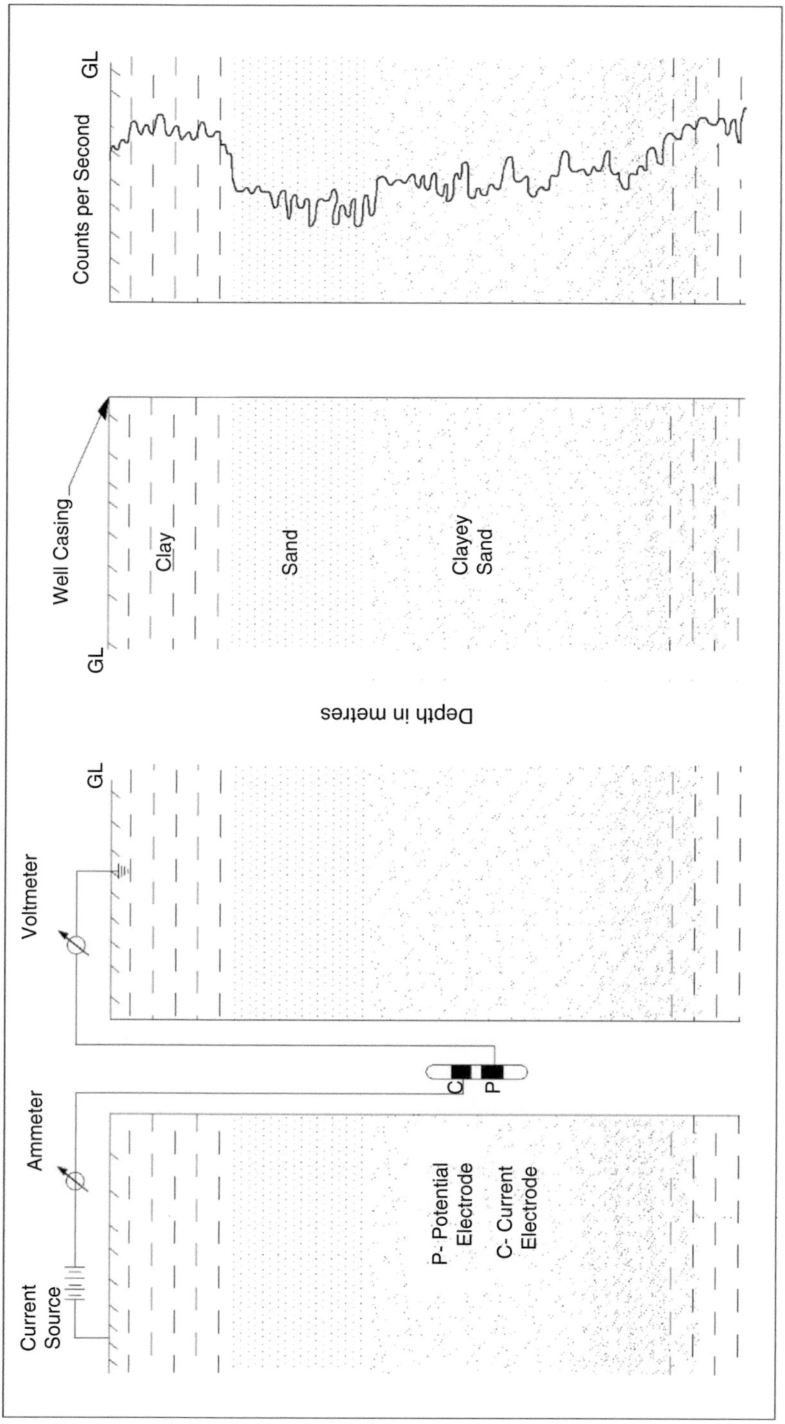

Fig. 10.1 A typical electrical log and a gamma log. (Adopted from CBPI, 1978)

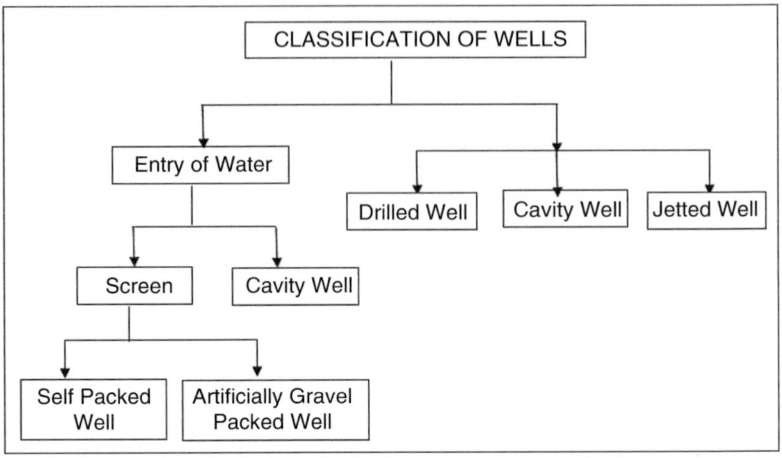

Fig. 10.2 Broad classification of water wells

3.1 Water Well

Water well is broadly a structure to extract groundwater but water wells are also used for water level monitoring purpose, domestic or stock use, community water supply, irrigation purpose, industrial purpose, injection purpose and dewatering in mines or construction site. Water wells can be classified on the basis of different parameters (Fig. 10.2).

In screen well a filtering device is provided in the intake portion of the well in unconsolidated or semi-consolidated aquifer which prevents in-rush of formation material in the well. Screen wells can be either artificially gravel-packed or self-packed. To prevent entering of finer particles of aquifer material in the tube well assembly, the annular space between drilled hole and tube well assembly is filled with coarse sand or gravel. These filler materials are smooth, uniform, clean, well-rounded and siliceous. Size of gravel is designed depending on average grain size of the aquifer material. In case average grain size of formation material is adequately large (>2 mm), no artificial packing is required. Larger particles of the aquifer material act as an envelope around the screen to prevent in-rush of finer particles of the aquifer material.

A cavity well is constructed by making an inner skin and an outer skin separated by a cavity. Both inner and outer skins are usually made out of brick or stone structures.

Drilled wells are those where formation material is cut or crushed manually or mechanically and cutting materials are taken out by some means to construct a well. Various methods of drilling and removal of cutting will be discussed in subsequent paragraphs. Drilled wells can be of any size and depth.

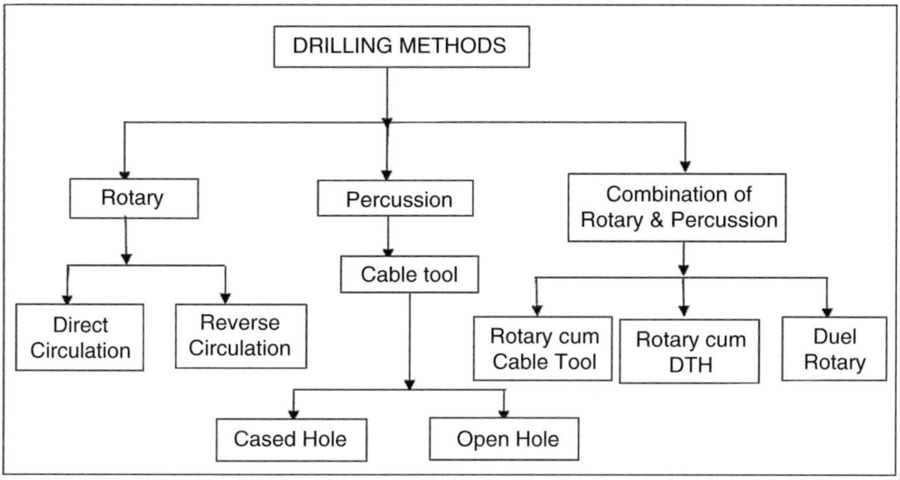

Fig. 10.3 Different types of drilling methods

Driven wells are suitable for unconsolidated sandy formations. Diameter and depth of driven wells are limited. Maximum 15 m can be driven mechanically by using very heavy hammer of 450 kg. In driven well, a hard conical shape metal is used to drive casing pipe. Casing pipe is used to protect well screen during driving. Once the screen is placed to the desired depth, the outer casing pipe can be unscrewed and removed.

Jetted wells are also suitable for unconsolidated formations. A 3–4 inch diameter borehole up to a depth of 50 m can be drilled by jetting method. A chisel-shaped bit is attached at the bottom of the drill pipe. Two nozzles are fitted on either side of the chisel. High pressure water is pumped through the drilled pipe which produces a jetting action through nozzle on the formation material. Cutting material comes out from the bore-hole along with the return flow of water.

Water wells can be drilled by different methods (Fig. 10.3) depending upon the following factors:

(a) Sub-surface geological formation to be encountered
(b) Expected yield from the well
(c) Size of well (depth as well as diameter)
(d) Accessibility of drilling machine and accessories.
(e) Available financial resource.

3.2 *Rotary Drilling*

In rotary drilling method, rotation is given to the bit attached at the bottom of the drill-string for cutting or crushing the formation material. Rotation is given to the bit either by means of Mechanical Rotary Table or Hydraulic Top-head Drive. Formation material is crushed or cut under heavy down pressure. Weight of the drill-string gives the required feed force to the bit. In top head drive rigs, weight can also be applied by hydraulic feed system. In rotary drilling, drilling fluid is used to remove the cutting materials with progress of bit in the bore-hole. Different types of drilling fluids are used in water well industry. The major types of drilling fluids are: (a) dry air, (b) clean water, (c) water with clay additive, (d) water with polymer additive, (e) foam etc. Combination of any of the above is also used as per requirement.

3.2.1 Direct Circulation Rotary

This method is commonly known as Direct Rotary drilling. In this method drilling fluid is pumped through drill pipe and the same drilling fluid comes out through the nozzles provided in the bit and removes the cuttings from the bottom of the bit. The spoil comes out of the bore-hole through the annular space between drill pipe and the bore-hole along with drilling fluid either by high return velocity of the drilling fluid or by suspending in drilling fluid due to its density and high viscosity depending upon the type of drilling fluid used.

In air rotary drilling, air compressor is a part of a drilling rig or a separate compressor unit is used where air, air-water mixture or air-water and foaming agent mixture are used as drilling fluid. Drilling with air has the advantage of reducing drilling cost by increased penetration rate, longer bit life and reduced well-development time. For effective drilling and removal of cuttings 20–30 m/sec up-hole velocity of return air is recommended. The up-hole air velocity is calculated by simple continuity equation (Eq. 10.2).

$$Q = A \times V \tag{10.2}$$

where Q is the volumetric capacity of the compressor, A is the area of annulus and V is the up-hole velocity.

For effective removal of drill cuttings, up-hole velocity of return air can be enhanced with same volumetric capacity compressor by reducing area of the annulus between the bore-hole and drill pipe. This can easily be done by reducing the size of drilling bit.

Direct circulation mud rotary drilling is very effective, hence popular drilling method. Almost all types of geological set-up can be negotiated by this method. Higher depth and larger diameter bore-holes can be drilled by this method (Fig. 10.4). Drilling process is similar to air rotary drilling. Only mud pump is used in place of air compressor. Water or water-bentonite clay mixture is used as drilling fluid in place of air. Water-bentonite clay mixture is commonly called mud. This circulating mud picks up the cuttings and flows upwards through the annular space between drill pipe and the bore-hole to the ground surface. At ground surface mud mixed with drill cuttings are allowed to pass through a settling pit. Drill cuttings

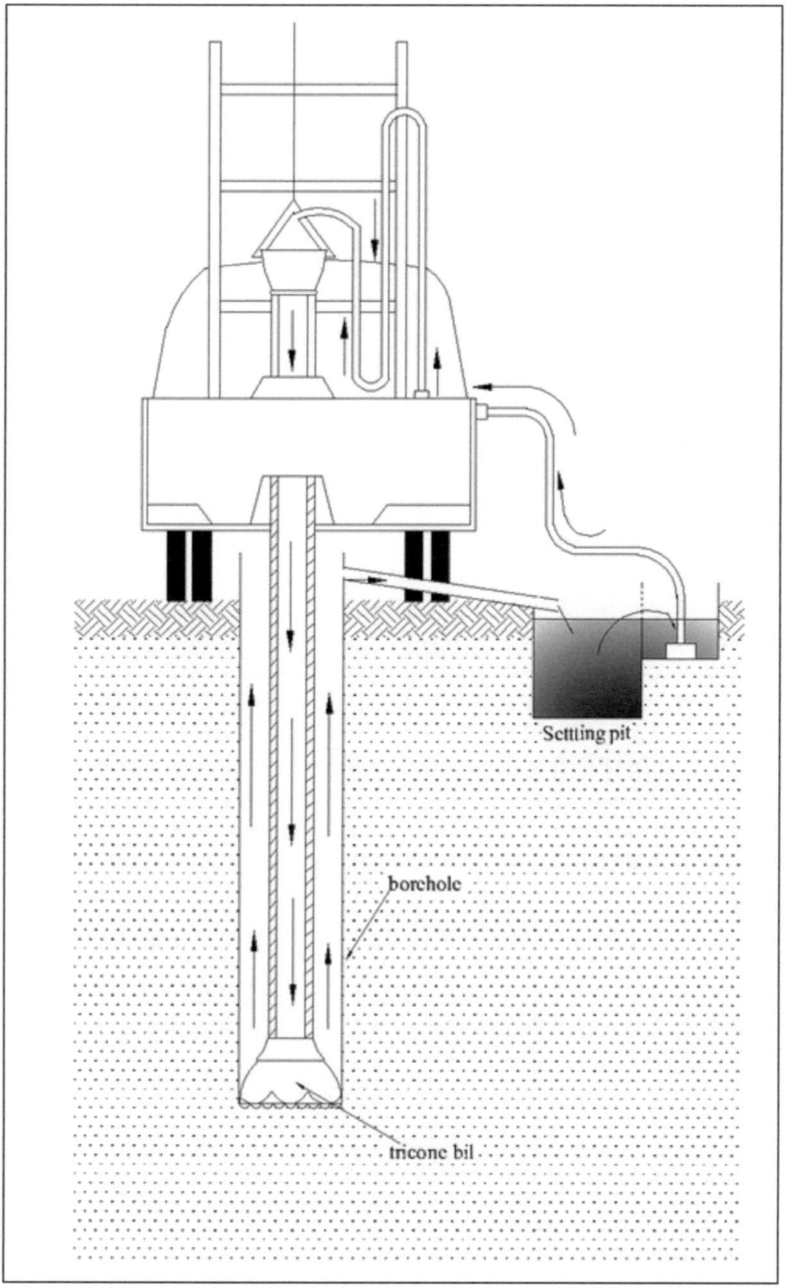

Fig. 10.4 Schematic diagram showing the process of rotary drilling. (Adopted from Dellur, 1998)

settle down at the bottom of the pit and only mud spills to the next pit from where it is again circulated through the mud pump.

In mud rotary drilling, selection of mud pump is very important. The major considerations are (i) mud volume requirement to develop required up-hole velocity for flushing cutting material and (ii) to develop adequate pressure to overcome pressure losses in drill string and surface equipment. Recommended up-hole velocity of drilling mud is 0.41–0.76 m/sec (80–150 fpm).

Next important factor in mud rotary drilling is preparation of mud and to maintain mud quality. Proper density and viscosity facilitate efficient removal of drill cutting. Formation water or formation clay makes the mud thin or thick in the process of circulation and effect the drilling process. Therefore a close observation on mud quality is must in mud rotary drilling.

3.2.2 Reverse Circulation Mud Rotary Drilling

This drilling method is commonly known as Reverse Rotary drilling. From the name itself it is clear that the direction of circulation of drilling fluid is reverse to that of Direct Circulation Mud Rotary drilling. In direct rotary, drilling mud is pumped through drill pipe to the bottom of the hole and the same mud comes out of the bore-hole along with drill cuttings through the annular space between drill pipe and the bore-hole. But in reverse rotary drilling fluid is allowed to enter the bore-hole through the annular space and reach the bottom of the hole by gravity. Subsequently the drilling fluid along with drill cuttings is pumped out through drill pipe. In reverse rotary drilling usually a centrifugal pump is used in place of mud pump used in direct rotary drilling. Reverse rotary drilling is suitable for unconsolidated formations. Bit and drill pipes are differently designed than that of direct rotary drilling so that drill cuttings are allowed to pass through them. In this method higher up-hole velocity can be achieved with a less capacity pump as the space through the drill pipe is much less than that of the annular space. Large diameter bore-holes can be achieved by this method. For drilling to a deeper depth compressor is used to airlift the fluid. When an airlift is used, the air is introduced in to the drill string at an appropriate depth by a small diameter pipe externally fitted with the drill pipe.

3.3 Cable Tool Drilling

Cable Tool Drilling is the oldest percussive drilling method developed in China about 4000 years ago. The method is still in practice with various technical and material improvements with the passage of time. Cable tool rig are also called Spudder Rig. Repeated lifting and dropping of heavy drilling tool into the bore-hole crushes formation material. This repeated lifting and dropping action is obtained mechanically by vertical motion of the spudding beam. The loosen material or drill cutting is mixed with the fluid in the bore-hole. Since no drilling fluid is circulated to remove drill cutting, the same is removed by bailing or by sand pump at relatively short drilling intervals. In this method, drilling and removal of drill cutting are alternative processes. Bailer or sand pump are mechanical devices to remove drill cutting at a regular interval (1–3 m of drilling). Bailer is a section of pipe with a one-way valve at the bottom. When the bailer is dropped, drill cuttings push the one-way valve and enter in to the bailer. While lifting the bailer, the one-way valve gets close due to the weight of the drill cutting accommodated in the bailer. The bailer is taken out and drill cuttings are removed from the bailer on the surface. Sand pump is a bailer fitted with a suction pipe and plunger. The plunger creates vacuum, which opens the one-way valve fitted at the bottom of the bailer and sucks drill cuttings into the bailer. Drill cutting size is a function of formation material and its hardness. They range from fine crushed rock to chips and particles. Small pebbles and gravels come out intact in the bailer.

Cable tool drilling method (Fig. 10.5) is a slow method as removal of cutting is done intermittently by stopping drilling action, but any geological formation can be negotiated by this drilling method. Since no foreign material is used as drilling fluid, most accurate sample of drill cutting can be obtained directly in this drilling method. Large diameter bore-hole can be obtained by this method. This method is still popular for drilling through unconsolidated boulders. Casing can be driven up to the bottom of the bore-hole simultaneously with drilling to prevent the bore hole from collapsing due to unconsolidated loose formation. Therefore this drilling is also termed as Cased Hole drilling**.** These casing pipes are retrieved after lowering the designed tube well assembly through the casing pipes by jacking. Since driving casing pipe up to the bottom and retrieving of entire quantity of casing pipe is a time taking process and more inventory is to be maintained at side, Open Hole drilling practice has been developed in cable tool drilling. In this process, casing is driven up to certain depth depending upon the characteristics and looseness of the formation and rest of the well is filled with water. Hydrostatic pressure of water in the bore-hole prevents from it collapsing. The hydrostatic pressure in the bore-hole increases directly with depth and takes care of the stability of bore-hole at higher depth. Circular hollow bits are used to minimize buoyancy effect in open-hole drilling, where bore-hole is filled with water.

3.4 Combination of Rotary and Percussion Drilling

Different drilling methods have got specific advantages and limitations over one another. To satisfy specific requirement, faster operations and better economy various improvises have been made in drilling methods. Therefore, combination of different drilling methods has been adopted popularly in drilling industry.

In rotary-cum-cable tool drilling method, rotary table is attached with the rig. Mud pump and drill pipes are also used to carry out direct rotary drilling in addition

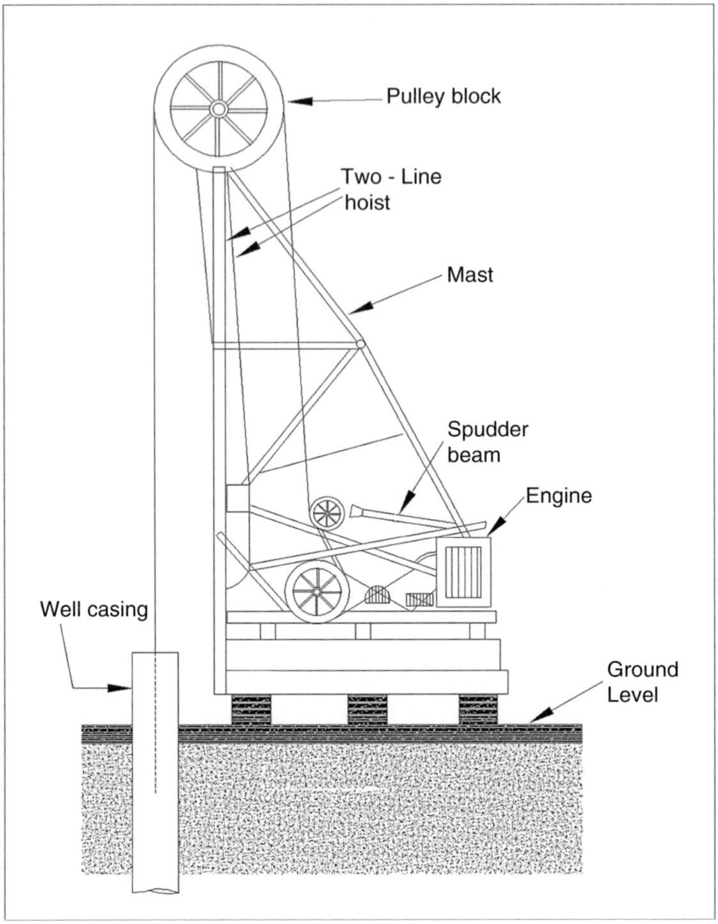

Fig. 10.5 Schematic diagram showing the process of cable tool drilling. (CBPI 1978)

to cable tool drilling. In such method, conventional cable tool drilling is initiated and drilling is continued to cross the boulder bed to case the boulder formation. Below that depth direct rotary drilling is resumed by engaging rotary table, drill pipes and mud pump. This method of drilling is much faster than that of conventional cable tool drilling. In this combination of drilling methods higher depths can be achieved.

Down the Hole Hammer (DTH) drilling is a popular drilling method for drilling through consolidated hard rock formations. Drill bits are known as DTH button bits. The buttons are usually made out of High Carbon Steel. Sometimes the buttons are made out of Tungsten-Carbide or diamond tipped to negotiate through harder materials like non-weathered quartz etc. In this drilling, the bit is attached with a pneumatically operated hammer and fitted at bottom of the drill string. Compressed air passes through the drill pipe to the hammer. The compressed air actuates the hammer and produce high frequency strokes by the bit to crush the rock formation. Drill cuttings are periodically flushed by blowing compressed air at the bottom of the bore-hole through drill pipe, hammer and bit so that bit can strike directly to the formation material. In DTH drilling, compressor is the most important component as it actuates the hammer and maintains adequate up-hole velocity to flush out drill cutting from the bore-hole. Though the hammers usually actuate at 100–110 psi pressure but for efficient removal of drill cutting and to maintain adequate up-hole velocity at least 200 psi pressure is required. Recommended up-hole velocity is 900 m/min or more. Volume of air requirement depends on the size of the bore-hole. Hammer produces 80–120 strokes per minute depending on their design factors. Button bit is attached with the hammer. The hammer imparts blows on the bottom of the bore-hole which is a percussive action. But at the same time the bit also rotates at the rate of 15–30 rpm to maintain the bore-hole annular. Therefore it is a combination of percussion and rotary action simultaneously.

Among all the drilling methods, DTH drilling is the fastest method. Various advancements and improvisations have been made in this method. Provision for bore-hole enlargement or simultaneous casing drive can be attached to this type of drilling machine to case the loose top soil over the compact rock formation.

Dual rotary drilling is a very advance technique getting popular for drilling through unconsolidated boulder. Drilling and casing driving through boulder by conventional cable tool drilling is a time taking method. After completion of well, retrieving of casing pipes by jacking is another time taking process. In this process, due to high friction with side wall of the bore-hole, casing pipes sometimes snaps from the joint. In such case, fishing out of remaining pipes from the bore-hole is further vigorous and time consuming process. These limitations have been taken care in dual rotary drilling method. In this method one rotary device is used to lower casing pipe. This device gives continuous rotation and up and down movement to the casing pipe. Due to this continuous movement casing pipes do not get stuck and become easy to retrieve. Retrieval of casing pipe can be done by the drilling machine instead of jacking which is a time taking process. Through the casing pipe DTH drilling is usually taken up to obtain faster penetration by using suitable air compressor. In this process return air comes out through the casing pipe which acts as a conduit and return air do not come in contact with the side wall of the bore hole. Hence there is no possibility of air loss in the bore hole and comparatively less capacity compressor can also perform. In this method, mud rotary drilling can also be resumed through the casing pipe by using a suitable capacity mud pump instead of air compressor. Therefore, it is obvious that Dual Rotary drilling method is more flexible and improvised method to obtain higher penetration rate and easy casing retrieval in unconsolidated boulder and gradually getting popular.

3.5 *Drilling Fluid*

Over the years, drilling fluid technology has developed to a great extent in water well industry. Beside water, other fluids like mud, air, air-foam mixture and polymers are popularly used in water well drilling. However the most commonly used drilling fluid is mud which is a colloidal solution of water and bentonite. Bentonite is finely ground clay free from any abrasive material or any toxic chemical. Circulation of drilling fluid is essential to remove drill cuttings from the bottom of the bore-hole to the surface and keep bottom of the hole clean so that bit can directly come in contact with the formation material for penetration.

Common functions of the drilling fluids are:

- Remove drill cuttings from bottom of the bore-hole and carry the same to the surface.
- Hold cutting material suspended in the fluid when circulation is stopped for the time being.
- Make temporary impermeable mud-cake on the wall of the bore-hole.
- Cooling and lubricating bit which suffers vigorous friction with formation material during drilling.
- Encounter sub-surface pressure.
- Support part of the weight of drill string and casing due to its buoyancy factor.

While performing the above duty, drilling mud should not cause the following adverse effects.

- Adversely affect the formation adjoining to the bore-hole
- Reduce rate of penetration
- Allow continuous suspension circulation of abrasive drill cuttings
- Require excessive pumping pressure for desired circulation
- More time to remove mud cake and excess washing or development of well.

Therefore, it is important to maintain proper drilling fluid properties for efficient completion of the well.

4 Well Construction

Tube well is an intake structure to draw groundwater from the aquifer. Tube wells can also be used for artificial recharge of non-committed surface water to the aquifer after proper treatment. Well completion basically comprises some distinct operations like: (i) drilling, (ii) installation of casing, (iii) placing of well screen and filter pack, if required, (iv) grouting of well to prevent dislocation of well assembly and to avoid seepage of contamination from surface and (v) development to ensure maximum intake of sand free water in the well. Well structure consists of two major elements (a) well casing and (b) intake portion.

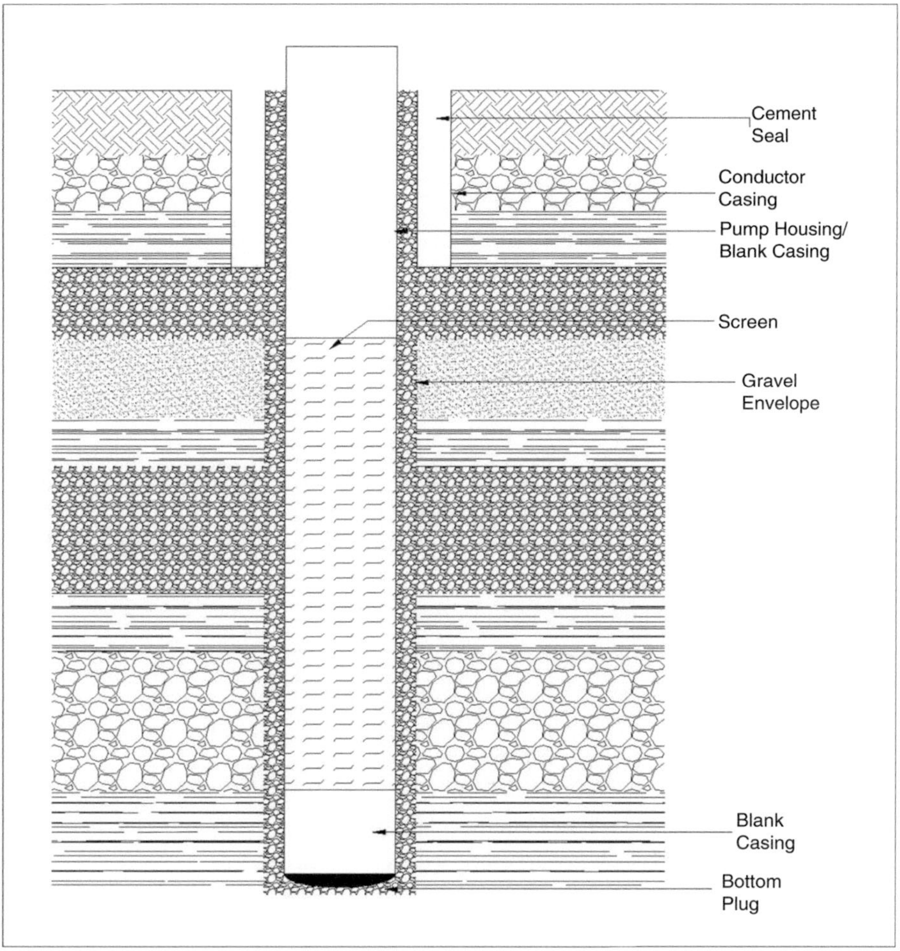

Fig. 10.6 Diagrammatic representation of a typical well assembly. (Adopted from Roscoe Moss 1990)

Well casing is again of two types, one is drive pipe and other one is well assembly casing pipe. The pipe used to support the bore-hole during drilling in case of cable tool drilling method is known as drive pipe. Drive pipe is taken out with the lowering of assembly pipe and shrouding of gravel. These drive pipes are re-used and not a part of the tube well assembly. Well assembly (Fig. 10.6) casing is part of the tube well structure which holds the intake portion of the well in proper position. Diameter of the casing pipe depends on the size of the bore-hole which in turn will depend on the capacity of the well and depth of aquifer tapped. The upper portion of the casing pipe used for pump setting is conventionally termed as Housing pipe.

Intake portion is the most important part of the well. Through this part of the tube well water enters the well from aquifer. In case of unconsolidated formations, the intake portion consists of screen or perforated pipes through which water flows into the well. Design of intake requires careful consideration of hydraulic parameters which influences performance of well. In case of consolidated hard rock formation only the top soil or over burden is cased and rest of the bore-hole is kept naked. Water from the fractures and fissures accumulates in the bore-hole. Due to consolidated nature of the formation, the hole remains intact. In case of unconsolidated water bearing formations, screen is provided to allow sand free water enter into the well freely and also to restrict the formation material to rush in. This part of the tube well requires proper design from various hydraulic considerations and that influence the performance of a well.

5 Well Design

A good water well design is to ensure an optimum combination of performance, longevity and economy (Driscoll 1986). The basic inputs required to design water well are:

- Expected yield and intended use of the well.
- Geological set up of the entire depth drilled.
- Character, thickness and sequence of potential aquifer.
- Size and gradation of aquifer materials.
- Water level condition and seasonal fluctuation.
- Quality of water.

In addition to the above basic inputs, some local information are also important like design and construction feature of previously constructed wells in that area and also operation and maintenance history of previously constructed well in that area.

Important steps for proper design of a well are:

- Determination of diameter and depth of casing.
- Mechanical analysis of formation material for screened well.
- Selection of water bearing strata to be screened and fixing the length of screen.
- Design of well screen i.e. diameter, length, percentage of screen opening and material.
- Selection of type of well i.e. naturally packed or artificially gravel packed well.
- Determination of gravel size.

5.1 Design of Casing

Diameter of casing is very important because it affects significantly the cost of structure. Housing part of the well casing must be large enough to accommodate the

Table 10.1 Recommended diameter of housing pipe

Anticipated yield of well (lps)	Nominal size of pump (mm)	Optimum size of housing pipe (mm)
< 6	100	150
5–11	125	200
10–25	150	250
22–40	200	300
37–56	250	350
53–82	300	400

Source: CGWB 2011

pump with proper clearance for installation and operation. The recommended diameter of the housing pipe is given in Table 10.1.

Diameter of the rest of the casing must be such as to ensure hydraulic efficiency. It can be calculated from the simple relation as follows (Eqs. 10.3 and 10.4):

$$a = Q/v \tag{10.3}$$

where a = Area of cross-section of casing pipe in meter, Q = Expected discharge of the well in cubic metre per second and v = Entrance velocity of water into the well (ideally, 3 m per second).

$$again, \quad a = \pi d^2/4 \tag{10.4}$$

where

d = Diameter of casing pipe in metre.

From the above relations diameter of casing pipe can be determined.

Length of casing is determined in such a way that most of the aquifer thickness is utilized by intake portion of the well, resulting in higher specific capacity (discharge per unit draw down). However, depth of the housing pipe is guided by the following relation (Eq. 10.5):

$$\begin{aligned} \text{Depth of housing pipe} = {} & \text{Water table below ground level} + \text{Length of the pump} \\ & + \text{Drawdown} + \text{Seasonal water level fluctuation} \\ & + \text{Allowance for submergence of pump} \end{aligned} \tag{10.5}$$

5.2 *Design of Well Screen*

Screen is the most important component of a well and hence must be carefully designed. Life of a well is mainly governed by the life of the screen. An efficient well screen is required to (i) offer minimum resistance to flow water in the well,

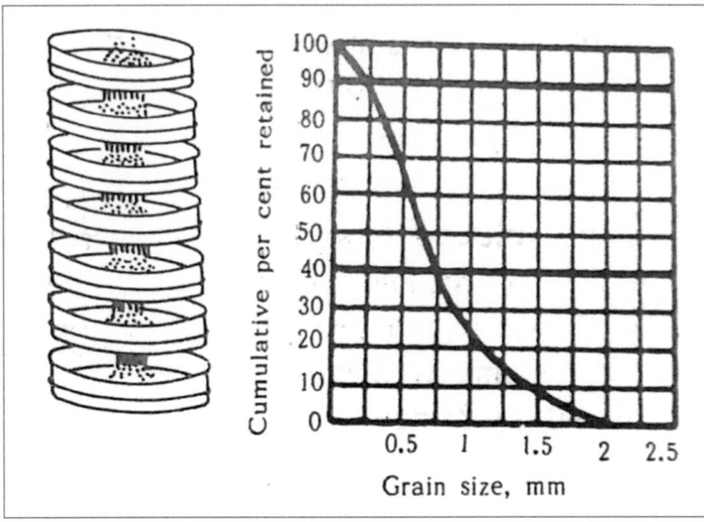

Fig. 10.7 An overview of sieve analysis. (Adopted from CBPI, 1978)

(ii) prevent movement of formation materials in the well, (iii) be strong enough to prevent collapse, and (iv) resist corrosion and incrustation.

However, for proper design of screen length, diameter, percentage opening and for determining corresponding gravel size, mechanical analysis of sample cuttings obtained from different depths is essential. Sieve analysis (Fig. 10.7) is one of such mechanical analysis of sample cuttings.

Preparation of sample is required before conducting sieve analysis. In this process, sample cuttings are properly washed and made free from any foreign material like mud etc. Then the sample is properly dried by heating up to 110 °C. In slandered procedure of sieve analysis, a set of sieve conforming to IS: 460–1962 or standard set by individual country is used. The weight retained in each sieve is measured. These weights are then expressed as a percentage of total weight of the sample. On the basis of the sieve size grain size of the sample can be determined. A graph is plotted by keeping grain sizes in X-axis and percentage retained in Y-axis (Fig. 10.7).

Uniformity Co-efficient (C_u) is a ratio expressing the variation in grain size (Eq. 10.6). It is usually measured by the sieve aperture that passes 60% of the material divided by the sieve aperture that passes 10% of the material.

$$\begin{aligned} C_u &= d_{60}/d_{10} \\ &= 60\%\text{pass or } 40\%\text{retained}/10\%\text{pass or } 90\%\text{retained} \end{aligned} \tag{10.6}$$

Generally, if the grain size is classified as uniform then the C_u is less than 2. For fairly distributed grain size the C_u lies between 2 and 3 and for heterogeneous grain size the value of C_u is greater than 3. This uniformity co-efficient is the guiding factor for selection of strata to be screened.

Effective Size (d_{10}) is defined as formation particle size, where 10% of the grains pass through the sieve and 90% of the grain is retained. This effective size of the grain is the guiding factor for selection of slot opening.

Selection of aquifer to be screened is carefully done after proper grain size distribution analysis. Permeability of aquifer is proportional to the square of the effective grain size $d_{10.}$ In case two sets of sample is having same effective size, the sample having lower Uniformity Coefficient is more permeable. The well screen length is determined on the basis of the characteristics of the aquifer and available drawdown. In any case drawdown should not be allowed up to the depth of screen. For different aquifer conditions different guidelines are followed.

For homogeneous confined aquifer, 70–80% of the aquifer thickness is screened depending upon thickness of aquifer. If aquifer thickness is less than 7.5 m, then 70% aquifer is tapped. If aquifer thickness is between 7.5 and 15 m, then 75% of aquifer is tapped. If aquifer thickness is more than 15 m, 80% of aquifer is screened. For non-homogeneous confined aquifer**,** it is best to screen most of the permeable or productive zone. This zone can be identified from the mechanical analysis of drill cutting sample as well as from electrical logging.

In case of homogeneous unconfined aquifer**,** the bottom one-third of the aquifer is usually screened. However, to obtain higher specific capacity (discharge per unit drawdown) more length, even up to 50% of the aquifer length can be screened. In case of non-homogeneous unconfined aquifer, the same principle is followed that is followed in the case of non-homogeneous confined aquifer. Only one consideration is taken care in such case, that is, the screen is placed at the lower most portion of the most productive zone. This is to obtain maximum available drawdown.

To maintain minimum turbulence at the entrance, the screen entrance velocity is maintained at 3 m/sec or less. This results in (i) minimum friction loss at the screen opening, and (ii) minimum rate of corrosion and incrustation. To determine the well screen diameter Eqs. (10.3) and (10.4) are used.

The size of the slot opening and percentage of open area of the screen is another important aspect of well screen design. As discussed earlier the ideal slot opening is the d_{10} size of the aquifer material. It means the slot size of the screen should be such that it can retain 90% of the formation material and may allow 10% of the formation material to pass through the slot. If formation material is of finer grade, it is recommended that V-wire screen is used in place of conventional slot pipe. It is difficult to cut slot less than 1.4 mm in normal MS pipe, whereas 0.5 mm opening can be obtained in V-wire screen.

For slotted pipes, the recommended maximum percentage of open area of the screen is 20%. It is not only difficult to design more slot per unit area but it is also not recommended from the strength point of view to have more percentage of open area in case of slotted pipe. However, for V-wire screen 70% of open area can be obtained.

5.3 Gravel Pack Design

In naturally packed wells (naturally developed wells), the fine material in the formation surrounding the screen is removed by development of the well to create more permeable zone around the screen. In such wells higher slot openings are used so that 60% of the formation materials are removed and 40% retained during development.

In artificially packed well, the zone immediately surrounding the screen is made more permeable by removing 10% of the formation material and replacing it by artificially graded coarser material (gravel).

In either case, the net hydraulic result is an increase in the effective diameter of the well and an increase in permeability around the screen. The mean size of the pack material is related to the mean gravel pack ratio (Fig. 10.8) which is given in Eq. (10.7) (Michael and Khepar 1989).

$$\text{PA Ratio} = 50\%\text{size of gravel} \div 50\%\text{size of formation} \tag{10.7}$$

The Central Board of Irrigation and Power (1967), based on a series of laboratory experiment, recommended the following criteria for PA Ratio:

(a) For uniform aquifer material ($C_u \leq 2$)

d_{50} of pack material ÷ d_{50} of formation material should lie between 9 and 12.5.

(b) For graded aquifer material ($C_u > 2$)

d_{50} of pack material ÷ d_{50} of formation material should lie between 12 and 15.5.

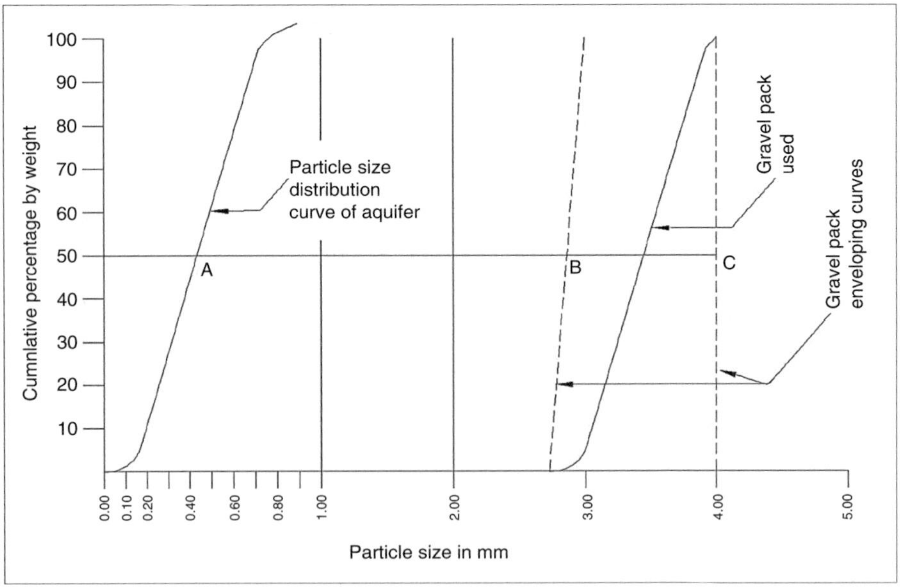

Fig. 10.8 Analysis for choosing appropriate gravel size based on grain size distribution of aquifer material

5.4 *Material for Construction of Well*

ERW Mild Steel (MS) pipes are commonly used for construction of well for any size and depth. But alternative materials are also used under different conditions such as size of formation material, water quality and last but not the least, the cost factor. Since the screen portion is the weakest portion of the well assembly, it is prone to fail early than any other part of the well. To avoid corrosion and incrustation Galvanized Iron pipes (GI) or High Density Polyvinyl Chloride (HD-PVC) pipes are used. But for large diameter and deeper wells MS pipe is recommended although stainless steel pipes have distinct advantages, but not very cost effective. To arrest fine formation material brass strainers or V-wire screens are used. Brass strainers are not cost-effective for deep and larger diameter bore holes and stainless steel V-wire screens are cost intensive than conventional low carbon galvanized steel (LCG Steel) V-wire screen.

6 Well Development

During drilling operation, normal setup of the geological formation is damaged or disturbed. Moreover, foreign materials like mud etc. are used during drilling which ultimately clog the water bearing aquifer. Mud cakes are also formed on the side wall of the bore-hole during drilling to prevent collapse of the unconsolidated formation. Therefore, after completion of the construction part, the tube wells are developed to remove those foreign materials from the well and repair the formation from the damage occurred during drilling.

The basic objectives of well development are:

- To remove the mud cake formed on the wall of the bore-hole.
- To clear the permeable zone.
- To remove desired quantity of formation material surrounding the screen for better yield of sand free water.
- To minimize skin effect and maximize well efficiency.

6.1 *Development Methods*

6.1.1 Back Washing

In this process fresh water is pumped in the well with pressure which comes out through the screen and agitates the area surrounding the screen. As a result not only the mud cake around the screen is broken, gravels and formation materials are also segregated. Back washing is usually done by mud pump and through drill pipe to the desired depth.

6.1.2 Hydraulic Jetting

In this method, high velocity fresh water is injected in the well. This water jet cleans the tube well assembly from inside. At the same time part of the jetting through the screen strikes the mud cake around the screen. For jetting, usually mud pump is used. Jetting tools are commonly fabricated from a section of blank pipe slightly less than the diameter of screen. Four straight holes are drilled at 90° apart. If nozzles are used in place of drilled hole, better hydraulic efficiency can be obtained. Simultaneous use of air lift pump with jetting is very effective development technique.

6.1.3 Mechanical Surging

In this process, fresh water is forced into the formation. The surge block moves up and down in the tube well assembly and acts as a plunger. This up and down movement of the plunger also creates suction inside the tube well assembly; as a result the fresh water forced into the formation is again sucked in along with mud and fine formation particles present around the screen.

6.1.4 Compressed Air Development

This is the most commonly used method. This is basically pumping by air-lift method. In air-lift pumping, compressed air is forced in the well and all the fine formation particles, mud etc. present in the tube well assembly is pumped out. Air-lift pump also creates suction around the screen portion and helps to clean the formation around the screen. In air-lift pump, air-line is placed inside a larger diameter eductor pipe. The eductor pipe is lowered against the bottom most screen. Air-line is lowered within the eductor pipe up to a maximum possible depth according to the capacity of compressor. But in any case air-line should not be lowered beyond the depth of eductor pipe. If air-lift pumping is combined with mechanical surging, better result can be obtained.

6.1.5 Chemical Treatment

If more time is consumed in drilling or there is a time gap between well construction and development, mud cake may become hard and conventional development methods may not break it. Also in case of hard-rock wells, sometime fractures may get clogged by clay or any softer material which restricts the hydraulic continuity and inflow of water in the well; in such cases chemical treatments are required. The commonly used chemicals – polyphosphates which act as dispersion agents for clay – are sodium tri-polyphosphate ($Na_5P_3O_{10}$), sodium pyrophosphate ($Na_4P_2O_7$) or sodium hexa-meta-phosphate {$(NaP_2)_6$}.

Apart from chemical treatment, acid treatments are also done to dissolve acid soluble material present in the formation for permitting higher flow rate in the well. Commonly used acids are hydrochloric acid (HCl), sulfamic acid (H_3NO_3S) etc.

A measured quantity and dilution of chemical or acid is poured in the well and a reaction time is given. But the most important function is to remove these foreign materials completely from the well and from the aquifer where these chemical/acid has propagated.

6.1.6 Hydro-Fracturing

This method of development is adopted in specific wells in hard rock formation. In this method of development a section of well is isolated by using packers and then fresh water is injected at high pressure (120–140 bars) to that isolated zone. In this process, drill cuttings or any softer material clogging the fractures will get cleared and due to better hydraulic continuity, yield of the well will increase. One hydro-fracturing unit consists of (a) one plunger type reciprocating pump to deliver about 3000 lph at about 180 bar pressure, (b) one water tanker of 9000–10,000 l capacity along with a pump to fill up the tanker from available source, (c) single and double packer units to isolate the required section of well where hydro-fracturing is to be conducted, and (d) different fittings and fixtures to fix packers to the desired depths and apply high hydraulic pressure to the selected area. This method of development is not recommended for all types of wells.

6.1.7 Over Pumping

This is the simplest method of development to remove the fine particles from the aquifer. In this method, a tube-well is pumped at a higher rate than the actual rate of pumping during its normal operation. Though a very effective method but usually adopted as a final development after using other development methods. This is because if the water is not fairly cleared by other development methods, the test pump may get easily worn out while pumping unclear water that contains formation sand which is abrasive in nature. The pump is normally set above the screen. When pump is in operation, water flows from formation to well but when pump is stopped the water in the column pipe gives a return pressure to the formation. Thus, by periodically running and stopping the pump, formation around the screen is stabilized. This method of development is not recommended for developing shallow wells with short length of screen.

References

CBPI (1978). Manual on Ground Water and Tube Wells. Central Board of Power and Irrigation, Govt. of India.

CGWB (2011). Water Well Construction: Techniques Equipment and Management. Rajiv Gandhi National Ground Water Training and Research Institute, Central Ground Water Board, Govt. of India.

Dellur, J.W. (1998). Hand Book of Ground Water Engineering. School of Civil Engineering, Purdue University, Indiana.

Driscoll, F.G. (1986). Ground Water and Wells. Johnson Screens, St. Paul, Minnesota.

Michael, A.M. and Khepar, S.D. (1989). Water Well and Pump Engineering. Tata McGraw- Hill, New Delhi.

Roscoe Moss Company (1990). Hand Book of Ground Water Development. 1st Edition. Wiley-Interscience.

Chapter 11
Pumping Test for Aquifers: Analysis and Evaluation

Pradip K. Sikdar

1 Introduction

A pumping test is a field experiment in which a well is pumped at a controlled rate and water-level response (drawdown) is measured in one or more surrounding observation wells and optionally in the pumped well (control well) itself. Response data from pumping tests are used to estimate the hydraulic properties of aquifers, evaluate well performance and identify aquifer boundaries. Aquifer test and aquifer performance test (APT) are alternate designations for a pumping test.

Although the terms pumping test and pump test are often invoked interchangeably, the use of pumping test is preferred (Woessner and Anderson 2002). If the performance of a pump is being tested then 'pump test' should be used; if the performance of an aquifer is being tested through the action of pumping a well then 'pumping test' should be used.

Pumping tests are carried out to determine: (i) the quantity of groundwater that can be extracted from a well based on long-term yield and well efficiency, (ii) the hydraulic properties of an aquifer or aquifers – transmissivity, hydraulic conductivity and storage coefficient, (iii) spatial effects of pumping on the aquifer – radius of influence and safe distance, (iv) the suitable depth of the pump and (v) water quality and its variability with time.

Common types of pumping tests that may be performed include the following:

Constant-rate tests maintain pumping at the control well at a constant rate. This is the most commonly used pumping test method for obtaining estimates of aquifer properties.

P. K. Sikdar (✉)
Department of Environment Management, Indian Institute of Social Welfare and Business Management, Kolkata, India
e-mail: p_sikdar@hotmail.com

P. K. Sikdar (ed.), *Groundwater Development and Management*,
https://doi.org/10.1007/978-3-319-75115-3_11

Step-drawdown tests proceed through a sequence of constant-rate steps at the control well to determine well performance characteristics such as well loss and well efficiency.

Recovery tests use water-level (residual drawdown) measurements after the termination of pumping. Although often interpreted separately, a recovery test is an integral part of any pumping test.

2 Preliminary Studies

To start with, it is essential to collect preliminary information about the subsurface geological and hydrogeological conditions such as lithological character and thickness of the aquifer, direction of groundwater flow, water table or piezometric surface gradients and regional groundwater level. The location of aquifer boundaries must be determined. Most pumping test formulae are based on the assumption that the aquifer is of infinite areal extent. In many cases the aquifer boundaries are too far away to vitiate the effect of this assumption. However, where barrier boundaries are close by, for example, buried impervious strata or lithological features like a fault, and where recharge boundaries are near by like a perennial river, a canal, a lake or the ocean, the effect of these must be considered. Sometimes barrier boundaries are hidden, and their presence is revealed only after a careful analysis of the drawdown plots.

The site of the test should be selected carefully so that the hydrogeological conditions are representative of the entire region and the gradient of the water table or the piezometric surface is low. The site should be easily accessible but away from railway lines or highways because the passing traffic may produce appreciable fluctuations in piezometric levels of confined aquifers.

The following concepts and reconnaissance work should factor into the design of a pumping test:

- time of year for pumping test
- natural agents of groundwater fluctuation such as barometric pressure changes, earth tides and tidal variations which may affect water levels in observation wells during the pumping test
- off-site groundwater use which may influence water levels during the pumping test
- effects of pumping test on surrounding water users
- depth setting and type of pump in control well
- pumping duration and rate
- pumping rate measurement and control
- water-level measurement and frequency
- disposal of pumped water
- collection of water quality samples during pumping

- location and orientation of streams, faults, lithologic contacts and other potential aquifer boundaries
- potential for salt water intrusion in coastal areas

3 Construction of Pumping Well and Piezometer

Several methods are available for construction of wells. The suitability of each depends upon the geology of the formation, the depth of groundwater and design of the well. Boring, jetting or driving can be employed for construction of shallow wells, and cable tool drilling; hydraulic rotary or reverse rotary are available for deep wells in soft sedimentary formations. In consolidated formations, a large pressure or impact has to be applied for loosening the formations. Methods for drilling in such formations are rotary-cum-percussion air drilling and core drilling. Accurate well logs should be prepared by collecting sediment samples from the drilled well and piezometers and performing sieve analysis. These help in taking decision about the best location of the well and piezometer screens.

The wells should be properly designed so as to draw clear water from the aquifer without excessive head loss and at the same time to keep the aquifer material out. Both are developed in place. Development removes the finer material from the aquifer surrounding the well so that only coarser material is left adjacent to the screen. The aquifer material around the well becomes more uniform in grain size and holds back the finer material of the aquifer further beyond so that it cannot clog the screen. When the well is gravel packed, much of the same purpose has been accomplished, although development is still beneficial. The choice of whether a well is to be provided with a screen or with a screen with gravel pack depends primarily upon the effective grain size d_{50} (50% finer) and the uniformity coefficient (d_{60}/d_{10}) of the aquifer material.

3.1 Design of Piezometers

Piezometers are pipes of small diameter; say up to 5 cm which acts as observation wells during pumping test. Large diameter piezometers should be avoided since they may cause a time lag in change of drawdown. Piezometer pipes end with screens generally 0.5–1 m long, and are placed about the same depth as the middle of the well screen in the pumped well. It is desirable to pack coarse uniform sand around the piezometer screens to facilitate quick entry of water. The rest of the annular space between the pipe and the hole should be filled by clay, very fine sand or concrete. Before use the piezometers should be flushed or pumped for a short time to remove the finer particles.

The distance of piezometers from the pumped well must be carefully fixed. The ultimate choice of the distance of piezometers from the pumping well depends upon

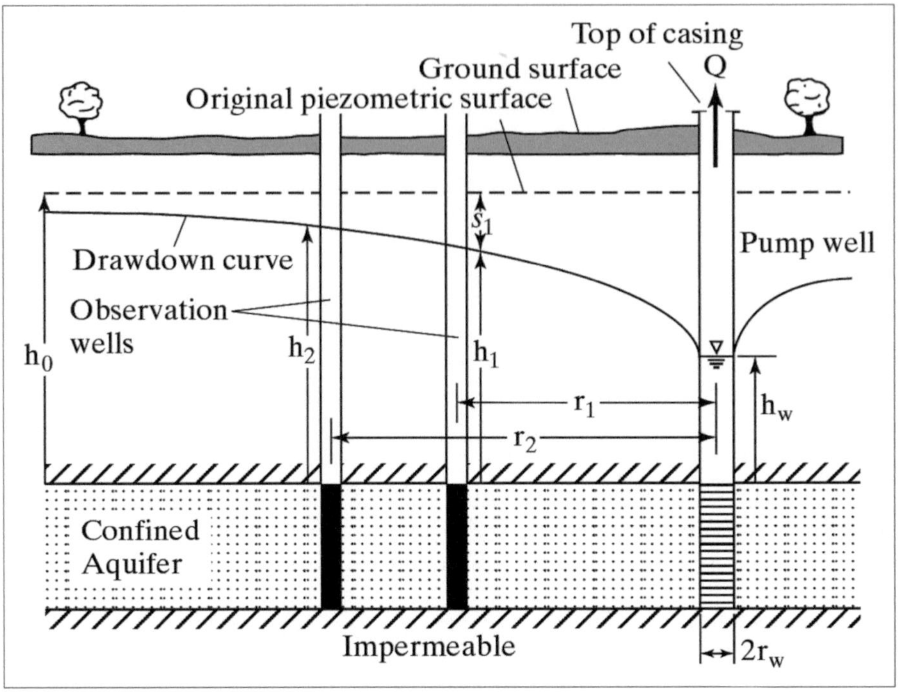

Fig. 11.1 Pumping well with observation wells in a confined aquifer

local conditions. Placing them between 10 and 100 m from the pumped well is satisfactory in most cases. In thick or stratified confined aquifers, the distance may have to be 100–250 m or more.

4 Pumping Test

The principle of a pumping test involves applying a stress to an aquifer by extracting groundwater from a pumping well and measuring the aquifer response to that stress by monitoring drawdown as function of time in one or more observation wells (Fig. 11.1). These measurements are then incorporated into an appropriate well-flow equation to calculate hydraulic parameters of the aquifer.

Water level measurements are taken at short intervals during the initial part of the test, and the time interval between measurements is gradually increased as the test continues. Pumping should be continued till the cone of depression has stabilised, that is, does not seem to expand as pumping continues.

During the test, observations that are made include measurement of water levels during pumping in the test well and observation wells, discharge rate and measurement of recuperation of water level in the test well and observation wells. All water level measurements should be done with reference to fixed measuring points in test and observation wells. The static water level should be measured for the test wells and observation wells prior to commencement of pumping.

Observation wells should not be too close to pumping well, for example it is best to have them at distances $\geq$5 m. They may be located on line parallel to any boundary or located on orthogonal line to identify any boundary.

Measuring the flow rate during a pumping test can be done in the following ways:

- container and stopwatch
- in-line flow meter
- V-notch
- weir or flume
- orifice meter

Flow rates should be recorded with sufficient frequency to demonstrate a constant rate or to monitor planned rate changes. In the event of temporary test interruption (e.g., power failure), pumping stop and restart times should be noted to allow for proper interpretation of the test. It should be kept in mind that the discharge rate often decreases with time as the water level in the control well drops. Kruseman and de Ridder (1994) recommend checking and, if necessary, adjusting the flow rate at least once every hour; however, more frequent measurements may be considered until it becomes evident how often rate adjustments are required (Stallman 1971).

Care should be exercised in the disposal of the pumped water. It is important to dispose of the water in accordance with any applicable laws and regulations. The following points should be kept in mind for the disposal of discharge water:

- Avoid direct discharge of water on the ground surface if the water is likely to recharge the pumped aquifer (e.g., a shallow unconfined aquifer or karst aquifer with sinkholes). Recycling of the pumped water through recharge can result in the false identification of a constant-head boundary or leakage.
- Discharge of water to a surface water feature such as a stream is a viable option if it is anticipated that the surface water body is hydraulically disconnected from the pumped aquifer or if it behaves as a constant-head boundary.

The decision to terminate a pumping test is best made on the basis of hydrogeologic conditions at the test site and the objectives of the test. Longer tests may be necessary to estimate specific yield in an unconfined aquifer or to observe boundary effects. In general pumping test should continue until steady state flow is attained. In some tests, steady state occurs within a few hours of pumping, in others it may never occur. However, 24–72 h testing is enough to produce the required data for analysis. Tests taking longer than 24 h may be required for large takes, such as community supplies, or situations where it may take longer to determine steady state.

Plotting and inspecting pumping test data as they are collected in the field can help to decide:

- When it's appropriate to end a pumping test.
- Have the objectives of the test been achieved?

- Have sufficient late-time data been collected to estimate specific yield in an unconfined aquifer?
- Has the test continued long enough to detect and locate aquifer boundaries?

Other factors may influence the planned duration of a pumping test including budgetary constraints and regulatory requirements. Often, applicable regulations may establish a minimum duration required for a test, but longer tests may be necessary to achieve other test objectives.

5 Analysis of Data

5.1 *Confined Aquifer*

Charles Vernon Theis (1935) developed, for unsteady flow to a well, a formula based on the assumption that flow of groundwater is analogous to the flow of heat, and that the mathematical theory of heat conduction is applicable to hydraulic theory. The formula was derived taking into account the time factor and also the removal of water from storage in the development of the cone of depressions. Known as the Theis formula or non-equilibrium formula, it shows the change in head, or drawdown as follows:

$$h_0 - h = s = \frac{Q}{4\pi T} \int_u^\infty \frac{e^{-u}}{u} du$$

where $u = r^2S/4Tt$, s = drawdown, in metres, T = transmissivity, in m^2/day, t = time, in days, since pumping started, S = storage coefficient, dimensionless, r = radial distance from discharge well to the point of observation, in metres, Q = constant rate of discharge of well in m^3/day, and e = base of natural logarithm.

In the above equation the exponential integral expression is symbolically expressed as $W(u)$ for 'well function of u'.

$$W(u) = -0.5772 - \ln u + u - \frac{u^2}{2.2!} + \frac{u^3}{3.3!} - \frac{u^4}{4.4!} + \cdots\cdots$$

The formula is based on the following assumptions:

1. The aquifer is isotropic and homogenous.
2. The aquifer has infinite areal extent.
3. The discharge well penetrates and receives water from the entire thickness.
4. The transmissivity is constant at all times and at all places.
5. The well has infinitesimal diameter.
6. Water removed from storage is discharged instantaneously with decline in the head.
7. The aquifer is pumped at a constant discharge rate.

Implied in the assumptions are other limiting conditions: the aquifer is horizontal and confined; has a constant coefficient of storage, is not recharged; the pumped well is fully penetrating and screened in the entire aquifer; the piezometric surface is horizontal and the storage in the well can be neglected.

The Theis equation enables to determine the formation constants hydraulic conductivity (K), transmissivity (T) and storativity (S). This equation cannot be solved directly. Four methods are available for using this equation. They are:

1. Theis' method
2. Chow's method
3. Cooper and Jacob's method
4. Theis' recovery method

The Cooper and Jacob's method is very useful to interpret pumping test data in the field condition and hence has been discussed in this chapter. For the other methods the readers may refer to any textbook on hydrogeology (Hantush 1964; Lohman 1972; Karanth 1987; Todd and Mays 2005).

5.1.1 Cooper and Jacob's Method for Evaluation of K, T and S

The Cooper and Jacob (1946) solution (or *Jacob's modified non-equilibrium method*) is useful for determining the hydraulic properties (hydraulic conductivity, transmissivity and storativity) of non-leaky confined aquifers. Analysis involves matching a straight line given by the solution to drawdown data plotted as a function of the logarithm time since the start of pumping.

The Cooper and Jacob solution is an approximation of the Theis non-equilibrium method which is given in compact notation as follows:

$$s = \frac{Q}{4\pi T} W(u)$$

$$\text{Or, } s = \frac{Q}{4\pi T}\left(-0.5772 - \ln u + u - \frac{u^2}{2.2!} + \frac{u^3}{3.3!} - \frac{u^4}{4.4!} + \cdots\cdots\right)$$

$$\text{Again, } u = \frac{r^2 S}{4Tt}$$

$$\text{Therefore, } u\,\alpha \frac{1}{t};\, u\,\alpha\, r^2$$

Jacob (1946, 1950) pointed out that for large values of "*t*" or small values of "*r*" the value of "*u*" becomes small enough so that in the above series the terms after the first two becomes negligible. The drawdown can then be expressed as

$$s = \frac{Q}{4\pi T}(-0.5772 - \ln\ u)$$

$$s = \frac{Q}{4\pi T}\left(-0.5772 - \ln\frac{r^2S}{4Tt}\right)$$

which after rewriting and changing into decimal logarithm reduces to

$$s = \frac{2.30Q}{4\pi T}\log\frac{2.25Tt}{r^2S}$$
$$s = \frac{2.30Q}{4\pi kb}\log\frac{2.25kbt}{r^2S}$$

Hence a plot of s vs. log t forms a straight line. If drawdown difference per log cycle of time is represented as Δs, the value of kb can be determined from

$$\Delta s = \frac{2.30Q}{4\pi kb} \quad \text{or} \quad kb = \frac{2.30Q}{4\pi\Delta s}$$

and if the line is extended to zero drawdown and the corresponding time t_0 determined, the value of S can be obtained from

$$\frac{2.25Kbt_0}{r^2S} = 1 \quad \text{or} \quad S = \frac{2.25Kbt_0}{r^2}$$

This method is valid for values of $u < 0.01$.

The following steps are involved in Cooper-Jacobs's method:

(i) The values of s vs. t for an observation well are plotted on a semi-logarithmic paper (t on log scale) and a straight line drawn through the plotted points (Fig. 11.2).

(ii) The line is extended to zero drawdown and the value of t_0 is determined.

(iii) The drawdown difference Δs per log cycle of time is determined.

(iv) The values of Q and Δs are substituted into the equation $Kb = \frac{2.30Q}{4\pi\Delta s}$ and Kb or T is found out.

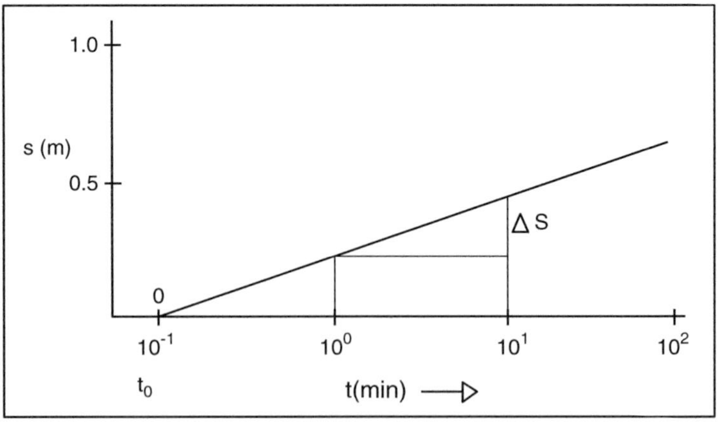

Fig. 11.2 Time-drawdown graph to compute aquifer parameters using Cooper-Jacob's method

(v) Substituting Kb and t_0 into $S = \frac{2.25Kbt_0}{r^2}$ the value of S is found out.

(vi) It is verified from $u = \frac{r^2}{4Tt}$ that the value of u is less than 0.01 for the portion of the data used.

(vii) This procedure should be repeated for all available observation wells.

5.2 *Unconfined Aquifer*

To find out the K, T and S values of an unconfined aquifer from pumping test data, Jacob's straight line method (Cooper and Jacob 1946) based on Theis (1935) equation for unsteady state flow condition in unconfined aquifer can be used. The 'third' segment of 'time-drawdown' curve (Fig. 11.3) of an observation well is used to determine transmissivity (T) and specific yield (S_Y) values because an unconfined aquifer reacts initially in the same way as does a confined aquifer. Gravity drainage is not immediate on pumping but water is released instantaneously from storage by compaction of aquifer material and by expansion of the water itself, that is, the early time response follows Theis equation with the confined "elastic" storage corresponding to storativity (S). This is the 'first' segment of the 'time-drawdown' curve. The second segment shows a decrease in slope because of the replenishment by gravity drainage from the interstices above the cone of depression and hence the flow is essentially vertical. The gravity drainage is controlled by the aquifer K_h/K_v ratio. The 'third' segment which starts after several minutes after pumping has started, represents the period during which the 'time-drawdown' curve conforms closely to the Theis type curve with gravity drainage providing storage corresponding to the specific yield (S_y).

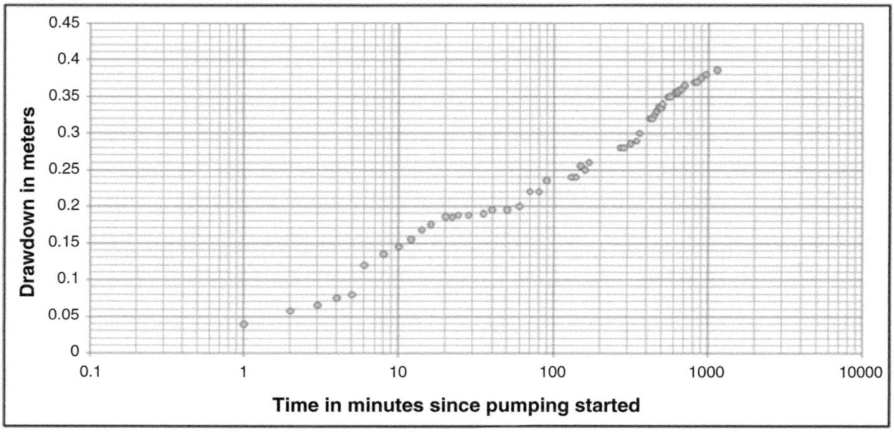

Fig. 11.3 Time-drawdown curve in an unconfined aquifer

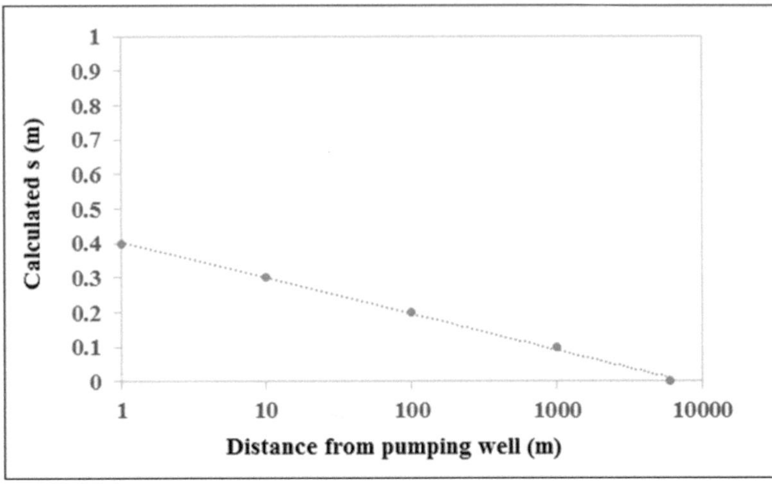

Fig. 11.4 Calculated drawdown-distance from pumping well to measure radius of influence

5.3 Determination of Safe Distance Between Two Pumping Wells

Safe distance ($2r_o$) is double the radius of cone depression or radius of influence (r_o). In unsteady state, safe distance is time dependent as the cone of depression expands with increase in pumping hours (Kruseman and de Rider 1994). The safe distance of wells depend upon a number of factors like pumping rate, and hydraulic parameters of the aquifer like transmissivity and storage coefficient. The safe distance is determined using the non-steady relation (Theis 1935). Since radius of influence depends on the balance between aquifer recharge and well discharge, the radius may vary from year to year.

The radius of influence of a well can be determined from a distance-drawdown plot. For all practical purposes, a useful comparative index is the intercept of the distance-drawdown graph on the distance axis (Fig. 11.4). After working out u, corresponding $W(u)$ values are taken from the table of Wenzel (1942). The drawdowns are worked out based on the Theis equation for 1, 10, 50, 100, 500 and 1000 m distances from pumping well. The calculated drawdown data versus distance from pumping well are plotted on semi-logarithmic graph paper with distance on the logarithmic scale. The best fit line plotted through the points is the configuration of the cone of depression (Fig. 11.4). The distance at which the best-fit line intersects zero drawdown line gives the radius of the cone of depression (r_o).

6 Conclusion

Groundwater is the most suitable source of drinking, irrigation and industrial water, supplies of which are brought to the surface by means of wells. Pumping tests are a practical way to obtain an idea of the cardinal aquifer parameters, the well's efficiency, its optimal production yield and safe distance between two adjacent wells. Groundwater levels and pumping rates measured during pumping tests provide information of the behaviour or state of 'health' of the aquifer system. Pumping test provides valuable information, helps to better understand the groundwater system and take decisions regarding optimal groundwater abstraction. However, decisions should be based on a wider understanding of the regional geology, hydrogeology and environment. It is no use blindly inserting data from pumping tests into equations without understanding the hydrogeological environment.

References

Cooper, H.H. and Jacob, C.E. (1946). A generalized graphical method for evaluating formation constants and summarizing well field history. *Am Geophys Union Trans*, **27:** 526–534.

Hantush, M.S. (1964). Hydraulics of Wells. *In:* Advances in Hydroscience. Vol. 1 (Ed.) V.T. Chow. Academic Press, New York.

Jacob, C.E. (1946). Radial flow in a leaky artesian aquifer. *Transactions, American Geophysical Union*, **27(2):** 198–205.

Jacob, C.E. (1950). Flow of ground water. *In:* Engineering Hydraulics. (Eds) Rouse, Hunter. John Wiley & Sons, New York.

Karanth, K.R. (1987). Ground Water Assessment Development and Management. Tata McGraw-Hill Publishing Co. Ltd., 720 p.

Kruseman, G.P. and de Ridder, N.A. (1994). Analysis and Evaluation of Pumping Test Data. Second Edition. Publication 47, International Institute for Land Reclamation and Improvement, P.O. Box 45, 6700 AA Wageningen, The Netherlands.

Lohman, S.W. (1972). Ground-Water Hydraulics. U.S. Geological Survey Professional Paper 708.

Stallman, R.W. (1971). Aquifer Test Design Observation and Data Analysis: Techniques of Water-Resource Investigation. U.S. Geological Survey Publication No. 03-B1.

Theis, C.V. (1935). The relation between the lowering of the piezometric surface and the rate and duration of discharge of a well using groundwater storage. *Am. Geophys. Union Trans*, **16:** 519–524.

Todd, D.K. and Mays, L.W. (2005). Groundwater Hydrology, 3rd Edition. John Wiley and Sons, 652 p.

Wenzel, L.K. (1942). Methods of determining permeability of water bearing materials with special references to discharging well methods. U.S. Geological Survey Water Supply Paper, 887.

Woessner, W.W. and Anderson, M. (2002). The hydro-malaprop and the ground water table. *Ground Water*, **40(5):** 465.

Chapter 12
Arsenic in Groundwater

John M. McArthur

1 Introduction

This chapter in not a review of arsenic (As) in groundwater. It is, rather, a brief summary of As pollution in the areas shown in Fig. 12.1. For a comprehensive account of the problem of As worldwide, the reader is referred to Ravenscroft et al. (2009).

Since 1993, the guideline value for the concentration of arsenic (As) in drinking water has been given as 10 g/L by the World Health Organization (WHO 2017). Prior to 1993, it was set at 50 µg/L. The national governments of many countries specify a concentration of As in drinking water that should not be exceeded and it is usually either 10 or 50 µg/L. In India, Pakistan and Bangladesh, it is 50 µg/L. The reason for limiting the concentration of As in drinking water is that it poses a hazard to health and, in much of the world, drinking water is groundwater. The hazard posed by As in drinking water is compounded by ingestion of As in food. Individuals at risk of As-poisoning from their drinking water may not wholly avoid the issue by switching to low-As water supplies, especially where As-rich groundwater continues to be used for irrigation and some of that As finds its way into the edible part of crops.

The adverse health effects of As in drinking water is particularly severe in southern Asia, where the degree and extent of As-pollution of groundwater is also most severe. Countries particularly badly affected are Bangladesh, northern India (especially West Bengal), southernmost Nepal and Pakistan (Fig. 12.1). In these areas, concentrations of As in groundwater typically exceed 50 g/L in a significant number of wells from which it is drawn. In Bangladesh, it is around 25% across the southern half of the country. In southern West Bengal, concentrations up to

J. M. McArthur (✉)
Department of Earth Science, University College London, London WC1E 6BT, UK
e-mail: j.mcarthur@ucl.ac.uk

P. K. Sikdar (ed.), *Groundwater Development and Management*,
https://doi.org/10.1007/978-3-319-75115-3_12

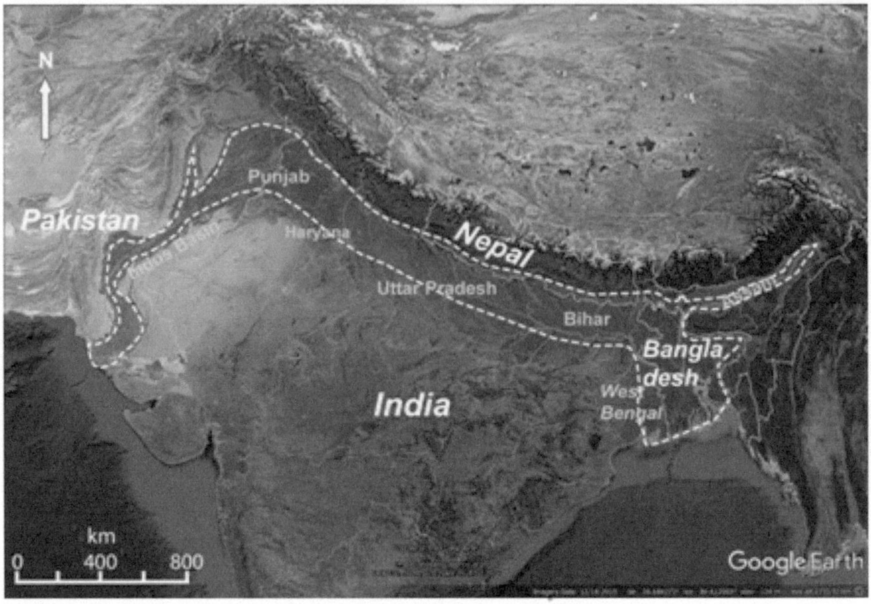

Fig. 12.1 Map of S Asia; the dotted line encloses the area prone to As-pollution of groundwater. Not all parts of those areas are As-affected. In Bangladesh, the pollution is mostly found in the southern and central regions of the country in shallow aquifers. In India and Pakistan, As-pollution is found in shallow groundwaters under alluvial plains flanking major rivers e.g. the active alluvial plains of the Brahamaputra River in Assam and the Ganga River in Bihar and upstream states, and the Terai region of southermost Nepal

4000 μg/L have been recorded. The affected (and unaffected) aquifers comprised fluvial and deltaic sands composed of quartz, mica, and feldspar that were derived from the Himalayas, Siwaliks (the southernmost foothills of the Himalayas), the Shillong Plateau that divides Assam from Bangladesh, and the Eastern Hills of Bangladesh.

2 Scale of the Problem

In this chapter, the term contamination refers to the presence in the environment of a substance at unusual concentrations. Pollution refers to the presence in the environment of a substance at a concentration that is both unusual and that causes environmental harm. The numbers of humans affected by As-pollution is not known with certainty because the four countries most affected, and discussed here, have largely rural economies with a high percentage of the population involved in agriculture and so dispersed in small communities for which accurate information on health is not

Table 12.1 Total population and percentage of rural population

Country	Total population (in million)	% of rural population
Bangladesh	158	65
Nepal	27	81
Pakistan	190	61
West Bengal (India)	93	61 in India total

Table 12.2 Summarised results of As testing in India

State	% >50 μg/L As	% >10 μg/L As
West Bengal	26	58
Assam (part)	6	26
Bihar (part)	11	29
Uttar Pradesh (part)	2.4	22
Jharkhand (part)	3.7	7.5

easy to obtain. For example, around 330 people live in the 'average' village in Bangladesh (DHS 2016) and about 65% of the inhabitants may be classified as living in rural settings in 2016 (http://data.worldbank.org/indicator/SP.RUR.TOTL.ZS; 13/07/2017). The problem of As-pollution is exacerbated by the fact that rural inhabitants seldom have access to piped supplies of water and so largely rely for their domestic, and drinking, water on private or communal hand-pumps tapping local groundwater, which is widely polluted by As. Finally, the problem is compounded again by the large numbers of consumers in these densely-populated countries. The approximate populations and the % classified as rural are shown in Table 12.1.

In Nepal, the Terai (the lowland alluvial plain) is home to about 12.5 million and some 90% of them depend on groundwater for their daily needs. It is this region where groundwater is polluted in patches by As. Shrestha et al. (2004) found 24% of groundwater wells in the Terai yielded water with >10 μg/L As, whilst 7% exceeded 50 μg/L, with wells <30 m deep being most contaminated (32% >10 μg/L As). Thakur et al. (2011) give lower figures for the Terai, with 10% of water wells giving groundwater with >10 μg/L As and only 2% having >50 μg/L As. More recent estimates appear to suggest that around 1 in 10 (1.4 million) are at risk from As-pollution of groundwater, according to UNICEF (https://www.unicef.org/wash/nepal_35975.html, 17/07/2017).

For Pakistan, as early as 2004, 36% of the population of Sindh Province in southern Pakistan was declared to be at risk from As in groundwater (Ahmad et al. 2004). For Punjab province in northern Pakistan, Ramay et al. (2006) reported that out of 34 districts, 18 have As polluted groundwater, with 40% of wells yielding groundwater with >10 μg/L As and 9% with >50 μg/L As. Similar figures were quoted in Government of Pakistan (2007).

In India, Nickson et al. (2007) summarised the results of extensive testing for As of groundwater from wells installed on lowland alluvial plains adjacent to major rivers in five states. The results were as shown in Table 12.2.

In Bangladesh, many testing programmes have been undertaken since Dipankar Chakraborti forced recognition of the problem by convening an international meeting on the issue in 1995. Early estimates by Peter Ravenscroft (DPHE 1999) based on 2020 well waters were that 51% exceeded 10 μg/L and 35% exceeded 50 μg/L. Further testing amended these figures to 46% and 27% respectively (DPHE 2001). Subsequent further testing of drinking water, as opposed to groundwater, revealed lower figures of 32% and 13% (Johnston and Sarker 2007), presumably reflecting the effects of both remediation efforts and avoidance strategies on the part of consumers.

Whilst these figures are alarming even now, some 20 years after awareness of the problem became widespread, an often overlooked facet of the figures is that whilst, for example in Bangladesh, roughly 25% of tested sources are As-polluted, it follows that 75% are not and are doing what they are designed to do; that is, supply low-As groundwater as a replacement for microbiologically-polluted surface waters which has been a traditional source of water and one that caused substantial harm, especially from diarrhoeal disease. The improvement in public health this has allowed should be balanced against the adverse effect of As on health.

3 Toxicity of Arsenic

For a detailed analysis of toxicology and wider aspects of the effects of As on human health, the reader is referred to the recent volume edited by States (2016). There is a common perception that inorganic AsIII is more toxic than inorganic AsV. For humans, the situation is more complex. Most inorganic arsenic ingested is excreted in urine as the mono- and di-methylated form, with minor amounts excreted unchanged (Thomas 2016; refs therein). For many years, methylation and excretion was viewed as a detoxification system. It is now recognised that these methylated forms are probably the more damaging forms of As in respect of human metabolism. Long-term ingestion of water containing more than the recommended amount of As can give rise to a variety of cancers e.g. of the skin, bladder, or kidneys, and diseases of the heart and lungs i.e. chronic As-poisoning. The display of symptoms lags ingestion by months to years, depending on the As concentrations in ingested food and water. The ill-effects are, to some degree, specific to individuals, with anecdotal evidence that well-nourished individuals suffer less than those on poor diets. No certain cures are known for chronic As-poisoning (as opposed to the acute form; Hughes 2016); avoidance of As in drinking water and in food (e.g. in local rice grown with As-rich irrigated water) can be recommended, but practicalities of life amongst the rural poor often precludes such action. Such avoidance strategies appear to ameliorate skin conditions but do not appear to ameliorate the long-term risks of contracting cancer.

4 Mechanism of As-Pollution of Groundwater

4.1 Anthropogenic Mechanisms

Human activity can add As to groundwater. Amongst the many potential sources are leaching from ash deposited from the smoke-plumes of brick-kilns, leaching of residues of arsenical pesticides and herbicides that are applied to crops; leaching of arsenical tick-control solutions from ponds used for dipping livestock; release of As from soils by application of phosphate fertilizer which promotes competitive exchange of PO_4 with As; the oxidation of gangue sulphides in mine-waste, in which rates of oxidation and weathering greatly exceed those that occur naturally in situ, and poor disposal practices at factories that manufacture As-pesticides.

In the areas under discussion (Fig. 12.1), these human impacts are not significant contributors to As-pollution of groundwater, except for one instance of industrial pollution in Kolkata (formerly Calcutta) by arsenical waste-products from manufacture of Paris Green, an As-rich pesticide (Chatterjee et al. 1993). That apart, invocation of some of these mechanisms results from misunderstanding. For example, it is well documented that arsenical pesticides and herbicides accumulate in soils – they sorb to soil particles, especially the iron oxyhydroxide component. This is typically not a problem until phosphate fertiliser is applied to those soils, usually when a new type of crop is grown (e.g. rice after cotton). The PO_4 in the fertiliser displaces much of the As that is sorbed on mineral surfaces and soil water becomes rich in As. The effect on crops can be devastating e.g. a rice-crop with straight head disease and no rice yield. This experience has been extrapolated, incorrectly, to suggest that, because As-polluted groundwater often contains 1–5 mg/L of PO_4, the As has been put into solution by competitive exchange with PO_4. In reality, because exchange is an equilibrium process, this amount of PO_4 could not liberate more than a few µg/L of As and cannot be responsible for the occurrence of groundwater containing tens to hundreds of µg/L as is commonly seen in alluvial and deltaic aquifers (see, for example, Ravenscroft et al. 2001; Kent and Fox 2004).

Migration of As-rich soil water downwards into aquifers is also unlikely to occur except in the most unusual of circumstances because sorption, dilution and dispersion, prevent it happening. As an end-member example, almost no migration of As away from a cattle-dip site was noted by Kimber et al. (2002), despite the soils of such sites being heavily polluted by As from its use as a tickicide at concentrations of around 1000 mg/L over many years.

Finally, it is worth mentioning one mechanism that is still invoked, incorrectly, as a cause of As-pollution: competitive exchange of HCO_3^- in groundwater for As sorbed on sediment grains. Were this mechanism important, As-pollution would be even more prevalent around the world than it is because HCO_3^- is usually the dominant anion in potable groundwaters worldwide. In the Bengal Basin of West Bengal, where As-pollution is patchy in distribution but severe where found, concentrations of HCO_3^- in groundwaters containing <10 µg/L As are around 440 ± 55 mg/L compared to around 480 mg/L ± 85 mg/L in groundwaters containing >10 μg/L As. The difference is trivial and accounted for by the fact that Fe-reduction generates HCO_3^-. Groundwaters with <200 mg/L HCO_3^- seldom contain detectable As because organic matter oxidation, a process that also generates HCO_3^-, has not proceeded sufficiently far to generate much HCO_3^-, nor cause anoxia in groundwater.

4.2 Natural Mechanisms

Arsenic finds its way into groundwater by five natural routes. These are:

1. From hydrothermal activity e.g. hot springs and geysers in volcanic areas (e.g. Yellowstone National Park, USA);
2. by oxidation of arsenic-bearing pyrite in fractured crystalline rocks and, rarely, in sediments;
3. By desorption from mineral surfaces in response to rising pH to values much above 8, the rise being driven by processes unconnected to As;
4. By microbially-mediated reductive dissolution of sedimentary iron oxyhydroxides, which strongly sorb As and PO_4 and liberate both to solution when destroyed; and
5. Evaporative concentration.

In the countries discussed in this chapter, only mechanisms 3 and 4 have been shown to operate. Reductive dissolution causes As-pollution everywhere except in northern Pakistan (Punjab Province around Lahore), where pH-induced desorption appears to be more common.

Reduction of iron oxyhydroxides occurs only in anoxic groundwater lacking both dissolved oxygen and dissolved nitrate i.e. anoxic groundwater (Gulens et al. 1979; Nickson et al. 2000). Anoxia tends to occur in stagnant or slowly moving groundwater. Such groundwater is found in low-lying, flat terrain – deltas and the alluvial plains of large rivers. Release of As to groundwater is accompanied by release of Fe, and usually by release of PO_4, which also strongly sorbs to sedimentary iron oxides.

In groundwater where reductive dissolution occurs, positive relationships are usually seen between As, Fe and PO_4, although they are often weak. The reason they are often weak is that during and after dissolution of sedimentary iron oxides, several secondary processes differentially affect As, Fe and PO_4, thereby degrading the strong positive relationships that would otherwise be seen. In the early stages of reduction, the liberated As will resorb to unreacted Fe-oxyhydroxides, as demonstrated by modelling (Welch et al. 2000) and visualised by McArthur et al. (2004). In anoxic groundwater, reduction of SO_4 can generate H_2S, which reacts with dissolved Fe and As (and undissolved Fe-oxides) to form, ultimately, pyrite containing hundreds to thousands of ppm of As, so removing both Fe and As from groundwater in a ratio different from that in groundwater (Kresse and Fazio 2003; Kirk et al. 2004). Formation of mixed-valence hydroxycarbonates (brown and green rusts) may remove Fe alone, whilst Fe and PO_4, but not appreciable As, may be removed into vivianite, an hydrated FeII-phosphate. Arsenic may be found in concentrations of

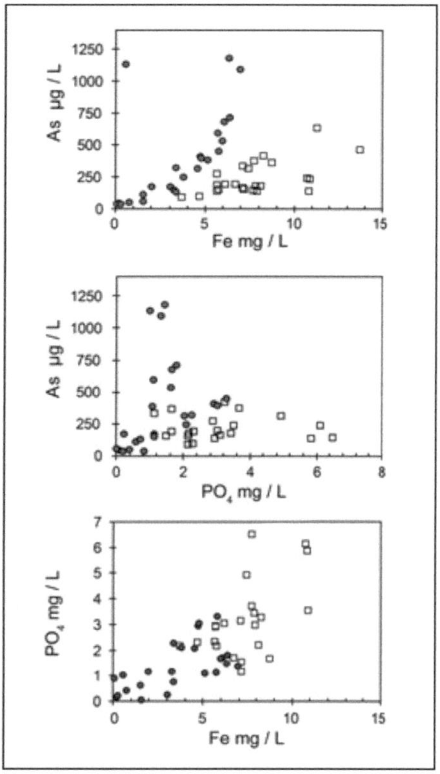

Fig. 12.2 Relation of As, Fe and PO_4, in groundwater from Moyna, West Bengal, showing the often-poor relationship in groundwater between the three. Open squares = wells 108 feet deep. Black circles = wells 120 feet deep

several hundred ppm in siderite (Mumford et al. 2012) but is not clear that the phase was pure and did not contain admixed Fe-oxyhydroxides, so its role in sequestering As in the study areas has yet to be defined. Finally, both Fe and As may be removed by sorption onto freshly precipitated Fe-oxyhydroxides as a result of anoxic oxidation of Fe(II) by NO_3 (Smith et al. 2017). Finally, sediments are not homogenous and will have a range of Fe/As ratio for their oxides; that heterogeneity will be reflected in the As/Fe of groundwaters. Given all of the above, it is surprising that As, Fe and PO_4 ever show co-variance in groundwater as they demonstrably sometimes do (Fig. 12.2).

The process applies everywhere in the regions discussed except northern Pakistan (of which exception see later). It is encapsulated in Fig. 12.3, which oversimplifies the process but captures its main elements. Glaciers in the Himalayas grind rocks into fine particles that are eventually carried downstream to accumulate in the delta. During grinding and transport, the grains are chemically weathered and gain a coating of iron-oxyhydroxide from the weathering of Fe-bearing minerals. The coating is patchy and thin because weathering is slow owing to the low temperatures and absence of vegetation in the source regions. Transport downstream is, in geological terms, very rapid, so preventing much weathering at the higher temperatures in the plains. During their passage downriver to the delta, the oxyhydroxides

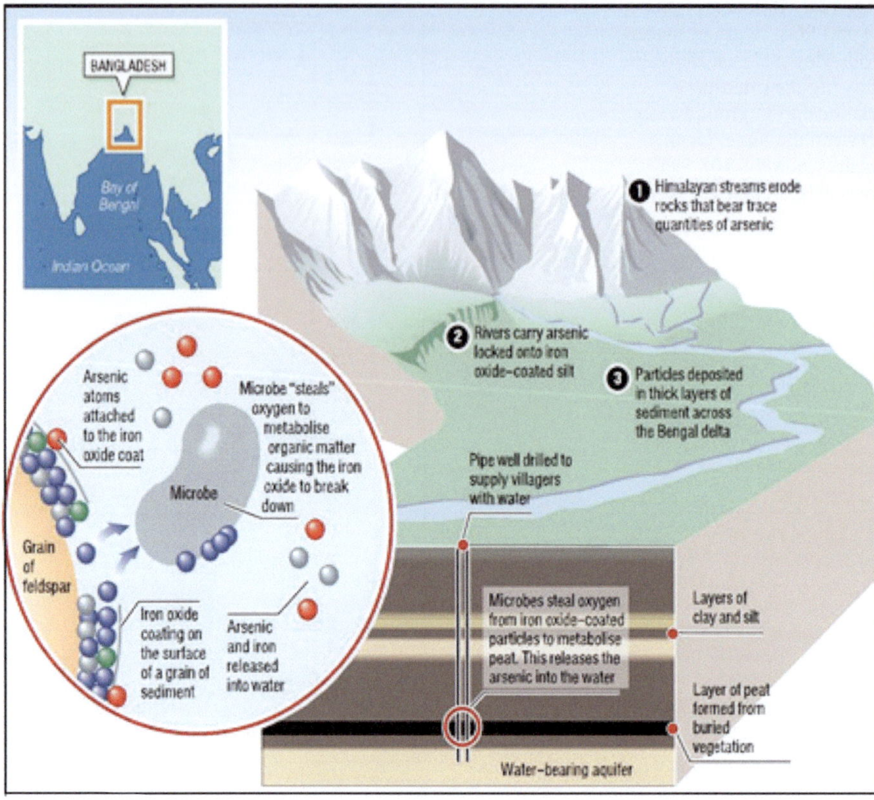

Fig. 12.3 Diagram of how As-pollution arises in the alluvial plains of the Ganga and Brahmaputra Rivers, and the rivers' delta, herein termed the Bengal Basin. The Indus River Basin in Pakistan shares the source elements of this process, but the drier climate in Pakistan, compared to further east, limited the growth of vegetation and so limited the amount of organic matter buried in the sedimets. In northern Pakistan, the amounts were too small to drive reduction of Fe-oxyhydroxides, so a different process drives As-pollution in that region – see the text for details

coatings strongly sorb As and PO_4 from river water. Once in the delta, they are rapidly buried by later sediments carried down the rivers. The hot, wet, climate of the delta allows lush growth of vegetation (dense jungles, swamps and mangroves, before humans removed them). The organic matter from this vegetation, particularly peatlands, was buried in the sediments along with the As-bearing Fe-oxyhydroxides. The sediments contain bacteria that use the organic matter as food; they oxidise it to gain energy for growth and to build their cells. To do so they need oxygen. The only oxygen available is that in the sedimentary Fe-oxyhydroxide, so they strip it from the Fe, turning Fe(III) into Fe(II), and leave the unwanted Fe(II) in the groundwater. The As and PO_4 that was sorbed to the Fe-oxyhydroxide has nowhere to go and ends up in the groundwater along with the Fe.

In northern Pakistan, a different situation applies. Whilst reduction of Fe-oxyhydroxides creates minor amounts of As-pollution in shallow groundwaters (e.g. Nickson et al. 2005; because some human activity releases organic matter to the surface and shallow subsurface), more prevalent as a mechanism of As-pollution appears to be release of As from sediments as pH rises (Farooqi et al. 2007a, b). This mechanism might be regarded as competitive exchange of As with hydroxide ions. It has, to date, been seen operating only in oxic environments (Robertson 1989; Welch et al. 2000, ibid.). The mechanism is invoked because laboratory studies of As sorption to iron-oxides shows that the strength of sorption decreases as pH increases. The characteristic signature of the process is a positive relation of pH to As concentration. Groundwaters that show this mechanism may also be rich in NO_3.

In Fig. 12.4 are compared the characteristic signatures of pH-induced desorption shown by groundwater from northern Pakistan, and Fe-reduction shown by groundwater from southern Pakistan. The northern groundwaters show a strong positive correlation between pH and As concentration, a correlation that is absent in groundwater from southern Pakistan. Concentrations of As and Fe are anti-correlated in groundwaters from northern Pakistan, but are positively correlated, although weakly, in groundwaters from southern Pakistan. Finally, groundwaters from northern Pakistan contain high concentrations of NO_3 whereas those from southern Pakistan contain little or none.

5 Drivers of As-Pollution

The previous section explained how As-pollution occurs, but not why the reactions causing it occur in the first place. This section does so.

5.1 Driver of Fe-oxyhydroxide Reduction

Reduction of iron oxyhydroxides is a by-product of the microbial metabolism of organic matter (Froelich et al. 1979; Gulens et al. 1979; Nealson 1997; Chapelle 2000). The oxygen used to convert organic carbon into bicarbonate is taken from iron oxide, thereby reducing the FeIII to FeII and releasing it to solution. The As and PO_4 sorbed to the Fe-oxyhydroxides are released to groundwater. As organic matter is oxidised, the organic-N and organic-P it contains are also released as NH_4 and additional PO_4. Both are present in the groundwaters of the alluvial aquifers of southern Pakistan, the Bengal Basin, and Nepal. The reaction, modified trivially

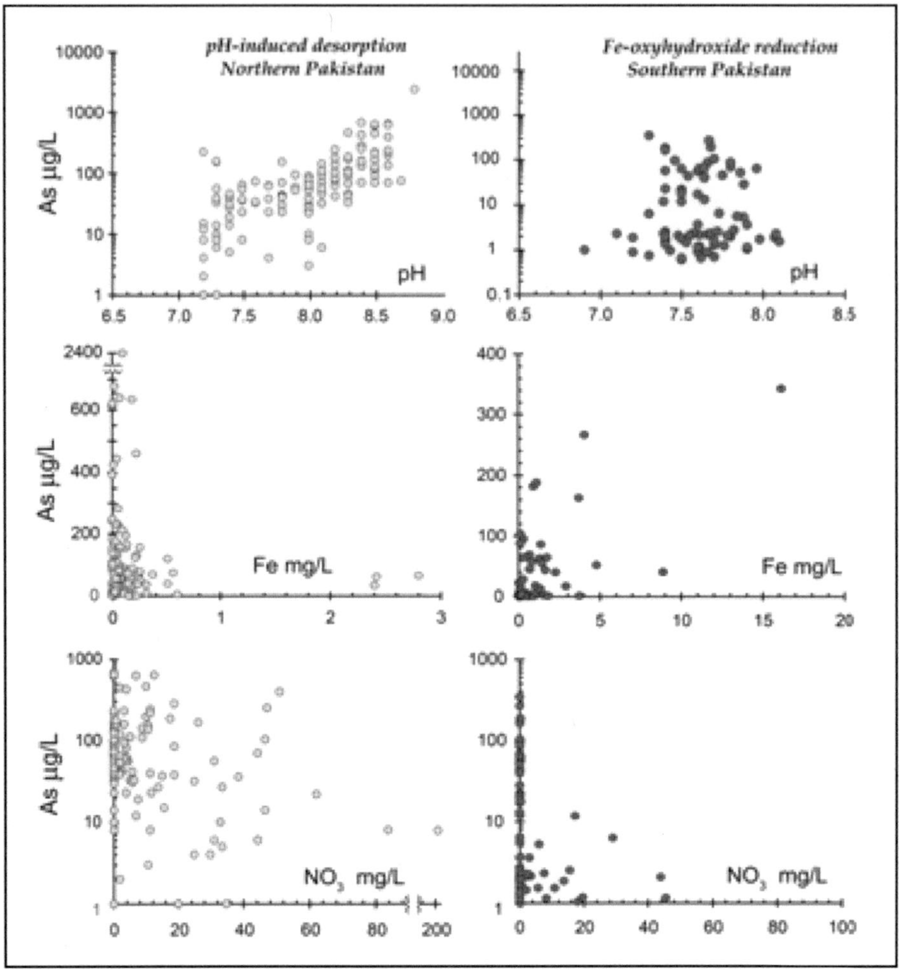

Fig. 12.4 Comparison of chemical signatures of two mechanisms of As-pollution: pH-induced desorption operates in northern Pakistan.and reduction of Fe-oxyhydroxide operates in southern Pakistan

from Froelich et al. (1979) and written to include calcite precipitation from the alkalinity generated, is:

$$\begin{aligned}&(CH_2O)_{106}(NH_3)_{16}(H_2PO_4) + 424\,FeOOH + 758\,Ca^{2+}\\&+652HCO_3^- = 758CaCO_3 + 16NH_4^+ + H^+ + H_2PO_4^-\\&+424Fe^{2+} + 636H_2O\end{aligned} \tag{12.1}$$

The source of the organic matter (OM) driving reduction in the regions is a matter of debate. There are only a few possible sources, but which are most important is still not clear, despite many years of research. It is likely that the contributions of each

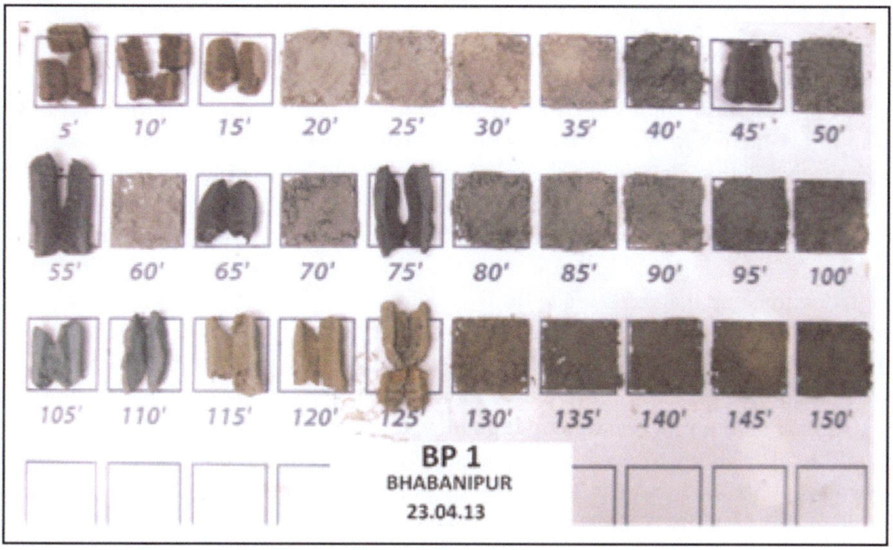

Fig. 12.5 Typical sediment profile in the shallow, post-LGM aquifer sands of the Bengal Basin. Depths are in feet below ground level. Note the red, oxidised, LGM palaeosol at 115–125 feet

source differ between localities. The possible sources were listed by Ravenscroft et al. (2001) as disseminated organic matter in sediments, OM in the upper silt/clay aquitard that covers much of the basin, buried peat, and OM leaching from latrines. Much research over the past 16 years allows a more informed discussion than before. To this list may now be added river water and pond water (natural and man-made), both of which carry OM and might recharge shallow aquifers, and OM in fine-grained deposits filling abandoned river channels (Desbarats et al. 2014; Sahu and Saha 2015; Donselaar et al. 2017). We discuss each in turn.

The aquifer sands of the post-LGM aquifers appear to contain almost no disseminated OM and this source can be discounted. To illustrate possible other aquifer sources, Fig. 12.5 shows a cored borehole from Bhabanipur, in southern West Bengal. Note the reduced, grey, colour of the post-LGM aquifer (above 105 feet), the red and oxidised LGM palaeosol, at 115–125 feet, and the underlying brown, oxidised, sands that cap the pre-LGM aquifer and which pass downward into grey, pre-LGM sands where drilling reaches sufficient depth, typically >100 m. At this site, the upper 15 feet are red and oxidised from weathering. The depth of the oxidation depends on the porosity of the upper aquitard; where it is less permeable than at Bhabanipur, it is less oxidised and sediment below the oxidation zone is often organic-rich. Leaching of this OM downward in recharge may well carry OM into the underlying post-LGM aquifer to drive As-pollution. At 40, 55, 65 and 75 feet, organic-rich, peaty, units occur. Organic matter leaking from these units might well drive reduction and As-pollution locally. Buried deposits of peat *sensu stricto* occur widely across the Bengal Basin and, where they occur, they are likely to be a major source of OM for Fe-reduction.

Pit latrines are common in rural areas worldwide, and the areas under discussion are no exception. Pit latrines provide a concentrated pool of OM which degrades via fermentation and generates OM-rich pollution plumes to groundwater that, in turn, generate reducing conditions. Pit latrines and water wells are often sited no more than a few metres to tens of metres apart because both, for convenience, are sited close to habitation. It follows that pit latrines should be driving As-pollution. The matter is, however, not quite that simple because pit latrines also generate plumes of NO_3-rich water, and NO_3 suppresses reduction of Fe-oxyhydroxides and so suppresses As-pollution. It may be that near the latrine, methanogenesis rules rather than Fe-reduction, and that as distance from the latrine source increases, successive redox zones (SO_4-reduction, Fe-reduction, NO_3-reduction) are generated and that only in the zone of Fe-reduction is As-pollution generated. Delineation of these redox zones around a pit latrine has yet to be undertaken in the regions under consideration.

Organic matter in river water and pond water should be able to drive reduction as it infiltrates. Within a few hundred metres of rivers, infiltration of river water in response to abstraction of groundwater by pumping is known to generate reduction fronts in aquifers; for a review, see Farnsworth and Hering (2011). Much of the As pollution in the Bengal Basin is far from rivers, however, so this influence can be of local importance only. In contrast, man-made ponds are made to hold water, not leak it to groundwater, so OM in pond water has been discounted as being of little importance for Fe-reduction in the Bengal Basin (Sengupta et al. 2008; Datta et al. 2011; McArthur et al. 2011; Desbarats et al. 2014), although Lawson et al. (2013) take an alternative view.

Finally, Desbarats et al. (2014), Sahu and Saha (2015) and Donselaar et al. (2017) note the abundance of abandoned river channels both in the Bengal Basin (Desbarats et al. 2014) and upstream in the Gangetic alluvial plain (Sahu and Saha 2015; Donselaar et al. 2017) and propose that the OM that drives As-pollution is sourced from the fine-grained sediments that accumulate in such channels after abandonment. Sahu and Saha (2015) suggest persuasively that it is fine-grained, organic-rich, capping to aquifer sands in abandoned palaeo-channels that drive Fe-reduction by leaking OM to the underlying aquifer. The widespread distribution of abandoned river-channels across the delta region and alluvial plains of the region suggests this source is worthy of concerted further study. The peaty, OM-rich, units in Fig. 12.5 may represent the remnants of such channel fills.

Attempts have been made to use compound-specific biomarkers to trace the source of the OM driving As-pollution. Differences in organic moieties between surface water and As-rich groundwater were noted by Mladenov et al. (2010, 2015) who interpreted the differences (cautiously) as possibly indicating a source of DOC in buried peaty sediments within the aquifer. In contrast, ^{14}C dating of microbial DNA was used by Mailloux et al. (2013) to suggest that most DOC driving Fe-reduction came from a *surface* source, interpreted to be shallow soils in this instance. The difficulty with invoking a surface source for the OM driving Fe-reduction is that DOC reacts with Fe-oxyhydroxides to cause As-pollution i.e. DOC is a reactant, not a product of reaction Eq. (12.1). Its concentration should decrease downward as Fe-reduction proceeds and DOC is consumed. Profiles of DOC in groundwater show it increasing with depth as As increases (Harvey et al. 2002; Mailloux et al. 2013).

5.2 *Driver of pH-induced Desorption*

The mechanism that drives pH-induced desorption is chemical weathering of sediments, especially Ca-poor sediments, ion exchange, and evaporative concentration of waters. pH-induced desorption probably occurs in the north of Pakistan because the hot and seasonally-dry climate promotes not only evaporative concentration, but also limits flushing of the evaporated water (Greenman et al. 1967). The climate is monsoonal and hot, so in the dry season (October–May) evaporation is strong. Recharge by rainfall of around 600 mm/year that occurs during the monsoon season (June–September) is insufficient to flush the evaporated groundwater from the aquifers. Evaporation occurs of water in which the ratio of alkalinity (mostly as HCO_3^-) is much greater than the concentration of Ca. Precipitation of calcite during evaporation reduces the concentrations of both.

$$Ca^{2+} + 2HCO_3^- = CaCO_3 + H_2O + CO_2 \quad (12.2)$$

If alkalinity is in excess, the ratio HCO_3^-/Ca increases as evaporation proceeds, thereby increasing pH as ever higher concentrations of alkalinity remove ever more H^+ from solution in order to maintain solution equilibria. Ion exchange removes Ca, allowing HCO_3 to increase, with the same effect. A side effect is to cause As to sorb less strongly to sediment particles.

The Punjab of Pakistan has been subject to irrigation with river water for hundreds of years, a trend that has increased greatly in the past 80 years. Whether the alkaline groundwater has evolved in response to this irrigation, or was present naturally beforehand, is under investigation (e.g. Naseem and McArthur, 2018).

6 Distribution of As-Pollution

6.1 *Distribution at the Regional Scale*

In the regions under consideration, the distribution of As-polluted groundwater appears highly heterogeneous (DPHE 1999, 2001; van Geen et al. 2003) but the apparent complexity reflects mostly the fact that most maps showing the aerial distribution of pollution use aggregate data from a range of well depths rather than a single depth. This compounds lateral changes with vertical changes, obscures patterns, and leads to confusion. In reality, the overall distribution of As-pollution is simple at the broad scale and reflects the geomorphological evolution of the region

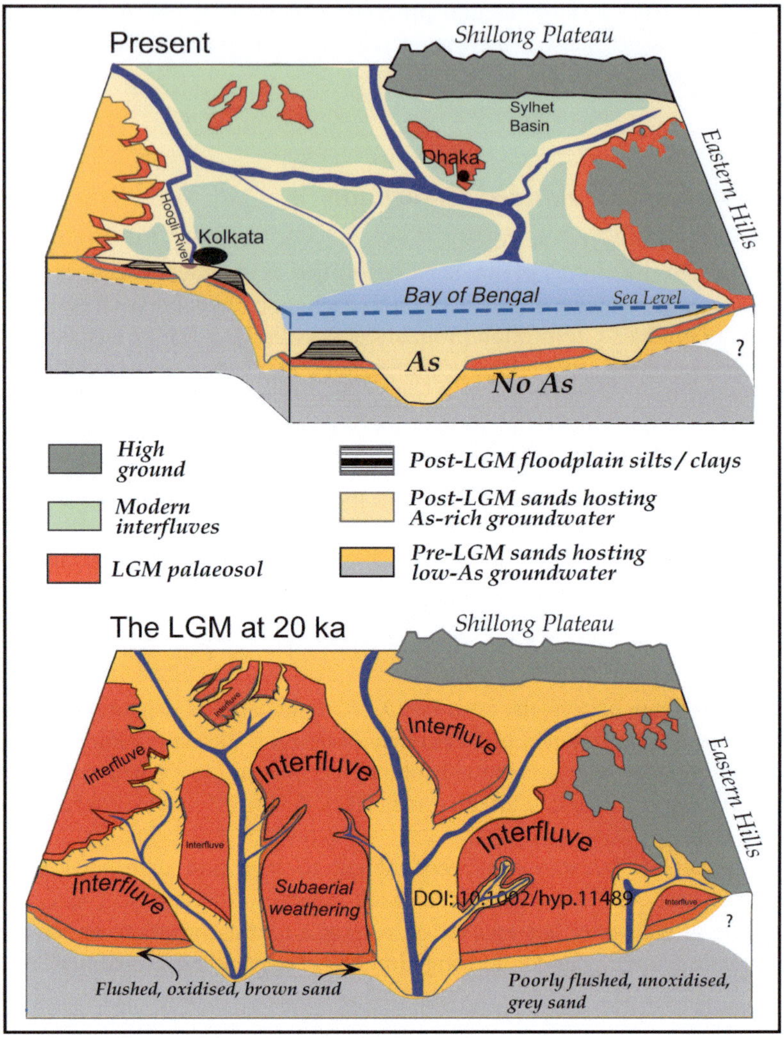

Fig. 12.6 Relation of sea-level to the distribution of As in groundwater of the Bengal Basin (adapted from McArthur et al. 2008). Pollution by As is found only in sediments (aquifers) deposited after the LGM. (After Umitsu 1993; DPHE 1999)

considered over the past 125 ka; that is, over the last full glacial climate-cycle. That evolution, and its relation to As, is summarised in Fig. 12.6. Around 125 ka, sea-level was close to its present level. Sea-level declined during the next 100 ka mostly in response to the growth of polar ice-caps, and to a small extent because of the global lowering of the snow-line and the growth of high-altitude glaciers. Around 20 ka the decline stopped as ice-caps stopped growing; this time is termed the last glacial maximum – often abbreviated to LGM. At that time, sea-level was around 120 m below its present level and ice-caps covered much of North America and northern Europe. Between 20 ka and 6 ka the climate warmed and much of the ice melted, leaving today's smaller ice-caps. As the ice retreated northwards, sea-level rose rapidly in response (Umitsu 1993); it stabilised at around its present level at about 6 ka and has varied only by a few metres since. It was the long decline in sea level by 120 m and the rapid rise thereafter that had a major influence on the distribution of As-pollution.

Whilst Umitsu (1993) was the first to highlight the effect of sea-level change on sediment distribution in the Bengal Basin, he made no mention of As-pollution in groundwater. The first to notice the connection between sea-level and As-pollution was Peter Ravenscroft, as main author of DPHE (1999), who noted that As-pollution was confined to groundwater in aquifers deposited since the sea-level lowstand at the LGM. He wrote "were the world in a low sea level stand there might not be an arsenic problem in the Bengal Basin" (DPHE 1999, Main Report, pp 4–10).

The As-polluted aquifers were formed between 20 and 6 ka as glaciers in the Himalayas retreated in response to global warming. As the climate warmed, the glacial deposits accumulated over the previous 100 ka were rapidly flushed down rivers to accumulate in the accommodation space created by rising sea-level as it flooded the previously-lowland areas. Today, groundwater in the aquifers deposited after 20 ka (post-LGM aquifers) are usually As-polluted. Groundwater in deeper aquifers (pre-LGM aquifers) usually contain <10 μg/L As and typically much less (DPHE 1999).

Upstream of the Bengal Basin, in Assam, Bihar and Uttar Pradesh, the direct effects of sea-level change were not felt, but the effects of climate-change that accompanied the sea-level changes were felt strongly. The period between 9 ka and 5 ka, known as the Holocene Climatic Optimum (also termed the hypsithermal), was warmer and wetter than the climate of the LGM. The wetter climate lead to stronger river flow, erosion, and carriage downstream to the delta of the glacial sediments stored in the catchment prior to and during the hypsithermal (Goodbred 2003). During this period, the alluvial plain was an erosional, rather than depositional, environment. After 7 ka, the climate became drier, whilst remaining monsoonal, so river flow decreased and the scoured rivers partially refilled their courses with fresh sediment. Arsenic pollution of groundwater has been reported in the alluvial plains of the rivers upstream of Bangladesh (Chakraborti et al. 2003; Sahu and Saha 2015; Kumar et al. 2016) and it seems to be largely within these post-hypsithermal sediments that groundwater is As-polluted (Sahu and Saha 2015; Donselaar et al. 2017). As in the Bengal Basin, deeper groundwater contains <10 μg/L of As. An approximation to this history is visualized in Fig. 12.7. The As is sourced by reduction of Fe-oxyhydroxides and the driver appears to be OM in infilled abandoned river channels where fine-grained and organic-rich sediment either caps underlying sand aquifers (Sahu and Saha 2015) or lies upflow of them (Desbarats et al. 2014).

On the northern fringes of the alluvial plain, approaching the Siwaliks, the depositional setting is again different, with rivers flowing north to south across outwash fans formed of both Himalayan and Siwalik sediments. Here, As pollution

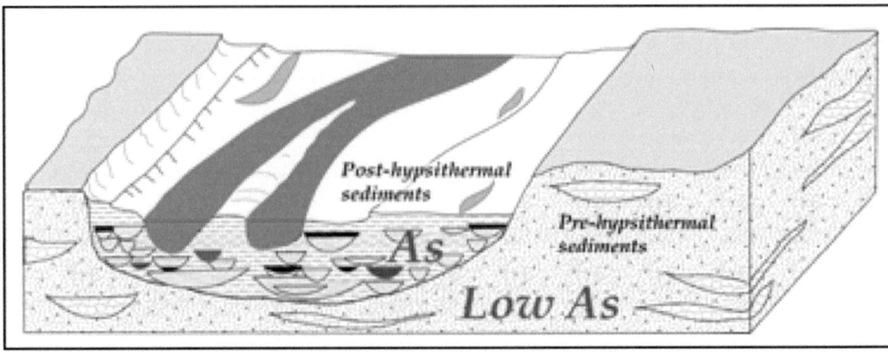

Fig. 12.7 Block diagram of the alluvial plain of the Ganga River in Bihar. During the hypsithermal, the environment was erosional and large river channels were cut into pre-hypsithermal sediments. After the hypsithermal, reduced river flow allowed the valley to partially fill with more recent sediments. The former form low-As aquifers, the latter form As-polluted aquifers. Black infill represent organic-rich fills to abandoned river channels. Modified from Goodbred (2003) and Sahu and Saha (2015)

appears to be confined to thin sand aquifers apparently comprising infills of abandoned river channels enclosed in thick sequences of silt and clay which is often organic-rich (Gurung et al. 2005; Brikowski et al. 2014). Leakage of OM from these overlying clays is the likely driver of As-pollution.

In Pakistan, the sediments in the northern part of the Indus Basin are As-polluted by the mechanism of pH-induced desorption, whilst those of the southern Indus Basin are polluted by Fe-reduction. The relation of the pollution to sediment architecture and nature is all but unknown, as little has been published on the matter. Nevertheless, similar considerations probably apply to the upper Indus Basin as outlined by Goodbred (2003; his Fig. 12.12) for the Ganges alluvial plain – roughly shown in Fig. 12.5. It is not known whether a buried palaeosol controls the distribution of As in the lower Indus Basin as it does in the Bengal Basin.

The broad-scale control of As-distribution by climate, and by its effect on sea level, has led to the vertical distributions of As-pollution seen in Fig. 12.8. Whilst the details of this distribution differ from place-to-place, As-pollution is broadly confined to the post-LGM sediments. In southern Pakistan, depth data is sparse but what there is suggests that As-pollution peaks around 30 m depth (Fig. 12.8). In northern Pakistan, a lack of published information precludes analysis of depth distributions, and it can be said only that pH-induced desorption can be seen in groundwaters ranging from 20 to 200 m in depth (Farooqi et al. 2007a, b). Depth information is available for the Bengal Basin (West Bengal and Bangladesh), and for the Terai. In both regions, concentrations of As increase downwards from essentially zero at the surface to a maximum between 20 and 50 m depth before decreasing again as depth increases (Fig. 12.8). In the Terai of Nepal, the maximum appears to occur around

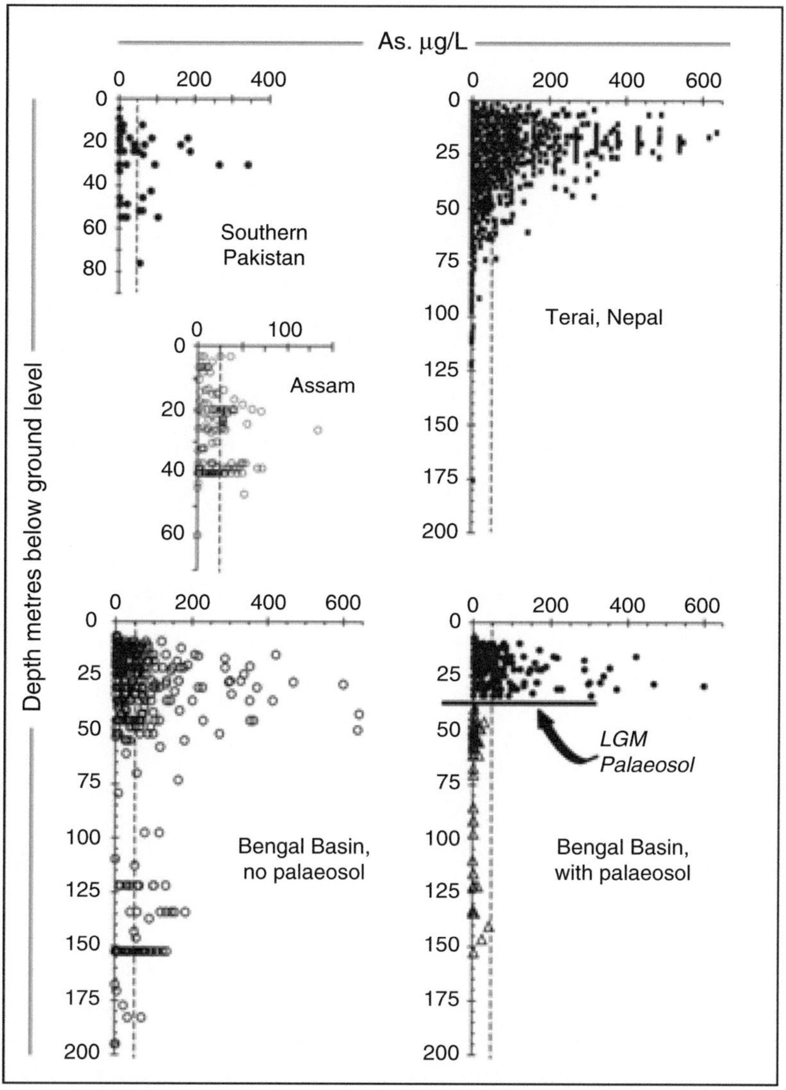

Fig. 12.8 Vertical distribution of As pollution in southern Pakistan, the Terai of southernmost Nepal, and the Bengal Basin, where profiles differ depending on whether the LGM palaeosol is absent or present. Assam profile from Choudhury et al. (2017)

20 m below ground level with values <50 μg/L occurring below 75 m depth (Shrestha et al. 2004).

In the Bengal Basin, some subtleties modify the vertical distribution of As-pollution. In many areas, a prominent palaeosol horizon is present at depths that range up to 40 m

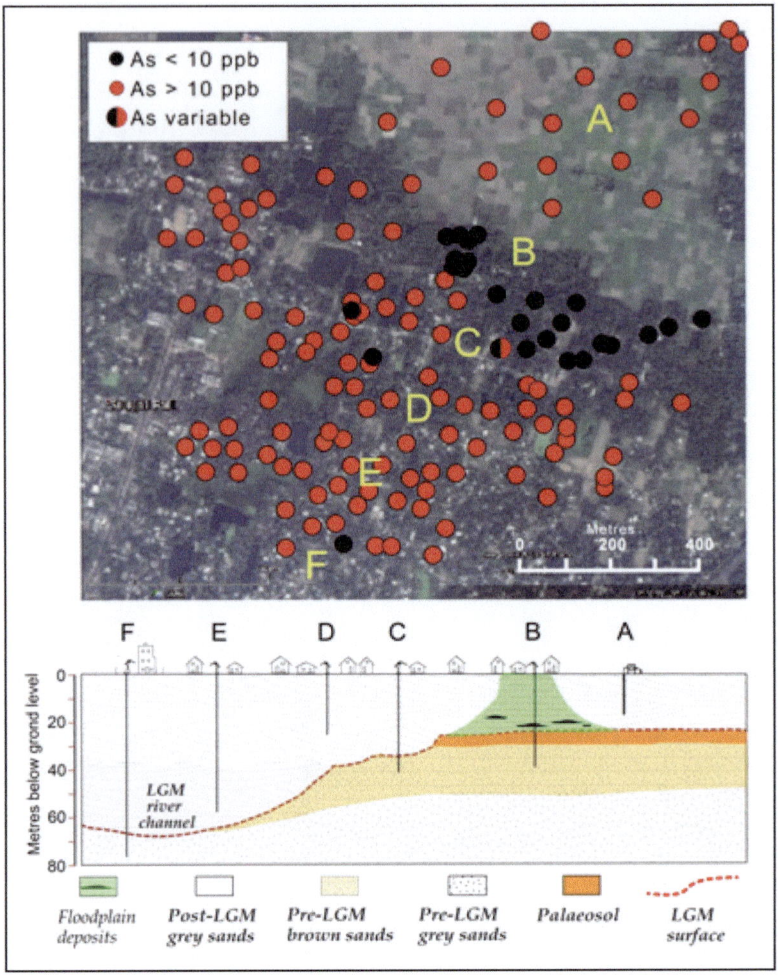

Fig. 12.9 Cross section at village scale through sediments of the Bengal Basin showing how the disposition of pre-LGM and post-LGM sediments interact with the depth of wells to provide a surface pattern of pollution that appears at face value to be complex, but is, in reality, of great simplicity. See accompanying text for an explanation of As pollution at Sites A to F

in the basin centre, and are progressively more shallow as the basin margins are approached, where it crops out (Fig. 12.6). The palaeosol is usually overlain by As-polluted post-LGM sand aquifers and underlain by unpolluted pre-LGM aquifers that contains <10 ppb of As and generally much less. The juxtaposition introduces a step-change in the vertical profile of As-pollution, and one normally disguised in plots of the vertical distribution of As-pollution (Fig. 12.8). Where the palaeosol is absent, the maximum concentrations of As occur around 30 m depth.

6.2 Distribution at the Local Scale

The sediment architecture presented above effects hydrochemistry and provides a framework within which to understand in outline both the patchy distribution of As laterally, its distribution vertically (already discussed; Fig. 12.8) and the likely effects of time on the As concentration in a well (discussed later). Figure 12.9 conceptualises this understanding by showing the distribution of As-pollution across a model village, together with a cross-section through the village, with typical wells shown at a range of depths. Although conceptual, the model quite closely represents the reality on the ground in Moyna, West Bengal (McArthur et al. 2004).

The cross section shows the landscape of the LGM now buried by post-LGM floodplain silts, clays and peats (at B) and by post-LGM river sands (at A, C-F). At LGM time, A and B would have been dry land, C would have been the banks of a river, and D, E and F would have been in the river. Where there was dry land at the LGM, weathering developed a thick soil, termed a palaeosol, and weathered and oxidised underlying pre-LGM sands to depths of around 30 m. The depth of this weathering was controlled by the depth of local rivers with respect to surrounding land; the palaeosol simply represents the extreme end-member result of that weathering and oxidation.

The distribution of As pollution across this generic model (and across Moyna) is easily understood in terms of sedimentology and well depth. At A, local farmers have installed motorised 4-inch wells in their fields to provide irrigation water. The wells are installed at a shallow depth because drilling deeper costs more. All found good flows at around 20 m depth, so did not drill deeper. As a consequence, all the wells are installed in the post-LGM aquifers and all are As-polluted. Had they drilled to 50 m, all would have obtained low-As water. Moreover, their pumping has not caused As-polluted groundwater to move down from the post-LGM aquifer because the LGM palaeosol, being all but impermeable, prevents it.

At B, on the northern fringes of Moyna, houses are on floodplain deposits – clays, silts and peats that are insufficiently permeable to form a useful aquifer. Wells must be installed beneath the floodplain deposits in the first aquifer encountered. This just happens to be the brown-sand, pre-LGM aquifer. Fe-reduction does not occur in this aquifer, so well waters are low in As – typically well below 10 g/L. Further into Moyna, at C, a well is installed 7 m into the brown sand of the pre-LGM aquifer. Initially low in As, concentrations increased with time after a few years of pumping as As-polluted groundwater from the overlying post-LGM aquifer was drawn in by pumping of the well. This increase is analysed in detail in Sect. 7.2.

Wells at D and E are installed in post-LGM grey sands so are As-polluted. The owner of well E installed it at 60 m depth in an effort to avoid As-pollution, having heard that it worked elsewhere. Unfortunately, that 'elsewhere' was at Site B. He was not to know that Site B was on floodplain deposits whilst he was living on an old river-channel that cut into the LGM landscape to 60 m depth and had subsequently been filled with post-LGM sands. The local drilling technology (using iron pipe in West Bengal) can drill only to 60 m (it is 90 m in Bangladesh where plastic pipe is

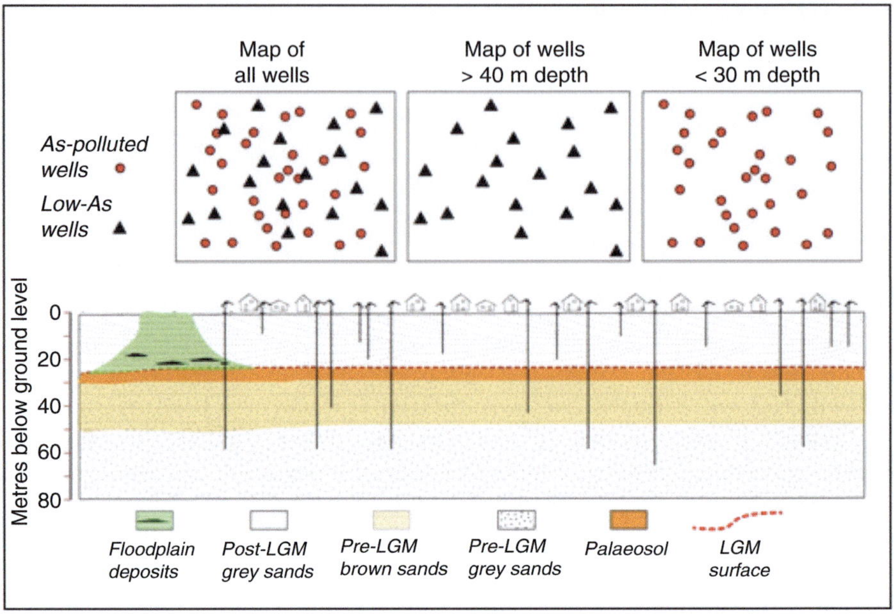

Fig. 12.10 Diagram of how well-depth can alter the apparent geographical distribution of As-pollution

used) so wells around and at E and F are installed at depth no greater than 60 m and, as a consequence, most are in post-LGM sands and are As-polluted. Finally, the well at F was drilled to 70 m depth because the owner was wealthy. He could afford to drill a deep well i.e. deeper than 60 m depth, a process that requires a drilling rig that is large and expensive compared to the drilling method used for shallow wells, so his well is in pre-LGM sediments and is (currently) free of As.

It is worth pursuing the aspect of depth on distribution further, because it is a concern wherever maps are made of As-pollution. Unless all wells are at the same depth, the map will give a false picture of the distribution of the pollution. As a consequence, remediation and avoidance measures may fail. This can be illustrated simply (Fig. 12.10). This figure shows an extended village located above a sedimentary sequence in which a pre-LGM aquifer is capped by the LGM palaeosol which is, in turn, overlain by post-LGM aquifer sands.

In the village, owners have installed wells at a range of depths. Deeper wells have penetrated the LGM palaeosol and have well-screens in the post-LGM aquifer, so all provide low-As groundwater. Wells at depths <30 m are all installed in post-LGM aquifers and all are As-polluted. If all wells are used to provide a map of As-pollution, a confusing picture is obtained. If only wells <30 m deep, or only wells >40 m deep, are mapped, completely different patterns are obtained for the distribution of As (Fig. 12.10). All three maps provide information on the distribution of As, but unless well depth is factored into an understanding of that distribution, false conclusions might be drawn e.g. that the area is free of As-pollution.

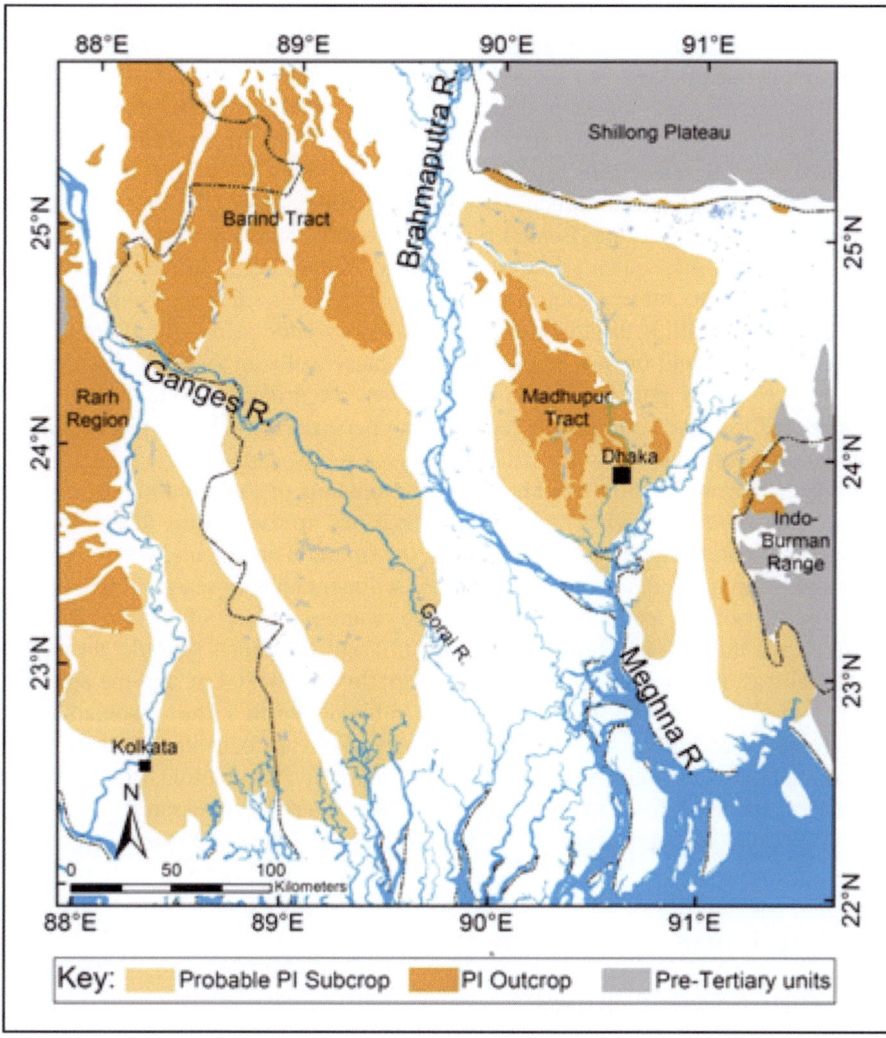

Fig. 12.11 In brown shading are shown areas of the Bengal Basin in which the situation in Fig. 12.8 pertains, and where drilling wells to 60 m depth would likely provide low-As water long-term. Modified from Hoque et al. (2014a)

The generic model shown in Figs 12.9 and 12.10 is applicable across the Bengal Basin. Some villages are entirely on floodplain deposits (Site B of Fig. 12.9), as is the city of Dhaka; the inhabitants are free of As-pollution in groundwater and have seldom heard about the matter. Some villages are over deep palaeo-channels (E, F of Fig. 12.9) and most wells are As-polluted, with dire consequences. A few villages, such as Moyna and Khaki, straddle floodplain and palaeo-channel deposits, leading to apparently puzzling distributions of As-pollution in well waters – until the underlying sedimentology is understood. Finally, the areas known as the Barind and Madhupur Tracts (Fig. 12.11) are outcrops of weathered palaeosols that have developed continuously over many tens of thousands of years, having not been below sea-level. Wells in these areas are free of As, as are the wells of Dhaka, which sits atop this thick palaeosol.

One of the tragedies of As-pollution in the Bengal Basin is that many villages occupy a position similar to that shown in Fig. 12.10. Wells are typically installed to a shallow depth because drilling deeper cost more. Simply by drilling 20 m deeper, to 50 or 60 m depth, well-owners would have avoided As-pollution entirely. An estimate of the regions of the Bengal Basin where this confusion may have arisen is shown in Fig. 12.11, which shows areas of the Bengal Basin in which the LGM palaeosol is present either as outcrop (darker brown) or as subcrop (lighter brown). In all these areas drilling through the LGM palaeosol reaches a low-As aquifer. In the areas coloured in light brown, where the LGM palaeosol is in the subsurface, many wells are needlessly polluted by As because they are drilled to a depth that is too shallow: drilling just 20 m deeper would have penetrated the LGM palaeosol and provided low-As water.

Whilst elements of this model, such as a relation of As-pollution to climate change, have some relevance to the fluvial regimes up-river in the alluvial plains of the Ganga, Brahmaputra and Indus rivers, the situation up-stream is both different and more heterogeneous. There the control of sedimentation by sea-level was absent and climate control paramount. It was the waning power of rivers after the hypsithermal that allowed renewed channel fill, and continued tectonic activity in the Himalayas, that created space for river migration of 10–20 km, and the accumulation of sediments within this shifting alluvial basin. The association of As-pollution with specific facies types, i.e. thick clay caps over thin sand aquifers, may provide guidance that enables prediction of where As-pollution will be found. Further exploration of this model requires immediate intensive research.

7 Migration of As-Pollution

7.1 Broad-Scale Changes

A major concern regarding natural As-pollution, as opposed to the industrial spill in Kolkata reported by Chatterjee et al. (1993), is whether As will migrate to pollute currently unpolluted aquifers and if so at what rate. In most of the areas where As-pollution is found, groundwater is pumped for both domestic use and irrigation during the dry season, so groundwater flow has been increased over natural flow (DPHE 1999, 2001; Harvey et al. 2002), a fact that should promote the migration of pollution.

Migration of As will be evidenced by changes in As-pollution in groundwater over time and there have been a number of such variations reported. In some cases, no data was presented to support the claims so they are not discussed here. In a few cases, variations are attributable to real changes in concentration of As in the groundwater, rather than attributable to analytical or sampling practice. The latter is a particular concern as it is established wisdom in hydrogeology that the chemical composition of water pumped from a well changes with the intensity of pumping; ensuring a reproducible pumping regime is seldom explicitly documented during reports of monitoring.

Recharge will always be variable in composition, so changes through seasons in very shallow wells (say 5–10 m depth) should be no surprise in the sand aquifers of the Bengal Basin, Indus Basin, and upstream alluvial plains. In sand aquifers, compositional change in recharge will be increasingly damped with increasing depth as hydrodynamic dispersion mixes groundwater, so variations in groundwater deeper than 15 m would not be expected on short timescales of months to a few years. Whilst early modelling studies (DPHE 1999, 2001) suggested the spread of As would be slow, a particular concern is whether pumping from the deep, pre-LGM, aquifer, as a mitigation measure to avoid As, will draw down into it As-rich groundwater from shallow aquifers. This appears not to be happening (yet) in Bangladesh, where major cities such as Dhaka and Khulna have drawn municipal supply from deep aquifers for decades without experiencing invasion of As. Indeed, in Dhaka, despite drawdown of 60 m, concentrations of As in groundwater remain <10 g/L (Hoque et al. 2014b). Furthermore, a survey of 927 deep wells spread across 180 km^2 of Bangladesh in and around Araihazar, some 25 km east of Dhaka, also showed only four with concentrations of As >50 g/L, a fact that suggests deep groundwater is not being polluted by drawdown of As-polluted groundwater from shallower levels (Choudhury et al. 2016).

The situation across the border in West Bengal is different; here, the deep, pre-LGM aquifer, at depths around 150 m, has been pumped on an industrial scale for irrigation since the 1960s. Mukherjee et al. (2011) reported drawdown of As resulting from this deep pumping. Although McArthur et al. (2016) showed that at least some of the deep As was probably generated in-situ, rather than by migration, they also showed that Cl from human sources had been drawn down into the pre-LGM aquifers beneath palaeo-channels by deep pumping and so As must be following, albeit at a much slower rate and one yet to be properly quantified.

7.2 *Local Changes*

In the Terai of Nepal, Brikowski et al. (2014) observed pronounced seasonal variations in the concentration of both As and other species in groundwater from shallow, thin, aquifers where irrigation pumping was all but absent. The variations represent variations of natural conditions. Studies over 5 years in Araihazar, Bangladesh (van Geen et al. 2007), showed some changes in water quality that

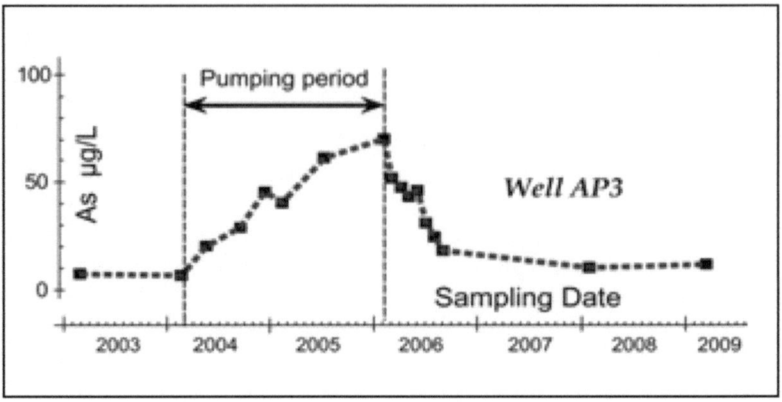

Fig. 12.12 Variation with time in the As concentration of well AP3 of McArthur et al. (2010), which is sited within a few metres of the edge of a palaeosol and has a screen 7–10 m below the top of the brown sands of the pre-LGM aquifer

were attributed to well failure, but some of which may have been attributable to local migration of As into pre-LGM brown sands from nearby As-polluted aquifers. Further studies of changing water quality at Araihazar (Dhar et al. 2008) documented both declines and increases in As concentrations, and other constituents over the monitoring period of 2–3 years. They noted that invasion of flood water either down a well, or its annulus, is one reason for changing groundwater composition. Other changes could not be related to details of flow or aquifer sedimentology.

Where the sedimentology was known in detail in relation to wells, McArthur et al. (2010) were able to relate changes in As concentrations in well water to both disturbance of flow from pumping, and general translocation downward through an aquifer of a peaky vertical profile in As concentration in groundwater. For example, their Well AP3, installed in exactly the position of Well C in Fig. 12.9, showed changes in As-concentration over time (Fig. 12.12). The screen was at 39–42 m depth and so was between 7 and 10 m below the interface of brown pre-LGM sands and overlying grey post-LGM sands, the LGMP having been eroded at this location. Initially around 7 g/L As, groundwater from AP3 remained low in As for a year whilst the well was used by one family only. Neighbouring wells were shallower, screened in grey, post-LGM sands and so As-polluted. When, out of charity, the owner allowed neighbours to draw water from AP3 (from Feb 2004), the increased abstraction drew in As-polluted water from the overlying post-LGM aquifer. After 2 years, the As concentration exceeded 50 g/L and use of the well for domestic supply was discontinued. On cessation of pumping, invasion of As-rich groundwater ceased and the As-concentration decreased to low values. This invasion of As occurred because the LGM palaeosol was not present and so did not protected the pre-LGM aquifers from downward invasion of As-polluted water from the overlying post-LGM aquifer. In contrast, wells near Site B (Fig. 12.9) stayed <5 g/ L As from 2002 until monitoring ceased in 2014 because the distance that invading,

As-rich, groundwater had to travel to get to the wells was much more than 7 m it had to travel to reach the top of the screen of Well AP3. The brown sand contains Fe-oxyhydroxides that scavenge As from groundwater and so As was removed before the groundwater reached Site B. Eventually, wells at B will be affected by As-pollution, but not for decades to centuries have passed, such is the sorptive capacity of Fe-oxyhydroxides in the brown, uppermost, pre-LGM aquifer.

8 Mixed Groundwaters

The main mechanism of As-pollution in the regions under discussion is reduction of Fe-oxyhydroxides. This adds As, PO_4, NH_4 and Fe(II), to groundwater. The sequence of redox-reactions known to occur in nature is well documented: O_2 consumption, NO_3 reduction, Mn-reduction, SO_4 reduction, CH_4/CO_2 production, CO_2 reduction. These reactions are sequential, so NO_3 should be removed from groundwater before Mn reduction occurs; Fe-reduction should occur after Mn reduction. Yet it is not uncommon for groundwaters across the region to contain redox-incompatible species such as NO_3 *and* As, Fe *and* Mn or to contain As *and* H_2S. Such occurrences might imply that Fe-reduction is not the whole story of As-pollution – until depth profiles of groundwater composition are examined and it is understood that the screens of domestic wells are generally 3.5 m long (irrigation-well screens are often 7 m long); which is long enough to span redox boundaries.

In most locations where it has been examined, groundwater is highly stratified; see, for example, Fig. 12.1 of Harvey et al. (2002). With increasing depth successive redox zones are encountered with O_2 consumption occurring in the uppermost few metres, followed by the other redox zones as depth increases. A well-screen that straddles a redox boundary will draw in water from more than one redox zone and give rise to a sample of 'mixed' waters containing redox-incompatible species. Even when a screen is wholly within one redox zone, it is possible to obtain mixed waters if abstraction is high, e.g. because a motorised pump is used or a well is pumped for a long time: the volume from which the water in the aquifer is drawn increases with increase in pumping (time and rate) and the expanding volume from which water is drawn may grow to intersect more than one redox zone. Reactions that will adjust the groundwater composition to equilibrate mixed redox species are generally slower than the time it takes to pump, sample, and bottle the groundwater for later analysis, hence preserving these mixed signals.

Finally, redox zones in groundwater need not be arranged in horizontal layers. Buried peat deposits, or organic-rich infills of abandoned channels, could create lateral redox gradients whereby at the core fermentation produces CH_4/CO_2, whilst SO_4 reduction and Fe reduction occur around it in a shell-like form that would be modified by groundwater flow. Future research into the details of redox zonation and As-release in aquifers needs to employ piezometer nests with very short well screens (0.5 to 1 m) long, and low-pumping rates for sampling, in order to avoid artefacts arising from mixed waters.

9 A Word About Data

Preparation of this account of As-pollution in India, Pakistan and Bangladesh, was hampered by the absence of data from many, if not most, papers published on the matter. This lack of data prevented comparisons being made and diagrams being drawn that would have illustrated useful points. It also means that many papers that could have been cited were not cited because of the absence of data. In some quarters, even the simple term 'data' appears to be misunderstood. Summary tables (mean, min, max etc.) are not 'data', they are summaries of data. Diagrams are not 'data' they are representations of that data.

Many papers contain no data. Some give summary tables in which means, medians, modes, maxima, minima concentrations for a data set are given, rather than the entire data specific to a well. Such practices deny the reader the opportunity to assess data quality by, for example, calculating ion-balance for comparison to measured EC, or calculating HCO_3, and checks on measured HCO_3. Such practices also prevent the claims made in a paper being tested against the data on which the claims are based. Such practices also make it impossible for the reader to mine the data by, for example, generating additional co-variation diagrams or grouping the data in different ways to those in the paper in order to obtain in sights over looked by the original authors. Giving full data not only avoids such problems but also allows future work to build on work already accomplished. For example, it was possible for Sultana et al. (2014) to assess changes in groundwater composition over time in northern Pakistan only because the data for groundwater from the same wells, originally reported on by Farooqi et al. (2007a, b), were available either in the paper or from the author herself.

A particular hazard of summary data arises when they are used in attempts to provide national assessments of groundwater quality as a guide to future aquifer development. If only summary data is available on which to base such assessments, risks may be overestimated because outlier data that may be well-specific (e.g. high NO_3 from a well next to a latrine) will skew conclusions. Some such data may not even represent groundwater composition e.g. Cu and Pb data, which mostly represents contamination from pump fittings.

The need is emphasised here for any report of any study of As-pollution or, indeed, any scientific study, to include in the paper all data relevant to the claims set out in the paper. Amongst the important data that should be given in any study of groundwater quality are the depth of wells sampled, their GPS locations, and the chemical data on groundwater analysed that is identifiable to each well. That is, data sufficient in detail to allow others to reproduce the results. To publish a paper without its supporting data is to make the statement: "trust us". The reader must take on trust that the data are correct, that no mistakes have been made in plotting data, or doing calculations etc. The best practice of science is not built on trust, it is built on proof. It can be encapsulated in the Popperian phrase "Here are my claims and here is the data on which they are based – falsify if you can".

References

Ahmad, T., Kahlown, M.A., Tahir, A. and Rashid, H. (2004). Arsenic an Emerging Issue: Experiences from Pakistan. 30th WEDC International Conference, Vientiane, Lao PDR.

Brikowski, T.H., Neku, A., Shresth, S.D. and Smith, L.S. (2014). Hydrologic control of temporal variability in groundwater arsenic on the Ganges floodplain of Nepal. *J Hydrol*, **518:** 342–353.

Chapelle, F.H. (2000). The significance of microbial processes in hydrogeology and geochemistry. *Hydrogeol J*, **8:** 41–46.

Chakraborti, D., Mukherjee, S.C., Pati, S., Sengupta, M.K., Rahman, M.M., Chowdhury, U.K., Lodh, D., Chanda, C.R., Chakraborti, A.K. and Basu, G.K. (2003). Arsenic groundwater contamination in Middle Ganga Plain, Bihar, India: A future danger? *Environ. Health Perspect*, **111:** 1194–1201.

Chatterjee, A., Das, D. and Chakraborti, D. (1993). A study of ground water contamination by arsenic in the residential area of Behala, Calcutta due to industrial pollution. *Environ Pollut*, **80:** 57–65.

Choudhury, I., Ahmed, K.M., Hasan, M., Mozumder, M.R.H., Knappett, P.S.K., Ellis, T. and van Geen, A. (2016) Evidence for elevated levels of arsenic in public wells of Bangladesh due to improper installation. *Groundwater*, **54(6):** 871–877.

Choudhury, R., Mahanta, C., Verma, S. and Mukherjee, A. (2017). Arsenic distribution along different hydrogeomorphic zones in parts of the Brahmaputra River Valley, Assam (India). *Hydrogeology Journal*, **25:** 1153–1163.

Datta, S., Neal, A.W., Mohajerin, T.J., Ocheltree, T., Rosenheim, B.E., White, C.D. and Johannesson, K.H. (2011) Perennial ponds are not an important source of water or dissolved organic matter to groundwaters with high arsenic concentrations in West Bengal, India. *Geophys Resear Letters*, **38:** L20404. doi:https://doi.org/10.1029/2011GL049301.

Dhar, R.K., Zheng, Y., Stute, M., van Geen, A., Cheng, Z., Shanewaz, M., Shamsudduha, M., Hoque, M.A., Rahman, M.W. and Ahmed, K.M. (2008). Temporal variability of groundwater chemistry in shallow and deep aquifers of Araihazar, Bangladesh. *J Contamin Hydrol*, **99:** 97–111.

DHS (2016). Bangladesh Demographic and Health Survey 2014. Ministry of Health and Family Welfare, Dhaka, Bangladesh. Mitra and Associates, Dhaka, Bangladesh; The DHS Program, Rockville, Maryland, U.S.A.

DPHE (1999). Groundwater Studies for Arsenic Contamination in Bangladesh. Final Report, Rapid Investigation Phase (6 vols); Department of Public Health Engineering, Government of Bangladesh. Mott MacDonald and British Geological Survey. Dhaka.

DPHE (2001). Arsenic Contamination of Groundwater in Bangladesh. Department of Public Health Engineering and British Geological Survey. BGS Technical Report WC/00/19 (4 vols).

Desbarats, A.J., Koenig, C.E.M., Pal, T., Mukherjee, P.K. and Beckie, R.D. (2014). Groundwater flow dynamics and arsenic source characterization in an aquifer system of West Bengal, India. *Water Resou Res*, **50L:** 4974–5002. doi: https://doi.org/10.1002/2013WR014034.

Donselaar, M.E., Bhatt, A.G. and Ghosh, A.K. (2017). On the relation between fluvio-deltaic flood basin geomorphology and the wide-spread occurrence of arsenic pollution in shallow aquifers. *Sci Total Environ*, **574:** 901–913.

Farnsworth, C.E. and Hering, J.G. (2011). Inorganic geochemistry and redox dynamics in bank filtration settings. *Environ Sci Technol*, **45:** 5079–5087. doi: https://doi.org/10.1021/es2001612.

Farooqi, A., Masuda, H. and Firdous, N. (2007a). Toxic fluoride and arsenic contaminated water in Lahore and Kasur districts, Punjab, Pakistan, and possible contaminant sources. *Environ Pollut*, **145:** 839–849.

Farooqi, A., Masuda, H., Kusakabe, M., Naseem, M. and Firdous, N. (2007b) Distribution of highly arsenic and fluoride contaminated groundwater from east Punjab, Pakistan, and the controlling role of anthropogenic pollutants in the natural hydrological cycle. *Geochem J*, **41:** 213–234.

Froelich, P.N., Klinkhammer, G.P., Bender, M.L., Luedtke, N.A., Heath, G.R., Cullen, D., Dauphin, P., Hammond, D., Hartman, B. and Maynard, V. (1979). Early oxidation of organic matter in pelagic sediments of the eastern equatorial Atlantic: Suhoxic diagenesis. *GCA*, **43:** 1075–1090.

Greenman, D.W., Swarzenski, W.V. and Bennett, G.D. (1967). Ground-water hydrology of Punjab, West Pakistan with emphasis on problems caused by canal irrigation. U.S. Geol. Surv. Water Supp. Pap., p. 1608-H.

Goodbred, S.L. Jr. (2003). Response of the Ganges dispersal system to climate change: A source-to-sink view since the last interstade. *Sedi Geol*, **162:** 83–104.

Government of Pakistan (2007). National Action Plan for Arsenic Mitigation, 2007–2011. Pakistan: Ministry of Science and Technology.

Gulens, J., Champ, D.R. and Jackson, R.E. (1979). Influence of redox environments on the mobility of arsenic in groundwater. *In:* Jenne, E.A. (Ed.), Chemical Modelling in Aqueous Systems: Speciation, Sorption, Solubility, and Kinetics. *Am. Chem. Soc. Symp. Ser.* 93.

Gurung, J.K., Ishiga, H. and Khadka, M.S. (2005). Geological and geochemical examination of arsenic contamination in groundwater in the Holocene Terai Basin, Nepal. *Environ Geol*, **49:** 98–113.

Harvey, C.F., Swartz, C.H., Badruzzaman, A.B.M., Keon-Blute, N., Yu, W., Ali, M.A., Jay, J., Beckie, R., Niedan, V., Brabander, D., Oates, P.M., Ashfaque, K.N., Islam, S., Hemond, H.F. and Ahmed, M.F. (2002). Arsenic mobility and groundwater extraction in Bangladesh. *Science*, **298:** 1602–1606.

Hoque, M.A., McArthur, J.M. and Sikdar, P.K. (2014a). Sources of low-arsenic groundwater in the Bengal Basin: Investigating the influence of the last glacial maximum using a 115-km traverse across Bangladesh. *Hydrogeol J*, **22:** 1535–1547.

Hoque, M.A., McArthur, J.M., Sikdar, P.K., Ball, J.D. and Molla, T.N. (2014b). Tracing recharge to aquifers beneath an Asian megacity with Cl/Br and stable isotopes – The example of Dhaka, Bangladesh. *Hydrogeol J*, **22:** 1549 pp. DOI: https://doi.org/10.1007/s10040-014-1155-8.

Hughes, M.F. (2016). History of arsenic as a poison and a medicinal agent. *In:* Arsenic: Exposure Sources, Health Risks, and Mechanisms of Toxicity. States, J.C. (Ed.). John Wiley & Sons, Inc. First Edition.

Johnston, R.B. and Sarker, M.H. (2007). Arsenic mitigation in Bangladesh: National screening data and case studies in three upazilas. *J Environ Sc Health*, Part A **42(12):** 1889–1896.

Kent, D.B. and Fox, P.M. (2004). The influence of groundwater chemistry on arsenic concentrations and speciation in a quartz sand and gravel aquifer. *Geochem Transac*, **5(1):** 12. DOI: https://doi.org/10.1063/1.1738211.

Kimber, S.W.L., Sizemore, D.J. and Slavich, P.G. (2002). Is there evidence of arsenic movement at cattle tick dip sites? *Australian J Soil Res*, **40(7)**: 1103–1114.

Kirk, M.F., Holm, T.R., Park, J., Jin, Q., Sanford, R.A., Fouke, B.W. and Bethke, C.M. (2004) Bacterial sulfate reduction limits natural arsenic contamination in groundwater. *Geology*, **32 (11):** 953–956.

Kresse, T.M. and Fazio, J.A. (2003). Occurrence of arsenic in groundwaters of Arkansas and implications for source and release mechanisms. Arkansas Dept. of Environmental Quality. Water Quality Report WQ03-03-01, Little Rock, AR.

Kumar, M.J., Ramanathan, AL., Rahman, M.M. and Naidu, R. (2016). Concentrations of inorganic arsenic in groundwater, agricultural soils and subsurface sediments from the middle Gangetic plain of Bihar, India. *Sc Tot Environ*, **573:** 1103–1114.

Lawson, M., Polya, D.A., Boyce, A.J., Bryant, C., Mondal, D., Shantz, A. and Ballentine, C.J. (2013). Pond-derived organic carbon driving changes in arsenic hazard found in Asian groundwaters. *Environ Sci Technol*, **47:** 7085–7094.

Mailloux, B.J., Trembath-Reichert, E., Cheung, J., Watson, M., Stute, M., Freyer, G.A., Ferguson, A.S., Ahmed, K.M., Alam, M.J., Buchholz, B.A., Thomas, J., Layton, A.C., Zheng, Y., Bostick, B.C. and van Geen, A. (2013). Advection of surface-derived organic carbon fuels microbial reduction in Bangladesh groundwater. *Proc Nat Acad Sci*, **110(14):** 5331–5335.

McArthur, J.M., Banerjee, D.M., Hudson-Edwards, K.A., Mishra, R., Purohit, R., Ravenscroft, P., Cronin, A., Howarth, R.J., Chatterjee, A., Talukder, T., Lowry, D., Houghton, S. and Chadha,

D.K. (2004). Natural organic matter in sedimentary basins and its relation to arsenic in anoxic ground water: The example of West Bengal and its worldwide implications. *App Geochem*, **19:** 1255–1293.

McArthur, J.M., Ravenscroft, P., Banerjee, D.M., Milsom, J., Hudson-Edwards, K.A., Sengupta, S., Bristow, C., Sarkar, A., Tonkin, S. and Purohit, R. (2008). How paleosols influence groundwater flow and arsenic pollution: A model from the Bengal Basin and its worldwide implication. *Water Resour Res*, **44:** W11411.

McArthur, J.M., Banerjee, D.M., Sengupta, S., Ravenscroft, P., Klump, S., Sarkar, A., Disch, B. and Kipfer, R. (2010). Migration of As, and $^{3}H/^{3}He$ ages, in groundwater from West Bengal: Implications for monitoring. *Water Res*, **44:** 4171–4185.

McArthur, J.M., Ravenscroft, P. and Srecek, O. (2011). Aquifer arsenic source. *Nature Geosci*, **4:** 655–656.

McArthur, J.M., Ghosal, U., Sikdar, P.K. and Ball, J.D. (2016). Arsenic in Groundwater: The Deep Late Pleistocene Aquifers of the Western Bengal Basin. *Environ Sci Technol*, **50:** 3469–3476.

Mladenov, N., Zheng, Y., Miller, M.P., Nemergut, D.R., Legg, T., Simone, B., Hageman, C., Rahman, M.M., Ahmed, K.M. and McKnight, D.M. (2010). Dissolved organic matter sources and consequences for iron and arsenic mobilization in Bangladesh aquifers. *Environ Sci Technol*, **44:** 123−128.

Mladenov, N., Zheng, Y., Simone, B., Bilinski, T.M., McKnight, D.M., Nemergut, D., Radloff, K. A., Rahman, M. and Ahmed, K.M. (2015). Dissolved organic matter quality in a shallow aquifer of Bangladesh: Implications for arsenic mobility. *Environ Sci Technol*, **49:** 10815−10824.

Mukherjee, A., Fryar, A.E., Scanlon, B.R., Bhattacharya, P. and Bhattacharya, A. (2011). Elevated arsenic in deeper groundwater of the western Bengal Basin, India: Extent and controls from regional to local scale. *Appl Geochem*, **26:** 600−613.

Mumford, A.C., Barringer, J.L., Benzel, W.M., Reilly, P.A. and Young, L. (2012). Microbial transformations of arsenic: Mobilization from glauconitic sediments to water. *Water Res*, **46:** 2859–2868.

Naseem S. and McArthur J.M. (2018). Arsenic and other water-quality issues affecting groundwater, Indus alluvial plain, Pakistan. Hydrological Processes, DOI: 10.1002/hyp.11489.

Nealson, K.H. (1997). Sediment bacteria: Who's there, what are they doing, and what's new? *Annual Reviews in Earth Planet Sc*, **25:** 403–434.

Nickson, R.T., McArthur, J.M., Ravenscroft, P., Burgess, W.B. and Ahmed, K.Z. (2000). Mechanism of arsenic poisoning of groundwater in Bangladesh and West Bengal. *Appl. Geochem*, **15:** 403–413.

Nickson, R.T., McArthur, J.M., Shrestha, B., Kyaw-Myint, T.O. and Lowry, D. (2005). Arsenic, and other drinking water quality issues, Muzaffargarh, Pakistan. *Appl Geochem*, **20:** 55–68.

Nickson, R., Sengupta, C., Mitra, P., Dave, S.N., Banerjee, A.K., Bhattacharya, A., Basu, S., Kakoti, N., Moorthy, N.S., Wasuja, M., Kumar, M., Mishra, D.S., Ghosh, A., Vaish, D.P., Srivastava, A.K., Tripathi, R.M., Singh, S.N., Prasad, R., Bhattacharya, S. and Deverill, P. (2007). Current knowledge on the distribution of arsenic in groundwater in five states of India. *J Environ Sc Health*, Part A **42:** 1707–1718.

Ramay, M.I., Ahmad, T., Shipin, O.V., Jezeph, D. and Kadushkin, A. (2006). Arsenic contamination of groundwater and its mitigation in the province of Punjab (Pakistan) in the light of the situation in South Asia. http://www.who.int/household_water/resources/Ramay.pdf

Ravenscroft, P., McArthur, J.M. and Hoque, B. (2001). Geochemical and palaeohydrological controls on pollution of groundwater by arsenic. *In:* Arsenic Exposure and Health Effects IV. W.R. Chappell, C.O. Abernathy & R. Calderon (Eds). Elsevier Science Ltd., Oxford (66 Cites April 1 2010).

Ravenscroft, P., Brammer, H. and Richards, K.S. (2009). Arsenic pollution: A global synthesis. Wiley-Blackwell, London.

Robertson, F.N. (1989). Arsenic in ground-water under oxidising conditions, south-western United States. *Environ Geochem Health*, **11:** 171–185.

Sahu, S. and Saha, D. (2015). Role of shallow alluvial stratigraphy and Holocene geomorphology on groundwater arsenic contamination in the Middle Ganga Plain, India. *Environ Earth Sci.*, **73:** 3523–3536.

Sengupta, S., McArthur, J.M., Sarkar, A., Leng, M.J., Ravenscroft, P., Howarth, R.J. and Banerjee, D.M. (2008). Do ponds cause arsenic-pollution of groundwater in the Bengal Basin? An answer from West Bengal. *Environ Sc Techno*, **42:** 5156–5164.

Shrestha, B.R., Whitney, J.W. and Shrestha, K.B. (Eds) (2004). The state of arsenic in Nepal – 2003. National Arsenic Steering Committee (Nepal) and Environment and Public Health Organization (Nepal).

Smith, R.L., Kent, D.B., Repert, D.A. and Böhlke, J.K. (2017). Anoxic nitrate reduction coupled with iron oxidation and attenuation of dissolved arsenic and phosphate in a sand and gravel aquifer. *Geochim Cosmochim Acta*, **196:** 102–120.

States, C.J. (Ed.) (2016). Arsenic Exposure Sources, Health Risks, and Mechanisms of Toxicity. John Wiley & Sons, Inc. First Edition. 560 p.

Sultana, J., Abida, F. and Usman, A. (2014). Arsenic concentration variability, health risk assessment, and source identification using multivariate analysis in selected villages of public water system, Lahore, Pakistan. *Environ Monit Assess*, **186:** 1241–1251.

Thakur, J.K., Thakur, R.K., Ramanathan, AL., Kumar, M. and Singh, S.K. (2011). Arsenic Contamination of Groundwater in Nepal – An Overview. *Water*, **3:** 1–20.

Thomas, D.J. (2016). The chemistry and metabolism of arsenic. *In:* Arsenic: Exposure Sources, Health Risks, and Mechanisms of Toxicity. States, J.C. (Ed.). John Wiley & Sons, Inc. First Edition.

Umitsu, M. (1993). Late Quaternary sedimentary environments and landforms in the Ganges Delta. *Sediment Geol*, **83:** 177–186.

van Geen, A., Zheng, Y., Versteeg, R., Stute, M., Horneman, A., Dhar, R., Steckler, M., Gelman, A., Small, C., Ahsan, H., Graziano, J.H., Hussain, I. and Ahmed, K.M. (2003). Spatial variability of arsenic in 6000 tube wells in a 25 km^2 area of Bangladesh. *Water Resour Res*, **39:** art. no. 1140.

van Geen, A., Cheng, Z., Jia, Q., Seddique, A.A., Rahman, M.W., Rahman, M.M. and Ahmed, K.M. (2007). Monitoring 51 community wells in Araihazar, Bangladesh, for up to 5 years: Implications for arsenic Mitigation. *J Environ Sc and Health*, Part A, **42(12):** 1729–1740.

Welch, A.H., Westjohn, D.B., Helsel, D.R. and Wanty, R.B. (2000). Arsenic in groundwater of the United States: Occurrence and geochemistry. *Ground Water*, **38:** 589–604.

WHO (2017). Guidelines for Drinking-Water Quality, 4th ed. World Health Organization, Geneva.

Chapter 13
Arsenic Contaminated Irrigation Water and Its Impact on Food Chain

Saroj K. Sanyal

1 Introduction

Arsenic (As), a toxic trace element, is of great environmental concern due to its presence in soil, water, plant, animal and human continuum. Its high toxicity and increased appearance in the biosphere has triggered public and political concern. Out of the 20 countries (covering Argentina, Chile, Finland, Hungary, Mexico, Nepal, Taiwan, Bangladesh, India and others) in different parts of the globe where groundwater arsenic contamination and human suffering there from have been reported so far, the magnitude is considered to be the highest in Bangladesh, followed by West Bengal, India (Sanyal et al. 2012, 2015; Sanyal 2016a, b). The scale of the problem is grave and unprecedented, exposing millions of people in the Bengal delta basin to risk. The widespread arsenic contamination in groundwater in different parts of West Bengal, located primarily in five districts adjoining the river Bhagirathi, as well as the contiguous districts in Bangladesh, is of great concern. Even beyond the Bengal delta basin, the widespread arsenic contamination in groundwater above the permissible limit (50 μg/L; WHO 2001; *see* below) has also been detected in several places in the country (Table 13.1, Fig. 13.1), for instance at Chandigarh (1976), Nepal (2001), Bihar (2002), Uttar Pradesh (2003), Jharkhand (2003–2004) (Sanyal et al. 2015; Sanyal 2016a, b), Chhattisgarh and Punjab (2006–2007).

S. K. Sanyal (✉)
Bidhan Chandra Krishi Viswavidyalaya, Mohanpur, Nadia, West Bengal, India
e-mail: sksanyalnaip@gmail.com

P. K. Sikdar (ed.), *Groundwater Development and Management*,
https://doi.org/10.1007/978-3-319-75115-3_13

Table 13.1 Groundwater arsenic contamination in the Indian subcontinent

State	Coverage	Level of contamination in groundwater (μg As. L^{-1})	Citation
West Bengal	12 districts (Maldah, Murshidabad, Nadia, North 24-Parganas, South 24-Parganas, Kolkata, Howrah, Hooghly, Bardhhaman, North Dinajpur, South Dinajpur, Coochbehar), 111 blocks	50–3700	http://www.soesju.org/arsenic/wb.htm
Assam	18 districts, 72 blocks	>50	Singh, A.K. (2007). Curr. Sci. 92 (11):1506–1515.
	5 districts (Barpeta, Dhemaji, Dhubgiri, Darrang and Golaghat)	100–200	
	4 districts (Jorhat, Lakhimpur, Nalbari and Nagaon)	228–657	
Bihar	12 districts (Bhagalpur, Khagaria, Munger, Begusarai, Lakhisarai, Samastipur, Patna, Baishali, Saran, Bhojpur, Buxar and Katihar), 32 blocks	>50	Acharya, S.K. and Shah, B.A. (2004). Environ. Health. Pers. 112 (1): 19–20.
Jharkhand	1 district (Sahibgunj)	>50	http://www.soesju.org/arsenic/jharkhand.htm
Uttar Pradesh	21 districts (Ballia, Lakhimpur, Kheri, Baharaich, Chandauli, Gazipur, Gorakhpur, Basti, Siddharthnagar, Balarampur, Sant Kabir Nagar, Unnao, Bareilly, Moradabad, Rae Bareli, Mirza-pur, Bijnore, Meerut, Sant Ravidas Nagar, Shahjahanpur and Gonda)	>50	http://www.nerve.in/news:253500133730
Madhya Pradesh	1 district (Rajnandgaon)	52–88	Press Trust of India, September 4, 1999.
Manipur	1 district (Thoubal)	798–986	Singh, A.K. (2007). Curr. Sci. 92 (11):1506–1515.
Tripura	3 districts (North Tripura, Dhalai and West Tripura)	65–444	Singh, A.K. (2007). Curr. Sci. 92 (11):1506–1515.
Nagaland	2 districts (Mokokchong and Mon)	50–278	Singh, A.K. (2007). Curr. Sci. 92 (11):1506–1515.

Source: Sanyal (2016a, b), Sanyal et al. (2015)

2 Guideline Value of Maximum Arsenic Concentration

The World Health Organization (WHO)-recommended provisional guideline value of *total* arsenic concentration in drinking water is 10 μg/L since 1993 (WHO 1993, 1996), mainly because lower levels preferred for protection of human health are not

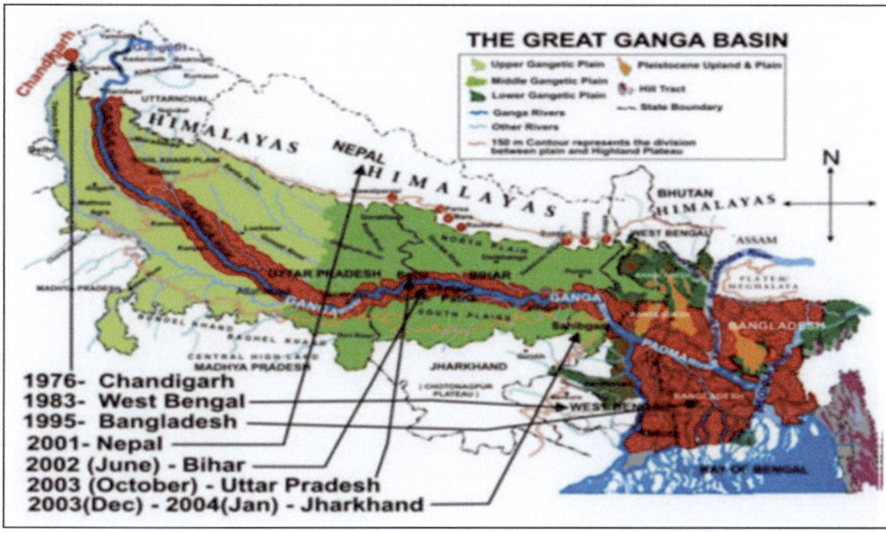

Fig. 13.1 Groundwater arsenic contamination in the Indian subcontinent. (Source: Sanyal 2005)

reliably measurable on a large scale. However, the National Standard for maximum acceptable concentration (MAC) of arsenic in drinking water is 50 μg/L in several countries including India and Bangladesh, based on an earlier WHO (1971) advice. The proposed new standard value of 5 μg/L is under consideration (WHO 2001). This is due mainly to the fact that inorganic Arsenic compounds are classified in Group 1 (carcinogenic to humans) on the basis of adequate evidence for carcinogenicity in humans and limited evidence for carcinogenicity in animals (IARC 1987). Adequate data on the carcinogenicity of organic Arsenic have not been generated. The joint FAO/WHO Expert Committee on Food Additives (JECFA) set a provisional maximum tolerable daily intake (PMTDI) of inorganic arsenic as 0.002 mg/kg of body weight for humans in 1983 and confirmed a provisional tolerable weekly intake (PTWI) as 0.015 mg/kg of body weight in 1989 (FAO/WHO 1989). Such guideline values for soil, plant and animal systems are not available.

Two major hypotheses, both of geogenic origin, have been proposed to account for such widespread arsenic contamination in the groundwater in parts of West Bengal and Bangladesh, confined within the delta bound by the rivers Bhagirathi and Ganga-Padma. Of these two hypotheses, namely the arsenopyrite oxidation hypothesis and the ferric oxyhydroxide reduction hypothesis, the latter is more consistent with the experimental observations reported for the aquifer sediments and aquifer water of the Bengal delta basin (Sanyal et al. 2012; Sanyal 2016a). According to this hypothesis, an anoxic condition of the aquifer causes arsenic mobilization from into the groundwater aquifer. The maintenance of such anoxic condition is proposed to be facilitated by the widespread practice of wetland paddy cultivation in the affected belt.

Table 13.2 Arsenic concentration in rocks and some other materials

Types of rocks/minerals	Arsenic content (mg/kg)
1. Rocks	
Igneous rocks	
Ultrabasic:	
Peridotite, Dunite, Serpentine	0.3–15.8
Basic:	
Basalt (extrusive)	0.18–113
Gabbro (intrusive)	0.06–28
Intermediate:	
Latite, Andesite, Trachyte (extrusive)	0.5–5.8
Diorite, Granodiorite, Syenite (intrusive)	0.09–13.4
Acidic:	
Rhyolite (extrusive)	3.2–5.4
Granite (intrusive)	0.18–15
Metamorphic rocks	
Quartzite	2.2–7.6
Slate/Phyllite	0.5–143
Schist/Gneiss	0.0–185
Sedimentary rocks	
Marine:	
Shale/Claystone(nearshore)	4.0–25
Shale/Claystone(offshore)	3.0–490
Carbonates	0.1–20.1
Phosphorites	0.4–188
Sandstone	0.6–9.0
Nonmarine:	
Shales	3.0–12
Claystone	3.0–10
2. Coal	Up to 2000
3. Crustal Average	2.0

Source: Sanyal et al. (2012)

3 Natural Abundance

Dissolved arsenic concentrations in natural waters (except groundwater) are generally low, except in areas characterized by geothermal water and/or mining activities. The sedimentary rocks generally have higher arsenic content (Table 13.2) than do igneous and metamorphic rocks, while suspended and bottom sediments in most aquatic systems contain more arsenic (Table 13.2) than most natural waters (Table 13.3). The capacity to retain arsenic is primarily governed by the sediment grain-size and the presence of surface coating composed of clays, clay-sized iron and manganese oxides and organic matter. Arsenic held by solid phases within the sediments, especially iron oxides, organic matter and sulphides may constitute the

Table 13.3 Arsenic concentrations in surface water sources

Source	Arsenic concentration (μg/L)
Rainwater and snow	<0.002–0.59
Rivers	0.20–264
Lakes	0.38–1.00
Sea water	0.15–6.00
Ponds (West Bengal, India)	4–70
Canals (West Bengal, India)	40–150

Source: Sanyal et al. (2012)

primary arsenic sources in groundwater under conditions conducive to arsenic release from these solid phases. These include abiotic reactions (oxidation/reduction, ion exchange, chemical transformations) and biotic reactions (microbial methylation) (Sanyal et al. 2012; Sanyal 2016a, b).

4 Arsenic Contamination in Ground Water in the Bengal Delta Basin

The groundwater arsenic concentration (50–3700 μg/L), reported from the affected areas of West Bengal, are several orders of magnitude higher than the stipulated Indian standard for the permissible limit in drinking water (50 μg/L, which is also the maximum acceptable concentration, MAC, for drinking water in Bangladesh, India and several other countries), as well as the WHO guideline value (10 μg/L). Further, the arsenic concentration in alluvial aquifers of Punjab varied from 4 to 688 μg/L (Sanyal et al. 2012; Sanyal 2016a, b). In West Bengal, the presence of arsenic in groundwater in concentrations exceeding the maximum acceptable concentration was first detected in 1978, while the first case of arsenic poisoning in humans was diagnosed at the School of Tropical Medicine in Calcutta in 1983 (Acharya 1997). The effect of ingestion of inorganic arsenic in drinking water and the health effects in adults have also been well established (Guha Mazumdar et al. 1998).

The main focus of attention, until recently, has been almost exclusively on arsenic contamination in groundwater-derived drinking water, *notwithstanding* the fact that the groundwater in the affected belt is predominantly used in the agricultural sector. Despite this, large number of systematic studies have been undertaken rather recently (during the last 15 years or so) to explore the influence of arsenic in groundwater, used as irrigation source, on soil-plant-animal continuum. Such interdisciplinary, inter-institutional studies, led by Bidhan Chandra Krishi Viswavidyalaya (BCKV), West Bengal, as well as others have given important leads as to the source of the given toxin in groundwater, its accumulation in soils and crops grown in the affected belt of West Bengal and in the animal tissues and products in the area of study (Sanyal et al. 2015; Sanyal 2016a, b). Such findings call for an immediate attention since what remains essentially a *point* and fixed source of arsenic contamination as for drinking water (e.g., a tube well discharging contaminated water), may well become a *diffuse* and uncertain source of contamination when arsenic finds its way into the food-web, accompanied with possible *biomagnification* up in the food chain. This assumes added significance in view of the reported finding of higher (than permissible) level of arsenic in the urine samples (an early biomarker of arsenic poisoning) of some people having no history of consuming arsenic contaminated drinking water (Guha Mazumdar Dr. DN, by private communication; *see* later). Interestingly, the surface water bodies, located in the affected belt, have remained largely free of arsenic. This tends to suggest that the soil, which receives the arsenic-contaminated water, acts as an effective *sink* to contain the toxin, thereby preventing the surface run-off to carry it to the adjoining water systems (Sanyal et al. 2012).

5 Health Implications of Arsenic Poisoning

Arsenic is a widely occurring toxic metal in natural ecosystems. Arsenic small as 0.1 g of arsenic trioxide can prove lethal to humans (Jarup 1992). Early symptoms of Arsenic poisoning include skin disorders, weakness, languor, anorexia, nausea and vomiting with diarrhoea or constipation. With the progress characteristic features, which include acute diarrhoea, edema (especially of the eyelids and ankles), skin pigmentation, arsenical melanosis and hyperkeratosis, enlargement of liver, respiratory diseases and skin cancer. In severe cases, gangrene in the limbs and malignant neoplasm are also observed (Guha Mazumdar et al. 1998; Sanyal et al. 2012). “Bell Ville Disease” (typical Arsenic-induced cutaneous manifestations among the people of Bell Ville) in Argentina, “Black Foot Disease” in Taiwan and “Kai Dam” disease in Thailand are well established as health disorders due to arsenic poisoning (Sanyal et al. 2012; Sanyal 2016a, b). Arsenic a matter of fact, the hair, nail, skin-scale and urine samples of a large number of people residing in the affected belt of West Bengal (India) and Bangladesh, have been analyzed by several workers. Many of these samples had arsenic loading more than the corresponding safe levels.

6 Chemistry of Arsenic in Groundwater-Soil Environment

Arsenic in groundwater and soil is present as dissolved oxyanions, namely arsenites ($As^{III}O_3^{3-}$, $H_nAs^{III}O_3^{(3-n)-}$, with $n = 1, 2$) or arsenate ($As^{V}O_4^{3-}$, $H_nAs^{V}O_4{}^{(3-n)-}$, with $n = 1, 2$), or both, besides the organic forms (Sanyal et al. 2015; Sanyal 2016a, b). The solubility, mobility, bioavailability and hence toxicity of arsenic in soil-crop system primarily depends on its chemical form, primarily the oxidation state of arsenic. Thus, the toxicity of arsenic compounds in groundwater/soil environment depends largely on its oxidation state, and hence on redox status and pH, as well as whether arsenic is present in organic combinations. The toxicity of arsenic compounds in groundwater/soil environment follows the order:

Arsine [AsH_3; valence state of As: 3] > organo-arsine compounds > arsenites (As^{3+} form) and oxides (As^{3+} form) > arsenates (As^{5+} form) > arsonium metals (+1) > native As metal (0).

The arsenites are much more soluble, mobile and toxic than arsenates in aquatic and soil environments. The organic forms, namely dimethylarsinic acid (DMA) or cacodylic acid, which on reduction (e.g., in anoxic soil conditions) forms di- and trimethyl arsines, are also present in soil. Another organic form present in groundwater and soil is monomethylarsonic acid (MMA). At pH 6–8, and in an aerobic oxidized environment (redox potential, $E_h = 0.2$–0.5 V), arsenic acid species and arsenate oxyanions ($H_nAs^VO_4{}^{(3-n)-}$ ions, with $n = 1, 2$; arsenic in pentavalent forms) occur in considerable proportions in most aquatic systems, whereas under mildly reducing conditions (such as one encounters in flooded paddy soils with $E_h = 0$–0.1 V), the arsenous acid, $H_3As^{III}O_3$, and arsenite oxyanion species ($As^{III}O_3{}^{3-}$, $H_nAs^{III}O_3{}^{(3-n)-}$, with $n = 1, 2$; As in trivalent form) are the predominant species. Furthermore, As (III) is more prevalent in soils of neutral pH range (and in most groundwater), as in the soils of the affected belt of West Bengal, India and Bangladesh, than otherwise thought, and hence is of concern. This is primarily because As (III) exists as a neutral, uncharged molecule, namely arsenous acid, $H_3As^{III}O_3{}^0$ ($pK_a = 9.2$), at the pH of the neutral soils and most natural groundwater as one would expect, based on the Henderson's equation, Eq. 13.1 (Sanyal et al. 2012, 2015; Sanyal 2016a, b), and is thus less amenable to retention by the charged mineral surfaces in soils and sediments.

$$pH = pK_a - \log(C_{acid}/C_{salt}) \tag{13.1}$$

There have been both direct and indirect evidence to suggest that arsenic (and selenium) is held in soils and sediments by oxides (e.g., of Fe, Al, Mn) through the formation of inner-sphere complexes via ligand-exchange mechanism. This is illustrated below by the following scheme of reactions.

$$[M - OH] + H_2O \rightleftharpoons [M - OH_2{}^+] + OH^- \tag{13.2}$$

$$\begin{aligned}&[M - OH_2{}^+] + As^VO_3{}^{3-} \rightleftharpoons \left[M - OH_2{}^+\cdots\cdots As^VO_4{}^{3-}\right]\\&\rightleftharpoons \left[M\text{-}As^VO_4{}^{2-}\right] + H_2O\end{aligned} \tag{13.3}$$

As shown above (Eq. 13.3), the said ligand exchange tends to increase the negative charge of the soil colloidal fraction, for instance, of iron oxides, and thus push the point of zero-charge (PZC) of the arsenic-laden soil to lower pH. Indeed, this was shown to be the case with concomitant increase in the negative magnitude of the variable-surface charge and the surface potential of the corresponding soil colloidal fraction (Sanyal et al. 2015). However, the non-specific adsorption (through electrostatic mechanism) of arsenic also occurs at pH values below the point of zero-charge for a given adsorbent (Sanyal et al. 2015; Sanyal 2016a, b).

It ought to be emphasized that the groundwater or the soil solution, which is subject to affluxes and influxes, as well as circulation, and also to man-made perturbations of groundwater due to its withdrawal, cannot be expected to remain in thermodynamic equilibrium, it being very much of an open system (thermodynamically speaking). Thus, more often than not, the ratio of concentrations of arsenic species, namely the ratio, [(As ^{III})/(As^{V})], in field soils (soil solutions) does *not* quite agree with the ones computed from the observed redox potential (E_h) and the application of the Nernst's equation (at 25 °C) to the equilibrium redox reaction, namely

$$As^{V}O_4{}^{3-} + 2H^+ + 2e = As^{IIII}O_3^{3-} + H_2O \quad (13.4)$$

$$E_h \cong E^0{}_h - 0.295 \log\left[\left(AsO_3{}^{3-}\right)\right] / \left(AsO_4{}^{3-}\right)\right] - 0.059\,pH \quad (13.5)$$

where the ( ) terms refer to the equilibrium concentrations of the respective ionic species in dilute soil solution (the ionic activity coefficients being assumed to be unity), and $E^0{}_h$ is the standard redox (reduction) potential of the [$As^VO_4{}^{3-}/As^{III}O_3{}^{3-}$] redox couple at 25 °C. It is evident from Eq. (13.5) that the proportion of As^{III}, and hence soluble arsenic level in soil, should increase substantially with diminishing E_h and increasing pH. Furthermore, at a high pH, the OH^- ion concentration would increase, causing displacement of As^{III} and As^V species from their binding sites through competitive ligand exchange reactions.

The dependence of arsenic sorption on pH of the sorption medium is governed largely by the nature of the soil colloidal fraction. A fall of arsenate adsorption was noted with increasing pH, but only at lower arsenate concentrations, which got reversed at a higher arsenate equilibrium concentration. This trend was explained in terms of the varying electrostatic potential of the variable-charge soil colloidal surfaces with pH, solubility product principles, and buffering action of the arsenic salt used (Majumdar and Sanyal 2003).

7 Selected Findings in Soil-Crop System

7.1 Arsenic Loading in Soils of West Bengal

Some of the research studies, conducted at the selected affected areas, revealed that the total and Olsen extractable arsenic (i.e., 0.5 M $NaHCO_3$, pH 8.5–extractable As, which constitutes the soil As pool amenable to plant uptake) varied from 8.4 to 24.3 mg/kg and from 2.90 to 15.8 mg/kg, respectively (Sanyal et al. 2012; Sanyal 2016a, b). The soil arsenic contents of these areas were generally higher than those reported for the soils of several other countries like Argentina, China, Italy, Mexico, France, Australia, etc. (Sanyal et al. 2012; Sanyal 2016a, b).

Inorganic soil arsenic fractions from the affected soils were also fractionated into different soil arsenic pools, namely water soluble arsenic (Ws–As), arsenic associated with Al compounds in soil, the so-called aluminium-bound arsenic (Al–As), iron-bound arsenic (Fe–As) and calcium-bound arsenic (Ca–As), by following the sequential extraction methodology. The findings suggested that these inorganic soil arsenic pools fell in the order: Ws–As < Ca–As < Al–As < Fe–As. In particular, the Fe–As fraction contributed 41%–74.7% towards the total soil arsenic sequential sum (Banik and Sanyal 2016).

7.2 *Interaction of Arsenic with Organics in Soil System*

As mentioned earlier, soil acts as an effective sink of arsenic present in the contaminated groundwater used for irrigating the crops. The soil organic fractions including humic acid (HA) and fulvic acid (FA) behave as effective accumulators of toxic heavy metals by forming the metal-humate complexes (chelates) with different degrees of stability (Datta et al. 2001; Mukhopadhyay and Sanyal 2004; Sinha and Bhattacharyya 2011; Ghosh et al. 2012; Sanyal et al. 2012, 2015). Besides, soil clays, Al oxides, Fe oxides, especially the amorphous Fe and Al oxides in soil also influence the arsenic retention by soils, soil minerals and sediments. The above mentioned organo-arsenic complexes were quite stable, even in the presence of competing oxyanions such as phosphate, sulphate and nitrate (Sanyal et al. 2015; Sanyal 2016a, b). Further, the moderating influence of the organic fractions from FYM, vermicompost, municipal sludge, mustard cake and surface soil of West Bengal was assessed in terms of the stability of the corresponding arseno-humic/fulvic complexes, formed in the organic manure-treated contaminated soils (Sinha and Bhattacharyya 2011; Ghosh et al. 2012). It was found that the organic manures, added as soil amendment, significantly reduced the accumulation (concentration) of arsenic in sesame seed to the maximum extent of 65.5% (vermicompost), 50% (phosphocompost), 42% (mustard cake) and 40% (FYM), compared with the control counterpart (Sinha et al. 2011). The risk associated with dietary exposure to arsenic-contaminated sesame oil reached a value of 15.6% of provisional tolerable weekly intake for arsenic at the maximum accumulation of arsenic in sesame oil. Thus, improving the soil organic matter stock in the tropical soils of the arsenic-affected belt, relatively poor in native organic matter, by adopting the appropriate management practices (such as recycling of crop residues, incorporation of the appropriate organic manure, etc.) will facilitate arsenic retention in the affected soils (Sanyal 2016a, b).

7.3 *Interaction of Arsenic with Phosphorus and Micronutrients*

Phosphorus (P) is one of the essential major plant nutrients for plant growth. Because arsenic and P are both placed in Group Vb of the Periodic Table, the interaction of arsenic and P in soil-plant system is an important issue in respect of arsenic mobilization. Indeed the indications are that these oxyanions would not be adsorbed independently in mixtures, but rather would tend to compete for some portion of the same type of adsorption sites (Sanyal et al. 2012; Das et al. 2014). Several workers showed that the presence of phosphate caused a reduction in arsenate adsorption, and that the reduction was much greater for the competitive effects of arsenate on phosphate adsorption by soil minerals, although a large variation in the degree of competition between these two oxyanions has also been reported (Mukhopadhyay et al. 2002; Sanyal et al. 2015).

On the other hand, certain micronutrient applications (such as application of zinc/iron salts) to the contaminated soil (where these micronutrients are deficient) help mitigate the arsenic toxicity in soil-crop system. As stated earlier, arsenic is released to a greater extent in soil solutions upon submergence, which can be countered by the presence of graded doses of applied Zn, possibly through the formation of the relatively less soluble zinc arsenate. Arsenic accumulation by rice crops, grown under submerged condition, was also reduced by the application of Zn to soil, while the latter led to the accompanying yield increment (Das et al. 2016). Zinc application (as zinc sulphate) in *boro* (summer) rice was especially helpful in reducing the plant accumulation of arsenic and its translocation to the plant biomass, whereas its residual impact influenced positively the arrest of soil arsenic, thereby bringing down its build-up in the succeeding crops. Similar effects were observed when such experiments were repeated with iron salts in place of zinc salts (Das et al. 2016).

8 Arsenic in Soil-Plant System

Several workers have reported accumulation and transformation of arsenic by a number of plant species grown in the arsenic affected areas. These crops (such as rice, elephant-foot-yam, green gram, cowpea, sesame, groundnut, etc.) tended to show a build-up of arsenic in substantial quantities in different plant parts. Indeed, pointed gourd, a vegetable creeper plant, has shown considerable arsenic loading when cultivated in the soils of West Bengal. A number of other vegetables, namely cauliflower, tomato and bitter gourd were also noted to accumulate arsenic in their economic produce. The distribution of arsenic content in plant parts generally followed the order:

$$\text{Root} > \text{stem} > \text{leaf} > \text{economic produce}$$

As mentioned earlier, reduction of arsenate to more toxic arsenite is facilitated by lowering of redox potential (E_h) which is encountered under anoxic soil conditions, with arsenite being more soluble and mobile than arsenate. Rice plant is thus rather susceptible to arsenic toxicity since it is grown under submerged soil conditions (low E_h). Further, the processing of rice (e.g., parboiling and milling, etc.) was found to increase the arsenic loading in rice for both the traditional and the high yielding cultivars (Sanyal et al. 2012). The toxicity of arsenic species in plant body is reported to follow the order: Arsine (AsH_3) > As^{3+} > As^{5+} > MMA (Monomethylarsonic acid) > DMMA (Dimethyl arsinic acid) (Sanyal et al. 2012).

Furthermore, screening of 200 rice genotypes showed a large variation of arsenic accumulation in grain. No significant correlation was found among the pattern of arsenic uptake by root, shoot and grain of the 200 rice lines examined. Initial analysis revealed that arsenic content in grain is controlled by more than one gene (BCKV, Unpublished work). It has also been found that crops like potato, pumpkin and sesame accumulate less arsenic than do others (Sanyal et al. 2012).

On storing the arsenic contaminated groundwater (from a shallow tube well, STW) in a pond, there was a gradual lowering (on standing) of arsenic loading of the stored pond water, while its progressive build-up in the corresponding sediment samples. Such decrease of arsenic content in the stored water might have arisen from the sedimentation of arsenic from the water to the pond sediment which obviously increases the arsenic loading in the latter. The dilution of the stored water by rainfall during the wet season (July to September) further decreased the contamination in the water. This opens up the possibility of the conjunctive use in agriculture of surface water and groundwater during the *lean period* (January-May) as a potential remedial option (Sanyal et al. 2012).

9 Hyperaccumulation vis-à-vis Detoxification of Arsenic by Plants/Microbial Species

The reported hyperaccumulation of arsenic from the contaminated soils by the brake-fern, *Pterisvittata,* and its subsequent translocation into the above-ground biomass suggests that the plant-accumulated arsenic was present almost entirely in the toxic inorganic forms with the proportion of highly toxic As (III) being in fact much greater in the plant body than that of the less toxic As (V) form, as compared to the distribution of these two forms in the contaminated soil in which the fern grows (Ma et al. 2001). Thus, it is worth noting in this context that such accumulation of arsenic does *not* necessarily lead to its detoxification per se, unless the plant-accumulated toxin is effectively detoxified (or *else* converted to less toxic forms) within the plant body by its metabolic processes. For this, a systematic search for phytoaccumulating or phytoexcluding plant species is necessary (Sanyal 2016a, b).

A scan of literature reveals a number of plant/microbial species, known for arsenic accumulation/or as bio-indicator, which can effectively remove arsenic (and other heavy metals) from the aquatic system, for instance, to the tune of 170 and 340 μg As. g^{-1} dry weight of water hyacinth in its stem and leaves, respectively (Chigbo et al. 1982), when grown in a pond containing 10 mg As. dm $^{-3}$. However, such accumulated arsenic in water hyacinth (*Eichornia crassipes*) is also liable to leaching out in the water body, particularly so on decomposition of such aquatic weed. Consequently, appropriate precaution has to be exercised while interpreting the arsenic status of aquatic environment by water hyacinth accumulation.

A number of microbial species (e.g., the bacterial species, namely *Proteus* sp., *Escherichia coli*, *Flavobacterium* sp.; *Corynebacterium* sp. and *Pseudomonas* sp., the fungus, namely *Candida humicola;* the freshwater algae, namely *Chlorella ovalis*, *Phaepdactuylum tricornutum*, *Oscillatoria rubescens*) have been reported to possess varying degrees of arsenic accumulating abilities. Several weed species, normally found along with crops like rice, potato, jute, mustard, etc., growing on arsenic contaminated soils (2–14 mg As. kg^{-1} soil), and subjected to irrigation (given to the desired crops) with arsenic contaminated groundwater, were noted to accumulate considerable amounts of arsenic in their biomass (Das et al. 2005).

10 Bio-Remediation of Arsenic in Soil-Plant System

In a study conducted with selected arsenic-contaminated soils of West Bengal, the arsenic-volatilizing indigenous soil bacteria, isolated from these soils, were tested for their ability to turn the toxic indigenous inorganic arsenic to less toxic volatile arsenicals. Approximately 37% of As (III) (under anaerobic condition) and 30% As (V) (under aerobic condition) were volatilized by these bacterial isolates in three days. In contrast to the genetically modified organism, the indigenous soil bacteria were capable of removing 16% of arsenic from the contaminated soil during the 60 days-incubation period in presence of FYM (Majumder et al. 2013a). Further, arsenic-oxidizing bacteria were isolated from selected arsenic-contaminated soils of West Bengal (Majumder et al. 2013b), which were closely related to various species of *Bacillus* and *Geobacillus*, based on their 16S rRNA gene sequences. They were found to be hyper-resistant to both As (V) (167–400 mM) and As (III) (16–47 mM). Elevated rates of As(III) oxidation (278–1250 μM. h^{-1}) and arsenite oxidase activity (2.1–12.5 nM. min^{-1} mg^{-1} protein) were observed in these isolates. Furthermore, among the superior arsenic-oxidizers, the AMO-10 completely (100%) oxidized 30 mM of As(III) within 24 h. The presence of the *aoxB* gene was confirmed in the screened isolates. Phylogenetic tree construction, based on the *aoxB* sequence, revealed that two strains, AGO-S5 and AGH-02, were clustered with *Achromobacter* and *Variovorax*, whereas the other two (AMO-10 and ADP-25) remained un-clustered. The increased rate of As(III) oxidation to the less toxic As (V) forms by these native strains might be exploited for the remediation of arsenic in the contaminated environments (Majumder et al. 2013b).

11 Arsenic in Soil-Plant System and Its Influence on Food Chain

It was noted that As (III) accounted for the major arsenic species recovered from grains of transplanted autumn paddy, while As (V) predominates in that from rice straw (Sinha and Bhattacharyya 2014a; Sanyal et al. 2015; Sanyal 2016a). Soil amendment through organic intervention reduced arsenic accumulation in rice grain and straw of autumn rice, as manifested through reduction of inorganic arsenic (Sinha and Bhattacharyya 2014a). Sinha and Bhattacharyya (2014b) also studied the arsenic toxicity profile in rice, grown in selected contaminated area of rural Bengal, and the possible risk of its dietary exposure. The unique character of the anaerobic rice ecosystem results in a significant build-up of inorganic arsenic (i-As) in soil and its concomitant accumulation in rice. The recoveries of i-As was dominated by As (III) in rice grain and As (V) in rice straw, thereby emphasizing the presence of higher levels of the more toxic As (III) in the edible portion. Recoveries of organic arsenic species in rice grain and straw further suggested the possibilities of methylation of i-As in the plant system. The risk of dietary exposure to i-As through rice, the staple food in the experimental area, poses an almost equal threat to human health as that posed by the contaminated drinking water. Organic amendments and augmented P fertilization showed considerable promise in reducing the total and inorganic arsenic accumulation in rice and the consequent dietary risk (Sinha and Bhattacharyya 2014a; Sanyal et al. 2015; Sanyal 2016a, b).

Few reports are available that characterize daily arsenic exposure through water and diet among the people living in groundwater contaminated regions and correlate the former with arsenic biomarkers. Demographic characteristics and the total daily arsenic intake through water and diet were determined in 167 participants (Group-1 participants selected from arsenic-endemic region) and 69 participants (Group-2 participants selected from arsenic-non-endemic region) in a study conducted in West Bengal. The findings showed significantly high dietary arsenic intake in people living in Nadia district of West Bengal, where contaminated groundwater was used for irrigation purpose, but significantly low in the region of Hoogly district, where groundwater was uncontaminated. Even after lowering the arsenic level in drinking water to <50 μg As. L^{-1} (the permissible limit in India), significant arsenic exposure occurred through water and diet, reflected by the elevated level of arsenic in the arsenic biomarker, namely urine, in people living in the arsenic-endemic region studied. Those with skin lesions were found to have a higher level of arsenic in urine and hair, compared to those without skin lesion (Guha Mazumder et al. 2013, 2014). In yet another study, the dose of daily arsenic intake from both water and diet was found to be significantly and positively associated with urinary arsenic levels in an arsenic-endemic region of West Bengal, even when people were using arsenic-safe water (<50 μg/L) for drinking and cooking purposes (Fig. 13.2). When arsenic levels in drinking water were further reduced to <10 μg/L (WHO safe limit), the dose from

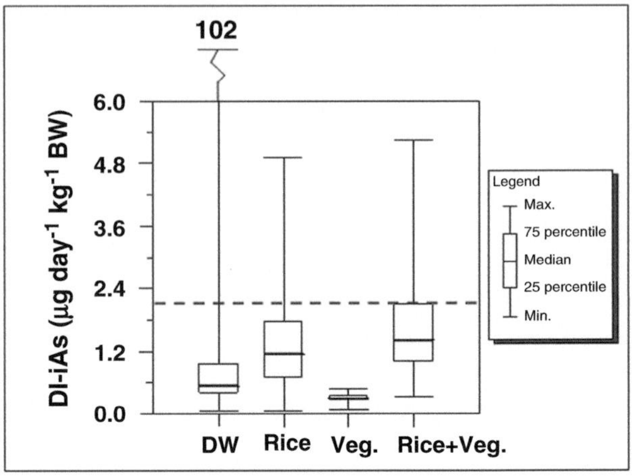

Fig. 13.2 Comparison among the daily intake of inorganic As (DI-iAs) due to consumption of drinking water (DW), rice and vegetables (Veg.). The dotted line indicates the WHO recommended provisional maximum tolerable daily intake (PMTDI) ($PMTDI_{As}$) value of 2.1 μg/kg body weight/day. (Source: Halder et al. 2012)

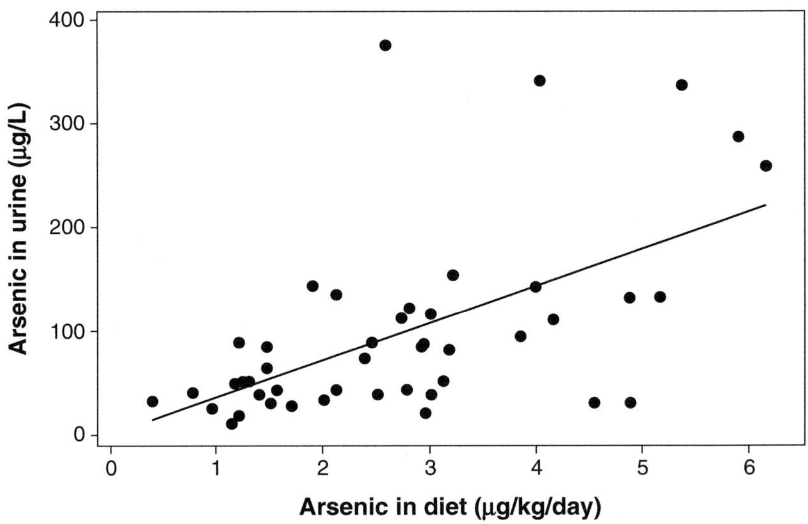

Fig. 13.3 Correlation of urinary As concentration (μg/L) with daily dietary As intake (μg/kg/day) for the participants drinking water with As level <0.3 μg/L having skin lesion (n = 45), West Bengal, India (r = 0.573). (Source: Guha Mazumder et al. 2014)

the diet was still found to be significantly associated with urinary arsenic excretion (Fig. 13.3). But no significant association was found with arsenic dose from drinking water in this group (Halder et al. 2012; Guha Mazumder et al. 2013; Sanyal et al. 2015; Sanyal 2016a, b). Further, when exposed to arsenic through only diet, urinary arsenic concentration was found to correlate positively with dietary arsenic intake in the participants, showing skin lesions (Fig. 13.3), while this correlation was insignificant in participants without skin lesion. Thus Figs. 13.2 and 13.3 amply demonstrate that supply of arsenic-safe drinking water (<10 μg/L) to the population in rural

Bengal *alone is not enough to reduce the risk of As poisoning*, consumption of *rice provides yet another potential pathway of i-As exposure*, that must also be considered for the effective remedial options. Indeed, any mitigation intervention of chronic arsenic toxicity in rural Bengal needed integrated approaches of attempting to reduce arsenic entry into the food-chain, on one hand, while reduction of arsenic in the drinking water below the safe limits, on the other (Halder et al. 2012; Guha Mazumder et al. 2014; Sanyal 2016a, b).

12 Risk Assessment of Metal-Contaminated Soil

There have been several attempts made to assess the risk of growing food crops in metal-contaminated soil and the entry of the hazardous metal in human food chain. Risk assessment of such metal-contaminated soil depends on how precisely one can predict the solubility of metals in soils. To ascertain this, simpler approaches like the integrated solubility and free ion activity model may be successfully used to predict the metal uptake through the edible portion of crops grown on contaminated soils (Datta and Young 2005). The guiding principle depends on the premises that the response of plants and soil organisms to metal toxicity is determined primarily by the variation in free metal ion activity in soil pore water. Indeed such approach has been extended very recently to cover the risk assessment arising from the entry of arsenic in the food chain and the subsequent intake of such contaminated food (especially rice) in a collaborative study of the soil scientists (of the Indian Agricultural Research Institute, New Delhi and BCKV, as well as the present author), on one hand, and the physicians (of the DNGM Research Foundation, Kolkata), on the other. The resulting findings have been summarized in a scientific article compiled by Golui et al. (unpublished).

Thus, Datta and Young (2005) developed the protocol for prescribing toxic limit of metals, based on extractable metals and soil characteristics, using the solubility and free-ion activity models. Solubility of metals in soil was predicted using the following pH-dependent Freundlich equation (Jopony and Young 1994) as per the free-ion activity of metal and metalloid (FIAM):

$(M^{2+}) = M_c/[k_M(H^+) - n_M]$where (M^{2+}) is the free metal ion activity in soil solution in soil pore, M_C is the labile pool of soil metal, assumed to be exclusively adsorbed on humus (mol. kg carbon^{-1}), and k_M and n_M are empirical constants which express the pH dependence of the metal distribution coefficient. It follows from detailed theoretical considerations (Datta and Young 2005; Meena et al. 2016):

$$p\left(M^{2+}\right) = [p(M_c) + k_1 + k_2\text{pH}]/n_F \tag{13.6}$$

where k_1 and k_2 are empirical, metal-specific constants, expressing the pH-dependence of metal distribution coefficient and n_F is the power term from the Freundlich equation. The metal transfer factor from soil solution to plant biomass is given as

$$\text{Transfer factor} = \log\left[\mathrm{M_{plant}}/\left(\mathrm{M}^{2+}\right)\right] \tag{13.7}$$

where M_{plant} is the metal loading of plant biomass.

Equations 13.6 and 13.7 can be combined to lead to Eq. 13.8 as follows:

$$p\left(\mathrm{M_{plant}}\right) = C + \beta_1 \mathrm{pH} + \beta_2 p(\mathrm{M_c}) \tag{13.8}$$

where C, β_1 and β_2 are empirical metal- and plant-specific constants. This model (FIAM) predicts the free-ion activity of trace metal and metalloid in soil solution as a function of labile soil extractable metal and pH with the simplifying assumption that the whole amount of metal (M_C) is adsorbed on soil humus.

This approach was recently used to predict the free-ion activity of Zn, Cu, Ni, Cd and Pb in metal-contaminated soil as a function of pH, soil organic carbon, and extractable metal content. It was found that the solubility model as a function of pH, organic carbon and EDTA-extractable metals was reasonably effective in predicting the metal ion activity in contaminated soils (Rang Zan et al. 2013).

Closely related to what is stated above, the risk to human health for intake of metal through consumption of green leafy vegetables has been computed in terms of what is known as the Hazard Quotient, HQ_{gv}. The latter is given as

$$\mathrm{HQ_{gv}} = (\mathrm{ADD}/\mathrm{R_fD}) \tag{13.9}$$

where ADD = average total daily dose of metal intake through diet and drinking water (mg metal.kg body weight^{-1}. day^{-1}) and R_fD = the corresponding reference dose which is defined as the maximum tolerable daily intake of the specific metal that does not lead to any deleterious health effects. Obviously, $HQ_{gv} > 1.0$ suggests hazard to human health. For arsenic, 2.1 μg/kg body weight/day was used as the corresponding PMTDI. As stated earlier, the joint FAO/WHO Expert Committee on Food Additives (JECFA) set a provisional maximum tolerable daily intake (PMTDI) of inorganic arsenic as 2.1 μg/kg body weight/day for humans in 1983 and confirmed a provisional tolerable weekly intake (PTWI) as 15 μg/kg body weight/day in 1988 (FAO/WHO 1989).

However, the point worth noting here is that ADD refers to the daily intake of a given metal *from all the food items and drinking water*. It is thus evident that for any one food item (e.g., rice or a vegetable), the limiting value of HQ_{gv} will be less than 1.0. In this context, Meena et al. (2016) argued that the permissible limits of metal and metalloid in soil was established based on (i) solubility of metal and metalloid in soil, (ii) metal and metalloid content in rice and wheat grain, and (iii) human health hazard, associated with intake of metal and metalloid through consumption of crops raised on metal-contaminated soils. For fixing the toxic limit of the extractable metal and metalloid in soils at a particular pH and organic carbon, the critical value of HQ_{gv} used by these authors was 0.5 for *any one given food item*, e.g., rice. These authors (2016) developed a ready reckoner to compute the permissible limit of the extractable metal and metalloid in soils, based on soil pH and organic carbon content of the soil. These permissible limits were based on the predicted HQ_{gv} by the aforesaid solubility-FIAM.

13 Remedial Options at a Glance

- Optimum conjunctive use of ground and surface water (e.g., harvested rainwater) and recharge of groundwater resource with harvested rainwater, free of arsenic.
- Development/identification of suitable low arsenic-accumulating high yielding crops/varieties, and preferring low-water requiring, farmer-attractive cropping sequences (especially for the lean period of January to May), suitable for the arsenic-contaminated areas.
- Irrigation with pond-stored groundwater in which partial decontamination is facilitated by sedimentation-cum-dilution with rain water.
- Enhancing the water use efficiency (through optimum water management) for groundwater irrigation, especially for summer (*boro*) paddy.
- Increased use of FYM and other manures + green manure crops, as well as application of appropriate inorganic amendments (zinc/iron salts as and wherever applicable).
- Identification/development of varieties/crops which accumulate less arsenic in the consumable parts and where the ratio of inorganic to organic forms of arsenic is low.
- Developing cost-effective phyto- and bio-remediation options.
- Creation of general awareness through mass campaigning, holding of farmers' day, field demonstrations, taking due cognizance of the socioeconomic factors (Sanyal et al. 2012, 2015; Sanyal 2016a, b).

14 Conclusions

More sustained research work is necessary to characterize the entire gamut of intricacies of arsenic contamination spectrum in water-soil-plant-animal system, as well as to evolve effective remedial measures to contain the toxin in such system. In addition to the remedial measures discussed above, these include, among others, identification of potential bio-remediating species, exploration of the genetic make-up of several important plant species, covering the varieties of such cultivars, commonly used in the arsenic-belt, vis-à-vis arsenic accumulation and tolerance by these species for identifying the relevant DNA markers and the enzyme systems of these plant species that are affected by arsenic. Some of these aspects are already being examined.

The immediate needs, among others, include the improvement of the field and the laboratory protocols for large-scale measurement of arsenic, and that for different forms/species of arsenic in groundwater-soil-plant-animal-human continuum in order to characterize the *net* toxicity due to the entry of arsenic in the food chain, strengthening of inter-institutional and inter-disciplinary action programme, long-term technical alternatives to reduce dependence on arsenic contaminated resources, and promotion of international networking in support of arsenic mitigation options.

References

Acharya, S.K. (1997). Arsenic in groundwater – Geological overview. Consultation on Arsenic in Drinking water and Resulting Arsenic Toxicity in India and Bangladesh. World Health Organization, New Delhi, India, 29 April to 01 May, 1997.

Banik, G.C. and Sanyal, S.K. (2016). Evaluation of inorganic fractions of arsenic in relation to soil properties in affected areas of West Bengal, India. *Curr. Sci.*, **111(8):** 1371–1377.

Chigbo, F.E., Smith, R.W. and Shore, F.I. (1982). Uptake of arsenic, cadmium, lead and mercury from polluted water by the water hyacinth (*Eichornia crassipes*). *Environ Pollution*, Ser A **27:** 31–36.

Das, I., Ghosh, K. and Sanyal, S.K. (2005). Phytoremediation: A potential option to mitigate arsenic contamination in soil-water-plant system. *Everyman's Sci*, **40(2):** 115–123.

Das, Indranil, Ghosh, Koushik, Das, D.K. and Sanyal, S.K. (2014). Transport of arsenic in some affected soils of Indian subtropics. *Soil Res*, **52:** 822–832.

Das, Indranil, Sanyal, S.K., Ghosh, Koushik and Das, D.K. (2016). Arsenic mitigation in soil-plant system through zinc application in West Bengal soils. *Bioremed J*, **20:** 24–37. doi.org/10.1080/10889868.2015.1124062.

Datta, A., Sanyal, S.K. and Saha, S. (2001). A study on natural and synthetic humic acids and their complexing ability towards cadmium. *Pl Soil*, **225:** 115–125.

Datta, S.P. and Young, S.D. (2005). Predicting metal uptake and risk to the human food chain from leaf vegetables grown on soils amended by long-term application of sewage-sludge. *Water, Air and Soil Pollut*, **163:** 119–136. doi: https://doi.org/10.1002/jsfa.6839.

FAO/WHO (1989). Joint FAO/WHO Expert Committee on Food Additives. Toxicological evaluation of certain food additives and contaminants. Cambridge University Press.

Ghosh, K., Das, I., Das, D.K. and Sanyal, S.K. (2012). Evaluation of humic and fulvic acid extracts of compost, oil cake, and soils on complex formation with arsenic. *Soil Res*, **50:** 239–248.

Golui, Debasis, Guha Mazumdar, D.N., Sanyal, S.K., Datta, S.P., Ray, P., Patra, P.K., Sarkar, S. and Bhattacharya, K. (Unpublished). Prescription of safe limit of As in soil in relation to dietary exposure of arsenicosis patients from Malda district, West Bengal – A case study.

Guha Mazumdar, D.N., Haque, R., Ghosh, N., De, B.K., Santra, A., Chakraborti, D. and Smith, A.H. (1998). Arsenic levels in drinking water and the prevalence of skin lesions in West Bengal, India. *Intern J Epidemio*, **27:** 871–877.

Guha Mazumder, D.N., Deb, D., Biswas, A., Saha, C., Nandy, A., Das, A., Ghose, A., Bhattacharyya, K. and Mazumdar, K.K. (2014). Dietary arsenic exposure with low level of arsenic in drinking water and biomarker: A study in West Bengal. *J Environ Sci Health*, Part A **49:** 555–564.

Guha Mazumder, D.N., Deb, D., Biswas, A., Saha, C., Nandy, A., Ganguly, B., Ghose, A., Bhattacharyya, K. and Mazumdar, K.K. (2013). Evaluation of dietary arsenic exposure and its biomarkers: A case study of West Bengal, India. *J Environ Sci Health*, Part A **48:** 896–904.

Halder, D., Bhowmick, S., Biswas, A., Mandal, U., Nriagu, J., Mazumdar, D.N., Chatterjee, D. and Bhattacharya, P. (2012). Consumption of brown rice: A potential pathway for arsenic exposure in rural Bengal. *Environ Sci Tech*, **46(7):** 4142–4148. doi: https://doi.org/10.1021/es204298a.

IARC (1987). International Agency for Research on Cancer. Overall evaluations of carcinogenicity: An updating of IARC Monographs vols. 1–42.

Jarup, L. (1992). Dose-response relation for occupational exposure to arsenic and cadmium. National Institute for Occupational Health, Sweden.

Jopony, M. and Young, S.D. (1994). The solid solution equilibria of lead and cadmium in polluted soils. *Eur J Soil Sci*, **45:** 59–70.

Ma, L.Q., Komar, K.M., Tu, C., Zhang, W., Cai, Y. and Kennelley, E.D. (2001). A fern that hyperaccumulates arsenic. *Nature*, **409:** 579.

Majumdar, K. and Sanyal, S.K. (2003). pH-dependent arsenic sorption in an Alfisol and an Entisols of West Bengal. *Agropedol*, **13:** 25–29.

Majumder, A., Bhattacharyya, K., Kole, S.C. and Ghosh, S. (2013a). Efficacy of indigenous soil microbes in arsenic mitigation from contaminated alluvial soil of India.Environ Sci Pollution Res, **20(8):** 5645–53. doi: https://doi.org/10.1007/s11356-013-1560-x.

Majumder, A., Bhattacharyya, K., Bhattacharyya, S. and Kole, S.C. (2013b). Arsenic-tolerant, arsenite-oxidizing bacterial strains in the contaminated soils of West Bengal, India. *Sci Total Environ*, **463–464:** 1006–1014.

Meena, R., Datta, S.P., Golui, D., Dwivedi, B.S. and Meena, M.C. (2016). Long-term impact of sewage irrigation on soil properties and assessing risk in relation to transfer of metals to human food chain. *Environ Sci Pollut Res*, **23:** 14269–14283. doi: https://doi.org/10.1007/s11356-016-6556-x.

Mukhopadhyay, D., Mani, P.K. and Sanyal, S.K. (2002). Effect of phosphorus, arsenic and farmyard manure on arsenic availability in some soils of West Bengal. *J Indian Soc Soil Sci*, **50:** 56–61.

Mukhopadhyay, D. and Sanyal, S.K. (2004). Complexation and release isotherm of arsenic in arsenic-humic/fulvic equilibrium study. *Austr J Soil Res*, **42:** 815–824.

Rang Zan, N., Datta, S.P., Rattan, R.K., Dwivedi, B.S. and Meena, M.C. (2013). Prediction of the solubility of zinc, copper, nickel, cadmium, and lead in metal-contaminated soils. *Environ Monit Assess*, **185:** 10015–10025. doi: https://doi.org/10.1007/s10661-013-3309-x.

Sanyal, S.K. (2016a). A textbook of soil chemistry (in press).

Sanyal, S.K. (2005). Arsenic contamination in agriculture: A threat to water-soil-crop-animal-human continuum. Presidential Address, Section of Agriculture &Forestry Sciences, 92nd Session of the Indian Science Congress Association (ISCA), Ahmedabad, January 3–7, 2005; Indian Science Congress Association, Kolkata.

Sanyal, S.K. (2016b). Arsenic toxicity in plant, animal and human and its alleviation strategies. *Indian J Fertilisers*, **12(12):** 74–83.

Sanyal, S.K., Gupta, S.K., Kukal, S.S. and Jeevan Rao, K. (2015). Soil degradation, pollution and amelioration. *In:* Pathak, H., Sanyal, S.K. and Takkar, P.N. (eds), State of Indian Agriculture-Soil, National Academy of Agricultural Sciences, New Delhi.

Sanyal, S.K., Jeevan Rao, K. and Sadana Upkar, S. (2012). Toxic elements and other pollutants – A threat to nutritional quality. *In:* Goswami, N.N. et al. (eds), Soil Science in the Service of Nation. Indian Society of Soil Science, New Delhi.

Sinha, B. and Bhattacharyya, K. (2011). Retention and release isotherm of arsenic in arsenic–humic/fulvic equilibrium study. *Biol Fert Soils*, **47(No.7):** 815–822. doi: https://doi.org/10.1007/s00374-011-0589-6.

Sinha, B. and Bhattacharyya, K. (2014a). Arsenic accumulation and speciation in transplanted autumn rice as influenced by source of irrigation and organic manure. *Intern J Bio-resource Stress Managmt*, **5(3):** 336–368.

Sinha, B. and Bhattacharyya, K. (2014b). Arsenic toxicity in rice with special reference to speciation in Indian grain and its implication on human health. *J Sci Food Agril*, **95(7):** 1435–1444.

Sinha, B., Bhattacharyya, K., Giri, Pradip and Sarkar, S. (2011). Arsenic contamination in sesame and possible mitigation through organic interventions in the lower Gangetic Plain of West Bengal, India. *J. Sci Food and Agril*, **91:** 2762–2767. Doi: https://doi.org/10.1002/jsfa.4519.

WHO (1971). International standards for drinking water, Third edition.

WHO (1993). Guidelines for Drinking-Water Quality, Second edition, Volume 1.

WHO (1996). World Health Organization guidelines for drinking water quality, Vol. II, Second Edition, WHO, Geneva.

WHO (2001). http://www.who.int/inf-fs/en/fact210.html

Chapter 14
Fluoride Pollution in Groundwater

Paulami Sahu

1 Introduction

With burgeoning population, the water demand of the world has increased by many folds and this increasing demand is met from surface water and groundwater resources. Though about 70% of the earth's surface is covered with water, only a tiny part of it is suitable for drinking. Surface water present in lakes, swamps and rivers (0.014%) and groundwater present in sub surface aquifers (1.7%) are the major source of water suitable for human consumption. But the quality of groundwater deteriorates due to natural or anthropogenic causes. Presence of arsenic, fluoride, nitrate, iron, manganese and heavy metals in groundwater, above permissible limit, may cause serious health problems. Among numerous contaminants, arsenic and fluoride are considered as the two main contaminants in India both in terms of the number of people affected and their areal extent of distribution. Present chapter attempts to summarize various aspects of fluoride contamination of groundwater with special emphasis on its source as well as its occurrence in South Asian countries, in general and in India, in particular.

2 Chemical Properties of Fluoride

Fluoride is the most electronegative and the most reactive chemical element in the periodic table (Gillespie et al. 1989). Fluoride is represented as "F" and this element is the lightest member of the halogen group. Fluorine is not found as fluorine in the

P. Sahu (✉)
School of Environment & Sustainable Development, Central University of Gujarat, Gandhinagar, India
e-mail: paulami_sahu@cug.ac.in

P. K. Sikdar (ed.), *Groundwater Development and Management*,
https://doi.org/10.1007/978-3-319-75115-3_14

environment, but it is always present in combined state as fluoride (F). It has an oxidation state of −1, atomic number 9 and atomic weight 18.998. Fluoride is the 13th most abundant element in the earth's crust (Weinstein and Davison 2003) and its abundance in the earth's crust is 0.06–0.09% (Fawell et al. 2006). Fluoride ions have the same charge and nearly the same radius as hydroxide ions and may replace each other in mineral structure (Hem 1985). Fluoride occurs in both organic and inorganic compounds.

3 Importance of Fluoride for Human Health

Fluoride is an essential micronutrient for human beings, serving to strengthen the apatite matrix of skeletal tissues and teeth (Maithani et al. 1998). In humans, about 95% of the total body F is found in bones and teeth. World Health Organization (WHO 1984) had recommended that F concentration in groundwater should be lower that 1.5 mg/L from the suitability for human consumption point of view. But this F guideline value that was set in 1984 was subsequently re-evaluated by WHO in 2006 and it was concluded that there was no evidence to suggest that it should be revised. The 1.5 mg/L guideline value of WHO is not a "fixed" value but is intended to be adapted to take account of local conditions (e.g. diet, water consumption, etc.) (Bailey et al. 2006). As far as Indian standard is concerned (IS 10500:2012) acceptable limit of F in groundwater is 1.0 mg/L and permissible limit in absence of alternate source of water is 1.5 mg/L. High F (>1.5 mg/L) results in dental and skeletal fluorosis; renal and neuronal disorders along with myopathy (Ayoob and Gupta 2006). The effects of F are best predicted by the dose and the duration of exposure. For example, the F requirements set for drinking water in temperate climates are not directly applicable to hot humid regions, where significantly more water is consumed (Fawell et al. 2006). Consumption of drinking water with F below or above the prescribed range is detrimental to human health. Therefore, regular monitoring of groundwater quality with respect to F is essential.

4 Environmental Occurrence and Source of Fluoride in Groundwater

Fluorine is an incompatible lithophile element that preferentially partitions into silicate melts during magmatic crystallization. The geochemical behaviour of F determines its accumulation in the upper continental crust, where its average abundance is 611 mg/kg (Wedepohl 1995). Fluorine always occurs in combined form in minerals as F because of its high reactivity and represents about 0.06–0.09% of the earth's crust (WHO 1984). Fluoride is found in all natural waters at some concentration. Seawater typically contains about 1 mg/L while rivers and lakes generally exhibit concentrations of less than 0.5 mg/L. In groundwater, however, low or high concentrations of F can occur, depending on the nature of the rocks and the occurrence of F-bearing minerals. Fluoride in groundwater is of either geogenic or anthropogenic origin. Predominantly F in groundwater is naturally present due to weathering of rocks rich in F. Fluorite occurs in igneous sedimentary and metamorphic rocks. Fluoride-rich groundwater is mostly found in sediments of marine origin and at the foot of mountainous areas (WHO 2001; Fawell et al. 2006; Brindha and Elango 2011).

4.1 *Geogenic Sources*

4.1.1 F-Bearing Minerals in Igneous, Sedimentary and Metamorphic Rocks

Granite with pegmatite layer, gneiss and schist rocks are the major geological formations that contribute to F concentration in groundwater. Granite is composed of quartz, feldspar and fluorite, whereas gneiss and schist are composed of quartz, potash-feldspar, hornblende, biotitic and fluorite. The F-bearing minerals with interaction of water provide a significant F concentration in groundwater. Fluoride may occur in a mineral as a primary constituent or as an impurity. The main sources of F include fluorine-bearing minerals such as fluorite (CaF_2), apatite [$Ca_5(Cl,F,OH)(PO_4)_3$], micas [$AB_{2-3}(X, Si)_4O_{10}(O, F, OH)_2$], amphiboles [$A_{0-1}B_2C_5T_8$-$O_{22}(OH, F,Cl)$], cryolite ($Na_3AlF_6$), sellaite ($MgF_2$) and topaz [$Al_2SiO_4(F,OH_2)$] (Hem 1985; Pickering 1985; Datta et al. 1996; Jadhav et al. 2015; Khan et al. 2016). Fluorite (CaF_2) is a common F-bearing mineral. This mineral has a rather low solubility and occurs in both igneous and sedimentary rocks. Most fluorides are sparingly soluble and are present in natural water in small amounts. As F-ions have the same charge and nearly the same radius as hydroxyl ions (OH^-), they may replace each other in the octahedral sheet of mineral structures (Brigatti and Guggenheim 2002). Concentration of F in the continental crust is 611 mg/kg. Various rock types contain F at different levels: basalt, 360 μg/g; granites, 810 μg/g; limestone, 220 μg/g; sandstone, 180 μg/gm; shale 800 μg/gm; oceanic sediments, 730 μg/gm and soils, 285 μg/gm (cgwb.gov.in). Minerals of the apatite group such as fluorapatite ($Ca_{10}(PO_4)_6F_2$) is one of the important F-bearing minerals associated mainly with igneous rocks and present as accessory phase (Pan and Fleet 2002).

Fluorite (CaF_2) is a common mineral that forms in hydrothermal systems usually with a cubic shape but can also form into octahedral or other sub-octahedral shapes, depending on the temperature and composition of the solution where the crystals grow (Zidarova 2010). Villiamite (NaF), an almost infinitely soluble mineral, may contribute considerably to F concentrations in groundwater associated with certain peralkaline intrusive bodies, such as the Lovozero Massif in Russia (Kraynov et al. 1969). Granitic rocks contain F ranging between 500 and 1400 mg/kg (Koritnig 1978; Krauskopf and Bird 1995), which is much higher than any other rock type. The weathering of these rocks and long residence time in aquifers with fractured

F-rich rocks enhances F levels in the groundwater. Fluoride may occur in sedimentary rocks as a cementing material in limestones and dolomites or as an accessory mineral in pegmatites. Other F-bearing minerals that are much less abundant include amphiboles such as hornblende and cryolite (Na_3AlF_6). Fluoride-bearing minerals present in metamorphic rocks are important source of groundwater pollution. Gneiss rocks are highly weathered which facilitate release of F from minerals into groundwater. Srinivasa Rao (1997) reported groundwater having F concentration of 3.4 mg/L in Vamsadhara river basin, Andhra Pradesh where metamorphic rock pyroxene amphibolites and pegmatites are the source rocks.

4.1.2 Sea Water

In sea water, F is mainly found as MgF^+ and F^- ions and they contribute about 47% and 51% respectively of the total F concentration in seawater. Fluoride that is removed from the seawater is incorporated into marine sediments, either as part of marine organisms, such as in shells or fish bones (Carpenter 1969) or by the precipitation of authigenic minerals, especially carbonate fluorapatite (Van Cappellen and Berner 1988).

4.1.3 Geothermal Fluid

Geothermal fluids comprise both hot water and steam containing dissolved solutes and gases. The most common types of geothermal water are alkali-chloride solutions with near-neutral pH values (Edmunds and Smedley 2013). The F concentration in geothermal fluid is closely linked to the solubility equilibria of F (Nordstrom and Jenne 1977), which depends on the fluid temperature.

4.1.4 Volcanic Sources

Fluoride is commonly associated with volcanic activity and fumarolic gases. Annual global emissions of gaseous F compounds from active volcanic sources ranges from 60,000 to 6 million tons (Symonds et al. 1988). Hydrogen fluoride is one of the most soluble gases in magmas and comes out partially during eruptive activity (D'Alessandro 2006). The emitted gases interact rapidly with the ash particles of the volcanic plume and form extremely thin salt coatings. This material is composed of relatively soluble sulfate and halide salts mixed with sparingly soluble fluorine compounds (Delmelle et al. 2007), such as CaF_2, AlF_3 and $Ca_5(PO_4)_3F$. Therefore, the water in contact with highly soluble volcanic ash deposits usually contains high concentrations of F (Ruggeri et al. 2010). Moreover, emission of F-rich lava is reported in the (calc-) alkaline volcanoes located at continental margins or island arcs (Rosi et al. 2003).

4.2 Anthropogenic Sources

Anthropogenic F is derived from aluminium smelters (Haidouti 1991), fertilizer factories, and industrial activities such as ceramic firing (WHO 2002), iron and steel production, fossil fuel burning, cement works and glass manufacture.

4.2.1 Atmosphere

Though small amount of F is contributed to the atmosphere by natural sources like rock dust, volcanic eruptions, or aerosol sea spray from the marine environment, major source of atmospheric F is anthropogenic. Industrial activities like aluminium smelting, brick and ceramic production, fossil fuel burning, iron and steel production etc. are the prime source of atmospheric F. These pollution sources release fluorine to the environment as gaseous (e.g., HF, SiF_4, F_2 and H_2SiF_4) or particulate (e.g., CaF_2, NaF and Na_2SiF_6) species (Ozsvath 2009). However, F concentrations in precipitation are generally low, typically <50 μg/L when compared to natural background levels. This rain water infiltrates into the unconfined aquifer and become the part of groundwater.

4.2.2 Fertiliser and Irrigation Water

Phosphate bearing fertilisers contribute to the F in groundwater (Farooqi et al. 2007b). Successive irrigation in agricultural lands also adds up F in groundwater (Young et al. 2010). Therefore, fertiliser and irrigation water are considered as potential threat for increase in F concentration in groundwater.

4.2.3 Burning of Fluoride-bearing Coal

Churchill et al. (1948) reported that coal contains 40–295 mg/kg of F. More than 100 million tons of fly ash is produced worldwide annually due to the combustion of coal especially from thermal power plants (Prasad and Mondal 2006) and their haphazard disposal will result in the leaching of F to groundwater. Thermal waters with high pH are rich in F (Edmunds and Smedley 1996). Jha et al. (2008) reported that brick kilns which use coal are also a source for F pollution.

5 Mechanism of Stabilization of Fluoride and Its Mobilization in Groundwater

The F in groundwater is mostly geogenic in origin. The occurrence, distribution pattern and mobilization of F in any aqueous medium such as surface water or groundwater, entirely depends on the combinations of different geochemical processes going on simultaneously in a natural aqueous system. Fluoride enrichment in any aqueous medium takes place due to weathering of the F-bearing minerals present in the host rocks and sediments and its leaching to the water source. Many environmental conditions or factors such as water pH (Genxu and Guodong 2001), temperature, origin of water, chemical composition of water, host rock lithology, amount of soluble and insoluble F in source rocks, duration of contact of water with F-bearing rocks, oxidation-reduction process (Mahapatra et al. 2005) etc. influence the geochemical processes or mechanism of F mobilization from its source to the groundwater reservoir. Ramesam and Rajagopalan (1985) reported that geological features such as dyke also play an important role in the concentration of F in groundwater. The presence of some doleritic dykes, acting as a natural barriers for groundwater flow and making the groundwater stagnated in fractures for longer period of time, give greater ionization facilities to F minerals in basic dyke rocks and as a consequence groundwater gets enriched with F.

Fluoride stabilizes in the rock through geochemical processes such as adsorption, precipitation and mobilizes to groundwater by dissolution and desorption processes.

5.1 *Stabilization*

5.1.1 Adsorption

Adsorption is defined as the transfer of ions (adsorbate) from the solution phase to the solid phase (adsorbent) via various mechanisms, such as physical and chemical adsorption and surface precipitation. Physiosorption is a nonspecific adsorption process in which outer-sphere complexes are formed between aqueous ions and the surface functional due to electrostatic interactions. On the contrary, chemisorption results in inner-sphere complexes through short range bonding between the solute complex and surface oxygens. Though the exact process of mobilization of F in groundwater is unclear, the most common mechanism for F mobilization is displacement of F^- by OH^- when F-rich minerals come in contact with water of high alkalinity (Edmunds and Smedley 2001; Khan et al. 2016). In the presence of water, the surface metal ions may first tend to coordinate with water molecules (Fig. 14.1a) and the dissociative chemisorption of water leads to the formation of surface of hydroxyl groups (Fig. 14.1b) which are highly reactive (Stumm 1992). If

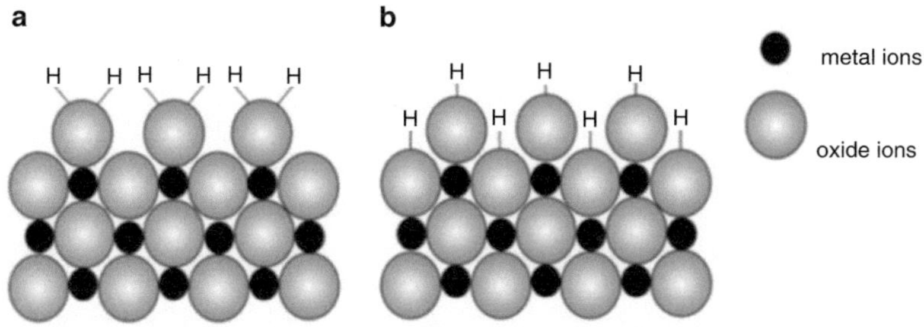

Fig. 14.1 (**a**) Coordination of surface metal ions with water molecules present at the water/solid interface of metal oxides; (**b**) formation of surface hydroxyl groups by dissociative chemisorption of water at the water/solid interface of metal oxides. (Source: García and Borgnino 2015)

the ion and surface have the same charge, the chemical affinity of an ion for the surface can supersede the electrostatic repulsion. In other words, if there is a very strong chemical interaction, an anion can adsorb onto a negatively charged surface (García and Borgnino 2015*)*.

The adsorption of F on metal hydroxides mainly depends on pH and higher adsorption occurs under strongly acidic condition. Farrah et al. (1987) first reported the interaction of F with aluminium hydroxides. This study reveals under strongly acidic conditions (pH ranging from 3 to 8) aluminium oxides dissolve and form F–Al complexes. To enhance the activity of aluminium oxide as absorbent, it was sufficiently heated. Many studies were carried out on adsorption of F onto iron hydroxides (Hiemstra and Van Riemsdijk 2000; Sujana and Anand 2010; García and Borgnino 2015). Here F is replaced by surface hydroxyl groups and this released F is adsorbed and form inner sphere complexes. Calcium carbonate or calcite acts as F sink in sediments due to the strong affinity of F for calcite sites. Robison and Edigton (1946) conveyed that the main source of F in ordinary soil is clay minerals present in it. Clay minerals for example, kaolinite, dickite etc. are hydrated alumina silicates having layered crystal structures (phyllosilicate). Due to negative electrical charge of clay minerals, generally they can't remove anions. However in soils, as clays are usually associated with iron hydroxides coatings (Zhuang and Yu 2002) they can adsorb F (Bia et al. 2012). Fluoride dissolution is also controlled by clay content in a soil as clay decreases the hydraulic conductivity and thus increases the residence time of water in the aquifer and contact duration between F-bearing minerals and groundwater (Rao et al. 2015).

5.1.2 Precipitation

The main authigenic F-bearing minerals that may precipitate from saturated solutions in the aquatic environments are fluorite, fluorapatite and carbonate fluorapatite (García and Borgnino 2015).

5.2 *Mobilization*

5.2.1 Dissolution

The important F-bearing minerals are muscovite, biotite, fluorite (fluorspar), fluorapatite, lepidolite, cryolite, topaz etc. Dissolution of fluorite (CaF_2) is the main cause of high concentrations of F in groundwater in many parts of the world (Edmunds and Smedley 2013). Handa (1975) reported that sodium bicarbonate-type facies are more prone to release F from fluorite mineral. The dissolution of CaF_2 takes place as:

$$CaF_2 + 2NaHCO_3 \rightarrow CaCO_3 + 2Na + 2F + H_2O + CO_2$$

High alkalinity, moderate electrical conductivity and bicarbonate concentration are important factors which controls F dissolution in groundwater (Saxena and Ahmed 2001). Other promoting factors are alkalinity and evaporation (Tirumalesh et al. 2007). If the groundwater is highly alkaline, minerals like mica and amphiboles displace F with OH^- (Hem 1985). As F ions have the same charge and nearly the same radius as OH ions, they may replace each other in the octahedral sheet of mineral structures (Brigatti and Guggenheim 2002). The replacement process in muscovite (Singh and Mukherjee 2014) is as follows:

$$KAl_3Si_3O_{10}(OH;F)_2 + CO_2 + 2.5H_2O \rightarrow 1.5Al_2Si_2O_5 + K + HCO_3^- + 2F^-$$

As presence of calcium creates fluorite saturation ceiling conditions (Frengstad et al. 2001) its removal from the system helps in F dissolution in groundwater. Study of Guo et al. (2007) that indicates sodium-rich groundwater has more chance of dissolution of F from the mineral into the aqueous solution than that of calcium-rich groundwater also support this theory.

High HCO_3 content of groundwater helps in dissolution of fluorite in groundwater (Guo et al. 2007) by the following reaction:

$$CaF_2 + 2HCO_3 \rightarrow CaCO_3 + 2F^- + H_2O + CO_2$$

5.2.2 Desorption

Under alkaline conditions, desorption from iron, aluminium and manganese oxyhydroxides has been considered as secondary source of F in groundwater (Borgnino et al. 2013). Chemical weathering of CaF_2 bearing rocks such as granite, gneiss, sandstone and shale is most effective under high temperature in arid to semiarid climatic conditions (Su et al. 2015; Khan et al. 2016) and it releases the adsorbed F.

6 Distribution of Fluoride in Groundwater

6.1 World Scenario

As per World Health Organisation (WHO) the permissible limit of F concentration in groundwater is 1.5 mg/L. Concentrations above the permissible limit have been recorded in several parts of the world. Fluoride is one such ion that causes health problems in people living in more than 25 nations around the world (Brindha and Elango 2011). The total number of people affected is not known, but a conservative estimate would number in the tens of millions. The occurrence of F contamination in groundwater has been reported by many researchers from all over the world working in the field of groundwater. Researchers from different countries like India, Sri Lanka, Pakistan, Kenya, Norway, North Jordan, Iran, Afghanistan, Turkey, China, Japan, Southern Algeria, Mexico, Korea, Italy, Brazil, Malawi, Ethiopia, Canada, Ghana, Kenya, South Carolina, Wisconsin, Ohio, South Korea reported the occurrence of F contaminated groundwater in their countries (Srinivasa Rao 1997; Dissanayake 1991; Amouei et al. 2012; Gaciri and Davies 1993; Banks et al. 1998; Rukah and Alsokhny 2004; Farooqi et al. 2007a; Mirlean and Roisenberg 2007; Msonda et al. 2007; Vivona et al. 2007; Davraz et al. 2008; Messaïtfa 2008; Moghaddam and Fijani 2008; Oruc 2008; Suthar et al. 2008; Looie and Moore 2010; Naseem et al. 2010; Derakhshani et al. 2014; Hassan Saffi and Jawid Kohistani 2015; Oruc 2003; Kim and Jeong 2005; Tekle-Haimanot et al. 2006; Khan et al. 2013; Valenzuela-Vásquez et al. 2006; Zheng et al. 2006; Chae et al. 2007; Yidana et al. 2010; Young et al. 2010; Desbarats 2009; Brindha et al. 2011; Keshavarzi et al. 2010; Li et al. 2009; Karthikeyan et al. 2010; Chae et al. 2007). Endemic fluorosis develops widely in many areas of the world, such as China (Guo et al. 2006), India (Gupta et al. 2005), Mexico (Carrillo-Rivera et al. 2002) and Africa (Gizaw 1996).

6.2 South Asian Scenario

The current territories of Afghanistan, Bangladesh, Bhutan, Nepal, India, Pakistan, Sri Lanka and Maldives, form the countries of South Asia. The South Asian Association for Regional Cooperation (SAARC) is an economic cooperation organisation in the region which was established in 1985 and includes all eight nations comprising South Asia. Out of these eight countries, F contamination in groundwater is reported in seven countries except Maldives. Any published data indicating occurrence of high concentration of F in groundwater of Maldives is not available.

6.2.1 Afghanistan

Elevated concentrations of F are found in groundwater of Faryab province, Afghanistan. Sixteen percent of the total 23,800 water samples showed concentrations above the WHO limit for F (Hassan Saffi and Jawid Kohistani 2013). 5174 chemical analysed data from UNICEF, ECHO, WASH project and National GMWs network reveals that the F concentration of Afghanistan varied between 0.01 mg/L (Shahri Naw, Kabul city) and 15 mg/L (Wara kalli, Gurbuz district of Khost province) (Hassan Saffi and Jawid Kohistani 2015).

6.2.2 Bangladesh

In general, the F concentration in groundwater of Bangladesh is low. But in some rural areas of Bangladesh the fluoride concentration in water is higher than 1.5 mg/L. Groundwater of Barisal, Faridpur, Jessore, Khulna and Rajshahi regions contain higher levels of F than any of the other regions (Hoque et al. 2003).

6.2.3 Nepal and Bhutan

No data are available for F in Nepal's groundwater. However, due to high rainfall, the presence of F at high concentrations is unlikely. Concentrations are likely to be much below 1 mg/L, i.e. less than the WHO guideline value for F in drinking water, in both the hill regions and the terai (Diwakar et al. 2015). Pant (2011) and Gurung and Oh (2012) reported occurrence of F in shallow wells (1.92 mg/L) and deep wells (1.76 mg/L) of Kathmandu Valley. There is no reported data of occurrence of F in groundwater >1.5 mg/L in Bhutan.

6.2.4 Pakistan

Rafique et al. (2008) reported that the groundwater of Thar Desert, Pakistan was highly contaminated with F. Study of Naseem et al. (2010) in Nagar Parkar Town, Thar Desert of Pakistan also supported the same. In Nagar Parkar Town mean F concentrations of 3.33 mg/L (maximum 7.85 mg/L) were found, with 78% of samples exceeding the WHO standard limits (Naseem et al. 2010; Azizullah et al. 2011). Fluoride concentration in most of the groundwater samples of Chachro and Diplo sub-districts of Tharparkar district, located in south-east edge of Sindh, Pakistan were higher than the permissible limits proposed by WHO (Brahman et al. 2013). Fluoride data of groundwater samples of Faisalabad (Kausar et al.

2003) and Karachi (Siddique et al. 2006), reveals that it was well below the standard limits with few exceptions from an industrial area of Karachi. The highest F concentrations have been reported from Khalanwala, East Punjab, where it reached as high as 21.1 mg/L (Farooqi et al. 2007a) and 22.8 mg/L, and 75% of samples were above the WHO normal standards (Farooqi et al. 2007b).

6.2.5 Sri Lanka

The problems associated with high F content in groundwater in the dry zone of Sri Lanka are well known (Dissanayake and Weerasooriya 1986; Lennon et al. 2004). It has been found that both shallow and deep groundwater exceeded the World Health Organization (WHO) recommended levels in drinking water (>1.5 mg/L). In the Kurunegala District in Northwestern Sri Lanka, the F problem is more acute in some parts. For instance, in the Galgamuwa region, nearly 15% of deep wells had more than 2.0 mg/L of F. Nikeweratiya, Nikawewa, Polpitigama and Ambanpola are among the other areas in the Kurunegala District that have significantly high F levels. The Anuradhapura district is another region where F levels are remarkably high. The Polonnaruwa district in the North Central Province of Sri Lanka is another region where the F problem is prevalent (Chandrajith et al. 2012).

6.2.6 India

Out of 85 million tons of F deposits present in the earth crust, 12 millions are found in India (Teotia and Teotai 1994). This fact indicates that the geogenic F contamination of groundwater is widespread in India (cgwb.gov.in). Endemic fluorosis is prevalent in India since 1937 (Shrott et al. 1937). As per CGWB record (cgwb.gov.in) F in groundwater beyond the permissible limit of 1.5 mg/L are reported in 184 districts of 19 states of India and it is a major health problem in India (Table 14.1, Fig. 14.2). Table 14.2 represents the literature value of concentration of F in groundwater in different states.

7 Technology for Fluoride Mitigation in Groundwater

The crise of F pollution is increasing rapidly as revealed by the world-wide survey. As per estimates of UNESCO (2007) more than 200 million people in the whole world rely on drinking water with F concentration more than the limit as prescribed by WHO guideline value. Though, it is better to avoid groundwater sources having

Table 14.1 Occurrence of fluoride in groundwater of India

Sl. No.	State	No. of districts affected	Districts
1.	Andhra Pradesh	16	All districts except Adilabad, Nizamabad, West Godavari, East Godavari, Vishakhapatnam, Srikakulam, Vizianagaram
2.	Assam	2	Karbi Anglong, Nagaon
3.	Bihar	5	Daltonganj, Gaya, Rohtas, Gopalganj, Paschim Champaran
4.	Chhatisgarh	2	Durg, Dantewara
5.	Delhi	7	Central, South, West, East, South-west, North West, North East zones
6.	Gujarat	18	All districts except Dang
7.	Haryana	11	Rewari, Faridabad, Karnal, Sonipat, Jind, Gurgaon, Mohindergarh, Rohtak, Kurukshetra, Kaithal, Bhiwani
8.	Jammu & Kashmir	1	Doda
9.	Jharkhand	4	Giridih, Palamau, Pakur, Sahabganj
10.	Karnataka	14	Dharwad, Gadag, Bellary, Belgaum, Raichur, Bijapr, Gulbarga, Chikmaglur, Mandya, Bangalore (rural), Mysore, Manglore, Kolar, Shimoga
11.	Kerala	2	Palghat, Allepy
12.	Maharashtra	8	Chandrapur, Bhandara, Nagpur, Jalgaon, Buldhana, Amravati, Akola, Yavatmal
13.	Madhya Pradesh	13	Shivpuri, Jhabua, Mandla, Dindori, Chhindwara, Dhar, Vidisha, Sehore, Raisen, Mandsour, Neemuch, Ujjain, Seoni
14.	Orissa	18	Phulbani, Koraput, Dhenkenal, Angur, Boudh, Nayagarh, Puri, Balasore, Bhadrak, Bolangir, Ganjam, Jagatpur, Jajpur, Kalahandi, Keonjhar, Kurda, Mayurbhanj, Rayagada
15.	Punjab	9	Mansa, Faridkot, Bhatinda, Muktsar, Moga, Sangrur, Ferozpur, Ludhiana, Amritsar
16.	Rajasthan	32	All districts
17.	Tamil Nadu	8	Salem, Erode, Dharmapuri, Coimbatore, Tiruchirapalli, Vellore, Madurai, Virudunagar
18.	Uttar Pradesh	7	Unnao, Agra, Meerut, Mathura, Aligarh, Raibareli, Sonbhadra
19.	West Bengal	7	Birbhum, Bardhman, Bankura, Purulia, Malda, U. Dinajpur & D. Dinajpur

Source: cgwb.gov.in, accessed on 5th March, 2017

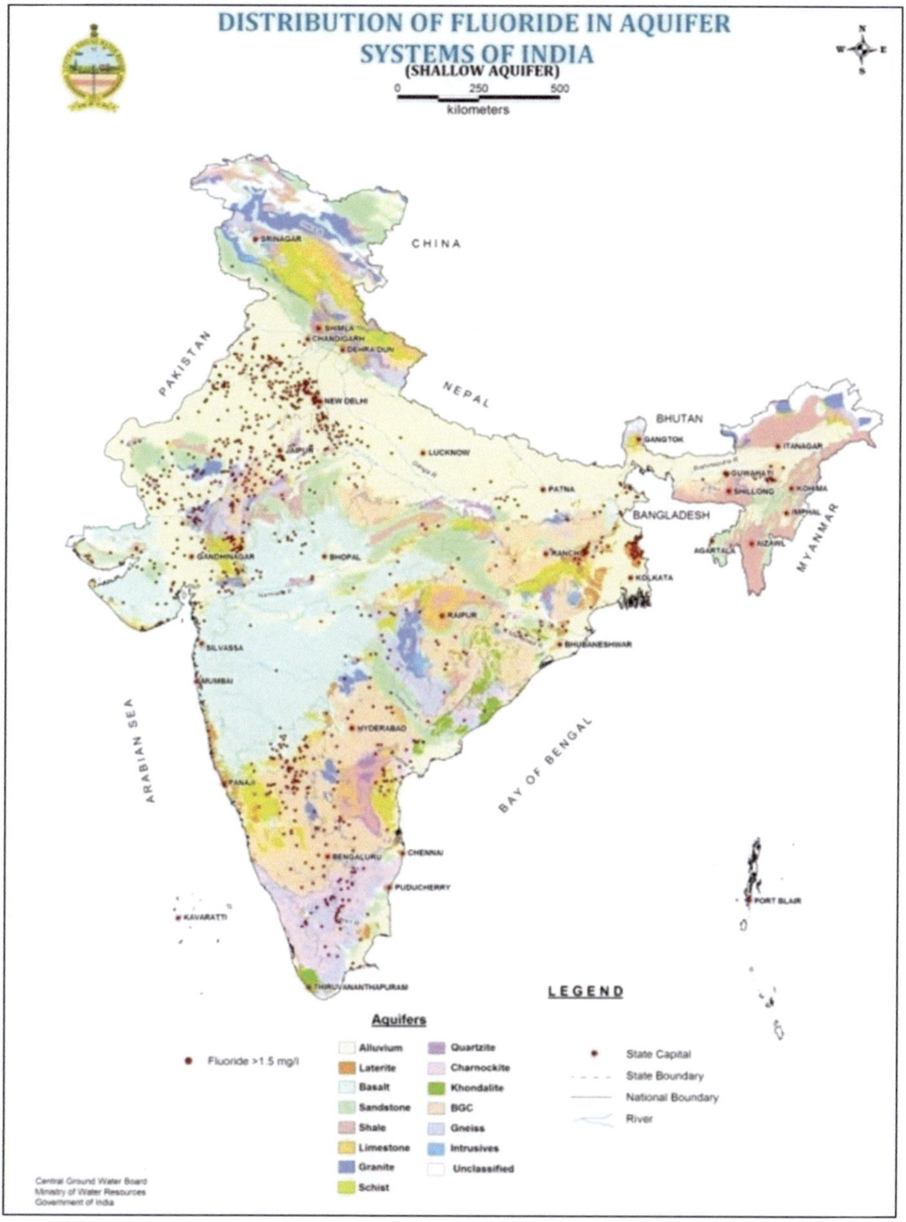

Fig. 14.2 Distribution of fluoride in shallow aquifer systems of India. (Source: cgwb.gov.in)

Table 14.2 Concentration of fluoride in groundwater of India

Sr. No.	Location	Fluoride concentration (mg/L)	Reference
1	Nalgonda District, Andhra Pradesh	Up to 7.6	Reddy et al. (2010)
2	Kurmapalli watershed, Andhra Pradesh	Up to 21.0	Mondal et al. (2009)
3	Kolar and Tumkur Districts of Karnataka State	0.36–3.34	Mamatha and Rao (2010)
4	Bellary, Karnataka	0.33–7.8	Wodeyar and Sreenivasan (1996)
5	Guwahati, Assam	0.18–6.88	Das et al. (2003)
6	Karbi Anglong, Assam	0.95–15.4	Kotoky et al. (2008)
7	Kamrup, Assam	0.4–8.92	Chakrabarty et al. (2009)
8	Delhi	0.1–16.5	Datta et al. (1996)
9	Mehsana, Gujarat	1.5–5.6	Dhiman and Keshari (2006)
10	Bhiwani, Haryana	0.14–86	Garg et al. (2009)
11	Palghat, Kerala	0.2–5.75	Shaji et al. (2007)
12	Yavatmal, Maharashtra	0.30–13.41	Madhnure et al. (2007)
13	Hanumangarh, Rajasthan	1.01–4.42	Suthar et al. (2008)
14	Churu District, Rajasthan	0.1–14	Muralidharan et al. (2002)
15	Erode, Tamil Nadu	0.5 and 8.2	Karthikeyan et al. (2010)
16	Kanpur, Uttar Pradesh	0.14–5.34	Sankararamakrishnan et al. (2008)
17	Sonbhadra, Uttar Pradesh	0.483–6.7	Janardhana Raju et al. (2009)
18	Hooghly, West Bengal	0.01–1.18	Kundu and Mandal (2009)
19	Nayagarh, Orissa	0.16–10.1	Kundu et al. (2001)
20	Birbhum, West Bengal	0.006–1.95	Gupta et al. (2006)

Note: Table modified from Brindha and Elango 2011

high F content, it is not always feasible. Therefore, several methods are developed by various researchers to treat groundwater with high concentration of F and reduce its F content. A detailed list of those F removal technology or methods is given in Table 14.3. However, the effective removal of F by these techniques depends on the initial concentration of fluoride, contact time, pH of target groundwater, type of absorbents and their size.

Table 14.3 Technologies for fluoride removal

Sr. No.	Technique	Chemical used	Reference
1.	Precipitation method	Alum or Aluminum sulphate	Culp and Stoltemberg (1958)
2.	Bone Char	Ground animal bones consisting of tricalcium phosphate and carbon	Sorg and Logsdon (1978), Ma et al. (2008)
3.	Nalgonda Technique	Alum or Aluminum sulphate or kalium aluminum sulphate and lime or calcium oxide	Dahi et al. (2013)
4.	Electro coagulation	Aluminum hydroxide	Mameri et al. (1998)
5.	Magnesia and Serpentine	Magnesia, Serpentine (Mg6Si4O10 (OH))	Chidambaram et al. (2003)
6.	Activated alumina	Aluminum trihydrate	Ghorai and Pant (2004)
7.	Ion Exchange Resins	Strong base exchange resins	Meenakshi and Viswanathan (2007)
8.	Red Mud	Granular red mud	Tor et al. (2008)
9.	Bentonite	Granular acid-treated bentonite	Ma et al. (2010)
10.	Hybrid Al_2O_3/Bio-TiO_2 Nanocomposite	Hybrid Al_2O_3/Bio-TiO_2 Nanocomposite impregnated Thermoplastic polyurethane (TPU) nanofibrous membrane	Suriyaraj et al. (2012)
11.	La-Al-Scoria adsorbent	Lanthanum-Aluminum loaded Scoria	Zhang et al. (2014)
12.	Bismuth oxide	Hydrous bismuth oxide (HBO_2)	Srivastav et al. (2015)

8 Conclusion

Fluoride pollution in groundwater is inevitable due to its geogenic origin. It is difficult to stop infiltration of agricultural runoff containing chemical fertilisers and industrial activities which add on anthropogenic F in groundwater. As small amount of F is essential for development of strong teeth and bones of human, total eradication of F from drinking water is not recommended. Instead of complete removal, F in groundwater is mitigated by using several in situ and ex situ defluorination techniques. Among several methods, though reverse osmosis has been considered as the most user friendly one and can be used for household purpose, they even remove the essential ions for human body. Some onsite methods which can reduce the concentration of F in groundwater by dilution within aquifer such as rain water harvesting using percolation tank or check dam, are also very effective. As all these techniques are very expensive, every technique or combinations of techniques should be customised for any target F affected area in such a way that in long run, that technique can supply low F-bearing groundwater in a sustainable and cost effective manner.

References

Amouei, A., Mahvi, A.H., Mohammadi, A.A., Fallah, S.H., Asgharnia, H.A. and Khafajeh, A.A. (2012). Physical and Chemical Quality Assessment of Potable Groundwater in Rural Areas of Khaf, Iran. *Wrld Appl Sc J*, **18(5):** 693–697.

Ayoob, S. and Gupta, A.K. (2006). Fluoride in drinking water: A review on the status and stress effects. *Critical Reviews in Environmental Science and Technology*, **36:** 433–487.

Azizullah, A., Khattak, M.N.K., Richter, P. and Häder, D. (2011). Water pollution in Pakistan and its impact on public health – A review. *Environ Inter*, **37:** 479–497.

Bailey, K., Chilton,J., Dahi, E. , Lennon,M., Jackson, P. and Fawell. J. (Eds) (2006) Fluoride in Drinking-water. World Health Organization, Geneva

Banks, D., Frengstad, B., Midtġard, A., Jan, R.K. and Strand, T. (1998). The chemistry of Norwegian groundwaters: I. The distribution of radon, major and minor elements in 1604 crystalline bedrock groundwaters. *The Sc Tot Environ*, **222:** 71–91.

Bia, G., De Pauli, C.P. and Borgnino, L. (2012). The role of Fe(III) modified montmorillonite on fluoride mobility: Adsorption experiments and competition with phosphate. *J Environ Manag*, **100:** 1–9.

BIS (Bureau of Indian Standards) (1992). Indian Standard Specifications for Drinking Water, IS:10500.

BIS (Bureau of Indian Standards). 2012. Specification for drinking water IS 10500: 2012, New Delhi, India

Borgnino, L., Garcia, M.G., Bia, G., Stupar, Y., Le Coustumer, P.H. and Depetris, P.J. (2013). Mechanisms of fluoride release in sediments of Argentina's central region. *Sc Tot Environ*, **443:** 245–255.

Brahman, K.D., Kazi, T.G., Afridi, H.I., Naseem, S., Arain, S.S. and Ullah, N. (2013). Evaluation of high levels of fluoride, arsenic species and other physicochemical parameters in underground water of two sub districts of Tharparkar, Pakistan: A multivariate study. *Water Research*, **47:** 1005–1020.

Brigatti, M.F. and Guggenheim, S. (2002). Mica crystal chemistry and the influence of pressure, temperature, and solid solution on atomistic models. *In:* Mottana, A., Sassi, F. P., Thompson, Jr. J. B. and Guggenheim, S. (ed.) Micas: Crystal Chemistry & Metamorphic Petrology. Mineralogical Society of America.

Brindha, K. and Elango, L. (2011). Fluoride in Groundwater: Causes, Implications and Mitigation Measures. *In:* Monroy, S.D. (ed.), Fluoride Properties, Applications and Environmental Management.

Brindha, K., Rajesh, R., Murugan, R. and Elango, L. (2011). Fluoride contamination in groundwater in parts of Nalgonda district, Andhra Pradesh, India. *Environ Monit and Assess*, **172:** 481–492.

Carrillo-Rivera, J.J., Cardona, A. and Edmundo, W.M. (2002). Use of abstraction regime and knowledge of hydrogeological conditions to control high fluoride concentration in abstracted groundwater San Luis Potosy basin, Mexico. *Hydrogeol J*, **261:** 24–47.

Carpenter, R. (1969). Factors controlling the marine geochemistry of fluorine. *Geochimica et Cosmochimica Acta*, **33:** 1156–1167.

Chae, G.T., Yun, S.T., Mayer, B., Kim, K.H., Kim, S.Y., Kwon, J.S., Kim, K. and Koh, Y.K. (2007). Fluorine geochemistry in bedrock groundwater of South Korea. *Sc Tot Environ*, **385:** 272–283.

Chakrabarty, S., Deka, J. and Sarma, H.P. (2009). Water quality issues in Baihata area of Kamrup (Rural) District, Assam, India. *Int J Chem Sci*, **7(4):** 2914–2920.

Chandrajith, R., Padmasiri, J.P., Dissanayake, C.B. and Prematilaka, K.M. (2012). Spatial distribution of fluoride in groundwater of Sri Lanka. *J Natn Sci Foundation Sri Lanka*, **40(4):** 303–309.

Chidambaram, S., Ramanathan, AL. and Vasudevan, S. (2003). Fluoride removal studies in water using natural materials. *Water SA*, **29(3):** 339–344.

Churchill, H.V., Rowley, R.J. and Martin, L.N. (1948). Fluorine content of certain vegetation in Western Pennsylvania area. *Analytical Chemistry*, **20(1):** 69–71.

Culp, R.L. and Stoltemberg, H.A. (1958). Fluoride reduction at La Crosse, Kansas. *J. American Water Works Association*, **50(3):** 427.

Dahi, E., Mtalo, F., Njau, B. and Bregnhj, H. (2013). Defluoridation using the Nalgonda Technique in Tanzania. *Ecological Engineering*, **52:** 211–218.

D'Alessandro, W. (2006). Human fluorosis related to volcanic activity: A review. *Environmental Toxicology Transaction: Biomedicine and Health*, **10:** 21–30.

Das, B., Talukdar, J., Sarma, S., Gohain, B., Dutta, R.K., Das, H.B. and Das, S.C. (2003). Fluoride and other inorganic constituents in groundwater of Guwahati, Assam, India. *Cur Sc*, **85(5):** 657–661.

Datta, P.S., Deb, D.L. and Tyagi, S.K. (1996). Stable isotope (^{18}O) investigations on the processes controlling fluoride contamination of groundwater. *J Contam Hydrol*, **24(1):** 85–96. doi: https://doi.org/10.1016/0169-7722(96)00004-6.

Davraz, A., Sener, E. and Sener, S. (2008). Temporal variations of fluoride concentration in Isparta public water system and health impact assessment (SW-Turkey). *Environ Geol*, **56:** 159–170.

Delmelle, P., Lambert, M., Dufrene, Y., Gerin, P. and Oskarsson, N. (2007). Gas/aerosol–ash interaction in volcanic plumes: New insights from surface analyses of fine ash particles. *Erth Planet Sc Let*, **259:** 159–170.

Derakhshani, R., Tavallaie, M., Raoof, M., Mohammadi, T.M., Abbasnejad, A. and Haghdoostf, A.A. (2014). Occurrence of fluoride in groundwater of Zarand region, Kerman province, Iran. *Fluoride*, **47(2):** 133–138.

Desbarats, A.J. (2009). On elevated fluoride and boron concentrations in groundwaters associated with the Lake Saint-Martin impact structure, Manitoba. *Applied Geochemistry*, **24:** 915–927.

Dhiman, S.D. and Keshari, A.K. (2006). Hydrogeochemical evaluation of high-fluoride groundwaters: A case study from Mehsana District, Gujarat, India. *Hydrol Sc J*, **51(6):** 1149–1162.

Dissanayake, C.B. and Weerasooriya, S.V.R. (1986). Fluorine as an indicator of mineralization hydrogeochemistry of a Precambrian mineralized belt in Sri Lanka. *Chemical Geology*, **56:** 257–270.

Dissanayake, C.B. (1991). The fluoride problem in the ground water of Sri Lanka – environmental management and health. *Inter J Environ Stud*, **38(2):** 137–155.

Diwakar, J., Johnston, S.G., Burtona, E.D. and Shresthab, S.D. (2015). Arsenic mobilization in an alluvial aquifer of the Terai region, Nepal. *J of Hydrol: Regional Studies*, **4:** 59–79.

Edmunds, W.M. and Smedley, P.L. (1996). Groundwater Geochemistry and Health: An Overview. *In:* Appleton, Fuge and McCall (Eds), Environmental Geochemistry and Health 113: Geological Society Special Publication, London.

Edmunds, W.M. and Smedley, P.L. (2001). Fluoride in natural waters. *In:* Selinus, O. (ed.), Essentials of medical geology. Springer, Netherlands.

Edmunds W. M. and Smedley P.L., (2013), Fluoride in Natural Waters. In: Essentials of Medical Geology, Second Edition. 2013. Eds: Selinus, O., Alloway, B., Centeno, J.A., Finkelman, R.B., Fuge, R., Lindh, U. and Smedley, P.L. Springer pp 311–336

Farooqi, A., Masuda, H. and Firdous, N. (2007a). Toxicfluoride and arsenic contaminated groundwater in the Lahore and Kasur districts, Punjab, Pakistan and possible contaminant sources. *Environ Pollut*, **145:** 839–849.

Farooqi, A., Masuda, H., Kusakabe, M., Naseem, M. and Firdous, N. (2007b). Distribution of highly arsenic and fluoride contaminated groundwater from east Punjab, Pakistan, and the controlling role of anthropogenic pollutants in the natural hydrological cycle. *Geocheml J*, **41:** 213–234.

Farrah, H., Slavek, J. and Pickering, W.F. (1987). Fluoride interactions with hydrous aluminum oxides and alumina. *Australian Journal of Soil Research*, **25:** 55–69.

Fawell, J., Bailey, K., Chilton, J., Dahi, E., Fewtrell, L. and Magara, Y. (2006) Fluoride in drinking water. WHO, IWA Publishing.

Frengstad, B., Banks, D. and Siewers, U. (2001) The chemistry of Norwegian groundwaters: IV. The pH-dependence of element concentrations in crystalline bedrock groundwaters. *Sci Total Environ* **277:** 101–117.
Gaciri, S.J. and Davies, T.C. (1993) The occurrence and geochemistry of fluoride in some natural waters of Kenya. *J Hydrology*, **143:** 395–412.
García, M.G. and Borgnino, L. (2015) Fluoride in the Context of the Environment. doi: https://doi.org/10.1039/9781782628507-00003.
Garg, V.K., Suthar, S., Singh, S., Sheoran, A., Garima, M. and Jain, S. (2009). Drinking water quality in villages of southwestern Haryana, India: Assessing human health risks associated with hydrochemistry. *Environ Geol*, **58:** 1329–1340.
Genxu, W. and Guodong, C. (2001). Fluoride distribution in water and the governing factors of environment in arid north–west China. *J Arid Environ*, **49(3):** 601–614. doi: https://doi.org/10.1006/jare.2001.0810.
Ghorai, S. and Pant, K.K. (2004). Investigations on the column performance of fluoride adsorption by activated alumina in a fixed-bed. *Chem Engineer J*, **98(1–2):** 165–173.
Gillespie, R.J., Humphries, D.A., Baird, N.C. and Robinson, E.A. (1989). Chemistry. Second ed. Allyn and Bacon, Boston.
Gizaw, B. (1996). The origin of high bicarbonate and fluoride concentration in waters of the main Ethiopian rift valley, East African rift system. *J of Afric Erth Sc*, **22:** 391–402.
Guo, Q., Wang, Y., Ma, T. and Ma, R. (2006). Geochemical processes controlling the elevated fluoride concentrations in groundwaters of the Taiyuan Basin, Northern China. *J Geochem Explor*, **93(1):** 1–12.
Guo, Q., Wang, Y., Ma, T. and Ma, R. (2007). Geochemical processes controlling the elevated fluoride concentrations in groundwaters of the Taiyuan Basin, Northern China. *J Geochem Explor*, **93(1):** 1–12. doi: https://doi.org/10.1016/j.gexplo.2006.07.001.
Gupta, S.K., Deshpande, D., Agarwal, M. and Raval, B.R. (2005). Origin of high fluoride in groundwater in the North Gujarat-Cambay region, India. *Hydrogeol J*, **3:** 596–605.
Gurung, A. and Oh, S. (2012). An overview of water pollution and constructed wetlands for sustainable wastewater treatment in Kathmandu Valley: A review. *Scientific Research and Essays*, **7(11):** 1185–1194.
Gupta, S., Banerjee, S., Saha, R., Datta, J.K. and Mondal, N. (2006). Fluoride geochemistry of groundwater in Nalhati-1 Block of the Birbhum district, West Bengal, India Research report. *Fluoride*, **39(4):** 318–320.
Haidouti, C. (1991). Fluoride distribution in soils in the vicinity of a point emission source in Greece. *Geoderma*, **49:** 129–138.
Handa, B.K. (1975). Geochemistry and genesis of fluoride-containing ground waters in India. *Groundwater*, **13(3):** 275–281.
Hassan Saffi, M. and Jawid Kohistani, A. (2013). Scientific Investigation report in Afghanistan: Water Resources potential, quality problems, challenges and solutions in Afghanistan. DACAAR report, Kabul, July 2013.
Hassan Saffi, M. and Jawid Kohistani, A. (2015). Study on water quality status with respect to fluoride contamination in drinking water sources of Balkh province, Afghanistan. Scientific Investigation Report. Editor Leendert Vijselaar.
Hem, J.D. (1985). Study and Interpretation of the Chemical Characteristics of Natural Water. 3rd Edition. US Geological Survey Water-Supply Paper 2254, University of Virginia, Charlottesville. 263 p.
Hiemstra, T. and Van Riemsdijk, W.H. (2000). Fluoride adsorption on goethite in relation to different types of surface sites. *Journal of Colloid and Interface Science*, **225:** 94–104.
Hoque, A.K.M., Fazlul Khaliquzzaman, M., Hossain, M.D. and Khan, A.H. (2003). Fluoride levels in different drinking Water sources in Bangladesh. *Fluoride*, **36(1):** 38–44.
Jadhav, S.V., Bringas, E., Yadav, G.D., Rathod, V.K., Ortiz, I. and Marathe, K.V. (2015). Arsenic and fluoride contaminated groundwaters: A review of current technologies for contaminants removal. *J Environ Manag*, **162:** 306–325. doi: https://doi.org/10.1016/j.jenvman.2015.07.020

Janardhana Raju, N., Dey, S. and Das, K. (2009). Fluoride contamination in groundwaters of Sonbhadra District, Uttar Pradesh, India. *Curr Sc*, **96(7):** 979–985.

Jha, S.K., Nayak, A.K., Sharma, Y.K., Mishra, V.K. and Sharma, D.K. (2008). Fluoride Accumulation in Soil and Vegetation in the Vicinity of Brick Fields. *Bulletin of Environmental Contamination and Toxicology*, **80:** 369–373.

Kotoky, P., Barooah, P.K., Baruah, M.K., Goswami, A., Borah, G.C., Gogoi, H.M., Ahmed, F., Gogoi, A. and Paul, A.B. (2008). Fluoride and endemic fluorosis in the Karbianglong district, Assam, India. *Fluoride*, **41(1):** 42–45.

Karthikeyan, K., Nanthakumar, K., Velmurugan, P., Tamilarasi, S. and Lakshmanaperumalsamy, P. (2010). Prevalence of certain inorganic constituents in groundwater samples of Erode district, Tamilnadu, India, with special emphasis on fluoride, fluorosis and its remedial measures. *Environ Monit and Assess* **160:**141–155.

Kausar, R., Ahmad, S., Rehman, K.U. and Ahmad, R. (2003). Fluoride Status of Underground Water of Faisalabad City Area. *International Journal of Agriculture & Biology*, **05(4):** 536–539.

Keshavarzi, B., Moore, F., Esmaeili, A. and Rastmanesh, F. (2010). The source of fluoride toxicity in Muteh area, Isfahan, Iran. *Environmental Earth Sciences*, doi: https://doi.org/10.1007/s12665-009-0390-0 doi:https://doi.org/10.1109/ICEET.2009.361.

Khan, N., Hussain, S.T., Saboor, A., Jamila, N., Khan, S.N. and Kim, K.S. (2013) Chemical Investigation of the Drinking Water Sources from Mardan, Khyber Pakhtunkhwa, Pakistan. *World Applied Sciences Journal*, **27(1):** 112–122.

Khan, S.A., Thakur, S.K., Sarkar, A. and Shekhar, S. (2016). Worldwide contamination of water by fluoride. *Environ Chem Lett.*, doi: https://doi.org/10.1007/s10311-016-0563-5.

Kim, K. and Jeong, G.Y. (2005), Factors influencing natural occurrence of fluoride rich groundwaters: A case study in the southeastern part of the Korean Peninsula. *Chemosphere*, **58(10):** 1399–1408.

Koritnig, S. (1978). Fluorine. *In:* Wedepohl, K.H. (ed.), Handbook of Geochemistry, vol. II/1. Springler-Verlag, Berlin. 9C1-B9O3.

Krauskopf, K.B. and Bird, D.K. (1995). An introduction to geochemistry. McGraw-Hill Int., Singapore.

Kraynov, S.R., Merkov, A.N., Petrova, N.G., Baturinskaya, I.V. and Zharikova, V.M. (1969). Highly alkaline (pH 12) fluosilicate waters in the deeper zone of the Lovozero Massif. *Geochemistry International*, **6:** 635–640.

Kundu, M.C. and Mandal, B. (2009). Assessment of potential hazards of fluoride contamination in drinking groundwater of an intensively cultivated district in West Bengal, India. *Environ Monit and Assess*, **152:** 97–103.

Kundu, N., Panigrahi, M., Tripathy, S., Munshi, S., Powell, M.A. and Hart, B. (2001). Geochemical appraisal of fluoride contamination of groundwater in the Nayagarh District of Orissa, India. *Environ Geol*, **41(3–4):** 451–460. doi: https://doi.org/10.1007/s002540100414.

Lennon, M.A., Whelton, H., Mullane, D.O. and Ekstrand, J. (2004). Rolling Revision of the WHO Guidelines for Drinking-water Quality: Fluoride. World Health Organization.

Li, X., Hou, X., Zhou, Z. and Liu, L. (2009). Distribution and Geochemical Evolution of Fluoride in Groundwater of Taiyuan Basin, China. ICEET '09 Proceedings of the 2009 International Conference on Energy and Environment Technology, **2:** 507–510.

Looie, S.B. and Moore, F. (2010). A study of fluoride groundwater occurrence in Posht-e-Koohe-Dashtestan, South of Iran. *World Applied Sciences Journal*, **8(11):** 1317–1321.

Ma, W., Ya, F., Wang, R. and Zhao, Y. (2008). Fluoride removal from drinking water by adsorption using bone char as a biosorbent. *Inter J Environ Techn Manag*, **9(1):** 59–69.

Ma, Y., Shi, F., Zheng, X., Ma, J. and Gao, C. (2010). Removal of fluoride from aqueous solution using granular acid-treated bentonite (GHB): Batch and column studies. *Journal of Hazardous Materials*. doi: https://doi.org/10.1016/j.jhazmat.2010.10.016

Madhnure, P., Sirsikar, D.Y., Tiwari, N., Ranjan, B. and Malpe, D.B. (2007). Occurrence of fluoride in the groundwaters of Pandharkawada area, Yavatmal district, Maharashtra, India. *Current Science*, **92(5):** 675–679.

Mahapatra, M.K., Mishra, A. and Das, B.P. (2005). Fluorosis first reported in Naupada district of Orissa India. *Ecology Environment and Conservation*, **11(2):** 277–280.

Maithani, P.B., Gurjar, R., Banerjee, R., Balaji, B.K., Ramachandran, S. and Singh, R. (1998). Anomalous fluoride in groundwater from western part of Sirohi district, Rajasthan and its crippling effect of human health. *Cur Sc*, **74(9):** 773–777.

Mamatha, P. and Rao, S.M. (2010). Geochemistry of fluoride rich groundwater in Kolar and Tumkur districts of Karnataka. *Environ Erth Sc*, **61:** 131–142.

Mameri, N., Yedou, A.R., Lounici, H., Belhocine, D., GRIB, H. and Bariou, B. (1998). Defluoridation of Septentrional Sahara Water of North Africa by Electrocoagulation Process using Bipolar Aluminium Electrodes. *Wat Res*, **32(1):** 1604–1612.

Meenakshi, S. and Viswanathan, N. (2007). Identification of selective ion-exchange resin for fluoride sorption. *J Colloid and Interface Science*, **308(2):** 438–450.

Messaïtfa, A. (2008). Fluoride contents in groundwaters and the main consumed foods (dates and tea) in Southern Algeria region. Special Issue: "Groundwater Flow – Selected papers from XXXIII IAH Congress, Zacatecas, Mexico (233–320)". *Environ Geol*, **55(2):** 377–383.

Mirlean, N. and Roisenberg, A. (2007). Fluoride distribution in the environment along the gradient of a phosphate-fertiliser production emission (southern Brazil). *Environ Geochem and Health*, **29(3):** 179–187.

Moghaddam, A.A. and Fijani, E. (2008). Distribution of fluoride in groundwater of Maku area, northwest of Iran. *Environ Geol*, **56:** 281–287.

Mondal, N.C., Prasad, R.K., Saxena, V.K., Singh, Y. and Singh, V.S. (2009). Appraisal of highly fluoride zones in groundwater of Kurmapalli watershed, Nalgonda district, Andhra Pradesh (India). *Environ Erth Sc*, **59:** 63–73.

Msonda, K.W.M., Masamba, W.R.L. and Fabiano, E. (2007). A study of fluoride groundwater occurrence in Nathenje, Lilongwe, Malawi. *Physics and Chemistry of the Earth*, Parts A/B/C, **32 (15–18):** 1178–1184.

Muralidharan, D., Nair, A.P. and Satyanarayana, U. (2002). Fluoride in shallow aquifers in Rajgarh Tehsil of Churu District, Rajasthan – An arid environment. *Curr Sc*, **83:** 699–702.

Naseem, S., Rafique, T., Bashir, E., Bhanger, M.I., Laghari, A. and Usmani, T.H. (2010). Lithological influences on occurrence of high-fluoride groundwater in Nagar Parkar area, Thar Desert, Pakistan. *Chemosphere*, **78:** 1313–1321.

Nordstrom, D.K. and Jenne, E.A. (1977). Fluorite solubility equilibria in selected geothermal waters. *Geochimica et Cosmochimica Acta*, **41:** 175–188.

Oruc, N. (2003). Problems of high fluoride waters in Turkey (hydrogeology and health aspects). The short course on medical geology – health and environment. Canberra, Australia.

Oruc, N. (2008). Occurrence and problems of high fluoride waters in Turkey: An overview. Issue entitled "Medical Geology in Developing Countries, Part 2". *Environmental Geochemistry and Health*, **30(4):** 315–323.

Ozsvath, D.L. (2009). Fluoride and environmental health: A review. *Reviews in Environ Sc and Biotech*, **8:** 59–79.

Pan, Y. and Fleet, M.E. (2002). Compositions of the Apatite-group Minerals: Substitution Mechanisms and Controlling Factors. *In:* Kohn, M.L., Rakovan, J. and Hughes, J.M. (eds), Phosphates. Mineralogical Society of America.

Pant, B. (2011). Ground water quality in the Kathmandu valley of Nepal. *Environ Monit Assess*, **178(1):** 477–485.

Pickering, W.F. (1985). The mobility of soluble fluoride in soils. *Environ Pollut Ser B Chem Phys*, **9(4):** 281–308. doi: https://doi.org/10.1016/0143-148X(85)90004-7.

Prasad, B. and Mondal, K.K. (2006). Leaching Characteristics of Fluoride from Coal Ash. *Asian J Water Environ and Pollut*, **4(2):** 17–21.

Rafique, T., Naseem, S., Bhanger, M.I. and Usmani, T.H. (2008). Fluoride ion contamination in the groundwater of Mithi sub-district, the Thar Desert, Pakistan. *Environ Geol*, **56:** 317–326.

Ramesam, V. and Rajagopalan, K. (1985). Fluoride ingestion into natural water of hard rock areas, Peninsular India. *Geol. Sco. India*, **26:** 125–132.

Rao, N.S., Rao, P.S., Dinakar, A., Rao, P.V.N. and Marghade, D. (2015). Fluoride occurrence in the groundwater in a coastal region of Andhra Pradesh, India. *Appl Water Sci.* doi: https://doi.org/10.1007/s13201-015-0338-3.

Reddy, D.V., Nagabhushanam, P., Sukhija, B.S., Reddy, A.G.S. and Smedley, P.L. (2010). Fluoride dynamics in the granitic aquifer of the Wailapally watershed, Nalgonda District, India. *Chem Geol*, **269(3–4):** 278–289.

Robison, W.D. and Edigton, G. (1946). Fluorine in soil. *Soil Sci*, **61:** 341–353.

Rosi, M., Papale, P., Lupi, L. and Stoppato, M. (2003). Volcanoes. Firefly Books Ltd., Spain.

Ruggeri, F., Saavedra, J., Fernandez-Turiel, J.L., Gimeno, D. and Garcia-Valles, M. (2010). Environmental geochemistry of ancient volcanic ashes. *J of Hazard Mater*, **183:** 353–365.

Rukah, Y.A. and Alsokhny, K. (2004). Geochemical assessment of groundwater contamination with special emphasis on fluoride concentration, North Jordan. *Chemie der Erde – Geochemistry*, **64(2):** 171–181.

Sankararamakrishnan, N., Sharma, A.K. and Iyengar, L. (2008). Contamination of nitrate and fluoride in ground water along the Ganges alluvial plain of Kanpur district, Uttar Pradesh, India. *Environ Monit and Assess*, **146:** 375–382.

Saxena, V. and Ahmed, S. (2001). Dissolution of fluoride in groundwater: A water-rock interaction study. *Environ Geol*, **40(9):** 1084–1087. doi: https://doi.org/10.1007/s002540100290.

Shaji, E., Viju, B.J. and Thambi, D.S. (2007). High fluoride in groundwater of Palghat District, Kerala. *Cur Sc*, **92(2):** 240–245.

Shrott, H.E., Merobert, G.R., Barnald, T.W. and Nayar, A.S.M. (1937). Endemic fluorosis in the Madras presidency. *Ind J Medical Research*, **25:** 553–561.

Siddique, A., Mumtaz, M., Saied, S., Karim, Z. and Zaigham, N.A. (2006). Fluoride concentration in drinking water of Karachi city (Pakistan). *Environ Monit Assess*, **120(1–3):** 177–185.

Singh, C.K. and Mukherjee, S. (2014). Aqueous geochemistry of fluoride enriched groundwater in arid part of Western India. *Environ Sci Pollut Res*, **22(4):** 2668–2678. doi: https://doi.org/10.1007/s11356-014-3504-5.

Sorg, T.J. and Logsdon, G.S. (1978). Treatment technology to meet the interim primary drinking water regulations for inorganics, part 2. *J. Am. Water Works Assoc.*, **70:** 379–393.

Srinivasa Rao, N. (1997). The occurrence and behaviour of fluoride in the groundwater of the Lower Vamsadhara River basin, India. *Hydrol Sc*, **42(6):** 877–892.

Srivastav, A.L., Singh, P.K. and Sharma, Y.C. (2015). Synthesis of a novel adsorbent, hydrous bismuth oxide (HBO_2) for the removal of fluoride from aqueous solutions. *Desalination Water Treatment*, **55(3):** 604–614.

Stumm, W. (1992). Chemistry of the Solid-Water Interface. Wiley Interscience Publication.

Su, C., Wang, Y., Xie, X. and Zhu, Y. (2015). An isotope hydrochemical approach to understand fluoride release into groundwaters of the Datong Basin, Northern China. *Environ Sci Process Impacts*, **17(4):** 791–801. doi: https://doi.org/10.1039/C4EM00584H.

Sujana, M.G. and Anand, S. (2010). Iron and aluminium based mixed hydroxides: A novel sorbent for fluoride removal from aqueous solutions. *Appl Surface Science*, **256:** 6956–6962.

Suriyaraj, S.P., Bhattacharyya, A. and Selvakumar, R. (2012). Hybrid Al_2O_3/Bio-TiO_2Nanocomposite impregnated Thermoplastic polyurethane (TPU) nanofibrous membrane for Fluoride removal from aqueous Solution. *RSC Adv.*, **5:** 26905–26912.

Suthar, S., Garg, V.K., Jangir, S., Kaur, S., Goswami, N. and Singh, S. (2008). Fluoride contamination in drinking water in rural habitations of Northern Rajasthan, India. *Environ Monit and Assess*, **145:** 1–6.

Symonds, R.B., Rose, W.I. and Reed, M.H. (1988). Contribution of Cl^- and F^- bearing gases to the atmosphere by volcanoes. *Nature*, **334:** 415–419.

Teotia, S.P.S. and Teotai, M. (1994). Dental Caries: A disorder of high fluoride and low dietary calcium interactions (30 years of personal research). *Fluoride*, **27:** 59–66.

Tekle-Haimanot, R., Melaku, Z., Kloos, H., Reimann, C., Fantaye, W., Zerihun, L. and Bjorvatn, K. (2006). The geographic distribution of fluoride in surface and groundwater in Ethiopia with an emphasis on the Rift Valley. *Sc Tot Environ*, **367:** 182–190.

Tirumalesh, K., Shivanna, K. and Jalihal, A.A. (2007). Isotope hydrochemical approach to understand fluoride release into groundwaters of Ilkal area, Bagalkot District, Karnataka, India. *Hydrogeol J*, **15(3):** 589–598.

Tor, A., Danaoglu, N., Arslan, G. and Cengeloglu, Y. (2008). Removal of Fluoride from Water by Using Granular Red Mud: Batch and Column Studies. *J Hazard Materls*, **164(1):** 271–278.

UNESCO (2007). (Internet source, status January 2007) Trace elements in Groundwater and public health. http://www.iah.org/briefings/Trace/trace.pdf. Accessed on 9th February, 2017.

Valenzuela-Vásquez, L., Ramírez-Hernández, J., Reyes-López, J., Sol-Uribe, A. and Lázaro-Mancilla, O. (2006). The origin of fluoride in groundwater supply to Hermosillo City, Sonora, Mexico. *Environ Geol*, **51:** 17–27.

Van Cappellen, P. and Berner, R. (1988). A mathematical model for the early diagenesis of phosphorous and fluorine in marine sediments: Apatite precipitation. *Ameri J of Sc*, **288:** 289–333.

Vivona, R., Preziosi, E., Madé, B. and Giuliano, G. (2007). Occurrence of minor toxic elements in volcanic-sedimentary aquifers: A case study in central Italy. *Hydrogeol J*, **15:** 1183–1196.

Wedepohl, K.H. (1995). The composition of the continental crust. *Geochimica et Cosmochimica Acta*, **59:** 1217–1232.

Weinstein, L.H. and Davison, A. (2003). Fluorides in the Environment – Effects on Plants and Animals. CABI Publishing, UK.

WHO. (1984). International Standards for drinking water. 3rd guidelines for drinking water quality. Vol. 2: Health criteria and other supporting information. Geneva, World Health Organization.

WHO (2001). Naturally Occurring Hazards; Water, Health and Human Rights; Floods and Droughts. World Water Day 2001.

WHO (2002). Fluorides. Environmental Health Criteria 227, 268. World Health Organization, Geneva.

Wodeyar, B.K. and Sreenivasan, G. (1996). Occurrence of fluoride in the groundwaters and its impact in Peddavankahalla basin, Bellary district, Karnataka – A preliminary study. *Curr Sc*, **70 (1):** 71–74.

Yidana, S.M., Yakubo, B.B. and Akabzaa, T.M. (2010). Analysis of groundwater quality using multivariate and spatial analyses in the Keta basin, Ghana. *J Afric Erth Sc*, **58:** 220–234.

Young, S.M., Pitawala, A. and Ishiga, H. (2010). Factors controlling fluoride contents of groundwater in north-central and northwestern Sri Lanka. *Environl Erth Sc*, doi: https://doi.org/10.1007/s12665-010-0804-z.

Zhang, S., Lub, Y., Linb, X., Sub, X. and Zhang, Y. (2014). Removal of fluoride from groundwater by adsorption onto La(III)-Al(III) loaded scoria adsorbent. *Applied Surface Science*, **303:** 1–5.

Zheng, L., Luo, H., Bandou, K., Kanai, F., Terasaki, K., Yoshimura, T., Sakai, Y., Kimura, S. and Hagiwara, H. (2006). High fluoride groundwater with high salinity and fluorite in aquifer sediments in Inner Mongolia, China. *Chinese Journal of Geochemistry*, **25(1):** 103.

Zhuang, J.I.E. and Yu, G.R. (2002). Effects of surface coatings on electrochemical properties and contaminant sorption of clay minerals. *Chemosphere*, **49:** 619–628.

Zidarova, B. (2010). Hydrothermal fluorite-forming processes in the Mikhalkovo deposit (Central Rhodopes, Bulgaria) field observation and experimental confirmation. *Neues Jahrbuch für Mineralogie*, **187:** 133–157.

Chapter 15
Human Health Hazards Due to Arsenic and Fluoride Contamination in Drinking Water and Food Chain

Gautam Ghosh and Dipak K. Mukhopadhyay

1 Introduction

Man interacts with the biosphere, atmosphere, hydrosphere and the lithosphere for sustaining himself on earth. This interaction ranges from the level of regional pattern of climate, geography and geology to different elements in rock, soil and water. In the universe, lighter elements are more frequently encountered than the heavier ones. In the Periodic Table of elements, the first 26 elements comprise, by weight, 99% of the continental crust. Again out of these 26, the first 20 make up more than 99% by weight of the human body. The living tissue of both animals and plants comprise 11 elements such as sodium, magnesium, potassium, calcium, hydrogen, carbon, oxygen, phosphorus, sulphur, nitrogen and chlorine. Iron is present in those species that have haemoglobin. Besides these 11 elements, the living tissue also requires certain other elements for functioning properly. These elements, available in traces, constitute the substances that help to regulate the different processes of life. Trace elements, necessary for nutrition, include fluorine, chromium, manganese, iron, cobalt, copper, zinc, selenium, molybdenum and iodine. Nickel, aluminum, arsenic and barium are observed to concentrate in human tissues with age. Concentration of trace elements generally is found to increase from rock to soil to water to plants to animals. Many trace elements, beyond certain levels of concentration, often cause serious health problems, when the human body is exposed to them (Keller 1985). Arsenic and fluoride have precipitated health issues of grave concern (Mathur 2004).

Chronic arsenic toxicity arising from drinking of arsenic contaminated water has been reported from many countries of the world. However cases of arsenic poisoning, related to ingestion of arsenic bearing groundwater, have reached an unprecedented dimension in West Bengal and Bangladesh. About six million people have

G. Ghosh (✉) · D. K. Mukhopadhyay
Geological Survey of India, Kolkata, India
e-mail: gautamghosh_28@rediffmail.com

P. K. Sikdar (ed.), *Groundwater Development and Management*,
https://doi.org/10.1007/978-3-319-75115-3_15

been estimated to have been exposed to arsenic contaminated water (arsenic level >0.05 mg/l) in eight districts of West Bengal alone. Large number of these people have manifested with hyperpigmentation, keratosis, peripheral vascular disease, neuropathy and skin cancer (Guha Mazumder 2004).

Fluoride has been extensively used in the treatment of dental caries since long. It has been observed, however, that ingestion of fluoride, through drinking water and other sources beyond a certain level of concentration, has been found to cause dental disfigurement and crippling skeletal and neurological disability especially in India and other South Asian countries (Gupta and Godbole 2004).

The impact of arsenic and fluoride on human health vis-a-vis drinking water and food chain is assessed and highlighted in the ensuing discussion.

2 Arsenic

2.1 General Aspect

Arsenic is a metalloid with Atomic Number 33 and Atomic Weight 74.91. It occurs in Group V of the Periodic Table of elements. Arsenic is present in nature in both inorganic and organic forms. It occurs in inorganic form in three states namely: metal [As(O)], trivalent Arsenite [As(+3)] and pentavalent Arsenate [As(+5)]. Its organic forms are dimethylarsenic and monomethylarsenic. It occurs as compounds of metals and sulphur and occasionally in the native state. Its chief minerals are realgar, orpiment, arsenopyrite and arsenolite. Native sulphur, iron pyrites and other sulphide minerals often contain arsenic. All arsenic compounds are poisonous and exposure to arsenic gives rise to various adverse impacts on human health. It is necessary for the arsenic, present in its natural geo-chemical environment, to come into contact with the human body, get lodged in human tissues and then access the vital organs of the body in order to achieve the adverse impacts.

2.2 Environment of Arsenic

Arsenic is widely distributed in natural environment. It comprises 0.00058% of the total mass of the earth's crust. However it exhibits preferential accumulation in certain environment. High concentration of arsenic is usually observed in the following geo-chemical environments:

- Basin fill deposits of alluvial-lacustrine origin specially in semi-arid areas
- Volcanic deposits
- Geothermal systems
- Uranium and gold mining areas

In the first two geo-chemical environments, arsenic is associated with sediments derived from volcanic rocks of intermediate to acidic composition. Mobilization of arsenic in sedimentary aquifers may be caused by changes in the geo-chemical environment near surface related to agricultural irrigation and at depth due to compaction consequent to groundwater withdrawal (Welch et al. 1988).

Arsenic concentrations in igneous, metamorphic and sedimentary rocks are the major sources of aqueous arsenic. Sedimentary rocks generally have higher concentrations than igneous and metamorphic rocks. Clay minerals, organic matter, iron and manganese oxides commonly adsorb trace elements such as arsenic. In soils, organic matter can concentrate arsenic near surface while adsorption onto ferric oxyhydroxide can concentrate arsenic at greater depths.

Dissolved arsenic concentrations in rain water and snow ranges from <0.002 to 0.50 μg/L (Andreae 1980). In river water the range is from 0.20 to 264 μg/L and in lake water the range is from 0.38 to 1000 μg/L (Benson and Spencer 1983). In sea water the concentration ranges from 0.15 to 6.0 μg/L (Kanamori 1965; Onishi and Sandell 1955). Suspended and bottom sediments, in most aquatic systems, contain much higher arsenic concentrations than the water. Arsenic in certain solid phases in sediments particularly such as iron oxides, organic matter and sulphides may constitute primary source of arsenic in groundwater. Major processes responsible for concentration of arsenic in groundwater comprise (a) mineral precipitation/dissolution, (b) adsorption/desorption, (c) chemical transformations, (d) ion exchange and (e) bio-mediated reactions. Higher than the average levels of concentration of arsenic may often be related to anthropogenic sources such as smelters and industrial waste and effluent.

2.3 Exposure to Arsenic

Arsenic, once it is available in soil, surface and groundwater, can have access to the human body either through drinking water, food chain and/or air. The impact of arsenic exposure depends on the concentration of arsenic in the concerned medium. Many countries have specified the permissible limit of arsenic for different uses similar to other water quality parameters. The United States Environmental Protection Agency (USEPA) had fixed the tolerance limit of arsenic in drinking water at 50 μg/L based on the consideration whether the impact is cancerous or systematic toxicity. Recently, however, the observation, that arsenic at even low values can have deleterious effect on human health, has led to reviewing the guidelines for tolerance limit of arsenic. The USEPA has revised the limit from 50 μg/L to 10 μg/L. The World Health Organisation has reduced the tolerance limit value of arsenic in drinking water from 50 μg/L to 10 μg/L.

Drinking water with high levels of arsenic, over a long period of time, causes chronic arsenic toxicity. In West Bengal, India and Bangladesh, high arsenic concentration in drinking water has given rise to keratosis, pigmentation and respiratory diseases. From studies carried out in the districts of Bardhaman, Murshidabad, Nadia, North 24 Parganas and South 24 Parganas in West Bengal, India,

Mukhopadhyay and Ghosh (2010) reported people being affected by hyper-pigmentation and keratosis by drinking groundwater containing arsenic ranging from 50 µg/L to as high as 3000 µg/L.

Consumption of food products like vegetables and rice, cultivated using arsenic contaminated groundwater, is found to be responsible for chronic arsenic poisoning as in the cases of reported arsenic toxicity in West Bengal, India. Rahman et al. (2003) have reported that the level of arsenic in groundwater, used to cultivate rice and vegetables, was found to range from 103 µg/L to 827 µg/L. The average arsenic levels in rice and vegetables were found to be 0.323 µg/g and 0.027 µg/g (Chakraborti et al. 2004). Their work has also confirmed that arsenic contaminated groundwater, used to cultivate vegetables and rice consumed by people, may form an important source of arsenic toxicity. This was because urinary arsenic concentration even in control subjects drinking “safe” water found to be higher than the norm. This led to the conclusion that consumption of food products contaminated by arsenic was the main cause for arsenic poisoning in the control subjects. There is uncertainty however about the bioavailability and associated toxicity of arsenic from different foods (Al Rmalli 2005), investigating arsenic levels in food imported from Bangladesh to the United Kingdom, showed that imported vegetables from Bangladesh have from 2 to 100-fold higher concentrations of arsenic than vegetables cultivated in the United Kingdom, European Union, and North America.

Inhaling fumes from burning of arsenic contaminated coal has been found to effect porphyrin metabolism. Porphyrin levels were reported to be higher in people spending more time indoors with increased arsenic exposure, in Guizhou province, China (Ng et al. 2005).

2.4 Accumulation of Arsenic in Tissues

People suffering from arsenic toxicity have significant accumulation of arsenic in their urine, hair and nail samples. Both trivalent and pentavalent arsenic in solution are readily absorbed after ingestion mostly from the respiratory and gastro-intestinal tract. In case of humans 75% of trivalent arsenic is eliminated through urine initially during the first week. In case of pentavalent arsenic, elimination is up to 80–90% during the first two days. In southwest Taiwan, patients having Black Foot Disease, showed significant accumulation of inorganic arsenic in urine, hair and fingernails thus suggesting that arsenic in urine, hair and fingernails are indicators of arsenic exposure in humans (Lin et al. 1998). Chakraborti et al. (2004) showed that individuals with arsenic associated skin lesions had high arsenic levels in hair, nail and skin tissue. Skin lesions and arsenic concentration in hair and nails therefore may help in early diagnosis of chronic arsenic poisoning. Lin et al. (1998) observed that Black Foot Disease patients in Taiwan, consuming groundwater contaminated by arsenic, excreted higher total urinary arsenic. Meza et al. (2004) established a positive co-relation between total arsenic in water and total arsenic eliminated in urine amongst patients in Mexico. It was observed that different communities had

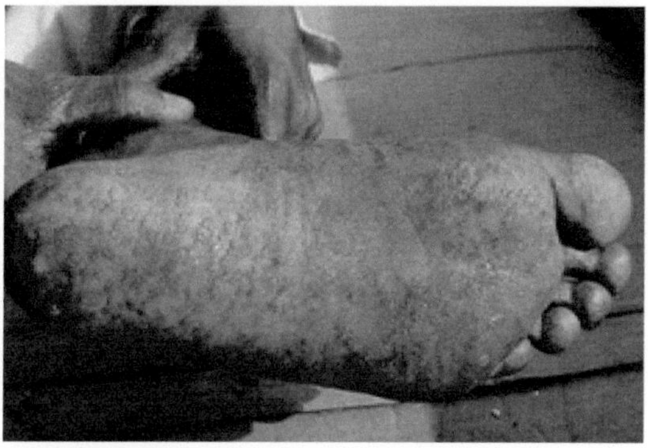

Fig. 15.1 Non-carcinogenic impact of arsenic exposure: Keratosis of foot. (Photograph by Dipak K. Mukhopadhyay)

different levels of arsenic metabolism. Ahsan et al. (2000) have emphasized the fact that skin lesions are three times more likely in people with the highest levels of urinary arsenic. This is due to the fact that urinary arsenic is a cumulative exposure indicator, and thus may be a good indicator for predicting negative health effects in humans.

2.5 *Impact of Arsenic Exposure on Human Health*

The impact of arsenic exposure on human health may be (A) non-carcinogenic and/or (B) carcinogenic in nature.

2.5.1 Non-carcinogenic Effects

Non-carcinogenic impact of arsenic exposure may be in the form of hyper-pigmentation, keratosis, blackfoot disease (Fig. 15.1), cardiovascular disease, neuropathy, degeneration of verbal IQ and long term memory, hormone regulation and hormone mediated gene transcription and increase in loss of foetus and premature delivery with decreased birth weights of infants. Guha Mazumder (2003) observed that chronic exposure to arsenic is associated with pigmentation, keratosis, skin cancer, weakness, anemia, dyspepsia, enlargement of the liver, spleen, and ascites (fluid in abdomen). Other symptoms include chest problems like cough, restrictive lung disease, polyneuropathy, altered nerve conduction velocity and hearing loss. In West Bengal, India, arsenic affected patients have been found to complain of irritability, lack of concentration, depression, sleep disorders, headaches, fatigue, skin itching, burning of eyes, weight loss, anemia, chronic abdominal pain, diarrhea, edema of feet, liver enlargement, spleen enlargement, cough, joint pain, decreased hearing, decreased vision, loss of appetite and weakness. In some cases liver enzymes were increased and liver histology showed fibrosis. Other effects included cirrhosis, hematemesis and melena.

Fig. 15.2 Advanced arsenicosis. [Photograph by Rajesh Chaturvedi, Geological Survey of India, Eastern Region (Source: GSI Portal)]

2.5.1.1 Hyper-Pigmentation, Keratosis and Blackfoot Disease

Arsenical toxicity takes six months to two years or more to develop depending on the quantity and level of concentration of arsenic in the water consumed. The first or early symptom is darkening of the skin (diffuse melanosis) giving rise to hyper-pigmentation of the body or palm. Spotted pigmentation may be observed on the chest, back or limbs. Again keratosis which is a growth of keratin in the form of red bumps on the skin or on mucous membranes, may develop. Guha Mazumder (2003) has also observed that the prevalence of keratosis and pigmentation also increases with increasing arsenic concentration in water. The association between exposure and response, and the prevalence of skin effects, were found to be co-relatable. Long-term ingestion of arsenic may be a cause for chronic respiratory diseases and skin lesions (Fig. 15.2). In Northern Chile, Smith et al. (2000) reported that chronic health effects of inorganic arsenic exposure included arsenic induced skin lesions despite good nutritional status. Guo et al. (2001) observed that the prevalence of arsenic dermatosis was highest in the regions that drank water with high concentrations of inorganic arsenic. They found that prevalence of skin lesions was greatest in people over 40 years of age. In a study carried out in West Bengal, it was found that patients who had arsenic-related skin lesions were consuming water with arsenic levels of 800 μg/L, and as a result, many of the patients with skin lesions also suffered from cancer (Rahman et al. 2003).

A severe form of peripheral vascular disease, known as Blackfoot disease in which the blood vessels in the lower limbs are severely damaged, resulting eventually in progressive gangrene, may occur. It has been observed mainly in the southwestern coast of Taiwan. Sporadic cases of Blackfoot were reported as early as early twentieth century, and peak incidence was observed between 1956 and 1960. Typical clinical symptoms and signs of progressive arterial occlusion are mainly found in the lower extremities; however in rare cases, the upper extremities might also be affected. Blackfoot disease ultimately results in ulceration, gangrene etc. requiring surgical amputation. Epidemiologic studies carried out since mid-twentieth century revealed that Blackfoot disease was associated with the consumption of inorganic arsenic in groundwater. Recent studies confirmed the existence of preclinical peripheral vascular disease, subclinical arterial insufficiency and defects in cutaneous micro-circulation in the residents of the endemic villages. It is suggested that the methylation capacity of arsenic can cause development of peripheral vascular disease among residents in areas with endemic Blackfoot disease. The incidences of Blackfoot disease were observed to decrease after replacing groundwater with tap water supplies to these areas. Although humic substances have also been suggested as a possible cause of Blackfoot disease, epidemiologic studies are required to confirm this (Tseng 2005).

2.5.1.2 Cardiovascular Diseases

Drinking of water contaminated with inorganic arsenic above a certain level of concentration can lead to risk of hypertension (Rahman et al. 1999). The prevalence of hypertension was found to increase in men and in women over the age of 60 years. Arsenic exposure is believed to trigger off cardiovascular disease. Lee et al. (2002) reported that arsenic ingestion affects the platelets and lead to increase of platelet aggregation. In the presence of thrombin, trivalent arsenic (arsenite) is observed to increase platelet aggregation. The authors indicated that platelet aggregation increased with long-term exposure to arsenic in drinking water, and this became one of the factors causing cardiovascular disease. In a study in southwest Taiwan, researchers evaluated a possible relationship between long-term arsenic exposure and Ischemic Heart Disease. Study, in a Black Foot Disease area, indicated that prevalence of Ischemic Heart Disease increased with increasing duration of consuming artesian well-water (Tseng et al. 2003). Wang et al. (2002) highlighted that long-term exposure to arsenic is associated with increased risk of carotid atherosclerosis.

2.5.1.3 Neuro-Behavioural and Neuropathic Effects

Arsenic toxicity often results in neuropathy. Peripheral neuropathy due to chronic arsenic exposure is one of the most common complications of the nervous system. Arsenic exposure may give rise to adverse neuropathic effects and negative neurobehaviour. Tsai et al. (2003), from a study in Taiwan, suggested that long-term accumulated arsenic may cause adverse neuro-behavioural effects in adolescence. Patients can suffer from constant pain, hypersensitivity to stimuli, muscle weakness, or atrophy (Tsai et al. 2003; Guha Mazumder 2003). Long-term exposure to arsenic and ultimate arsenic poisoning are believed to adversely affect memory and intellectual functioning. Calderon et al. (2001) and Wasserman et al. (2004) have found that children's intellectual function may decrease due to increased arsenic exposure. Children, who had more than 50 μg/L arsenic exposure, had lower performance scores than children with less than 5.5 μg/L exposure. In addition, Watanabe et al. (2003), evaluating the effects of arsenic at different ages, found that age is a very important factor when evaluating effects. In younger generations, clinical manifestations are not always obvious and, as a result, can be missed or underestimated.

2.5.1.4 Respiratory System Diseases

Guha Mazumder (2003) has observed that chronic respiratory diseases increase significantly with increasing arsenic concentrations in drinking water. Clinical manifestations may be cough and shortness of breath. People subjected to chronic arsenic exposure have been found to suffer from respiratory diseases. In another study, Milton et al. (2003) have observed that patients, with chronic arsenic exposure, have skin manifestations associated with weakness, conjuctival congestion, redness of the eyes, chronic cough and chronic bronchitis (inflammation of the respiratory tract). Milton and Rahman (2002) showed that chronic bronchitis increased with increasing arsenic exposure.

2.5.1.5 Effects on Hormonal System

Different doses of arsenic can affect hormone regulation in cells at different levels. Arsenic is considered to be an endocrine disruptor at doses as low as 0.4 μg/L. It is suggested that arsenic effects on gene expression may depend on internal conditions in the human body. Different organs in the body will respond differently to arsenic exposure (Bodwell et al. 2004). Long-term arsenic ingestion has been co-related positively with Diabetes Mellitus-type two. Tseng et al. (2000, 2002) suggest that inorganic arsenic is diabetogenic in humans and people exposed to arsenic may suffer from type two-diabetes.

2.5.1.6 Effects on Reproductive System

Researches have indicated that arsenic ingestion can adversely affect human health during pregnancy. Chakraborti et al. (2003) found that chronic exposure from groundwater arsenic gave rise to complications during pregnancy. They established that increased arsenic exposure caused increased fetal loss and premature delivery in women. Toxic effects on the fetus were also suggested by Hopenhayn et al. (2003a). A separate study by Hopenhayn et al. (2003b) found that women, exposed to arsenic in drinking water during pregnancy, have changes in urinary excretion and metabolite distribution that can cause toxic effects on the developing foetus. Study by Milton et al. (2005) indicated a strong link between chronic arsenic exposure and spontaneous abortion and stillbirth.

2.5.1.7 Steatosis (Fatty Liver)

Chen et al. (2004a) concluded from experimental studies on mouse, that chronic oral inorganic arsenic exposure causes steatosis.

2.5.2 Carcinogenic Effects of Arsenic Exposure

Chronic exposure to arsenic may cause cancer of skin, bladder and lungs. The International Agency for Research on Cancer has considered arsenic as a carcinogen for humans (IARC 1980). Centeno et al. (2006) in fact reported that arsenic is a unique carcinogen. It is the only known human carcinogen for which there is adequate evidence of carcinogenic risk by both inhalation and ingestion. Rahman et al. (2003) indicated that arsenic affected patients in West Bengal, who died prematurely due to cancer, had serious arsenical skin lesions prior to that. Also, in follow-up visits, people, who were exposed to high levels of arsenic from drinking water and/or food for many years, were frequently developing cancer.

2.5.2.1 Skin Cancer

Arsenite can play a role in the enhancement of UV-induced skin cancers. The mechanism of action may involve effects on DNA methylation and DNA repair (Rossman et al. 2004). Luster and Simeonova (2004) reported epidemiological evidence indicating that arsenic is associated with cancers of skin and internal organs, as well as with vascular disease.

2.5.2.2 Bladder Cancer

Steinmaus et al. (2003) observed that there was an increased risk of bladder cancer in smokers who were exposed to arsenic concentration of nearly 200 μg/L in drinking water compared to smokers consuming lower arsenic levels. This suggests that arsenic is synergistic with smoking at relatively high arsenic levels (200 μg/L). They highlighted that lag between arsenic exposure and development of bladder cancer can be very long (more than 40 years). However, Kurttio et al. (1999) reported significant increase in the risk of bladder cancer at levels of arsenic

>0.5 μg/L in Finland. Thus this correlation suggests adverse effect of exposure to arsenic at concentrations many times lower than any current drinking water quality guideline.

2.5.2.3 Lung Cancer

Recent study of residents in areas of endemic arsenic toxicity in Taiwan by Chen et al. (2004b) indicated increased risk of lung cancer with high levels of arsenic exposure through drinking water. The authors suggested that reduction in arsenic exposure should reduce the lung cancer risk in cigarette smokers. Study by Chiu et al. (2004), in south-west Taiwan, indicated that the mortality from lung cancer declined after the levels of arsenic in the well water were reduced. Xia and Liu (2004) opined that in arsenic endemic areas in China, cancer incidences may increase over the next 10–20 years mainly due to previous exposures. In China, often areas of chronic arsenic toxicity have high levels of fluoride in drinking water. This suggests that the combination of the two could increase the risk to human health due to potential synergism. It is already known that cigarette smoking is a main risk factor for lung cancer and in fact Ferreccio et al. (2000) found from a study in Chile that cigarette smoking plus ingestion of inorganic arsenic from drinking water had a synergistic effect in precipitating lung cancer.

3 Fluoride

3.1 General Aspects

Fluorine is a common element with Atomic Number 9 and Atomic weight 19.0 in Group VII B of the Periodic Table of elements. It is a trace element widely distributed in nature. It comprises about 0.3 g/kg of the earth's crust. It is found in the form of fluorides in a number of minerals, such as fluorspar, cryolite and fluorapatite.

3.2 Environment of Fluoride

Fluoride is present in air, surface water, sub-surface water, various types of food and also in toothpaste. Fluorine may be released to the environment during the production of phosphate fertilizers, bricks, tiles and ceramics.

Natural background concentration of fluoride in air is 0.5 μg/m^3. However considering anthropogenic emissions, the background concentration comes to 3 μg/m^3. Cao and Li (1992) found that fluoride concentrations may range from

16 to 46 μg/m^3 in indoor air, where high fluoride bearing coal is used for cooking, drying and curing food indoor as reported from some provinces of China.

Fluoride is present in traces in both surface and sub-surface water. Fluoride has been reported from seawater. It may also occur as a contaminant in river water due to industrial discharges. In groundwater, fluoride concentrations vary with the type of rock the water flows through but do not usually exceed 10 mg/L (USEPA 1985a, b). In areas rich in fluoride-containing minerals, groundwater may contain up to about 10 mg of fluoride per litre. Sarmah (2004) reported fluoride ranging in concentration from 1 to 19.89 mg/L in water in parts of Nagaon and Karbi Anglong districts of Assam. He observed that people of the area suffered from stiff neck, joint pain, dental caries and skeletal deformities. He further observed that while water samples from streams, springs and ring wells showed fluoride concentration to be <0.5 mg/L, samples from unlined dug wells and one pond gave values above 3 mg/L. Sarmah (2004) found that deep tube wells gave much higher values of fluoride. Fluoride concentrations in the groundwater of some villages in China have been reported to be greater than 8 mg/L (Fuhong and Shuqin 1988). In Canada, fluoride levels in drinking-water have been reported in municipal waters to range from <0.05 to 0.2 mg/L (non-flourinated) and from 0.6 to 1.1 mg/L (fluorinated); in drinking water prepared from groundwater, concentrations up to 3.3 mg/L have been reported. In the USA, 0.2% of the population is exposed to more than 2.0 mg/L (USEPA 1985a, b). In some African countries where the soil is rich in fluoride-containing minerals, levels in drinking water can be very high (e.g., 8 mg/L in the United Republic of Tanzania) (US EPA 1985a, b). Normally therefore the concentration of fluoride in sub-surface water is higher than in surface water.

Fluoride is absorbed from soil and water by all vegetation. As such, vegetables such as betel leaf, betel, tobacco, green garlic, onion, cabbage, soya bean, carrot and potato contain traces of fluoride. Fluoride may be present in egg, cow liver and kidney (Gupta and Gupta 2004). Other foods containing high levels of fluoride include fish (0.1–30 mg/kg) and tea (USEPA 1985a, b). High concentration in tea is caused by high natural concentrations in tea plants or by the use of additives during growth or fermentation. Dry tea contain 3–300 mg/kg (average 100 mg/kg) of fluoride, so 2–3 cups of tea contain approximately 0.4–0.8 mg (IPCS 1984). Thus in areas where water with high fluoride content is used to prepare tea, the intake via tea can be several times greater.

3.3 *Exposure to Fluoride*

Water, food, air, medicines and cosmetics utilizing fluoride preparations constitute major sources of fluoride for human exposure. However 60% of the total intake is through drinking water (Gupta and Gupta 2004).

Intake values ranging from 0.46 to 3.6–5.4 mg/day have been reported in several studies (IPCS 1984). However intakes of fluoride, in areas where high fluoride coal is used indoors or where there is elevated fluoride in drinking water, can be significantly higher as in some areas of China (IPCS 2002). Daily exposure in volcanic areas (e.g., the United Republic of Tanzania) with high fluoride levels in drinking water may be up to 30 mg for adults, mainly from drinking water intake. The guideline value is 1.5 mg/L for fluoride was set in 1984. Concentrations above this value may be responsible for dental fluorosis. Higher fluoride concentrations may cause skeletal fluorosis. The value, for artificial fluoridation of water supply, is usually 0.5–1.0 mg/L (Murray 1986). In setting national standards or local guidelines for fluoride or in evaluating the possible health consequences of exposure to fluoride, it is essential to consider the intake from water, air and food. Where the intakes are likely to approach, or be greater than 6 mg/day, it would be appropriate to consider setting a standard or local guideline at a concentration lower than 1.5 mg/L.

3.4 Accumulation of Fluoride in Human Tissues

Water-soluble fluorides are rapidly and almost completely absorbed in the gastrointestinal tract. The rate of absorption depends on the gastric pH, dietary calcium, plasma fluoride levels and other dietary cations (Whitford 1994). Absorbed fluoride is transported through blood and the distribution of fluoride is a rapid process. It is incorporated into teeth and bones. There is virtually no storage in soft tissues. Bone, teeth and cartilage are the main sites for retention of fluoride in human tissues. High concentration of flouride is reported in trans- iliac bone biopsies of patients with skeletal fibrosis (Boivin et al. 1989). Bone-fluoride concentration increases with duration and dose of fluoride (Guo et al. 1988). 50% of fluoride is excreted through urine (Waterhouse et al. 1980), faeces and sweat. Incorporation into teeth and skeletal tissues is reversible: after cessation of exposure, mobilization from these tissues takes place (IPCS 1984; USEPA 1985a, b; Janssen et al. 1988). Fluoride in inhaled particles is also absorbed, the extent of absorption depends on the size of the particles and the solubility of fluoride compounds present (IPCS 2002).

3.5 Impact of Fluoride Exposure on Human Health

Fluoride is considered to be an essential element for animals and humans. For humans, there is no data indicating the minimum requirement for nutrition. Janssen et al. (1988) opined that minimum oral dose of at least 1 mg of fluoride per kg of body weight may cause acute fluoride intoxication. Fluoride induces structural and functional changes in bone cells, matrix protein and mineral components. Many epidemiological studies of possible adverse effects of the long-term ingestion of fluoride via drinking water clearly establish that fluoride primarily produces effects on skeletal tissues (bones and teeth). Fluoride affects several systems and tissues. Low concentrations provide protection against dental caries, especially in children. However, fluoride can also have an adverse effect on tooth enamel and may give rise

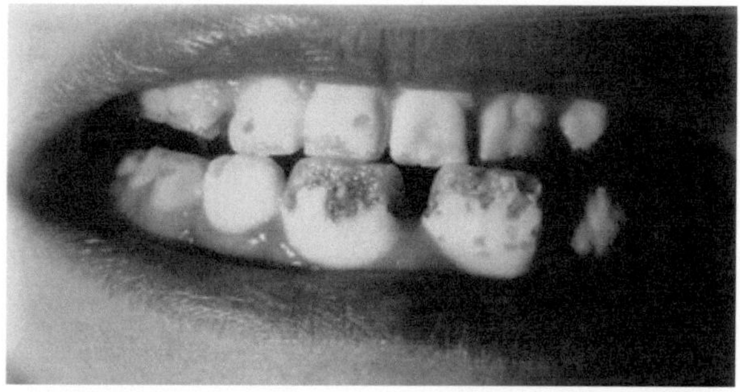

Fig. 15.3 The brown pitting and staining of teeth indicative of dental fluorosis. (Photograph by Dipak K. Mukhopadhyay)

Fig. 15.4 Skeletal fluorosis. [Photograph by Rajesh Chaturvedi, Geological Survey of India, Eastern Region (Source: GSI Portal)]

to mild dental fluorosis (Fig. 15.3) at drinking water concentrations between 0.9 and 1.2 mg/L (Dean 1942). In general, dental fluorosis does not occur in temperate areas at concentrations below 1.5–2 mg of fluoride per litre of drinking water. In warmer areas, dental fluorosis can occur at lower concentrations in the drinking water due to greater consumption of drinking water (IPCS 1984; USEPA 1985a). It is possible that in areas where fluoride intake other than drinking water (e.g. air and food) is elevated, dental fluorosis will develop at concentrations in drinking water below 1.5 mg/L. Elevated fluoride intakes can also have more serious effects on skeletal tissues. Skeletal fluorosis, with adverse changes in bone structure (Fig. 15.4) may be observed when drinking water contains 3–6 mg/L of fluoride. Crippling skeletal fluorosis usually develops only where drinking-water contains over 10 mg of fluoride per litre (IPCS 1984). Skeletal involvement in fluorosis may initially manifest as mild non–specific musculoskeletal symptoms.

Advanced stages may manifest as polyarthralgia, bone pain, stiffness of joints, restricted movement of joints and ultimate development of bow legs, knock knee, sabre shin and spinal deformities. In many parts of the world including India, symptoms of fluorosis may occur below these concentrations due to climatic conditions, food habit and malnutrition. A study in China has reported an increased prevalence of skeletal fluorosis above the level of 1.4 mg fluoride per litre in drinking water (Xu et al. 1997). However, other studies (Liang et al. 1997) estimate that, at least in some regions of China and India, the contribution from food can greatly exceed that from water. Therefore, one cannot rule out that high rates of skeletal fluorosis associated with a level greater than 1.4 mg/L in drinking water are due to other exposures. While there is a clear excess of skeletal fluorosis in these studies for a total intake of 14 mg/day, the quantitative relationship between total intake of fluoride from different sources and the risk of skeletal fluorosis cannot be estimated because of substantial uncertainties in the prevalence of effects in the range of intakes between 3 and 14 mg/day. In the concentration range of 1.45–2.19 mg/L of fluoride in drinking water, corresponding to a total intake of 6.54 mg/day, there is a relative risk for all fractures.

On the issue of possible association between fluoride in drinking-water and cancer rates among the population, IPCS (2002) considered all data and concluded that overall evidence of carcinogenicity in laboratory animals is inconclusive and that the weight of evidence does not support the hypothesis that fluoride causes cancer in humans; however, the data on bone cancer are relatively limited.

Studies on the possible adverse effects of fluoride in drinking water on pregnancy indicate that there is no apparent relationship between the rates of Down Syndrome or congenital malformation and the consumption of fluoridated drinking-water (IPCS 1984, 2002; USEPA 1985a, b; Janssen et al. 1988). It is known that persons suffering from certain forms of renal impairment have a lower margin of safety for the effects of fluoride than the average person. The data available on this subject are, however, too limited to allow a quantitative evaluation of the increased sensitivity to fluoride toxicity of such persons (USEPA 1985a, b; Janssen et al. 1988).

4 Conclusion

Many trace elements, beyond certain levels of concentration, cause serious human health problems, when the human body is exposed to them. Arsenic and fluoride have precipitated health issues of grave concern.

Chronic arsenic toxicity related to drinking of arsenic contaminated water has been reported from many countries of the world. However cases of arsenic poisoning, related to ingestion of arsenic bearing groundwater, have reached an unprecedented dimension in West Bengal and Bangladesh. About six million people have been estimated to have been exposed to arsenic contaminated water (arsenic level

>0.05 mg/L) in eight districts of West Bengal alone. Large number of these people have manifested with hyperpigmentation, keratosis, peripheral vascular disease, neuropathy and skin cancer.

Fluoride has been extensively used in the treatment of dental caries since long. It has been observed, however, that ingestion of fluoride through drinking water and other sources, beyond a certain level of concentration, has been found to cause dental disfigurement and crippling skeletal and neurological disability as reported from especially India and other South Asian countries.

Arsenic comprises 0.00058% of the total mass of the earth's crust but it exhibits preferential accumulation in certain environment. High concentration of arsenic is usually observed in the following geo-chemical environments:

- Basin fill deposits of alluvial-lacustrine origin specially in semi-arid areas
- Volcanic deposits
- Geothermal systems
- Uranium and gold mining areas

Major processes responsible for concentration of arsenic in groundwater comprise: (a) mineral precipitation/dissolution, (b) adsorption/desorption, (c) chemical transformations, (d) ion exchange and (e) bio-mediated reactions.

Drinking of water with high levels of arsenic, over a long period of time, gives rise to chronic arsenic toxicity. Consumption of food, from areas where food products like vegetables and rice are cultivated using arsenic contaminated groundwater, is another source of chronic arsenic poisoning. Different foods have different arsenic concentrations.

It has been observed, in case of people suffering from arsenic toxicity, that they have significant accumulation of arsenic in their urine, hair and nail samples.

All arsenic compounds are poisonous and exposure to arsenic gives rise to various adverse impacts on human health. Exposure of humans to arsenic may give rise to non-carcinogenic effects such as hyper-pigmentation, keratosis, blackfoot disease, cardiovascular disease and neuropathy. Arsenic toxicity may adversely affect verbal IQ and long-term memory and suppress hormone regulation and hormone mediated gene transcription and also cause increase in loss of foetus and premature delivery with decreased birth weights of infants. Arsenic is a unique carcinogen. It is the only known human carcinogen for which there is adequate evidence of carcinogenic risk by both inhalation and ingestion. Arsenic toxicity may also result in cancer of lungs, urinary bladder and skin.

Fluorine is a trace element widely distributed in nature. It comprises about 0.3 g/kg of the Earth's crust. It is found in the form of fluorides in a number of minerals, such as fluorspar, cryolite and fluorapatite.

Fluorine is present in air, surface and sub-surface water and also in various types of food. Natural background concentration of fluoride in air is 0.5 $\mu g/m^3$. Fluoride is present in traces in both surface and sub-surface water. It is found in both sea water and river water where it may reach as discharge from industrial sources. In groundwater, fluoride concentrations vary with the type of rock the water flows through but do not usually exceed 10 mg/L. Fluoride is absorbed from soil and water by all vegetation, and all foodstuffs contain at least traces of it.

Concentrations above 1.5 mg/L of fluoride may be responsible for dental fluorosis. Still higher fluoride concentrations may cause skeletal fluorosis. The value for artificial fluoridation of water supplies is usually 0.5–1.0 mg/L.

Water-soluble fluorides are rapidly and almost completely absorbed in the gastrointestinal tract. Absorbed fluoride is transported via the blood. Distribution of fluoride is a rapid process. Bone, teeth and cartilage are the main sites for retention of fluoride in human tissues.

Fluoride is considered to be an essential element for animals and humans. For humans, there is no data indicating the minimum nutritional requirement. Fluoride induces structural and functional changes in bone cells, matrix protein and mineral components. Many epidemiological studies of possible adverse effects of the long-term ingestion of fluoride via drinking water clearly establish that fluoride primarily produces effects on skeletal tissues (bones and teeth). Fluoride can have an adverse effect on tooth enamel and may give rise to mild dental fluorosis. Elevated fluoride intakes can also have more serious effects on skeletal tissues. Skeletal fluorosis (with adverse changes in bone structure) may be observed when drinking water contains 3–6 mg/L of fluoride. Crippling skeletal fluorosis usually develops only where drinking water contains over 10 mg/L of fluoride. Skeletal involvement in fluorosis may initially manifest as mild non-specific musculoskeletal symptoms. Advanced stages may manifest as polyarthralgia, bone pain, stiffness of joints, restricted movement of joints and ultimate development of bow legs, knock knee, sabre shin and spinal deformities.

References

Ahsan, H., Perrin, M., Rahman, A., Parvez, F., Stute, M., Zheng, Y., Milton, A.H., Brandt-Rauf, Geen, A.V. and Graziano, J. (2000). Associations between drinking water and urinary arsenic levels and skin lesions in Bangladesh. *J. Occup. Environ. Med.*, **42:** 1195–1201.

Al Rmalli, S.W. (2005). A survey of arsenic in foodstuffs on sale in the United Kingdom and imported from Bangladesh. *The Sci. of the Tot Environ*, **337(1–3):** 23–30.

Andreae, M.O. (1980). Arsenic in rain and the atmospheric mass balance of arsenic. *J Geophys Res*, **85:** 4512–4518.

Benson, L.V. and Spencer, R.J. (1983). A hydrochemical reconnaissance study of the Walker River Basin, California and Nevada. US Geological Survey open file report.

Bodwell, J.E., Kingsley, L.A. and Hamilton, J.W. (2004). Arsenic at very low concentrations alters glucocorticoid receptor (GR)-mediated gene activation but not GR-mediated gene repression: Complex dose-response effects are closely correlated with levels of activated GR and require a functional GR DNA binding domain. *Chem. Res. Toxicol.*, **17:** 1064–1076.

Boivin, G., Chavassieux, P., Chapuy, M.C., Baud, C.A. and Meunier, P.J. (1989). Skeletal Fluorosis: Histomorphometric analysis of bone changes and bone fluoride contents in 29 patients. *Bone*, **10:** 89–99.

Calderon, J., Navarro, M.E., Jimenez-Capdeville, M.E., Santos-Diaz, M.A., Golden, A., Rodriguez-Leyva, I., Borja-Aburto, V. and Diaz-Barriga, F. (2001). Exposure to arsenic and lead and neuropsychological development in Mexican children. *Environ. Res.*, **85:** 69–76.

Cao, S.R. and Li, Y.F. (1992). The evaluation of indoor air quality in areas of endemic fluorosis caused by coal combustion. *In:* Proceedings of the XIX Conference of the International Society for Fluoride Research, Kyoto, Japan, 1992. Kyoto, Department of Hygiene and Public Health, Osaka Medical College.

Centeno, J.A., Tchounwou, P.B., Patlolla, A.K., Mullick, F.G., Murakata, L., Meza, E., Gibb, H., Longfellow, D. and Yedjou, C.G. (2006). Environmental Pathology and Health Effects of Arsenic Poisoning: A Critical Review. *In:* Managing Arsenic in the Environment: From Soil to Human Health. Naidu, R., Smith, E., Owens, G., Bhattacharya, P. and Nadebaum, P. (Eds). CSIRO Publishing, Melbourne, Australia. (ISBN: 0643068686). Review of Chronic Arsenic Poisoning 2425.

Chakraborti, D., Mukherjee, S.C., Pati, S., Sengupta, M.K., Rahman, M.M., Chowdhury, U.K., Lodh, D., Chanda, C.R., Chakraborti, A.K. and Basu, G.K. (2003). Arsenic groundwater contamination in Middle Ganga Plain, Bihar, India: A future danger? *Environ. Health Perspect.*, **111:** 1194–2201.

Chakraborti, D., Sengupta, M.K., Rahman, M.M., Ahamed, S., Chowdhury, U.K., Hossain, M.A., Mukherjee, S.C., Pati, S., Saha, K.C., Dutta, R.N. and Quamruzzaman, Q. (2004). Groundwater arsenic contamination and its health effects in the Ganga-Meghna-Brahmaputra plain. *J. Environ. Monit.*, **6(6):** 74–83.

Chen, H., Li, S., Liu, J., Diwan, B.A., Barrett, J.C. and Waalkes, M. (2004a). Chronic inorganic arsenic exposure induces hepatic global and individual gene hypomethylation: Implications for arsenic hepatocarcinogenesis. *Carcinogenesis*, **25:** 1779–1786.

Chen, C.L., Hsu, L.I., Chiou, H.Y., Hsueh, Y.M., Chen, S.Y., Wu, M.M. and Chen, C. (2004b). Ingested arsenic, cigarette smoking, and lung cancer risk: A follow-up study in arseniasis-endemic areas in Taiwan. *JAMA*, **292(24):** 2984–2990.

Chiu, H.F., Ho, S.C. and Yang, C.Y. (2004). Lung cancer mortality reduction after installation of tap-water supply system in an arseniasis-endemic area in Southwestern Taiwan. *Lung Cancer*, **46(3):** 265–270.

Dean, H.T. (1942). Epidemiological studies in the United States. *In:* Moulton, F.R. (ed.), Fluorine and dental health, Washington, DC. American Association for the Advancement of Science (AAAS Publication No. 19).

Ferreccio, C., Gonzalez, C., Milosavjlevic, V., Marshall, G., Sancha, A.M. and Smith, A.H. (2000). Lung cancer and arsenic concentrations in drinking water in Chile. *Epidemiology*, **11(6):** 673–679.

Fuhong, R. and Shuqin, J. (1988). Distribution and formation of high-fluorine groundwater in China. *Environmental Geology and Water Science*, **12(1):**3–10.

Guha Mazumder, D.N. (2003). Chronic arsenic toxicity: Clinical features, epidemiology, and treatment: Experience in West Bengal. *J. Environ Sci Health*, **A38:** 141–163.

Guha Mazumder, D.N. (2004). Health Hazard caused by Chronic arsenic toxicity in West Bengal. Proceedings: Workshop on Medical Geology (IGCP-454), Special Publications No. 83, Geological Survey of India.

Guo, M.K., Nopakun, H., Messer, H.H., Opaug, R. and Singer, L. (1988) Retention of skeletal fluoride during bone turnover in rats. *Journal of Nutrition*, **118:** 326–366.

Guo, X., Fujino, Y., Kaneko, S., Wu, K., Xia, Y. and Yoshimura, T. (2001). Arsenic contamination of groundwater and prevalence of arsenical dermatosis in the Hetao plain area, Inner Mongolia, China. *Mol. Cell. Biochem.*, **222:** 137–140.

Gupta, A.B. and Gupta, S.K. (2004). Recent advances in fluorosis and defluoridaton with special reference to Rajasthan. Proceedings Workshop on Medical Geology, IGCP – 454. Special Publication No. 83, Geological Survey of India.

Gupta, S.A. and Godbole, M.M. (2004). Endemic Fluorosis and Its Medical Implications. Proceedings Workshop on Medical Geology, IGCP – 454. Special Publication No. 83, Geological Survey of India.

Hopenhayn, C., Ferreccio, C., Browning, S.R., Huang, B., Peralta, C., Gibb, H. and Hertz-Picciotto, I. (2003a). Arsenic exposure from drinking water and birth weight. *Epidemiology*, **14:** 593–602.

Hopenhayn, C., Huang, B., Christian, J., Peralta, C., Ferreccio, C., Atallah, R. and Kalman, D. (2003b). Profile of urinary arsenic metabolites during pregnancy. *Environ. Health Perspect.*, **111:** 1888–1891.

IARC (International Agency for Research on Cancer) (1980). Arsenic and arsenic compound. 23. http://www-cie.iarc.fr/htdocs/monographs/vol23/arsenic.html (accessed July 2005).

IPCS (1984). Fluorine and fluorides. Geneva, World Health Organization, International Programme on Chemical Safety (Environmental Health Criteria 36).

IPCS (2002). Fluorides. Geneva, World Health Organization, International Programme on Chemical Safety (Environmental Health Criteria 227).

Janssen, P.J.C.M., Janus, J.A. and Knaap, A.G.A.C. (1988). Integrated criteria document fluorides—effects. Bilthoven, National Institute of Public Health and Environmental Protection (Appendix to Report No. 75847005).

Kanamori, S. (1965). Geochemical study of Arsenic in natural waters: III. The significance of ferric hydroxide precipitate in stratification and sedimentation of arsenic in Lake Waters. *J of Erth Sci*, 13.

Keller, Edward A. (1985). Environmental Geology, Fourth Edition. Charles E. Merril Publishing Company.

Kurttio, P., Pukkala, E., Kahelin, H., Auvinen, A. and Pekkanen, J. (1999). Arsenic concentrations in well water and risk of bladder and kidney cancer in Finland. *Environ. Health Perspect*, **107:** 705–710.

Lee, M.Y., Bae, O.N., Chung, S.M., Kang, K.T., Lee, J.Y. and Chung, J.H. (2002). Enhancement of platelet aggregation and thrombus formation by arsenic in drinking water: A contributing factor to cardiovascular disease. *Toxicol. Appl. Pharmacol.*, **179:** 83–88.

Liang, C.K., Ji, R. and Cao, S.R. (1997). Epidemiological analysis of endemic fluorosis in China. *Environmental Carcinogenesis and Ecotoxicology Reviews*, **C15(2):** 123–138.

Lin, T.H., Huang, Y.L. and Wang, M.Y. (1998). Arsenic species in drinking water, hair, fingernails, and urine of patients with blackfoot disease. *J. Toxicol. Environ. Health*, **A53:** 85–93.

Luster, M.I. and Simeonova, P.P. (2004). Arsenic and urinary bladder cell proliferation. *Toxicol. Appl. Pharmacol.*, **198:** 419–423.

Mathur, K.N. (2004). Health Hazard and Geochemical Attributes. Proceedings: Workshop on Medical Geology (IGCP-454), Special Publications No.83, Geological Survey of India.

Meza, M.M., Kopplin, M.J., Burgess, J.L. and Gandolfi, A.J. (2004). Arsenic drinking water exposure and urinary excretion among adults in the Yaqui Valley, Sonora, Mexico. *Environ. Res.*, **96:** 119–126.

Milton, A.H., Smith, W., Rahman, B., Hasan, Z., Kulsum, U., Dear, K., Rakibuddin, M. and Ali, A. (2005). Chronic arsenic exposure and adverse pregnancy outcomes in Bangladesh. *Epidemiology*, **16:** 82–86.

Milton, A.H. and Rahman, M. (2002). Respiratory effects and arsenic contaminated well water in Bangladesh. *Int. J. Environ. Health. Res.*, **12:** 175–179.

Milton, A.H., Hasan, Z., Rahman, A. and Rahman, M. (2003). Non-cancer effects of chronic arsenicosis in Bangladesh: Preliminary results. *J. Environ. Sci. Health*, **A38:** 301– 305.

Mukhopadhyay, D.K., Ghosh, Gautam (2010). Arsenic pollution of ground water in parts of West Bengal – A case study. *Ind J of Geosc*, **64(1–4):** 41–48.

Murray, J.J. (ed.) (1986). Appropriate use of fluorides for human health. World Health Organization, Geneva.

Ng, J.C., Wang, J., Zheng, B., Zhai, C., Maddalena, R., Liu, F. and Moore, M.R. (2005). Urinary porphyrins as biomarkers for arsenic exposure among susceptible populations in Guizhou province, China. *Toxicol. Appl. Pharmacol.*, **206(2):** 176–184.

Onishi, Y. and Sandell, E.B. (1955). Geochemistry of arsenic. *Geochimica at Cosmochimica Acta*, **7**.

Rahman, M.M., Mandal, B.K., Chowdhury, T.R., Sengupta, M.K., Chowdhury, U.K., Lodh, D., Chanda, C.R., Basu, G.K., Mukherjee, S.C., Saha, K.C. and Chakraborti, D. (2003). Arsenic groundwater contamination and sufferings of people in North 24-Parganas, one of the nine arsenic affected districts of West Bengal, India. *J. Environ. Sci. Health*, **A38:** 25–59.

Rahman, M., Tondel, M., Ahmad, S.A., Chowdhury, I.A., Faruquee, M.H. and Axelson, O. (1999). Hypertension and arsenic exposure in Bangladesh. *Hypertension*, **33:** 74–78.

Rossman, T.G., Uddin, A.N. and Burns, F.J. (2004). Evidence that arsenite acts as a cocarcinogen in skin cancer. *Toxicol. Appl. Pharmacol.*, **198:** 394–404.

Sarmah, G.C. (2004). Fluoride contaminated water and its implications on Human Health in parts of Nagaon and Karbi Anglong Districts, Assam. Proceedings Workshop on Medical Geology, IGCP – 454. Special Publication No. 83, Geological Survey of India.

Smith, A.H., Arroyo, A., Guha Mazumder, D.N., Kosnett, M.J., Hernandez, A.L., Beeris, M., Smith, M.M. and Moore, L.E. (2000). Arsenic-induced skin lesions among Atacameno people in Northern Chile despite good nutrition and centuries of exposure. *Environ. Health Perspect*, **108:** 617–620.

Steinmaus, C., Yuan, Y., Bates, M.N. and Smith, A.H. (2003). Case-control study of bladder cancer and drinking water arsenic in the western United States. *Am. J. Epidemiol.*, **158:** 1193–1201.

Tsai, S.Y., Chou, H.Y., The, H.W., Chen, C.M. and Chen, C.J. (2003). The effects of chronic arsenic exposure from drinking water on the neurobehavioural development in adolescence. *Neurotoxicology*, **24:** 747–753.

Tseng, C.H. (2005). Blackfoot disease and Arsenic: A never-ending story. *Journal Environ Sci. Health*, **23(1):** 55–74.

Tseng, C.H., Chong, C.K., Tseng, C., Hsueh, Y.M., Chiou, H.Y., Tseng, C.C. and Chen, C.J. (2003). Long-term arsenic exposure and ischemic heart disease in arseniasis hyperendemic villages in Taiwan. *Toxicol. Lett.*, **137:** 15–21.

Tseng, C.H., Tseng, C., Chiou, H.Y., Hsueh, Y.M., Chong, C.K. and Chen, C.J. (2002). Epidemiologic evidence of diabetogenic effect of arsenic. *Toxicol. Lett.*, **133:** 69–76. Review of Chronic Arsenic Poisoning 2427.

Tseng, C.H., Tai, T.Y., Chong, C.K., Tseng, C., Lai, M.S., Lin, B.J., Chiou, H.Y., Hsueh, Y.M., Hsu, K.H. and Chen, C.J. (2000). Long-term arsenic exposure and incidence of non-insulin-dependent diabetes mellitus: A cohort study in arseniasis-hyperendemic villages in Taiwan. *Environ. Health Perspect.*, **108:** 847–851.

USEPA (1985a). Drinking water criteria document on fluoride. Washington, DC, US Environmental Protection Agency, Office of Drinking Water (TR-823-5).

USEPA (1985b). National primary drinking water regulations; fluoride; final rule and proposed rule. US Environmental Protection Agency. Federal Register, **50(220):** 47142–47171.

Watanabe, C., Inaoka, T., Matsui, T., Ishigaki, K., Murayama, N. and Ohtsuka, R. (2003). Effects of arsenic on younger generations. *J. Environ. Sci. Health*, **A38:** 129–139.

Welch, A.H., Lico, M.S. and Hughes, J.L. (1988). Arsenic in ground water of the Western United States. *Groundwater*, **26(3):** 333–347.

Wang, C.H., Jeng, J.S., Yip, K., Chen, C.L., Hsu, L.I., Hsueh, Y.M., Chiou, H.Y., Wu, M.M. and Chen, C.J. (2002). Biological gradient between long-term arsenic exposure and carotid atherosclerosis. *Circulation*, **105:** 1804–1809.

Wasserman, G.A., Liu, X., Parvez, F., Ahsan, H., Factor-Litvak, P., van Geen, A., Slavkovich, V., LoIacono, N.J., Cheng, Z., Hussain, I., Momotaj, H. and Graziano, J.H. (2004). Water arsenic exposure and children's intellectual function in Araihazar, Bangladesh. *Environ. Health Perspect*, **112(13):** 1329–1333.

Waterhouse, C., Taves, D. and Munzer, A. (1980). Serum inorganic fluoride changes related to previous fluoride intake, and bone resorption. *Clinical. Science*, **58:** 145–152.

Whitford, G.M. (1994). Effects of plasma fluoride and dietary calcium concentrations on GI absorption and secretion of fluoride in the rat. Calcif Tissue Int.

Xia, Y. and Liu, J. (2004). An overview on chronic arsenism via drinking water in PR China. *Toxicology*, **198:** 25–29.

Xu, R.Q., Wu, D.Q. and Xu, R.Y. (1997). Relations between environment and endemic fluorosis in Hohot region, Inner Mongolia. *Fluoride*, **30:** 26–28.

Chapter 16
Arsenic and Excess Fluoride Removal in Public Water Supply: Key Issues and Challenges

Arunabha Majumder

1 Introduction

Water is essential for our life, livelihood, food security and sustainable development. It is a scarce natural resource but renewable through cyclic process of the climate. India having one-sixth of world population and 2.4% of world's land area is to satisfy with only 4% of world's renewable water resources. The per capita availability of water is decreasing in the country and it has reduced to one-third today compared to water availability during independence. The demand of water is increasing at a faster rate due to growing population, growing agriculture, food production, rapid industrialization and economic development. Again climate change in recent time has resulted in alteration of rainfall pattern. It has been predicted that the climate change is likely to increase the variability of water resources affecting livelihood and human health. Spatial rainfall pattern of varying characteristic has resulted in uneven water resources and thereby developing water stress and water-scarce situation in many stretches of the country. The situation has aggravated due to natural and anthropogenic pollution of both groundwater and surface water sources.

Water is a public good and every person has the right to demand drinking water. To increase economic productivity and improve public health, there is an urgent need to enhance access to safe and adequate drinking water. Accordingly every water supply agency must ensure water security and safety for the consumers addressing the issue of potability, reliability, sustainability, convenience and equity.

A. Majumder (✉)
School of Water Resources Engineering, Jadavpur University, Kolkata, India

Arsenic & Fluoride Task Force, Government of West Bengal, Kolkata, India

All India Institute of Hygiene & Public Health, Government of India, Kolkata, India
e-mail: arunabhamajumder@hotmail.com

P. K. Sikdar (ed.), *Groundwater Development and Management*,
https://doi.org/10.1007/978-3-319-75115-3_16

In India 85% rural population depend on groundwater to fulfil their domestic requirement. The groundwater is extensively used for agriculture as well as to meet the industrial requirement. Over-abstraction of groundwater has resulted in depletion of groundwater level in many parts of the country. The quality of groundwater in certain regions has undergone a change to such an extent that the use of such water could be risky and hazardous. Increase in overall salinity of the groundwater and/or presence of high concentration of fluoride, arsenic, iron, nitrate, total hardness and few toxic metals have been noticed in large areas of several states in India. The problems associated with chemical constituents of drinking water arise primarily from their ability to cause adverse health effects after prolonged periods of exposure; of particular concern are contaminants that have cumulative toxic properties, such as heavy metals, and substances that are carcinogenic (WHO 1996).

Both arsenic and fluoride contamination of groundwater have emerged as a serious problem for public water supply in the country. Though surface water supply may be a good alternative to groundwater supply in arsenic and fluoride affected areas but it may not be feasible in many areas due to non-availability of acceptable surface water sources as well as cost. Removal of arsenic and fluoride from groundwater by adopting appropriate technology may be another solution for public water supply in affected areas. Today, installation of sustainable decentralized arsenic and fluoride removal units for the benefit of the rural community is a challenge to all stakeholders.

2 Arsenic in Groundwater

Arsenic is a chemical that is distributed widely in air, water, soil, rocks, plants and animals in variable concentrations. The source of arsenic in soil is mainly parent (or rock) materials from which it is derived (Majumder 2001). It gets introduced in water through dissolution of minerals and ores and through erosion from natural resources. Anthropogenic activities resulting from combustion of fossil fuels, mining, discharge of effluents from metallurgical, ceramic, dye and pesticides manufacturing industries, petroleum refinery, etc. may cause increased arsenic levels in surface water and groundwater. The cycling of arsenic in the environment is regulated by natural processes and human activities. Thus, humans all over the world are exposed to small amounts of arsenic, mostly through food and water.

Arsenic pollution in groundwater, used for drinking purposes has been envisaged as a problem of global concern. High arsenic concentrations recognized in many parts of Asia and elsewhere are dominantly found in groundwater, and many of the health consequences encountered have emerged in relatively recent years as a result of the increased use of groundwater from bore-wells for drinking and irrigation. In terms of numbers of groundwater sources affected by arsenic contamination and population at risk, the problem is of considerable proportion in several States of

India. The water quality monitoring study carried out by All India Institute of Hygiene and Public Health in 32 arsenic affected blocks in West Bengal ($n = 7680$ hand-pump fitted bore-wells) revealed presence of arsenic beyond 50 μg/L in 20% of bore-wells, arsenic up to 50 μg/L in 44% bore-wells. Arsenic was found to be absent in 36% of bore-wells (Ananthanarayanan and Majumder 2003). In West Bengal arsenic has been found in 83 blocks situated in Malda, Murshidabad, Nadia, North 24-Parganas, South 24-Parganas, Bardhaman, Hooghly and Howrah. However, most of the arsenic affected blocks are situated along eastern part of River Ganga. The groundwater in arsenic affected areas is characterized by high iron, calcium, magnesium, bicarbonate with low chloride, sulphate and fluoride. The surface water sources including dug-wells are not affected by arsenic contamination.

Long-term exposure to arsenic via drinking water may cause arsenicosis showing symptom of keratosis, hyperkeratosis, melanosis, disorder of digestive, urinary and nervous system. It may even cause cancer of skin, lungs, liver, urinary bladder and kidney. Arsenic has been detected as natural pollutant in groundwater in Ganga-Bramhaputra-Meghna River Basin. Inorganic arsenic can occur in the environment in several forms but in groundwater in India, it is mostly found as trivalent arsenite as well as pentavalent arsenate. The water quality monitoring result highlights presence of arsenic along with iron in groundwater. Millions of people in the States of West Bengal, Bihar, Uttar Pradesh, Jharkhand, Chhattisgarh, Assam and Manipur are at risk since they reside in arsenic affected areas.

Arsenic has been detected in paddy and vegetables cultivated in arsenic affected areas. Arsenic contaminated groundwater is used in these areas for cultivation. Presence of arsenic in rice grains and vegetables is considerably high causing risk of arsenicosis in human body (Ananthanarayanan and Majumder 2003).

In order to mitigate arsenic problem, safe water supply needs to be ensured for the people. Accordingly, alternative sustainable water sources free from arsenic contamination must be explored. Surface water sources are normally free from arsenic contamination and as such the water can be supplied after necessary purification. The arsenic contaminated groundwater can also be supplied after removal of arsenic from the water. The arsenic removal process should be technically feasible, economically viable and socially acceptable. The rejects from arsenic removal process should not cause deleterious effect to the environment and accordingly it must be disposed through eco-friendly manner causing no risk to the environment and public health.

2.1 Action Plan Suggested for Mitigation of Arsenic Problem

Arsenic mitigation action plan is to be adopted for undertaking several activities so that people are prevented from the exposure of arsenic ingestion as well as extending curative medical services to the people who are under the threat of arsenicosis disease. The following action plan is suggested for mitigation of arsenic problem (Ananthanarayanan and Majumder 2003):

(a) Supply of river water through piped water supply system after conventional treatment.
(b) Adoption of roof-top rainwater harvesting system and supply of safe water for drinking and cooking.
(c) Collection of rainwater in ponds or impounding reservoirs as surface run-off and supply of safe water for drinking and cooking after appropriate treatment.
(d) Supply of groundwater from arsenic-free deeper aquifers by installing bore-wells.
(e) Supply of arsenic-safe water after removal of arsenic from contaminated groundwater.
(f) Regular water quality monitoring of all water sources including mapping and making the information available in public domain.
(g) Development of infrastructure and services for proper diagnosis and treatment of arsenicosis patients.
(h) Provision of nutritional support to the villagers in arsenic affected areas.
(i) Organize mass awareness and motivation campaign, and development of Information, Education and Communication (IEC) materials

2.2 *Treatment Technologies for Arsenic Removal*

Arsenic occurs in aquifers in trivalent (arsenite) or pentavalent (arsenate) form and these forms are considered to be most important in selecting removal methodology (Majumder 2001). The following criteria may be considered for selection of technology for removal of arsenic from contaminated groundwater (Majumder 2014):

(i) Higher arsenic removal efficiency
(ii) Ensuring arsenic in treated water below 10 μg/L
(iii) Simple operation and maintenance
(iv) Economic viability
(v) Least risk from arsenic-rich rejects

2.2.1 Technology Options for Arsenic Removal

Arsenic can be removed from groundwater by the application of following techniques:

(i) Oxidation and co-precipitation
(ii) Adsorption
(iii) Ion-exchange
(iv) Reverse osmosis

Technology Park for demonstrating various mechanisms for arsenic removal was initiated by All India Institute of Hygiene & Public Health under ICEF Project in Baruipur, South 24 Parganas. In all, 17 hand-pumps attached arsenic removal units were installed in the villages and the performance of the units were monitored on weekly basis. Different technologies were followed with the use of different chemicals and media for arsenic removal. Based on field performance and evaluation of arsenic removal units working under the principle of co-precipitation, adsorption and ion-exchange in Technology Park, critical review has been presented hereunder.

2.2.1.1 Oxidation and Co-precipitation

In groundwater arsenic is present in both arsenite and arsenate forms in varying proportion. Arsenic removal by co-precipitation process necessitates oxidation of arsenite to arsenate. This can be achieved by prolong aeration, chlorination, application of U-V ray etc. In groundwater, arsenic is present with iron of varying proportion but mostly having iron content more than 1.0 mg/L. In natural process with overnight detention period, a considerable quantum of arsenic gets removed after standard filtration process. In this process while dissolved iron gets oxidized in contact with oxygen to ferric hydroxide the arsenate gets adsorbed over ferric hydroxide facilitating removal of arsenic from contaminated groundwater. The arsenic removal process by co-precipitation thus gets strengthened by adding proper dose of oxidizing agent (average chlorine dose @ 0.5 mg/l) and chemical coagulant (aluminum sulphate/ferric sulphate/poly aluminium chloride etc.). The process requires rapid and slow mixing of chemicals, settling and filtration. The reject sludge requires proper treatment and disposal.

The oxidation and co-precipitation process is simple and economically viable. This process can be used in developing household model (Fig. 16.1), hand-pump attached model (Fig. 16.2) and also power-pump attached piped water supply

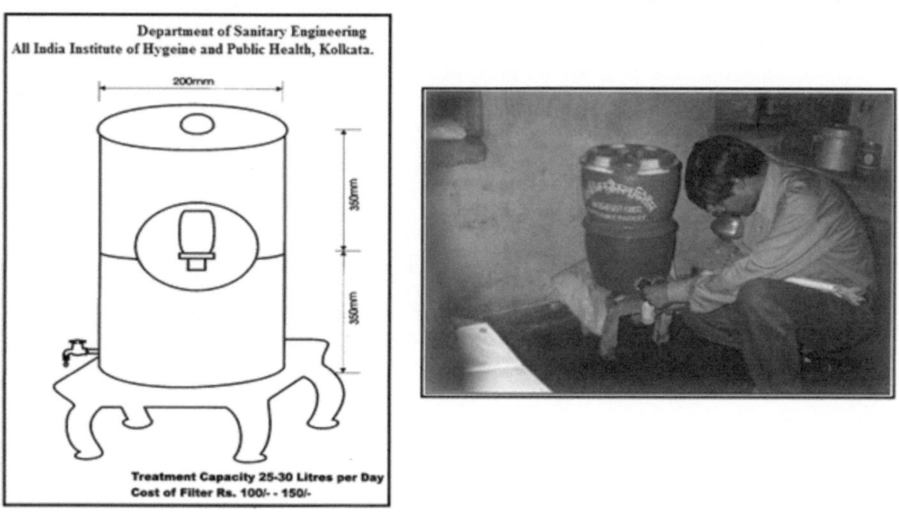

Fig. 16.1 Domestic arsenic removal unit operated with co-precipitation and filtration process

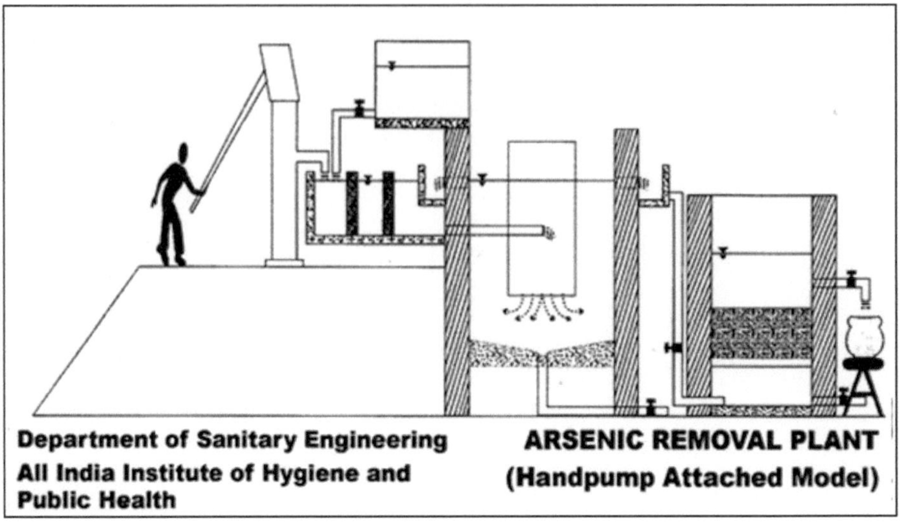

Fig. 16.2 Hand-pump attached arsenic removal unit operated with co-precipitation and filtration process

scheme. Since it is a single-stage treatment unit, it cannot bring arsenic in treated water below 10 μg/L. All India Institute of Hygiene and Public Health developed both domestic as well as hand-pump attached arsenic removal units and distributed in the villages for use (Figs. 16.1 and 16.2) (Ananthanarayanan and Majumder 2003). Critical review on the performance of both domestic and hand-pump attached models are furnished below.

(i) Availability of chemicals to the users. Addition of chemicals as per prescribed dose is to be ensured.
(ii) No mechanized system; so does not require power.
(iii) Both arsenic and iron are removed in the process.
(iv) 90% treated water ensures arsenic <50 μg/L.
(v) Iron in treated water remains to be less than 0.3 mg/L.
(vi) Arsenic in treated water ranges between 20 and 45 μg/L.
(vii) Sludge is to be stored, treated and properly disposed.
(viii) Risk of arsenic re-entering in the environment if not taken proper care.
(ix) Domestic candle filter is to be cleaned once a week.
(x) Hand-pump attached arsenic removal unit is to be cleaned once a month. It may require 3/4 h for cleaning.
(xi) Less recurring cost.

Fig. 16.3 Arsenic removal unit operated with adsorption process

(xii) Community participation is to be ensured for operation and maintenance of hand-pump attached units.

(xiii) Institutional development with capacity building is to be ensured for extending support to the users.

2.2.1.2 Adsorption

Arsenic from groundwater can be removed by the adsorption process. The process is simple and user-friendly. Because of its ease of handling, regeneration capability of the media and sludge-free operation, the fixed bed operation using adsorption technique has secured place as one of the popular methods of arsenic removal. Activated alumina and ferric hydroxide have been found to be excellent adsorbent for removal of arsenic. Both the media have been extensively used in the development of domestic model as well as hand-pump attached model (Fig. 16.3). Activated alumina can be regenerated 4–5 times by washing with acid and alkali solutions. Iron oxide coated sand, iron-nail, bauxite, hematite, laterite etc. can also be used for removal of arsenic.

Removal of arsenic is higher in adsorption process than co-precipitation. Different adsorbents have different capacities of arsenic adsorption; so, arsenic removal efficiency varies with different adsorbent. It is difficult to achieve arsenic <10 μg/L in 100% of treated water after single stage adsorption process. Presence of iron in groundwater causes clogging of fixed bed of adsorbent and accordingly regular cleaning of the media by backwashing is a necessity for proper operation of the arsenic removal units. Often two columns in series filled with media are installed with the objective to remove iron in the first column and arsenic in the second column packed with adsorbent. Critical review of the performance of arsenic removal units operated with adsorption process is presented below.

(i) Granular activated alumina/ferric hydroxide/hematite were used as adsorbent.
(ii) Better performance compared to co-precipitation method.
(iii) 90% treated water showed arsenic <50 μg/L.
(iv) 75% treated water showed arsenic <10 μg/L.
(v) Twice a week backwashing was necessary.
(vi) On an average 45 min time was necessary for backwashing.
(vii) Risk of arsenic present in backwash water.
(viii) In general, arsenic removal units were user-friendly.
(ix) Operation and maintenance (O&M) cost was more than O&M cost of units operated with co-precipitation method.
(x) O&M cost was affordable to the consumers.
(xi) Media needed change/regeneration after exhaustion.
(xii) Community participation was necessary for O&M.
(xiii) Technical support and well-knit infrastructure was the necessity for proper O&M of the units.

2.2.1.3 Ion Exchange

Ion exchange is a physico-chemical process in which ions are exchanged between solution phase and solid resin phase. The solid resin is typically an elastic hydrocarbon network containing a large number of ionized groups electro-statically bound to the resin. These groups are exchanged for ions of similar charge in solution that have a stronger exchange affinity for the resin. Critical review of arsenic removal units packed with ion exchange resins are presented below.

(i) Bucket of resin was used as media.
(ii) Resin oxidized as well as exchanged arsenic during treatment.
(iii) Performance of arsenic removal units was not satisfactory.
(iv) 65% of treated water showed arsenic <50 μg/L.
(v) 25% of treated water showed arsenic <10 μg/L.
(vi) Resin was required to be disposed in a eco-friendly manner after being exhausted.
(vii) Twice a week cleaning was necessary.
(viii) Community participation was necessary for O&M.

(ix) O&M cost appeared higher than the units operated with adsorption process.
(x) Infrastructure development was a necessity for extending technical support to the community.

2.2.1.4 Reverse Osmosis

Reverse osmosis process removes most of the impurities from water. Membrane is used in the process and water is forced through the membrane resulting in elimination of dissolved solids (impurities) including arsenic from water. However, reject management is very important in the application of reverse osmosis.

2.2.2 Two-Stage Arsenic Removal Unit

Field monitoring study revealed that single stage arsenic removal process could not produce treated water having less than 10 μg/L. arsenic at all times for any input arsenic concentration. Two-stage arsenic removal processes have an advantage to overcome such problem. It has been found that oxidation-co-precipitation process followed by adsorption could ensure to conform to desirable limit (<10 μg/L) of arsenic in treated water at all times for any arsenic input concentration. Public Health Engineering Department, Government of West Bengal has adopted two-stage arsenic removal processes for installation of arsenic removal units in different places in West Bengal. The single stage arsenic removal unit (oxidation, co-precipitation-filtration) developed by All India Institute of Hygiene and Public Health has been modified successfully to two-stage arsenic removal unit (oxidation, co-precipitation, adsorption, filtration) by School of Water Resources Engineering (SWRE), Jadavpur University under DST Project (Figs. 16.4 and 16.5).

2.3 *Sludge/Exhausted Media Disposal*

In co-precipitation process considerable quantum of sludge is generated. The sludge should not be disposed in drain or open land. The sludge must be stored in a leak proof chamber or container. The arsenic-rich sludge is hazardous. There may be health risk as well as environmental risk if the same is not handled, stored and disposed properly. Similarly, the exhausted media used in the adsorption process for removal of arsenic must be disposed safely. Following disposal options are suggested for adoption (Majumder 2014):

(i) Disposal of sludge in on-site sanitation pit or septic tank for anaerobic digestion.

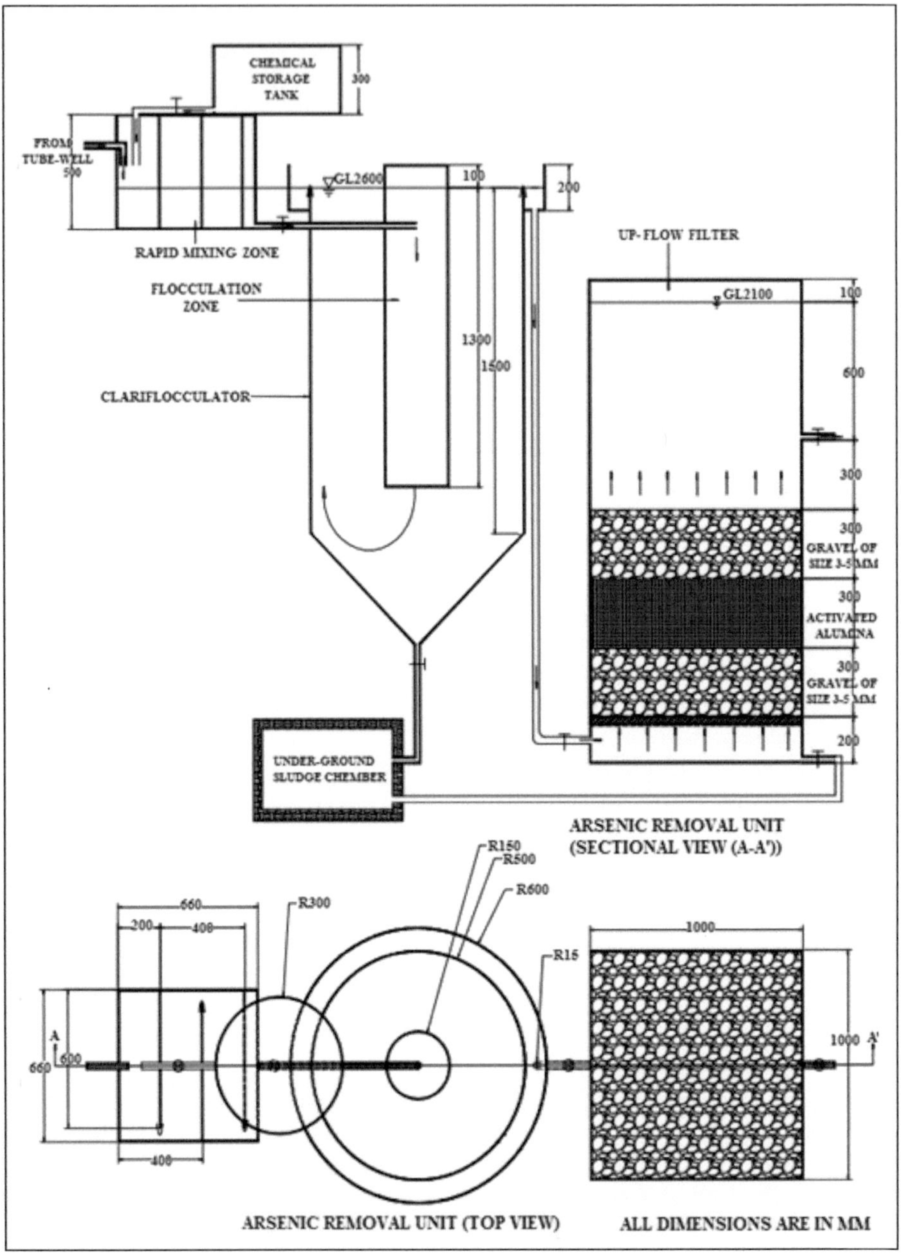

Fig. 16.4 Process diagram of two-stage arsenic removal unit of SWRE, Jadavpur University (DST Project)

Fig. 16.5 Arsenic removal unit operated with two-stage treatment at Malatipur village, Lalgola block, Murshidabad district, West Bengal of SWRE, Jadavpur University (DST Project)

(ii) Mixing of arsenic-rich sludge/grinded media with concrete in prescribed proportion.
(iii) Mixing of arsenic-rich sludge/grinded media with clay for preparation of brick and burning in kiln.

Environmental monitoring plan must be adopted around arsenic removal units during operational phase. This includes monitoring of arsenic in nearby surface water sources, sub-surface groundwater and soil.

2.4 *Performance of Arsenic and Iron Removal Plants Attached to Big-diameter Borewells: Case Study*

A study carried out by the All India Institute of Hygiene & Public Health, Government of India indicated that around 64% hand pump fitted wells ($n = 7680$) were contaminated by arsenic (>10 μg/L) in arsenic affected areas. In the contaminated tube-wells concentration of arsenic was mostly between 10 and 50 μg/L and sometimes even more. Based on this study Public Health Engineering Department (PHED), Government of West Bengal has implemented many surface water supply schemes based on the River Ganga water after conventional treatment to supply arsenic-free water in arsenic affected areas. Many more are in the process of execution. But in some affected areas, which cannot be covered under the existing or proposed surface water based piped water supply schemes, groundwater based piped water supply schemes were envisaged. Accordingly a good number of groundwater based mini piped water supply schemes have been implemented. Since groundwater in such areas contain

Fig. 16.6 Jasaikathi arsenic and iron removal plant

concentration of arsenic and iron beyond the desirable limit as per BIS 10500 (arsenic: >10 μg/L; iron: >300 μg/L), PHED has installed arsenic and iron removal plants (AIRP) to provide potable water to the village communities. The treatment plants have been installed by different agencies and are working under different principles and procedures with different operation and maintenances approaches. Mostly the implementing agencies are operating and maintaining the AIRP.

There are three AIRPs in North 24 Parganas District in West Bengal. They are (i) Jasaikathi Water Supply Scheme (Zone II) in Baduria Block; (ii) Bajitpur Water Supply Scheme in Baduria Block; and (iii) Kola Water Supply Scheme in Bagdah Block. A brief description of each of these AIRPs is given below.

2.4.1 Jasaikathi Water Supply Scheme

In Ramchandrapur village of Baduria block groundwater contains arsenic and iron which are more than the permissible limit prescribed by BIS. The arsenic concentration ranges between 37 and 80 μg/L. Since surface water is not available PHED implemented a groundwater based water supply scheme known as Jasaikathi Water Supply Scheme in February, 2015 to serve a population of 16315 (Fig. 16.6). This

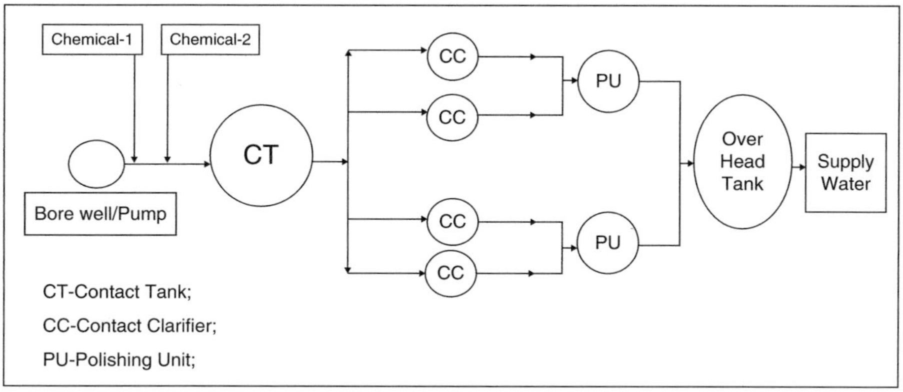

Fig. 16.7 Process flow diagram for arsenic and iron removal in Jasaikathi Water Supply Project. The plant was evaluated by School of Water Resources Engineering, Jadavpur University, and manufactured and installed by Odissi Innovation

water supply scheme has an AIRP attached to a borewell to remove arsenic and iron from the groundwater. The techniques used to remove arsenic and iron are chemical coagulation, precipitation, filtration and adsorption.

The process flow diagram of the plant is given in Fig. 16.7. The contact tank (CT) facilitates mixing of sodium aluminate and sodium sulphide with the raw water. The contact clarifier (CC) is packed with sand and gravel. Part of the arsenic is removed in the contact clarifier. Remaining part of arsenic is removed in the polishing unit (PU) where ferric hydroxide is packed to adsorb arsenic. Monochloramine is added for disinfection of water. Underground sludge tank allows storage of arsenic-rich sludge. The contact tank and contact clarifier is cleaned regularly (twice a week) by backwashing.

The treatment capacity of the plant is 110 m^3/hr and the plant is operated for 7 hr/day. After treatment the arsenic concentration reduces to <10 μg/L in 70% occasion. In the rest 30% occasion the concentration ranged between 10 and 17 μg/L. The iron concentration was reduced to <0.3 mg/L. This technology is suitable for low arsenic concentration in groundwater (<100 μg/L) but is doubtful for higher concentration of arsenic in raw water (>300 μg/L) in bringing down arsenic below 10 μg/L in treated water. Sludge management is also not satisfactory. Mixing of the sludge in nearby pond has been detected.

2.4.2 Bajitpur Water Supply Scheme

In Bazitpur village of Baduria block groundwater contains arsenic and iron which are more than the permissible limit prescribed by BIS. The arsenic concentration ranges between 300 and 400 μg/L. Since surface water is not available PHED implemented a groundwater based water supply scheme known as Bajitpur Water

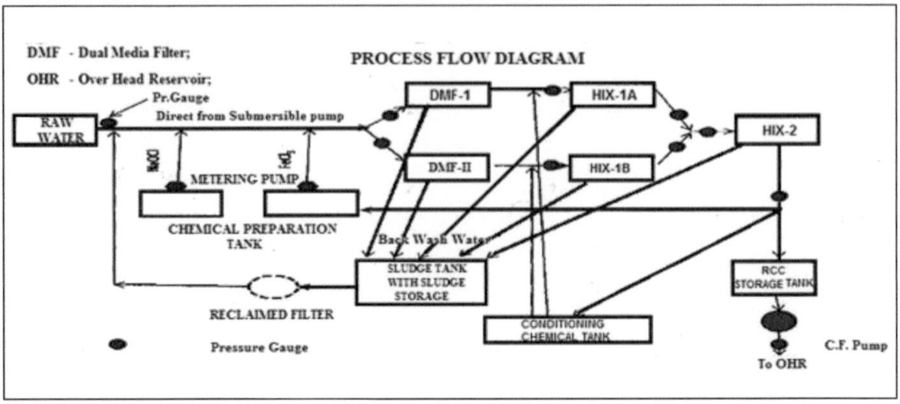

Fig. 16.8 Process flow diagram of arsenic and iron removal plant in Bajitpur Water Supply Project. The plant was evaluated by School of Water Resources Engineering, Jadavpur University, and manufactured and installed by Rites Water Solution (India) Pvt. Ltd

Supply Scheme in April, 2016 to serve a population of 16,335. This water supply scheme has an AIRP attached to a borewell to remove arsenic and iron from the groundwater. The AIRP is operated under the principle of chemical addition, oxidation, co-precipitation, dual-media filtration, two-stage ion-exchange for removal of arsenic and iron. The DMFs are required to be cleaned daily by backwashing. The back-washed water is re-circulated in the system for treatment. Media (ion-exchange) regeneration is done through chemical treatment as and when required. Sludge is stored in separate tank at AIRP site.

The process flow diagram of the plant is given in Fig. 16.8. The treatment capacity of the plant is 1464 m^3/day but the present operational rate is 91.5 m^3/hr and the plant is operated for 6 hr/day. The volume of water treated per day is 546 m^3. Performance of physico-chemical treatment is found to be satisfactory. Sodium hypo-cholorite, ferric chloride and HAIX (adsorbent) are showing encouraging result in arsenic removal process. After treatment the arsenic concentration reduces to <10 μg/L. Sludge management is not satisfactory.

2.4.3 Kola Water Supply Scheme

In Kola village of Bagdah block groundwater contains arsenic and iron which are more than the permissible limit prescribed by BIS. The arsenic concentration is <100 μg/L. Since surface water is not available PHED implemented a groundwater based water supply scheme known as Kola Water Supply Scheme in March, 2015 to serve a population of 2515 (Fig. 16.8). This water supply scheme has an AIRP attached to a borewell to remove arsenic and iron from the groundwater. The AIRP is operated under the principle of oxidation, flash mixing, coagulation and settling. The unit consists of oxidation chamber (OC), slow mixing flocculation tank (SMFT),

Fig. 16.9 Kola arsenic and iron removal plant

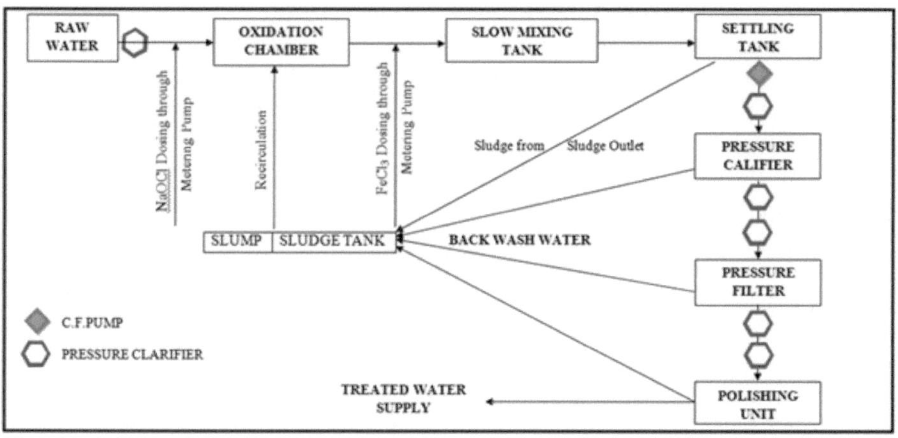

Fig. 16.10 Process flow diagram of arsenic and iron removal plant in Kola Water Supply Scheme. The plant was evaluated by School of Water Resources Engineering, and manufactured and installed by M/s Puspsa Enterprise

settling tank (ST), clarifier tank (CT), pressure filter (PF) and polishing unit (PU) (Fig. 16.8). Initially sodium hypochlorite is added for oxidation of arsenite. Ferric chloride as coagulant is added with flash mixing. The settling tank facilitates co-precipitation for removal of arsenic and iron. The polishing unit is packed with activated alumina which removes remaining arsenic from water by adsorption. The activated alumina will be regenerated as and when required. The AIRP also has sludge separation bed and sludge tank (Figs. 16.9 and 16.10). The treatment capacity of the plant is 240 m^3/day but the present operational rate is 30 m^3/hr and the plant is operated for 4 hr/day. The volume of water treated per day is 120 m^3. Performance of physico-chemical treatment is found to be satisfactory and the treated water contains <10 μg/L. As sludge management is not satisfactory.

2.5 Sustainability of Arsenic Removal Units

The feasibility of arsenic removal units is to be assessed before installation. The feasibility depends on existing basic water supply system in the area, amount of arsenic in groundwater and percentage of arsenic to be removed, level of managerial and technical capacity to install and maintain the arsenic removal units, community participation in managing the units, willingness and level of income to contribute towards O&M. Public Health Engineering Department installed around 2400 hand-pump attached arsenic removal units in arsenic affected habitations as per recommendation of All India Institute of Hygiene and Public Health based on study of different arsenic removal processes in Technology Park. But majority of hand-pump attached arsenic removal units did not function properly due to lack of participation of the community for O&M. It was observed that beneficiaries were not motivated for O&M of the units, lack of initiative for O&M from Gram Panchayet, funds were not collected from the beneficiaries for O&M, often units remained idle for want of minor repair, no-availability of media regeneration facility and sense of ownership of units was not generated amongst the beneficiaries.

But there are instances where arsenic removal units are functioning very well with community involvement for O&M. In each of these, users committee was formed at beginning with the leadership of women to look after the operation and regular maintenance of arsenic removal unit. Subscriptions were collected from the user-families and the money was kept in Post Office or Bank. The success story indicates that sustainability of arsenic removal units depend on how efficiently the units are managed by the community with affordable resources of the community.

AIRPs attached to big-diameter bore-well are operated and maintained by the agency that manufactured and installed the unit. Safe water is supplied through pipeline system via over-head reservoir. PHED is looking after the water supply scheme with support from the local Gram-Panchayet. Departmental monitoring and surveillance strategy has been developed towards sustainability of the AIRPs.

3 Fluoride in Groundwater

In India drinking water supply is provided from groundwater and surface water sources. The rural water supply is predominantly dependent on groundwater sources. The selections of aquifers are based on geo-hydrological conditions of the area. The aquifers below thick impervious clay layers are usually free from bacteriological contamination. Generally, groundwater contains higher dissolved solids than surface water. While recharging, rain water passing through soil grains are enriched with different dissolved solids (remaining as molecules or ionic forms). Higher concentrations of calcium and magnesium salt (causing hardness) and also iron are very common in groundwater sources. In coastal areas higher concentration chloride are found in aquifers. Fluoride is usually present in groundwater, but presence of fluoride beyond permissible level (1.5 mg/L) has been found in 20 states. In West Bengal, fluoride in groundwater has been detected beyond permissible level (1.5 mg/L) in 43 blocks of seven districts.

In India, fluorosis is considered to be one of the major environmental health problems. Though prevalence of fluorosis was detected more than 60 years ago, the progress of understanding fluorosis and its clinical manifestation was slow (Susheela 2001). Further rural water supply agencies (mostly Public Health Engineering Department) focused only on quantitative aspect with little emphasis on bacteriological quality and the aspect of chemical quality was ignored.

The risk of fluorosis in the country was found to be increasing rapidly in 70's and 80's. The medical professionals working with fluorosis diagnosis and treatment did not interact with water supply agencies and vice versa. The two groups worked in isolation and as a result suffering of the people were not properly attended. It was only in 1986 that National Drinking Water Mission, Government of India launched Mini-Mission Programme on control of fluorosis, removal of excess iron, desalination of water, eradication of guinea-worm, and conservation of water and recharging of groundwater (RGNDWM 1993).

In order to mitigate problem of fluorosis one has to ensure fluoride-safe water supply to the community. This involves selection of area specific appropriate methodology for fluoride- safe water supply as well as sustainability of the programme through active participation of the user community.

During the last 25 years with active support of Central Government, various State Governments have taken up action programmes for mitigation of excess fluoride problem in drinking water. UNICEF, WHO, UNDP etc. have also stretched their hands to supplement governmental efforts. Various non-governmental organizations are also in the fray to mitigate fluoride problem.

3.1 Extent and Magnitude of the Problem

All groundwater contain varying concentrations of fluoride due to universal presence of fluorides in earth's crust. However, there can be major differences within a relatively small area and different depths of boreholes. Fluoride in untreated groundwater is one of the most chronic toxic substances which have affected a considerable number of rural populations causing dental, skeletal and non-skeletal fluorosis. Contamination of groundwater with fluoride is severe in Rajasthan, Andhra Pradesh, Gujarat, Haryana, Karnataka and Punjab. In addition, such problems also exist in certain areas of Madhya Pradesh, Chhattisgarh, Maharashtra, Orissa, Tamil Nadu,

Uttar Pradesh, Kerala, Bihar, Jharkand, Delhi and Jammu & Kashmir. In 90s fluoride concentrations beyond permissible limit in groundwater have been reported from Assam and West Bengal. In affected areas, majority of fluoride concentration varies between 1.5 and 8 mg/L. However, fluoride concentrations in much higher proportion have also been reported from many areas (10–18 mg/L). In India, 200 districts in 20 states are endemic to fluorosis due to excess fluoride in drinking water. It has been estimated that around 66 million rural populations are residing in excess fluoride prone risk areas of the country (Susheela 2001).

3.2 Action Plan to Mitigate Excess Fluoride Problem

A holistic approach should be taken to mitigate fluoride problem in the country. The following action plan is suggested in excess fluoride affected areas (Majumder 2004):

(i) Supply of fluoride-safe water to the community
(ii) Extensive water quality monitoring
(iii) Diagnosis and treatment of fluorosis affected people
(iv) Extending nutritional support
(v) Awareness, motivation and training

3.3 Alternate Fluoride-Safe Water Supply System

There would be various systems for supply of fluoride-safe water to the community. Such water supply system must be appropriate to suit area specific condition. The alternate fluoride-safe water supply system could be as follows (Majumder 2004):

(i) River water based piped water supply
(ii) Big-diameter tube well based piped water supply
(iii) Installation of deep tube well
(iv) Handpump attached excess fluoride removal unit
(v) Excess fluoride removal unit attached with big-diameter tubewell for piped water supply
(vi) Use of traditional water sources pond/lake water after treatment
(vii) Rain water harvesting
(viii) Household treatment for excess fluoride removal

3.3.1 River Water Based Piped Water Supply

River water usually does not contain excess fluoride. This water can be treated by conventional process (coagulation-flocculation-sedimentation-filtration-disinfection) and can be supplied in the excess fluoride affected villages through piped network system. This alternative water supply system has following advantages and constraints:

(i) Permanent/long term solution
(ii) Very expensive (capital cost)
(iii) Execution process lengthy
(iv) Difficulty in operation and maintenance.

3.3.2 Big Diameter Tubewell Based Piped Water Supply

Water of aquifers free from excess fluoride can be drawn through big-diameter tubewell and supplied through piped network system. Such type of tube wells need to be installed with required skilled by providing proper sealing avoiding risk of contamination of excess fluoride through vertical leaching. The following are the advantages and constraints:

(i) Comparatively easier to execute
(ii) Comparatively less expensive than river water based piped water supply
(iii) Operation and maintenance require skill
(iv) Risk of contamination (vertical leaching)

3.3.3 Installation of Deep Tubewell

Deep tube well for tapping excess fluoride safe aquifer can be installed as a spot source for supplying water. Installation of deep tube well requires meticulous skill for sealing against vertical leaching. Following are the advantages and constraints:

(i) Easier to execute and operate
(ii) Less expensive
(iii) Skilled driller required to seal contaminated aquifer zone
(iv) Risk of contamination
(v) Appearance of excess fluoride after a considerable time

3.3.4 Handpump Attached Excess Fluoride Removal Unit

The other alternative to supply fluoride safe water is by adopting defluoridation techniques. The standard methods for fluoride removal are:

(i) Co-precipitation
(ii) Adsorption
(iii) Ion-exchange
(iv) Reverse Osmosis

The three important factors that are to be considered for implementing defluoridation technique are (i) simplicity of application, (ii) applicability in case of small water supplies, and (iii) cost effectiveness.

In Nalgonda technique, co-precipitation principle is adopted for removal of excess fluoride. Nalgonda technique is a combination of several unit operations and the process incorporates rapid mixing, chemical interaction, flocculation, sedimentation, filtration and disinfection. Lime, bleaching and alum are added in this process for defluoridation. The cost of the process is very cheap but the process is not user-friendly. The process can be installed with handpump attached excess fluoride removal model or domestic model. Such units are usually non-mechanized and no electricity is required to run the unit.

Activated alumina can be used as a media for adsorption of fluoride. Capacity of activated alumina depends upon the basicity of water and decreases considerably with increasing basicity. Following information are essential for manufacturing of defluoridation unit with activated alumina:

- Operating characteristics
  - (i) Capacity curves
  - (ii) Exhaustion curves
  - (iii) Exchange capacity and corresponding regenerating chemicals requirement with concentration
  - (iv) Number of times regeneration of the adsorbent should be considered
  - (v) Back wash and rinse water quantity and rate of application
  - (vi) Loading rate
  - (vii) Attrition losses and replacement requirement
  - (viii) Anticipated cycles of operation
  - (ix) Impact of variation in Ca, Mg, Sodium bicarbonate in groundwater
- Waste water (wash) disposal
  - (i) Activated alumina regenerated waste water quantity and quality
  - (ii) Method and cost of effluent treatment and/or disposal
- Cost aspect
  - (i) Cost of activated alumina
  - (ii) Capital cost of the plant for various treatment rates
  - (iii) Life expectancy of plant and activated alumina
  - (iv) Operation and maintenance requirement and cost

Appropriate treatment and safe sludge disposal technique need to be adopted if defluoridation techniques are used by the community. The hand pump attaches unit needs to be operated and maintained by the community. A community group is to be formed for each of the unit for operation and maintenance. A subscription will be required from the user family to maintain the plant.

3.3.5 Excess Fluoride Removal Unit Attached with Big Diameter Tubewell for Piped Water Supply

Excess fluoride removal unit working with the principle of either co-precipitation or adsorption or both can be installed with big diameter tubewell for piped water supply. Such unit however needs to be operated and maintained meticulously by skilled operators and technicians.

3.3.6 Use of Traditional Water Sources (Pond and Lake Water) After Treatment

Traditional water sources like lake, pond etc is free from excess fluoride. The water can be upgraded by using horizontal roughing filter (HRF) – slow sand filter (SSF). Such unit can be maintained by the community at an affordable cost. The following are the advantages and constraints:

(i) Availability of sufficient quantum of water
(ii) The ponds should not be used for bathing and pisiculture
(iii) The pond must be free from external pollution
(iv) Sound technology available for up-gradation of water quality
(v) Protection of source difficult
(vi) Disinfection is necessary after treatment of water; ensuring regular disinfection often becomes difficult
(vii) Community response sometimes not positive
(viii) Total community participation needs to be ensured for the success of the program

3.3.7 Rain Water Harvesting

Rain water harvesting by roof top collection can provide fluoride safe water. However, sufficient quantity of water needs to be stored for use during dry period. The water is to be used for drinking and cooking only.

Rain water can also be stored in impounding reservoir or pond in village. However, such water needs up-gradation by installing HRF-SSF. The following are the advantages and constraints:

(i) Technology available
(ii) Strong awareness and motivation are required
(iii) Community based schemes are to be ensured
(iv) For roof top collection individual family members are to be motivated
(v) Expensive for individual household

3.3.8 Household Treatment for Excessive Fluoride Removal

Both co-precipitation and adsorption techniques can be used for household treatment unit. The cost of the unit must be affordable to the people. The co-precipitation technique is cheaper than the adsorption technique. The following are the advantages and constraints:

(i) Easy to maintain and handle
(ii) Different models are available
(iii) Efficient in removal of excess fluoride
(iv) Availability of chemical often becomes difficult in the villages
(v) Sludge disposal strictly to be followed

3.4 Case Study

A fluoride removal unit (installed jointly by WIST Inc. US and Rite Solutions Pvt. Ltd) is functioning at Laxmi Narayanpur A. J. Adibasi High School, Nalhati, Birbhum, West Bengal. Public Health Engineering Department, Government of West Bengal, had arranged for the site for demonstrating the technology for excess fluoride removal from groundwater as per the recommendation of the Fluoride Task Force, Government of West Bengal.

3.4.1 De-Fluoridation Unit: Technological Aspects

The de-fluoridation unit has been developed primarily focusing selective removal of fluoride through effective sorbent-based treatment technology. HIX–NanoZr in which nano particles of zirconium oxide are loaded within a polymeric support of ion-exchange resin offers selective fluoride removal from a background of other competing ions. The schematic diagram of fluoride removal unit installed at school is presented in Fig. 16.11. The sequence of water treatment may be as follows:

(i) Column 1 with appropriate ion-exchange (cations) materials for partial desalination and pH adjustment/reduction
(ii) Column 2 with HIX-NanoZr for fluoride removal
(iii) Column-3 with specific ion-exchange materials for passive pH control
(iv) Column-4 with granular activated carbon (GAC) media as a polishing unit
(v) Additional arrangement for application of UV irradiation for disinfection of treated water

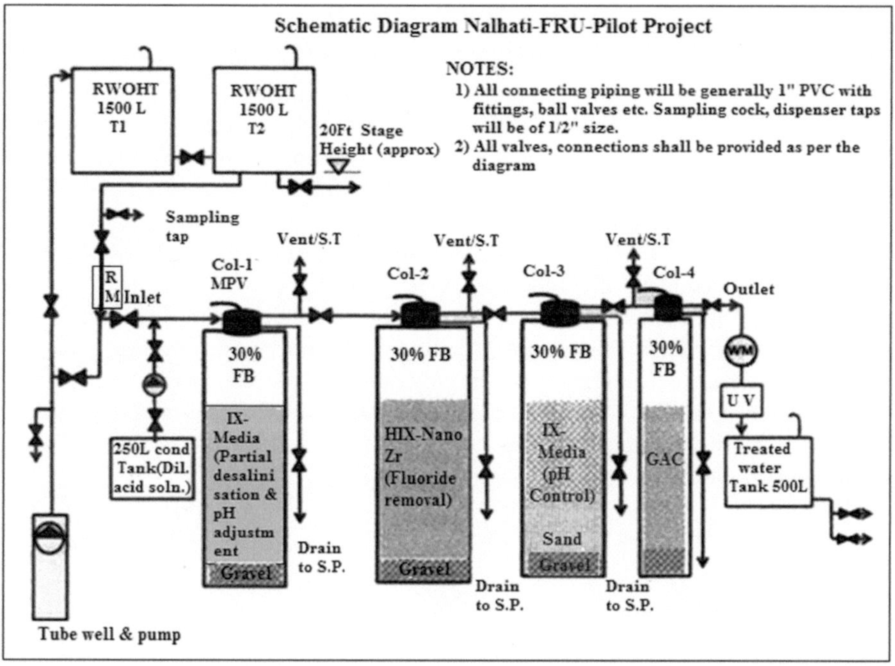

Fig. 16.11 A schematic diagram for the HIX-NanoZr fluoride removal plant installed at Nalhati, Birbhum

3.4.2 Operation

During de-fluoridation unit operation, column 1 and column 2 will need reconditioning. Ion-exchange media in column 1 needs reconditioning periodically depending on quantity and quality of groundwater. On an average, reconditioning interval may be seven days (weekly). The reconditioning is done by passing dilute hydrochloric acid. Reconditioning of HIX-NanoZr media in column 2 will be done at six monthly intervals (depends on raw water quality). Alkaline solution will be required for reconditioning of media in column 2 followed by rinsing with low pH/acid water.

In addition to above, regular back washing is required for column 1 (daily), column 2 (alternate day) and column 3 (twice a week).

The treatment capacity of the installed de-fluoridation unit is 500 L/h. Initially a 3000 L capacity tank is filled up with excess fluoride contaminated groundwater by pumping. Thereafter, the unit is operated in continuous mode till the raw water tank gets exhausted. At the most the de-fluoridation unit can treat 5000 L water daily. At present 400 students of the school are provided with the treated water. The treated water is also used for preparation of mid-day meal. The installed water-meter reading indicated that so far 57,500 L water has been treated.

3.5 Sustainability of Defluoridation Units

The technology to be adopted for removal of fluoride must be appropriate to actual groundwater quality of the site where the defluoridation unit will be installed. Presence of certain parameters on higher concentration or wide variation of pH may interfere with the operation of the unit and accordingly removal efficiency may vary. It is normally recommended to examine the groundwater quality before the installation of the defluoridation units. If necessary, treatability test may be carried out and modification in the model may be considered before installation. The manufacturer must assess the running time or bed-volume for the unit before regeneration of media or change of media. Regular surveillance as well as evaluation of the performance of the units will certainly help in achieving sustainability of the defluoridation units for the benefit of the rural community. Economic viability and social acceptability are the two other key issues to be considered for the success of defluoridation units.

4 Conclusions

Very many technologies have been developed for removal of arsenic and excess fluoride from contaminated groundwater. But in most of the cases treatability studies for different technologies at varying groundwater quality have not been carried out and as a result performance of different arsenic and excess fluoride removal units in the field condition are varying widely. Chemical dosages, quantity of adsorbents and quantity of ion-exchangers are not considered at field condition by the manufactures resulting in wide variation in the expected time for regeneration of the media. Again interference of certain water quality parameters present in excess in groundwater is posing problem in the contaminant removal process. Sludge management is often found to be poor in many water treatment plants. Thus, water quality monitoring and surveillance along with performance evaluation of arsenic and excess fluoride units should be planned and followed to achieve sustainability of the units.

References

Ananthanarayanan, P.H. and Majumder, A. (2003). State of Art on Arsenic Contamination of Groundwater in India. All India Institute of Hygiene & Public Health.

Majumder, A. (2001). Arsenic in groundwater: Technical options and methodology for removal. National Workshop on Drinking Water Quality Surveillance, Nov. 6-8, Bangalore (Organized by CPHEEO, Ministry of Urban Development, Govt. of India).

Majumder, A. (2004). Fluoride-safe Water Supply. National Workshop on Control & Mitigation of Excess Fluoride in Drinking Water, Jaipur.

Majumder, A. (2014). Sustainable Arsenic Removal System—A Challenging Task. Arsenic in Groundwater: Complexities and Challenges Ahead in West Bengal. Silpa Nagari Prakashani.

RGNDWM (Rajiv Gandhi National Drinking Water Mission) (1993). Water Quality & Defluoridation Techniques. Vol. II.

Susheela, A.K. (2001). Treatise on Fluorosis. Fluorosis Research and Rural Development Foundation, India.

WHO (World Health Organization) (1996). Guidelines for Drinking Water Quality, 2nd Edition. Volume 2.

Chapter 17
Impact of Climatic Stress on Groundwater Resources in the Coming Decades Over South Asia

Rajib Chattopadhyay, Surajit Chakraborty, and Atul K. Sahai

1 Introduction: The Problem of Climate Change in the Perspective of Groundwater Conservation

Adequate availability of groundwater resource is an indispensable component for the sustainable growth of society and human civilization. Groundwater is the largest source of accessible freshwater resources and as per estimates made, it supplies drinking water to ~36% of world population and ~42% of water used for irrigation (Taylor et al. 2013). According to recent estimates by various surveys (e.g. Shiklomanov 1993), of the 1.4 billion cubic km water available on earth, 2.5% of it is available as freshwater. Of all the freshwater available, 10 million cubic km is estimated to be groundwater which is essential for food security for both developing and developed countries. According to International Panel on Climate Change (IPCC) reports documented since last 20 years (IPCC Report 1996, 1998, 2001, 2008, 2014), the impact of climate change on water cycle including groundwater as an indispensable component is largely uncertain. Very few studies have taken the matter in a holistic way that includes groundwater as an explicit variable. Hence the question: "*What is the impact of climate change on groundwater resource*?" is yet to be quantified for societal impact studies e.g. chapter 3 of IPCC Report (2014). Any natural resource can be extracted for continuous usage only when the supply is replenished adequately. Hence it is undeniable that extensive usage of groundwater for drinking, agricultural as well as industrial use is sustainable only if the recharge of groundwater is maintained at a constant pace.

R. Chattopadhyay (✉) · A. K. Sahai
Indian Institute of Tropical Meteorology, Pune, India
e-mail: rajib@tropmet.res.in; sahai@tropmet.res.in

S. Chakraborty
Department of Environment Management, Indian Institute of Social Welfare and Business Management, Kolkata, India

P. K. Sikdar (ed.), *Groundwater Development and Management*,
https://doi.org/10.1007/978-3-319-75115-3_17

The natural recharge of groundwater and its sustenance is a complicated process involving many natural processes on multiple time scales (Balek 1988; Kumar 1997). The net groundwater level could be maintained in equilibrium for a given period when the recharge, discharge, and variable boundary conditions depending on surface soil characteristics, natural vegetation, river network and subsurface geology (lithology) of the region are in balance with each other. Given a source, these entire components would play a role in recharging an aquifer. The precipitation on the surface defines the major climatic component that acts as a source of freshwater that is available for recharge. The rainwater or the snow can infiltrate through the surface and serve as a steady source. The precipitation as rain is the major component of water cycle reaching the surface in the tropics especially in the Indian subcontinent and South Asian countries. Once rainwater reaches the surface, the recharge flow below the land surface takes place due to the process of soil infiltration and percolation. The soil will not get completely saturated with water unless the surface rainwater supply is sustained for prolonged time intervals. Thus, if water is applied only intermittently, there may be no recharge possible. Considering the monsoon climate over the Indian subcontinent and similar regions where rainfall takes place for a significant number of days during the rainy season, the constant rainwater supply can act as a primary source of groundwater.

One of the most direct results of climate change will be the increased variability of the global water cycle related to climate feedbacks. This directly will impact on the quantity and quality of regional water resources (Gleick and Adams 2000; Xu 2000; IPCC Report 2014 (ch. 3)). Such regional modulations of hydrological cycles are primarily related to variability in precipitation where rainfall becomes high intensity with a shorter irregular period; it is bound to impact the recharge process of local groundwater resources which is extensively sensitive to the duration of availability of surface water source.

The surface water source, in addition to precipitation, is dependent on other meteorological parameters such as evapotranspiration and hydrological parameters like river run-off, etc. It is expected that the strategic importance of groundwater for global water and food security will probably intensify under climate change as even conservative model projections show more frequent and intense climate extremes (droughts and floods) associated with increased variability in precipitation, soil moisture and surface water.

This chapter will highlight the relation between the climatic variability associated with climate change and the future uncertainty of groundwater recharge process assuming a large climate change scenario in the next century (or more precisely the *representative concentration pathways* (RCP) 8.5 scenario as defined in IPCC reports). Climatic variables are forced by the greenhouse gases which can be easily understood based on Planck's blackbody radiation law. The net input solar radiation on the earth's surface is absorbed and redistributed by land and atmosphere. The atmospheric redistribution of the radiation depends on the gases like oxygen, nitrogen, carbon-dioxide, water vapour, methane, etc. Some of these gases (e.g. carbondioxide, methane, water vapour) have excellent ability to absorb and redistribute the radiations which are called greenhouse gases. The heat/radiation thus released can act as an additional heat source to earth's atmosphere and are known to have the ability to alter the circulation, precipitation, etc. The RCPs are formulated to generate the estimated radiative stress scenarios that the future greenhouse gas concentration (due to anthropogenic activity) would impart on the climate system. The four standard RCPs are defined: RCP2.6, RCP4.5, RCP6 and RCP8.5. They are named after a possible range of radiative stress values in the year 2100 relative to pre-industrial values (+2.6, +4.5, +6.0 and +8.5 W/m^2, respectively). The RCP8.5 scenario is the strongest one that is assumed in this study.

These RCP scenarios are used to force the climate models or more generally, the earth system models (ESM) of which response of climate is an integral part. ESMs and climate models are of various types, and its use is common in several fields of meteorology and climatology. The next section would give a brief overview of earth system models that is used in climate change studies.

2 Earth System Models: Working Principle to Generate Climate/Earth System Simulations

The earth system modelling for climate change studies has two components: (a) Climate system and (b) Human/Anthropogenic system. The two elements together are called an earth system modelling (ESM) framework. The ESM framework is shown in Fig. 17.1. Human component forces the climate part of ESM. Climate change scenario modelling framework such as CMIP5 framework as used here computes the response of climate component as a function of the anthropogenic component. In doing so, an estimate of anthropogenic forcing is made (such as RCP8.5 scenario) for the future period. To generate a future period view of the climate state considering it as a dynamical system (i.e. it follows Newton's second law), meteorologists often rely on results from climate models. Climate variables such as winds (horizontal and vertical), temperature, precipitation, humidity of the atmosphere, etc. are simulated through prognostic mathematical models. These prognostic models are essentially a set of fluid-dynamic equations (more specifically, the Navier-Stokes equation) which are solved numerically using several added assumptions and statistical relations (known as parameterization). Regarding logic flow, it may be described as an input-output system with feedback loops. These feedback loops are referred to as coupling of two components (e.g. Ocean and Atmosphere, Land and Atmosphere, etc.). Considering, all parts of the earth e.g. land, ocean, atmosphere, biosphere, greenhouse gases, clouds, aerosols in the atmosphere, solar forcing, etc. can be coupled to each other via feedback coupling of one component with another forming a complex network of coupled equations solved at a gridded structure. The final solution would depend on the dynamical forcing components and feedback between the components. A set of dynamical and thermodynamical forcing to an input state would lead to a new finite climatic state where output climate variables would give the effect of forcing and feedback. It is

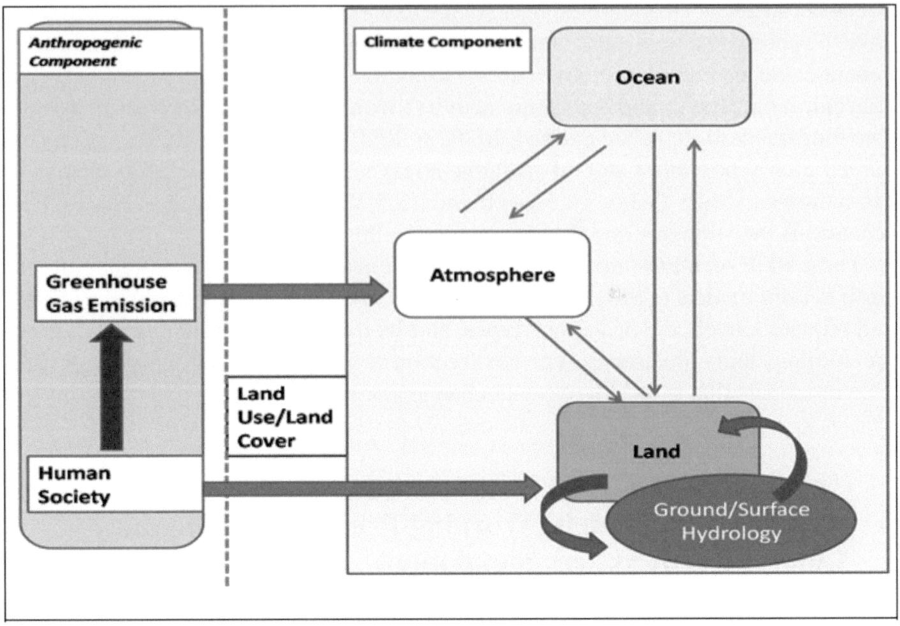

Fig. 17.1 Schematic diagram of an Earth System Model

easy to follow that the radiative forcing imparted by the future concentration of greenhouse gases, in the same way, would be able to alter the climate output from a state when there is no such forcing. The effect of this new altered state in response to radiative forcing could be studied to understand the effect of greenhouse radiative forcing which could be further used for computation of groundwater feedback. This is systematically done using several methods. The next section would discuss the state-of-the-art framework adopted by IPCC.

3 Standard Practice to Study the Impact of Climate Change: The Coupled Model Inter-comparison Project (CMIP5)

It was discussed earlier that the modelling studies are required to estimate the net stress on any natural resources due to increase in irradiative stresses related to climate changes. Since climate variables like precipitation, evapotranspiration (evaporation from soil surface and transpiration from leaves of vegetation cover over a particular area) and the river runoff are the primary variables that take part in groundwater recharge, it is important to have an intelligent guess and scenario-based projections on how these variables are affected by climate change through a scientifically consistent framework. The Coupled Model Intercomparison Project5

(CMIP5) is one such framework (Taylor et al. 2011) for numerically simulating the earth's climate through a set of dynamical equations which is solved through numerical coding of fluid dynamic equations, commonly referred to as the *ocean-atmosphere coupled general circulation models* (CGCMs). Ocean-atmosphere coupled models are the latest state-of-art tool which gives realistic estimation/simulation of the climate variables. CMIP began in 1995 under the auspices of the Working Group on Coupled Modeling (WGCM), which is in turn under auspices of CLIVAR (http://www.clivar.org/) and the Joint Scientific Committee for the World Climate Research Program (WCRP). Several leading climate modelling institutes participated in CMIP5 comparison which follows a consistent framework to run the model. The *Lawrence Livermore National Laboratory, US (http://cmip-pcmdi.llnl.gov/cmip5/)* is the nodal agency that collects output from the participating CGCMs and stores it for public use. Among various purposes, such models are employed both to detect anthropogenic effects in the climate record of the past century until near past (1850 A.D. i.e., pre-industrial era) to present and to project future climatic changes due to the anthropogenic production of greenhouse gases and aerosols in the coming century (present to 2100 A.D.).

To study the impact of climate change, the CMIP5 framework has archived output from both constant radiative forcing ("*control runs*") and perturbed (e.g. 1% per year increasing atmospheric carbon dioxide) simulations ("*perturbed runs*"). The estimate of additional stress can be made through the difference term: $\mathbf{\Delta}$ = *perturbed run–control run.* The current study will use this $\mathbf{\Delta}$ measure as an estimate of change for any climatic variables in the perturbed scenario (also known as *bias*). For climate change scenarios which use RCP8.5 radiative forcing for the coming century (present-to-2100), we will term it as ***F*** runs while the run representing the past to present scenario we shall term it as *Historical or **H** runs.* Thus, $\mathbf{\Delta} = \mathbf{F} - \mathbf{H}$ is the estimate of change of a climate variable under the RCP8.5 scenario used in this chapter. The $\mathbf{\Delta}$ could be utilized for understanding on how the process of groundwater recharge would be affected by climate change. This is discussed in next section.

4 Climate Stress and Groundwater Recharge

4.1 Groundwater Recharge and Resources

Various types of aquifers will be recharged in different ways. The main types of aquifers are unconfined and confined. An unconfined aquifer is directly recharged by local rainfall, and other sources like rivers and lakes. The rate of recharge will be primarily influenced by the permeability of overlying rocks and soils. Recharge also occurs where the underlying geology is highly fractured or is characterized by numerous sinkholes. Such recharge can be critical in some semi-arid areas. In principle, "rapid" recharge can occur in any region with favourable geology whenever it rains, and where recharge is limited by this process, it will be affected more by changes in the quantum of rainfall amount than by the seasonal cycle of soil moisture variability. Superficial unconfined aquifers along floodplains, which are most common in semi-arid and arid environments, are almost always recharged by annual streamflows. It can be depleted directly by the evaporation process. Changes in recharge, therefore, will be determined by changes in the duration of flow of these streams, and the permeability of the overlying beds, but increased evaporative demands would tend to lower groundwater storage. The thick layer of sands significantly reduces the impact of evaporation.

It will be noted from the preceding discussion that unconfined aquifers are sensitive to local climate change, abstraction, and seawater intrusion. However, the quantification of recharge is complicated by several factors including the characteristics of the aquifers themselves as well as overlying rocks and soils. A confined aquifer, on the other hand, is described by an overlying bed that is impermeable, and local rainfall does not control the recharge properties of these aquifers. It is regularly recharged from water bodies like lakes, rivers and rainfall that may take place at spatial distances ranging from a few kilometres to thousands of kilometres. Beside from the control of climate, recharge to aquifers is very much reliant on the characteristics of the aquifer media and the properties of the overlying soils. Several empirical approaches are available to estimate recharge based on surface water, unsaturated zone, and groundwater data. Among all these approaches, numerical modelling is the only tool that can predict recharge. Modelling study is also particularly useful for identifying the relative importance of different controls on recharge, provided that the model realistically accounts for all the processes involved. However, the accuracy of recharge estimates are largely based on the availability of high quality hydrogeologic and climatic data which is often not available.

To determine the potential impact of climate change on groundwater resources, in particular, is difficult due to the sparse data locations and the incomplete understanding of the intricacy of the recharge process, and the discrepancy of recharge within and between different climatic zones. Earlier, several attempts have been made to estimate the rate of recharge by using carbon-14 isotopes and other modelling techniques. It is possible to some extent for aquifers that are recharged from short spatial distances and after short temporal durations. However, recharge that takes place from long distances and after decades or centuries has been challenging to calculate with accuracy, making an estimation of the impacts of climate change involved. The medium through which recharge takes place often is poorly known and is very heterogeneous. This is again a challenging problem in recharge modelling. In general, there is a need to intensify research on modelling of aquifer characteristics, recharge rates, and seawater intrusion, as well as observation of groundwater abstractions.

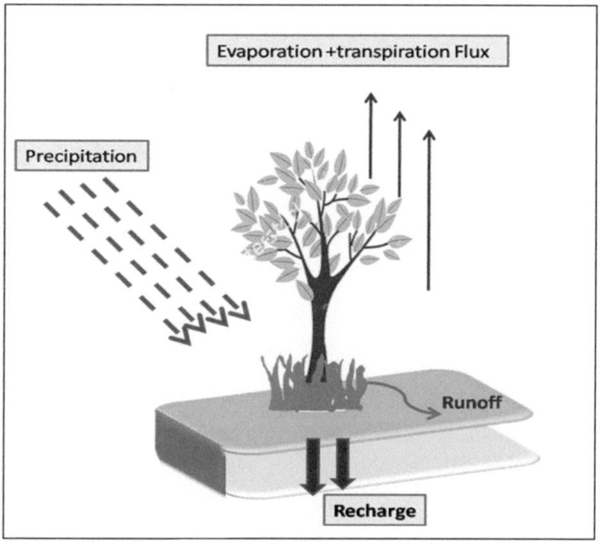

Fig. 17.2 Schematic diagram of storage in surface and water available for groundwater recharge

4.2 Climate Sensitivity to Groundwater Recharge

The recharge of freshwater aquifers is possible through surface recharge, which connects the water cycle to the groundwater storage (refer Fig. 17.2). The groundwater recharge may be explained by a simple equation that connects the climatic and hydrological variable with the net surface source or storage term that act as a source of groundwater recharge:

$$\Delta S = \Delta P - \Delta E - \Delta R$$

where for any time interval ΔT, ΔS is the change in net surface source or storage, ΔP is the net accumulated precipitation at the surface, ΔE is the net change in evapotranspiration and ΔR is the net change in river runoff. It is clear that for any sustainable groundwater use, the net storage or the source (ΔS) has to be stable.

In the backdrop of global warming, several studies have estimated a substantial variability or uncertainty in the precipitation process. The precipitation process is dependent on atmospheric moisture holding capacity. With the warming atmosphere, the moisture holding capacity of atmosphere increases (~7% per °C increase in temperature). The atmosphere adjusts to a newer equilibrium with change (or increase) in temperature due to change in atmospheric radiative forcing associated with altered greenhouse gas concentration as discussed in the previous section. What is the expected change in precipitation in a warming scenario? Although answers differ for different regions and different climate systems, a general agreement is that the frequency and intensity of extreme precipitation events show an increase in several climatic regions while the low intensity and moderate rainfall events would decrease in the near future. A significant change in monsoon

Table 17.1 Agriculture and GDP in the countries of South Asia

Country	Agriculture contribution to GDP (%)	Rural population (%)	Labour force employed in agriculture (%)	Agricultural Area (%)	Irrigated area (%)
Afghanistan	31.6	77	70.0	58	3.4
Bangladesh	18.6	72	48.0	65	35.1
Bhutan	17.4	65	59.4	15	1.0
India	19.0	70	56.0	55	18.9
Maldives	5.6	60	12.0	30	–
Nepal	32.8	81	66.0	30	8.0
Pakistan	21.2	64	45.0	33	25.0
Sri Lanka	12.8	85	33.0	40	8.9

precipitation pattern with an increase in extreme precipitation is observed by using data of last 50–60 years. Also, another important observation is the wet place to go wetter and dry place to go drier over different climatic zones.

Since the storage term (ΔS) is easily seen to be sensitive to the precipitation mean and its variability over time, its recharge potential to aquifers would also get affected in a global warming scenario. For example, an increase in extreme events implies pouring of a significant amount of rainfall in a short time interval which goes away as a surface runoff instead of recharging the groundwater reserve. If the drier region gets less rainfall, it would add to the stress in groundwater reserve, which is used for agricultural or industrial activity in this area.

5 Societal Impact of Precipitation Variability and Water Storage: A Case Diary for South Asia

In all the countries in South Asia, it is estimated that the percentage of surface and groundwater withdrawal as a proportion of internal renewable water resources is 58% (UNEP 2008). Therefore, this statistics is close enough to show that the countries of South Asia are near the threshold level of 60% which is defined as "water scarcity is approaching" situation. Under stress, it could easily touch the threshold of 75%, above which the sustainable limits of water withdrawal are exceeded. Agriculture contributes a significant part of India's GDP. The same is the case for all the countries in South Asia (refer Table 17.1; adopted from IGES-GWP 2012). It is the livelihood of a vast number of people in this region where agriculture is the chief source of income for the rural population. The farm activity over a vast region of India is strongly dependent on either monsoon rainfall or groundwater reserves. In either case, the precipitation plays a significant role in maintaining and sustaining the water availability for agricultural activity, industry and drinking water supply to cities and villages. So a substantial variability in precipitation in the form of drought and floods over Indian region and most countries

Table 17.2 Climate, population and water resources

Country	Area (km^2)	Population (millions)	Population density (per km^2)	GDP/ Capita US$ (2009)	Adult literacy 2004	Annual per capita water availability (m^3)	Area covered by forests (%)
Afghanistan	652,230	26.1	40	1352	28.0[a]	2709	2.1
Bangladesh	144,000	146.2	1015	1593	43.0[a]	8370	11.1
Bhutan	38,390	0.7	18	5167	–	114,134	69.1
India	3,287,260	1182.1	360	3266	61.0	1603	23.0
Maldives	300	0.3	1000	6730	96.3	95	3.3
Nepal	147,180	28.3	192	1261	48.6	7642	25.4
Pakistan	796,100	166.5	209	2700	49.9	1064	2.2
Sri Lanka	65,610	20.7	316	4747	90.7	2200	28.8

[a]For 2000–2004

of South Asia in general, which has a similar economic dependence on irrigation, agricultural resources and industrial activity, would impact the whole socio-economic process and increase the uncertainty and vulnerability. Several studies link this aspect in clear way (IGES-GWP 2012; refer Table 17.2 adopted from IGES-GWP 2012).

However, the studies identifying the impact on groundwater recharge in different climate change scenarios and projections are still unclear. Especially, studies cannot quantitatively specify to what extent the stress and strains on the available groundwater resources in a warming scenario would be affecting the society in the coming century. The withdrawal estimate of groundwater over Indian region is shown in Fig. 17.3a–c as a percentage of recharge (Ministry of Water Resources 2006). It is evident from the figure that a significant fraction of withdrawals are occurring over north-western states of India while most of the places of north India shows more than 50% withdrawal rate. In the scenario of prolonged drought-like situations or higher variable scenario of precipitation, the withdrawals are likely to increase. The next section would show some results from climate models which will highlight the uncertainty of the groundwater recharge in the climate change scenario.

6 Examples of Future Projections and Uncertainties: Net Surface Stress Over Indian Region

In this section, a brief overview of future projections would be presented. The "*future scenario*" (*F runs*) is generated from two models: the MIROC5 model (Watanabe et al. 2010) and the MPI earth system model, which have participated in the CMIP5 evaluations. We select the RCP8.5 scenario in which the maximum radiative forcing of 8.5 W/m^2 is assumed due to highest increase in greenhouse gas concentration. Such extreme scenarios are helpful to assess the impact in a stress-

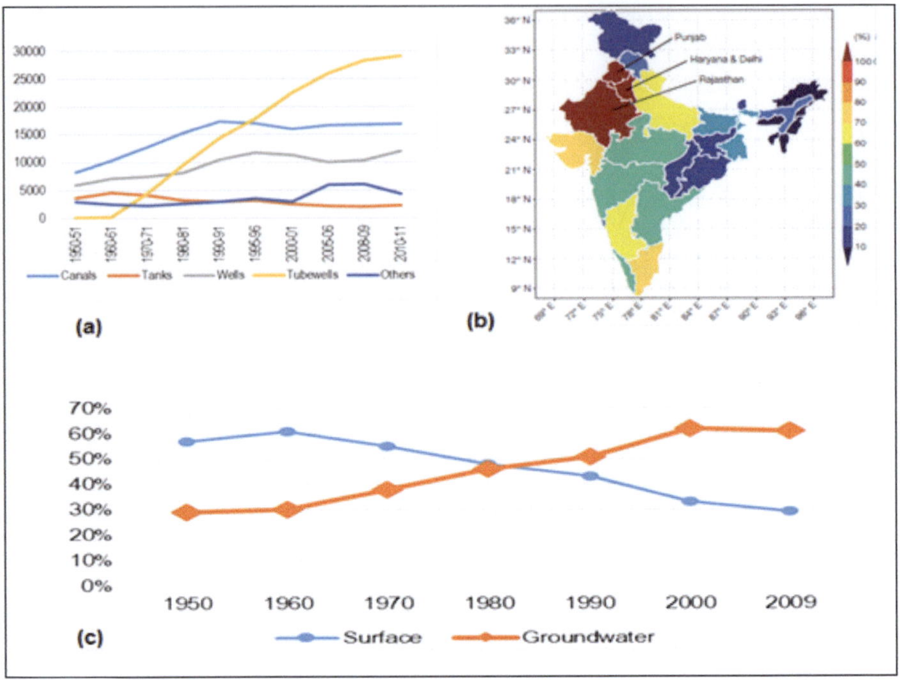

Fig. 17.3 (**a**) Sources of irrigated land in India in '000 hectares. (Agricultural Statistics at Glance 2014; Ministry of Agriculture 2014) (**b**) Groundwater withdrawals as a percentage of recharge. (**c**) Increase in groundwater in agriculture as compared with surface water. (Sources: Agricultural Statistics at Glance 2014, Ministry of Agriculture). Fig (**a**) and (**c**) directly adapted from Figs. 17.4 and 17.5 of: "Overview of ground water in India" by Roopal Suhag (2016) 4 pp. also available at: http://www.prsindia.org/administrator/uploads/general/1455682937~~Overview%20of%20Ground%20Water%20in%20India.pdf (last accessed on 31st March 2017). Fig (b) is directly adapted from Fig. 17.1 of Rodell et al. (2009) "Satellite-based estimates of groundwater depletion in India" (Nature) Vol 460|20 August 2009| doi:https://doi.org/10.1038/nature08238. Also available at www:https://www.nasa.gov/topics/earth/features/india_water.html (last accessed on 31st March 2017)

limiting eventuality. These two models are selected out of several models as these models have a proper monsoon as we see today (Sharmila et al. 2015) which is the driver of the hydrological cycle over the Indian subcontinent and contributions from monsoon precipitation dominate the climatic component of groundwater recharge. The basic elements of MIROC5 and MPI Earth System Model (ESM) are given in Table 17.3.

To have a first-hand idea of the current state of climate we briefly show the observed evidence of climate change that will serve as a basis for the model study of the future scenario.

Table 17.3 Different components of MIROC5 and MPI coupled models which participated in the CMIP5 framework

Model name	Atmospheric component model	Oceanic component model	Land component model	River model	Sea ice component model	Output variables used
MIROC5	Spectral dynamical core, σ coordinates with 64 vertical levels. Explicit treatment of cumulus convection, aerosol, radiation and turbulence	COCO version 4.5	Minimal Advanced Treatments of Surface Interaction and Runoff (MATSIRO; which predicts the temperature and water in six soil layers down to a 14-m depth, one canopy layer, moreover, three snow layers	River network for the T85 resolution and explicitly calculating the river discharge	The sea ice concentration, ice thickness, snow thickness, and energy of ice melting are predicted for multiple categories in a grid cell	T, Pr, E, R
MPI-ESM	ECHAM6	MPIOM/ HAMOCC (biogeochemistry)	JSBACH land	JSBACH land	MPIOM	T, Pr, E, R

Sources: 1. For MIROC5 – Watanabe et al. 2010
2. For MPI-ESM – http://www.mpimet.mpg.de/en/science/models/mpi-esm/

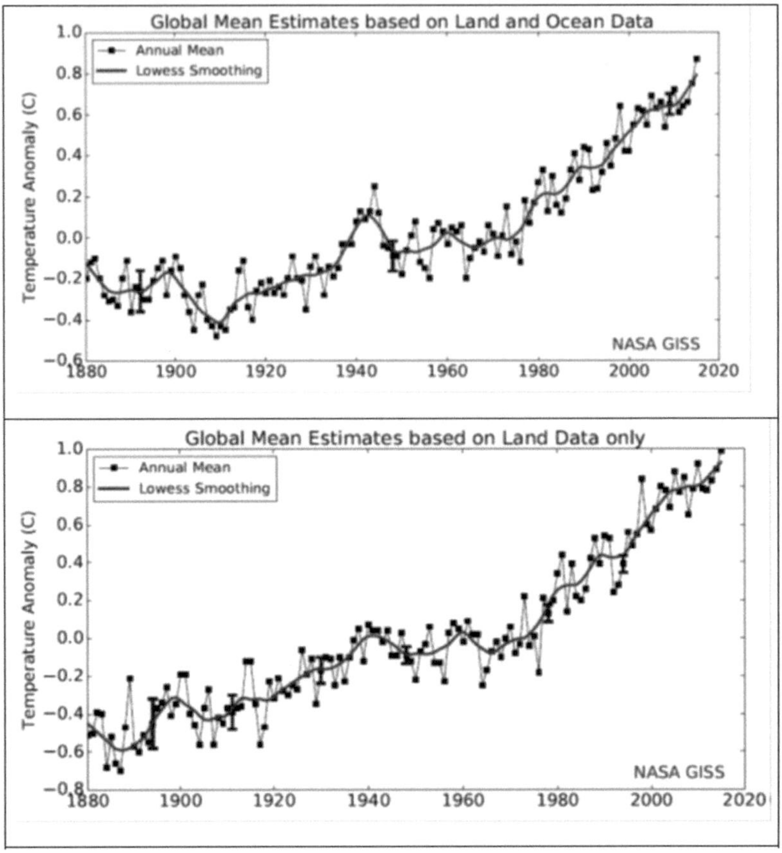

Fig. 17.4 The temperature trend for the last 150 years. Plot courtesy: NASA/GISS interactive plotter

6.1 Current Trends in Temperature and Precipitation

The current trends in temperature are shown in Fig. 17.4. The top panel shows the NASA/GISS (Hansen et al. 2010; GISSTEMP Team 2016) plots of earth's surface temperature averaged over land and ocean. The bottom panel shows the surface temperature averaged over the land regions only excluding the ocean. While both the panels show an exponential increase in surface temperature, the land data points show a larger increase in surface temperature. Assuming the earth as a (nearly) blackbody, exponential growth in surface temperature implies a shift in atmospheric radiation as the earth's surface long-wave radiation acts as the chief driver of atmospheric energy balance. This change in energy balance would indicate a change in the precipitation and circulation as documented in several studies. The plot in Fig. 17.5 shows the trend of precipitation extreme events based on last hundred years of India Meteorological Department's (IMD) $0.25° \times 0.25°$ rainfall data, which is

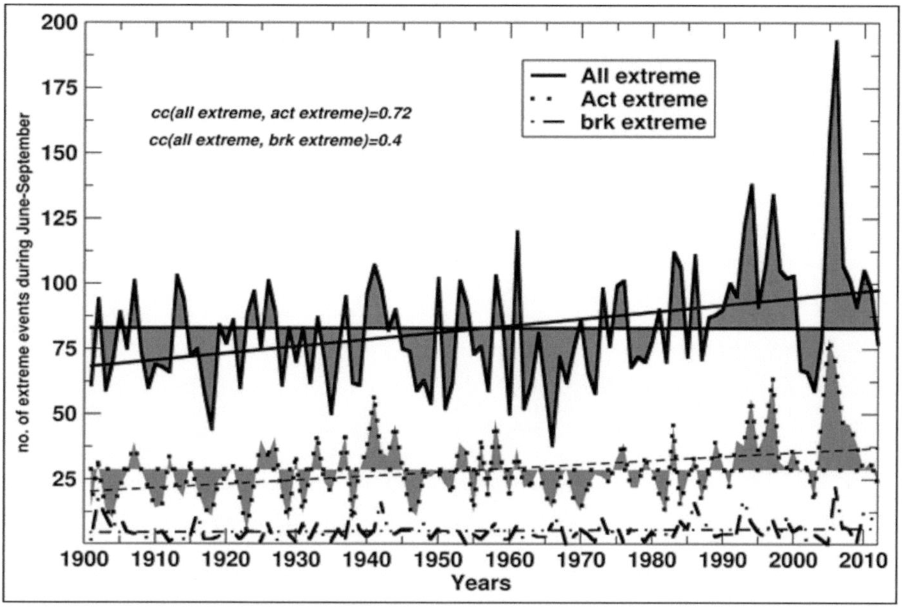

Fig. 17.5 The plots showed for the precipitation extremes. Extremes are defined when rainfall at a particular grid over Indian region is greater than 120 mm/day. "Act" extremes are the extremes during the active phase, "brk" extreme are the extreme during the break phase. "All" extremes are the extremes for all types of cases (i.e., active, break and normal cases). The horizontal line is the 110 years average number of extreme events for each instance. The trend lines are also shown

longest available time series that is representative of the Indian subcontinent. The plot shows the yearly number of extreme events counted over Indian Region during the June–September monsoon season. Extreme events are defined when the rainfall over a particular grid location over Indian land region is greater than 120 mm/day. It is clear that extreme rainfall events show a strong trend, and most of the extreme event occurs when the monsoon is active (intense precipitation phase) over Indian region rather than when it is in break (weak precipitation phase) condition. Although the total precipitation does not show a strong trend (Goswami et al. 2006) over Indian region, the slow to moderate precipitation rate shows a steadily decreasing trend (Goswami et al. 2006), consistent with several earlier studies. Since medium intensity rainfall events are of large importance for groundwater recharge rather than high-frequency intermittent rains, the increase in extreme events and the declining trend of low to medium rainfall events indicate that for the last 100 years, rate of recharge is likely to decrease over Indian region in the present period as compared to earlier period.

In the light of the above result, it would be important to understand the recharge potential over different regions of India when the concentration of greenhouse gas is increased to a large extent in the RCP8.5 scenario.

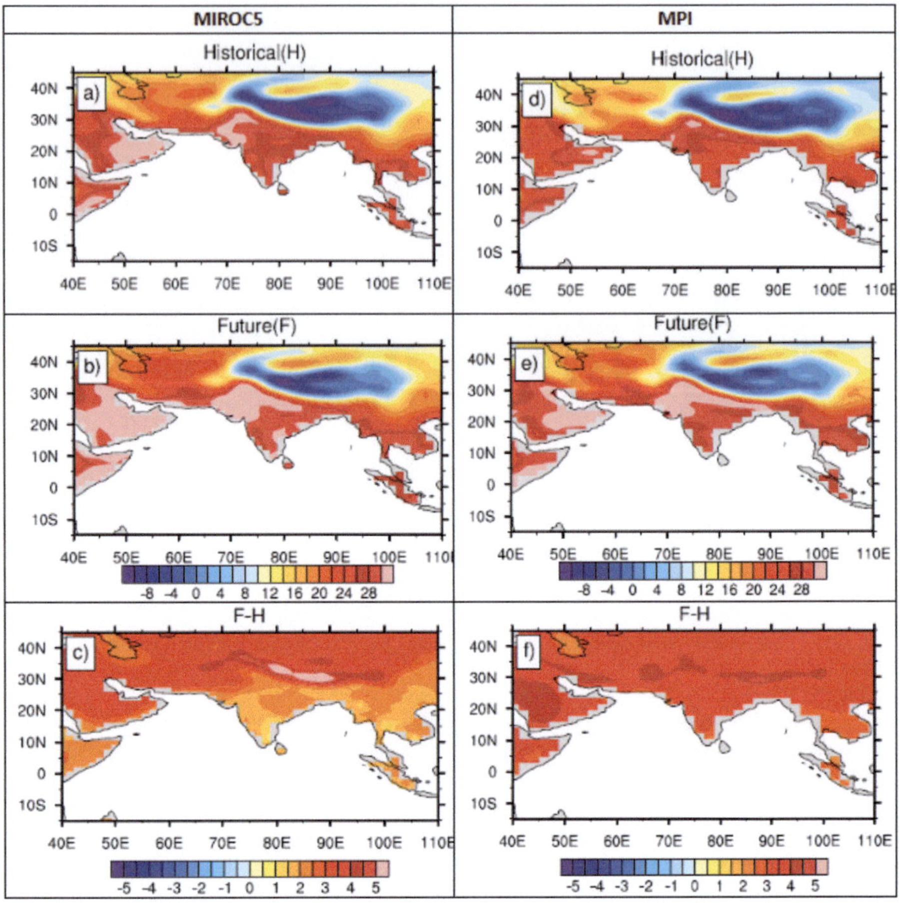

Fig. 17.6 (**a**) Plot of temperature (°C) over Indian Region for the *H* run, (**b**) *F* run and (**c**) $\Delta = F - H$ bias for the MIROC5 model. (**d–f**) same as (**a–c**) but for MPI-ESM run

6.2 *The Model-derived Result for the Next 100 Years and Computing the Difference of Future Versus Present*

As discussed earlier, the historical runs or the H runs represent the past 100/150 year climate after industrial era started from 1850. The historical runs for the models are generated based on an observed estimate of greenhouse radiative forcing and other boundary conditions that the model requires for running. Similarly, starting from the current period, model is run for future 100 years with the RCP8.5 scenario-projected-increase in greenhouse radiative forcing. This is the F runs also discussed earlier. The temperature plot for the larger region of South Asia incorporating the Indian subcontinent is shown for MPI-ESM and MIROC5 model in Fig. 17.6. The top panel is for H run, the middle panel is for F run, and the bottom panel represents

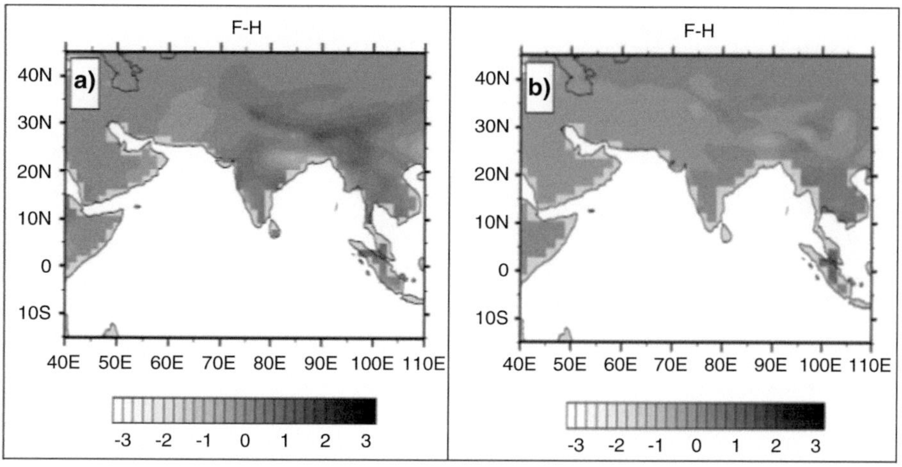

Fig. 17.7 The mean precipitation Δ bias (mm/day): (**a**) for MIROC5 run and (**b**) for MPI-ESM runs

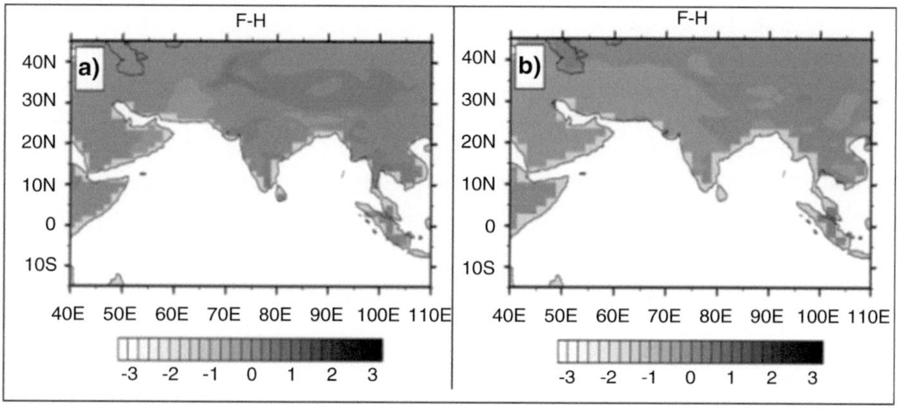

Fig. 17.8 The mean soil surface evaporation flux (unit: 10^{-5} kg m^{-2} s^{-1}) Δ bias: (**a**) for MIROC5 run and (**b**) for MPI-ESM runs. The variable includes sublimation flux

the difference (bias) $\Delta = F - H$. The other figures in this section would follow the same convention. It is evident from the Δ plot (bottom panel) of both the models that all the regions of South Asia show a significant increase in temperature in the coming century consistent with the earlier figure (Fig. 17.4). The temperature increase is primarily larger in the central Indian plain and the Himalayan region. The change in temperature would naturally lead to change in other climatic variables. For groundwater recharge, the precipitation, evapotranspiration and river runoff are the chief contributing factors.

Hence, the **Δ** biases for precipitation, evapotranspiration and river runoff are shown in Figs 17.7, 17.8 and 17.9. The precipitation scenarios (Fig. 17.7) for both

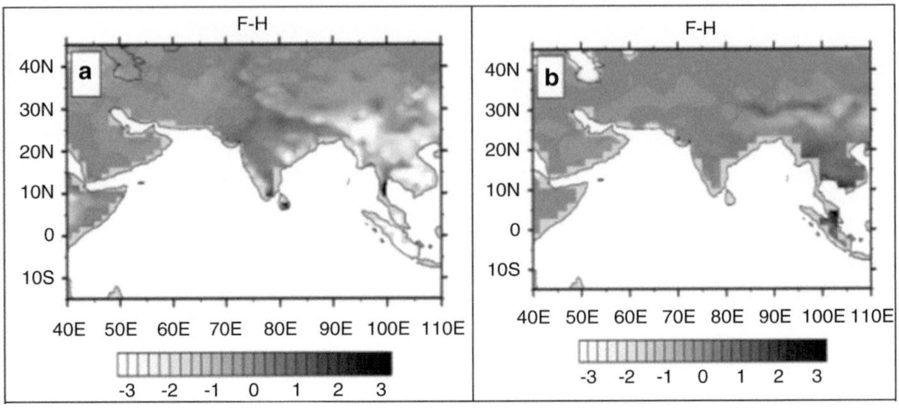

Fig. 17.9 The net ΔS flux (unit: 10^{-6} kg m^{-2} s^{-1}) bias: (**a**) for MIROC5 run and (**b**) for MPI-ESM runs

the models are interesting as the MIROC5 shows strong amplitude increase or decrease, while MPI-ESM shows weak amplitude increase and decrease over a larger region. The evapotranspiration pattern (Fig. 17.8) shows a steady growth over Indian region in MIROC5, while it shows an increase in MPI towards the east of 80°E and a decrease in future towards the west. A similar conclusion is valid for river runoff (Fig. 17.9) where the models disagree on the amplitude and spatial pattern. Although there is disagreement, the fallout of this fact is that the uncertainty in groundwater recharge has to be carefully monitored and regions which show large amplitude departures have to be carefully evaluated with realistic models in the near future.

6.3 *Stress in Groundwater Storage*

The change in the temperature, precipitation and other allied climatic factors as indicated in the last section would also imply a change in the storage measure ΔS. The storage pattern difference $\Delta = F - H$ for the ΔS is shown in Fig. 17.9. The spatial patterns for the two models show the quite similar distribution of anomalies. A large part of Indian landmass shows positive values while the east coast and north-eastern foothills and north-eastern state show negative anomalies. Over regions east of 90°E both the models differ in result indicating substantial uncertainty.

6.3.1 Variability in Storage Spatial Pattern: Results from Empirical Orthogonal Function Analysis

Because of much uncertainty in a model simulation in small spatial scale which is expected as the models differ in the statistical representation of climatic features; a multivariate data analysis identifying the large scale features (where errors are less) can be used to identify and study the dominant spatial pattern of ΔS and its temporal variability. The empirical orthogonal functions (EOFs) are the data-driven patterns or spatial maps which identify and orders the spatial maps according to the dominant variance patterns existing in the data domain selected for analysis. Empirical orthogonal function or EOF method is a widely used method applied to the study of various fields and takes different names (e.g. Karhunen-Louve transform, principal component analysis, etc.). In the area of multivariate data processing, the method of EOF analysis is a decomposition of a signal or multivariate data set (e.g. a time series data in several spatial locations) regarding orthogonal basis functions which are determined from the data. This is similar to harmonic analysis except that, in the harmonic analysis, the *basis* functions are fixed in harmonic analysis (e.g. Legendre polynomials, sinusoidal Fourier *basis* functions, etc.), and it is data driven in EOF analysis. The space-time dataset is used to construct a covariance matrix. Then an Eigen analysis of the covariance matrix is carried out which gives the output as a sum of space-time series:

$$\text{Data}(\text{space}, \text{time}) = \sum_{i=1}^{i=N} EOF_i(\text{space}) \times PC_i(\text{time})$$

This step is similar to Fourier or any other linear harmonic analysis. However, the basis functions in EOF analysis are arranged in such a way that the first basis functions explain maximum variances in the data and so on. The basis functions are orthogonal to each other; hence the construct is strictly geometrical and only first few basis functions e.g. EOF1, EOF2 are supposed to have physical significance. The time series of the PCs give an idea of the temporal evolution of the dominant spatial structure. Thus PC1 provides a temporal evolution of EOF1 and so on.

In climate data analysis, EOF is used to isolate dominant spatial and temporal mode of variability in data. These dominant modes (or spatial maps) would be useful to study future and current climate states. For example, the change in model data for present and future climate states can be derived by applying EOF analysis separately to future and current climate data. The first EOF of ΔS for the *H* runs is shown in Fig. 17.10 (a, b) for both the models. Similarly, the same EOF for the F runs is shown in Fig. 17.11 (c, d). One aspect is very evident from the plots that the first EOF has increased (in fact doubled) its variance share in both MIROC5 and MPI run in the future RCP8.5 scenario. For the MIROC5 run, *H* run shows a uniform pattern of negative shading to the east of 80°E and a pattern of positive shading to the west of 80°E. This east-west variability pattern is destroyed in F runs. The F run shows north-south asymmetry i.e. to the north, EOF1 has variability of one sign, and to the

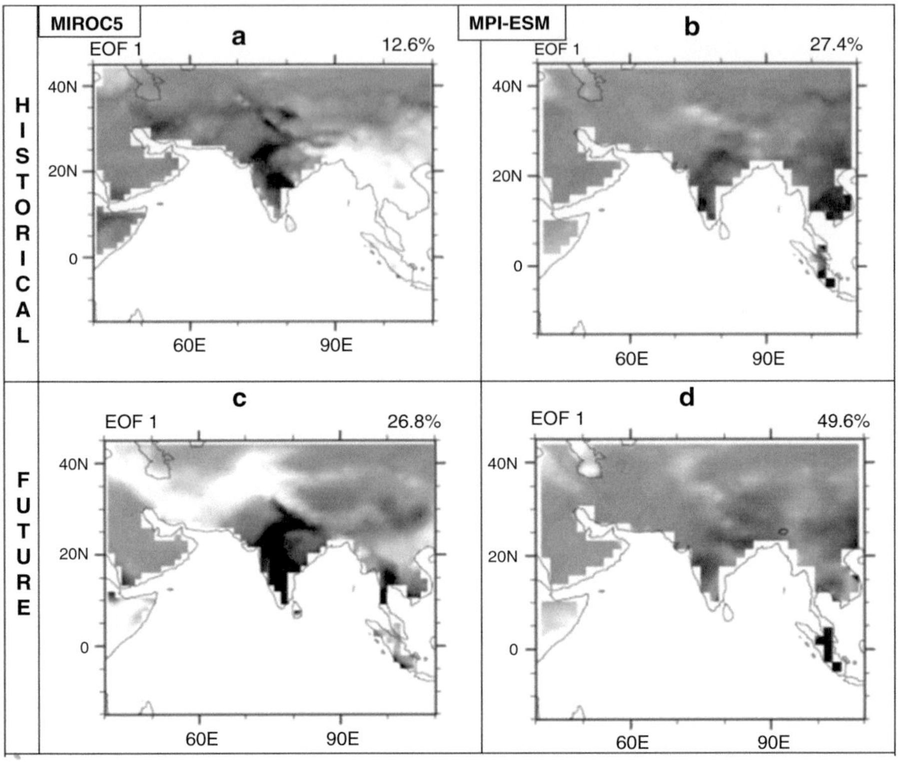

Fig. 17.10 The first EOFs of ΔS for the historical run (top panels) and the future run (bottom) panels. Figures (**a**) and (**c**) correspond to MIROC5 data while (**b**) and (**d**) correspond to MPI-ESM run. Top right-hand corner of each panel denotes the percentage variance explained by the each pattern. Shadings are arbitrary with shades in red indicates positive (one type) and blue indicates negative (other type) pattern. Deeper red and blue shades indicate stronger variability

south, the variability has another sign. Thus the east-west pattern becomes north-south pattern indicating a change in the whole physical process that controls groundwater recharge. The *H runs* from the MPI model similar kind of east-west pattern (except over Indian region) which becomes more of a north-south pattern in the future scenario. Thus the future scenario indicates an increase in variance in both runs and a north-south spatial pattern instead of an east-west pattern.

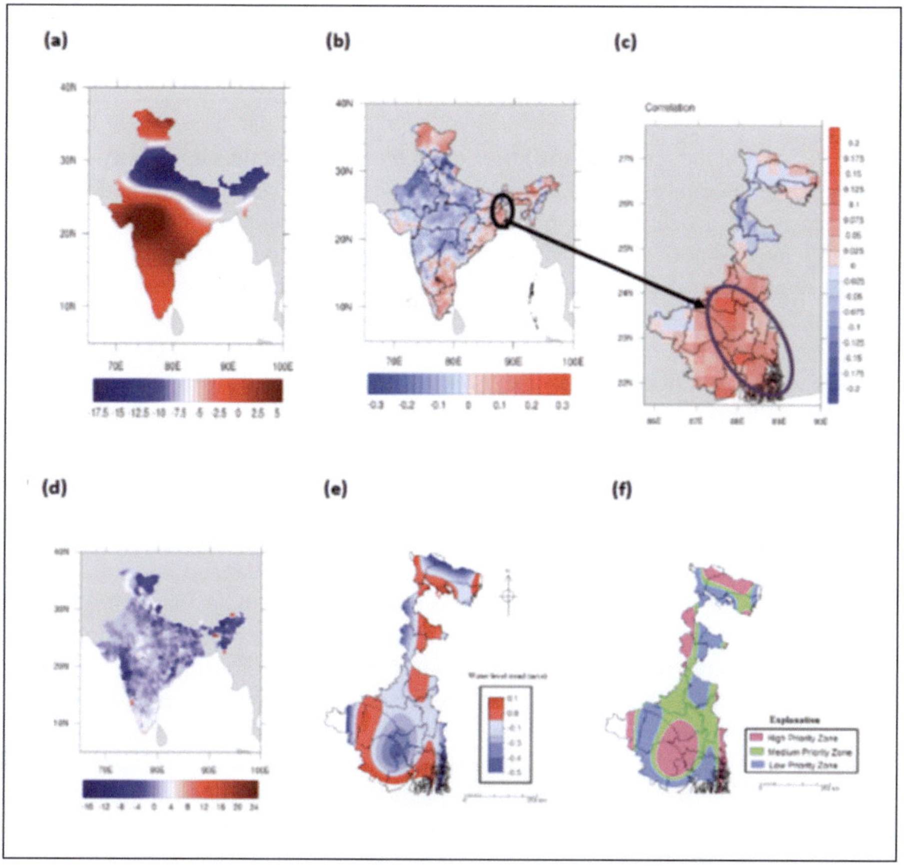

Fig. 17.11 (**a**) Plot showing a trend of GRACE TWS (mm/year) calculated during 2002–2015 when data is available. (**b**) The linear correlation coefficient (cc) between GRACE TWS and IMD RF computed for each grid during 2002–2015 based on monthly data. (**c**) Same as (**c**) but zoomed in for West Bengal State, India with cc values within ± 0.1 is masked out to identify potential source region (hotspots) for artificial rainwater harvesting. The shaded region encompassed by the ellipse could be one such hotspot for artificial rainwater harvesting. (**d**) The trend of IMD RF (rainfall, mm/year). Both data are monthly data and has a grid resolution of $0.25° \times 0.25°$. (**e**) Spatial distribution of the difference between average water table depth from ground level (in m) for the period 1996–2015 and yearly water table depth for the same period for 38 locations in West Bengal. Negative values (blue shades) indicate falling trend of water table and positive values (red shades) indicate rising trend of water table. Water table data for analysis have been taken from Central Ground Water Board website. (**f**) Priority of zones for artificial recharge of groundwater in West Bengal

7 Discussion

7.1 The Uncertainty and the Way Forward

It is evident from the results in the last section that (i) the models invariably show an increase in temperature of the land regions in the coming century, (ii) models show an increase and a decrease in precipitation, evapotranspiration, and river runoff. Although they do not coincide with each other regarding spatial location, or in other words, regional projections are quite uncertain for quantification, and (iii) the groundwater recharge storage term in the surface similarly share substantial uncertainty, even though they show regions of increased stress/decreased ΔS stress.

Despite the significant uncertainty, the results from the climate models clearly indicate one thing: variability in ΔS is bound to increase to high proportions over several parts of Indian region especially the eastern part of India. In the scenario of nearly constant or increased groundwater demand for irrigation and industrial uses in arid or semi-arid regions, the increased variability in net storage—that is potentially available for groundwater recharge in a climate change scenario—is definitely adding to the groundwater recharge uncertainty due to unplanned agricultural and industrial activities with groundwater over many parts of India. In the regions of climate stress, it is, therefore, mandatory to frame a formal policy to maintain the groundwater reserve for sustainable use in near future.

Since groundwater recharge of the aquifer is a slow process, adequate policy measures have to be formulated for the protection and sustainable usage of groundwater. These steps include but may not be limited to (a) detection and monitoring of groundwater depletion as well as recharge pattern related to regional climate change, (b) involving local community to restrict and judiciously use groundwater resources, (c) conjunctive management of surface water and groundwater recharge pattern, (d) adopt scientific procedures to increase recharge, (e) use of climate information services to maximize use of precipitation for agriculture, and (f) improve regional cooperation for surface water and groundwater recharge and management.

In the context of the Indian subcontinent, the large amount of rainfall the region receives through seasonal mean monsoon rainfall would provide an opportunity to recharge groundwater through rainwater harvesting and other artificial recharge methods. Rainwater harvesting is beneficial in many ways, but it is recommended as it is essentially the easiest choice for building up fresh water resources and groundwater recharge in urban as well as rural areas. An undeniable advantage of rainwater over other water sources is that it is one of the purest sources of water available. The superiority of rainwater is a prime incentive for people to choose rainwater as their primary water source for daily main uses, or for specific uses such as watering gardens. Rainwater quality exceeds that of groundwater or surface water as it is derived from natural distillation process and it does not come into contact with soil and rocks where it dissolves salts and minerals. The rainwater is not exposed to most of the pollutants that often are freely discharged into surface waters such as rivers, and which can contaminate groundwater. The rainwater harvesting can be easily implemented through precise execution and is especially useful in the context of climate change scenarios where it is important to use efficient recharge methods when rainfall variability is increasing.

Central Ground Water Board has stressed the importance of rainwater harvesting and also recommended a particular set of guidelines for such rooftop rainwater harvesting and artificial recharge (CGWB 2007). Several other methods of artificial recharge are also recommended here: (i) surface spreading techniques (e.g. flooding, and ditch and furrow method), (ii) runoff conservation structure (e.g. check dams, modification of village tanks etc.), (iii) subsurface techniques (e.g. injection wells or recharge wells), and (iv) indirect methods (e.g. induced recharge, aquifer modification techniques). Scientific implementation of each of these methods is helpful to augment the uncertainty arising due to depletion caused by overuse and indirect effect of climate stress in a climate change scenario.

7.2 *The Near-future Scenario and Potential Hotspots for Groundwater Recharge on Regional Scale*

Such scenarios based regional guidance maps could be easily developed based on climate model runs. However, as we have seen, the projections could be uncertain for smaller domain or local user or when the projection is required for shorter (e.g. decadal) time scale, it is necessary to have a practical understanding of the surface recharge potential of groundwater for local domains. For Indian regions and state level applications there is a need to develop guidance maps which could be identified as a potential hotspot for rainwater recharge of groundwater. For example, a recent study has emphasized the link of surface rainfall and groundwater abstraction for irrigation (Asoka et al. 2017). In this chapter, we develop an empirical method based on observed rainfall and JPL-NASA GRACE-project-derived monthly satellite data (Swenson and Wahr 2006; Landerer and Swenson 2012) which estimates the change in net terrestrial water mass storage (TWS), which could be expressed as:

$$TWS = GW + q_{\text{soil}} + w_{\text{surf}} + S + I$$

where GW = groundwater, q_{soil} = soil moisture, w_{surf} = surface water, S = snow and I = ice.

Now based on an estimation of ΔS as given earlier, it is clear that $\Delta S \propto \Delta(GW + q_{soil} + w_{surf})$, neglecting S and I for Indian region. This implies that TWS is a measure of ΔS. The spatial pattern of trend of TWS for the period during which data is available (January 2002–July 2016) is shown in Fig. 17.11a and the pattern of precipitation for the same period is shown in Fig. 17.11d. We can assume that where a statistically coherent (significant) correlation is present, it will provide a first-hand estimate of hotspots where an artificial recharge through rainwater harvesting could potentially be done with more efficiency for urban and rural planning, development of industries, etc. A correlation map of monthly precipitation anomaly (i.e. actual rainfall minus climatology) and TWS is shown in Fig. 17.11b for such regional applications. Regions having stronger positive correlations are identified as "source" or "hotspots," where rainfall recharge of groundwater would be preferred. Zooming over to the state of West Bengal (India), as an example, where groundwater is often known to be contaminated with arsenic posing a severe health hazard in several districts, we identify the regional hotspots over West Bengal which is shown in Fig. 17.11c.

We estimated changes (using linear trend) in groundwater level depth (m) using 38 well observation data (~2 wells from each district) from the CGWB (Central Ground Water Board) for 1996–2015 with groundwater storage anomalies from GRACE data. However, GRACE-based estimates of trends are lower than those of observation wells, as GRACE examines larger spatial domains (~100 km grid), whereas well observations are for point scale and represent very local depletion, which is not visible at GRACE resolution. Although the GRACE data shows uniform decreasing trend, the well data shows regions with patches of increasing trend (red shades). While it is an artifact of the data source or it is real is not clear. Nevertheless, the majority of the region shows decreasing trend, a result matching with the satellite-based data. For shorter real-time purpose when well data is not available, the satellite data can serve as a useful proxy indicator of the large scale behaviour. These zones with high rainfall, where precipitation is positively correlated with a surface component of recharge, are also operationally important for water managers as shown in Fig. 17.11e. The Kolkata region and the southern part of Bengal have an abundance of rainfall and show a positive trend with the surface source of groundwater (rainfall). This region must use rainfall as a sustainable and reliable source of groundwater recharge. A priority zone map for artificial recharge of groundwater for West Bengal (Fig. 17.11f) is generated based on negative (falling) and positive (rising) trends of water level. The high priority zones demand immediate attention for artificial recharging of groundwater. An interesting opposite scenario arises when we consider extreme northern parts of Bengal which have high priority zones (Fig. 17.11f) where rainfall, as well as groundwater, is showing falling trend (Fig. 17.11a, d, e) and rainwater cannot serve as a reliable component of recharge due to lack of correlation (Fig. 17.11c). Water management in this region, therefore, has to be tied to river system and distribution of river water with the efficient distribution network.

Figure 17.11 provides a reasonable way to estimate an (empirical) relationship between $\mathbf{\Delta}S$ and $\mathbf{\Delta}P$ (change in precipitation). Based on groundwater estimation and trend in usage pattern, this relation could further be exploited in designing bore wells for industrial purpose, the home supply of drinking water and irrigation purposes, etc. Similar relationships could be utilized for seasonal to sub-seasonal predictions of locally available source term if the predicted rainfall is accessible. Since now-a-day's operational rainfall predictions from daily to decadal timescale is possible, a strong collaborative multiagency modelling effort could be undertaken to enhance water security over Indian subcontinent or in regional scale anywhere in the world. However, it is to be remembered that these empirical statistics are for short term management purpose. For longer term planning (more than a decade), it is required that climate model based projections also has to be taken into account, albeit remembering its uncertainty.

8 Conclusion

Sustainable groundwater resource is a prime necessity for a country like India and several countries in South Asia where agriculture and industry both depend on ample water supply throughout the year. Groundwater could be considered as reliable source as its availability can be estimated in advance with short and long-term planning possible for industrial and agricultural uses. On the other hand, unplanned misuse would reduce its availability to different climatic zones of South Asia. Owing to its slow recharge, the arid and semi-arid regions where surface rainfall is less or do not have perennial surface water sources would be affected most. It is projected that the hydrological cycle would be under stress as the precipitation variability would be increased in the coming century. This chapter uses a simplistic view of groundwater recharge based on precipitation variability and has highlighted the uncertainty in the link through inter-comparison of CMIP5 models results for the next 100 years. While water stress is likely to increase over many regions of South Asia, in several areas, there will be surplus. It is therefore recommended that a judicial planning of artificial groundwater recharge has to be implemented with the participation of central and provincial governments across the countries of South Asia. An empirical method has been devised to identify "hotspots" where rainwater harvesting could act as a source of groundwater recharge.

Acknowledgements RC acknowledges the research support from IITM, an autonomous institute fully supported by the Ministry of Earth Sciences, Govt. of India. Plots are made using NCL-NCAR software and Xmgrace. The work is carried out with facilitation from India-UK Joint water centre (IUKWC). CMIP5 data is directly downloaded from https://pcmdi.llnl.gov/search/cmip5/. GRACE land grid data are available at http://grace.jpl.nasa.gov, supported by the NASA MEaSUREs Program. For information on GRACE NetCDF data used here, website documentation following is referred http://podaac.jpl.nasa.gov/dataset/TELLUS_LAND_NC_RL05. India Meteorological Department, Pune office (IMD) has provided the gridded rainfall data used in this study. The authors would also like to thank Prof. P.K. Sikdar, Department of Environment Management, IISWBM, Kolkata for providing constructive comments and suggestions which improved the manuscript.

References

Asoka, A., Gleeson, T., Wada, Y. and Mishra, V. (2017). Relative contribution of monsoon precipitation and pumping to changes in groundwater storage in India. *Nat Geosci*, **10:** 109–117. doi: https://doi.org/10.1038/ngeo2869.

Balek, J. (1988). Estimation of Natural Groundwater Recharge, I. Simmers (Ed.). Springer.

CGWB (2007). Manual on Artificial Recharge of Ground Water. Government of India, New Delhi.
GISSTEMP Team (2016). http://data.giss.nasa.gov/gistemp/. In: NASA GISS. http://data.giss.nasa.gov/gistemp/.
Gleick, P.H. and Adams, D.B. (2000). Water: The Potential Consequences of Climate Variability and Change for the Water Resources of the United States. USGS, Washington D.C.
Goswami, B.N., Venugopal, V., Sengupta, D. et al. (2006). Increasing Trend of Extreme Rain Events over India in a Warming Environment. *Science*, **314:** 1442–1445. doi: https://doi.org/10.1126/science.1132027.
Hansen, J., Ruedy, R., Sato, M. and Lo, K. (2010). Global Surface Temperature Change. *Rev Geophys*, **48:** RG4004. doi: https://doi.org/10.1029/2010RG000345.
IGES-GWP (2012). Technical Report on Issues related to Water and Agriculture in South Asia. IGES-GWP, Hamaya (Japan).
IPCC Report (1996). Climate Change 1995, Impacts, adaptations and mitigation of climate change: Scientific-technical analyses. Cambridge University Press, UK, London.
IPCC Report (1998). The Regional Impacts of Climate Change: An Assessment of Vulnerability. Cambridge University Press, UK, London.
IPCC Report (2001). Climate Change 2001, Impacts, Adaptation and Vulnerability. Cambridge University Press, UK, London.
IPCC Report (2008). Climate Change and Water. IPCC Technical Paper VI. Cambridge University Press, UK, London.
IPCC Report (2014). Fifth Assessment Report - Impacts, Adaptation and Vulnerability. Cambridge University Press, UK, London.
Kumar, C.P. (1997). Estimation of Natural Groundwater Recharge. *ISH J Hydraul Eng*, **33:** 61–74.
Landerer, F.W. and Swenson, S.C. (2012). Accuracy of scaled GRACE terrestrial water storage estimates. *Water Resour Res*, **48:** W04531. doi: https://doi.org/10.1029/2011WR011453.
Ministry of Water Resources (2006). Central Ground Water Board. Dynamic Ground Water Resources of India (as on March 2004). Government of India, New Delhi.
Rodell, M., Velicogna, I. and James, S.F. (2009). Satellite-based estimates of groundwater depletion in India. *Nature*, **460:** 999–1002; doi: https://doi.org/10.1038/nature08238.
Sharmila, S., Joseph, S., Sahai, A.K. et al. (2015). Future projection of Indian summer monsoon variability under climate change scenario: An assessment of CMIP5 climate models. *Glob Planet Change*, **124:** 62–78. doi: https://doi.org/10.1016/j.gloplacha.2014.11.004.
Shiklomanov, I.A. (1993). World Freshwater Resources in Water in Crisis. Oxford University Press, UK.
Swenson, S. and Wahr, J. (2006). Post-processing removal of correlated errors in GRACE data. *Geophys Res Lett*, **33:** L08402. doi: https://doi.org/10.1029/2005GL025285.
Taylor, K.E., Stouffer, R.J. and Meehl, G.A. (2011). An Overview of CMIP5 and the Experiment Design. *Bull Am Meteorol Soc*, **93:** 485–498. doi: https://doi.org/10.1175/BAMS-D-11-00094.1.
Taylor, R.G., Scanlon, B., Döll, P. et al. (2013). Ground water and climate change. *Nat Clim Change*, **3:** 322–329. doi: https://doi.org/10.1038/nclimate1744.
UNEP (2008). Freshwater under Threat, South Asia. UNEP-AIT, Nairobi, Kenya.
Watanabe, M., Suzuki, T., O'ishi, R. et al. (2010). Improved Climate Simulation by MIROC5: The Mean States, Variability, and Climate Sensitivity. *J Clim*, **23:** 6312–6335. doi: https://doi.org/10.1175/2010JCLI3679.1.
Xu, C. (2000). Modelling the Effects of Climate Change on Water Resources in Central Sweden. *Water Resour Manag*, **14:** 177–189. doi: https://doi.org/10.1023/A:1026502114663.

Chapter 18
Rainwater Harvesting and Artificial Recharge of Groundwater

Ajoy Kumar Misra

1 Rainwater Harvesting

1.1 Introduction

Close to three fourths of our planet is made of water, but not all of it is suitable for use. The water in the oceans and seas cannot be used as drinking water and little of it can be utilized for other purposes. As a result, there is a constant shortage of water that is suitable for drinking, domestic or industrial uses. Areas on the planet that have long faced water shortage were able to combat this problem by harvesting what little rain water they received. This slowly started spreading to areas where there was plenty of rainfall. As a result, the modern day rainwater harvesting system was brought into place.

The concept of rainwater harvesting involves 'tapping the rainwater where it falls'. A major portion of rainwater that falls on the earth's surface runs off into streams and rivers and finally into the sea. An average of 8–12% of the total rainfall is considered to recharge the aquifers. Rainwater harvesting is a technique of collection and storage of rainwater into natural reservoirs or tanks, or the infiltration of surface water into subsurface aquifers before it is lost as surface runoff (Fig. 18.1). The harvesting of rainwater simply involves the collection of water from surfaces on which rain falls, and subsequently storing this water for later use. Normally water is collected from the roofs of buildings and stored in rainwater tanks (Fig. 18.2). Water can also be collected in dams from rain falling on the ground and producing runoff. Either way, the water collected can be considered to be precious. Rainwater is collected when it falls on the earth, stored and utilized at a later point. It can be purified to make it into drinking water, used for daily applications and even utilized

A. K. Misra (✉)
Central Ground Water Board, Kolkata, India
e-mail: misraajoy@yahoo.co.in

P. K. Sikdar (ed.), *Groundwater Development and Management*,
https://doi.org/10.1007/978-3-319-75115-3_18

Fig. 18.1 Rainwater harvesting through dug well

Fig. 18.2 Roof-top rainwater harvesting in rural areas

in large scale industries. In short, rainwater harvesting is a process or technique of collecting, filtering, storing and using rainwater for various purposes.

To reduce the consumption of groundwater, many people around the world are using rainwater harvesting systems. This practice has been around for thousands of years and has been growing at a rapid pace. Till today, rainwater is used primarily as a source of drinking water in several rural areas. The best thing about rainwater is that it is free from pollutants as well as salts, minerals and other natural and man-made contaminants. In areas where there is excess rainfall, the surplus rainwater can be used to recharge groundwater through artificial recharge techniques.

In an urban setting, the simplest method of rainwater harvesting is usually done with the help of storage tanks. In this, a catchment area for the rainwater is directly linked to cisterns, tanks and reservoirs. Water can be stored in these structures until needed or used on a daily basis. The roofs of our homes are the best catchment areas, provided they are large enough to harvest daily water needs. Other than that, large bowls and tarps can also fulfill the function (Behzadian and Kapelan, 2015).

The reasons for practicing rainwater harvesting are to: (i) improve water quality in aquifers, (ii) conserve surface water runoff during monsoon, (iii) reduce soil erosion, and (iv) inculcate a culture of water conservation.

1.2 Roof-Top Rainwater Harvesting

A major portion of rainwater that falls on the earth's surface runs off into streams and rivers and finally into the sea. An average of 8–12% of the total rainfall is considered to recharge the aquifers. The technique of rainwater harvesting involves collecting the rain from localized catchment surfaces such as roofs, plain/sloping surfaces etc., either for direct use or to augment the groundwater resources depending on local conditions. Construction of small barriers across small streams to check and store the running water also can be considered as water harvesting.

Among the various techniques of water harvesting, harvesting water from rooftops (Figs. 18.2 and 18.3) need special attention because of the following advantages:

- (i) Roof-top rainwater harvesting is one of the appropriate options for augmenting groundwater recharge/storage in urban areas where natural recharge is considerably reduced due to increased urban activities and not much land is available for implementing any other artificial recharge measure. Roof-top rainwater harvesting can supplement the domestic requirements in rural areas as well.
- (ii) Rainwater runoff which otherwise flows through sewers and storm drains and is wasted, can be harvested and utilized.
- (iii) Rainwater is bacteriologically safe, free from organic matter and is soft in nature.
- (iv) It helps in reducing the frequent drainage congestion and flooding during heavy rains in urban areas where availability of open surfaces is limited and surface runoff is quite high.
- (v) It improves the quality of groundwater through dilution.
- (vi) The harnessed rainwater can be utilized at the time of need.
- (vii) The structures required for harvesting rainwater are simple, economical and eco-friendly.

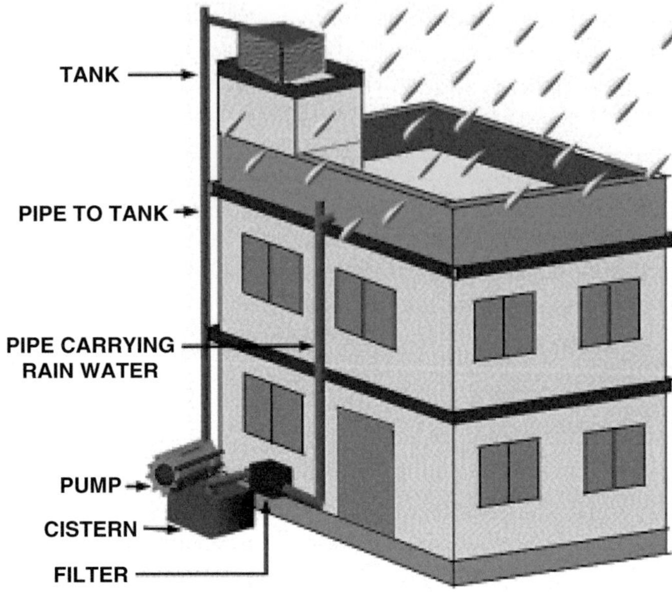

Fig. 18.3 Roof-top rainwater harvesting in urban areas

(viii) Roof catchments are relatively cleaner and free from contamination compared to the ground level catchments.

(ix) Losses from roof catchments are much less when compared to other catchments.

1.3 Rainwater Harvesting Through Water Detention Structures

Construction of small barriers across small streams to check and store the running water also can be considered as water harvesting. Water Resources Investigation and Development Department, Government of West Bengal, has started the project called Accelerated Development of Minor Irrigation (ADMI) in West Bengal since 2012. The objective of the project is to enhance agricultural production of small and marginal farmers in 10 blocks of South 24 Parganas district of West Bengal through conservation of surface water resources. So far about 9600 ha area has been brought under irrigation facility through water detention structures (Fig. 18.4). It is expected that construction of more water detection structures in the area will facilitate to create

Fig. 18.4 Water detention structure at Paruldaha, Baruipur block, South 24 Parganas district, West Bengal, India

additional irrigated area where more crops could be grown during winter and summer periods. In course of time agriculture based social livelihood in the area will be enhanced and the area will be gradually economically uplifted (Nag Choudhury et al. 2016).

1.4 Advantages and Disadvantages of Rainwater Harvesting

The advantages of rainwater harvesting are: (i) the system is easy to construct, operate and maintain, (ii) reduce water bills, (iii) suitable for irrigation, (iv) reduces demand on groundwater, (v) reduces floods and soil erosion, (vi) can be used for several non-drinking purposes such as flushing toilets, washing clothes, watering the garden, washing cars etc. (vii) provides self-sufficiency to water supply, (viii) reduces the cost for pumping of groundwater, (ix) provides high quality water, soft and low in minerals, (x) improves the quality of groundwater through dilution when recharged, (xi) in desert environment rainwater harvesting can provide some relief in non-rainy periods, and (xii) in coastal areas and islands rainwater provides good quality water.

The disadvantages of rainwater harvesting are: (i) unpredictable rainfall, (ii) initial high cost ranging between Rs. 12,000 and Rs. 1,20,000 and benefit from it cannot be derived until it is ready for use, (iii) requires regular maintenance, and (iv) limitation to volume of water stored.

Rainwater harvesting is a system that is gaining importance with time. Areas that experience high amounts of rainfall will benefit the most and will be able to distribute water to dry lands with ease.

1.5 Components of Rainwater Harvesting

Rainwater harvesting systems have three principal components, namely, catchment area, collection device, and conveyance system.

1.5.1 Catchment Area

Catchment area can be either roof top or land surface. Roof top can be made of galvanized corrugated iron, aluminum or asbestos cement sheets, tiles and slates, although thatched roofs tied with bamboo gutters and placed in proper slopes can produce almost the same amount of runoff (Gould, 1992). However, bamboo roofs are least suitable because of possible health hazards. Rainwater harvesting using ground or land surface as catchment areas is less complex way of collecting rainwater. Compared to roof-top catchment techniques, ground catchment techniques provide more opportunity for collecting water from a larger surface area (Wall and McCown, 1989). More rainwater may be collected from the land surface by clearing or altering vegetation cover, increasing the slope of the land, and compaction of the soil by physical or chemical treatments.

1.5.2 Collection Device

Collection device of rainwater is storage tanks which may be either above or below the ground (Figs. 18.5 and 18.6). Precautions required in the use of storage tanks include provision of an adequate enclosure to minimise contamination from human, animal or other environmental contaminants, and a tight cover to prevent algal growth and the breeding of mosquitoes. Open containers are not recommended for collecting water for drinking purposes.

1.5.3 Conveyance System

Conveyance system is required to transfer the rainwater collected on the roof-tops to the storage tanks. This is usually done by making connections to one or more down-pipes connected to the roof-top gutters. The pipes used for the collection of rainwater, wherever possible, should be made of plastic, PVC or other inert substance as the pH of rainwater can be low (acidic) and could cause corrosion, and mobilization of metals in metal pipes.

Fig. 18.5 A rainwater collector tank

Fig. 18.6 A rainwater collection tank made of ferro-cement. *Source:* Dolman and Lundquist (2008)

1.6 Installation and Maintenance

Rainwater harvesting technologies are simple to install and operate. Local people can be easily trained to implement such technologies, and construction materials are also readily available. Depending upon household capacity and needs, both the water collection and the storage capacity may be increased as needed within the available catchment area.

Maintenance is generally limited to the annual cleaning of the tank and regular inspection of the gutters and down-pipes. Maintenance typically consists of the removal of dirt, leaves and other accumulated materials. Such cleaning should take place annually before the start of the major rainfall season.

1.7 Suitability

The augmentation of municipal water supplies with harvested rainwater is suited to both urban and rural areas. The construction of cement tanks or provision of gutters does not require very highly skilled manpower.

1.8 Effectiveness of Technology

The feasibility of rainwater harvesting in a particular locality is highly dependent upon the amount and intensity of rainfall. Other variables, such as catchment area and type of catchment surface, usually can be adjusted according to household needs. As rainfall is usually unevenly distributed throughout the year, rainwater collection methods can serve as only supplementary sources of household water. The viability of rainwater harvesting systems is also a function of the quantity and quality of water available from other sources, household size, per capita water requirements, and budget available.

If rainwater is used in household appliances such as washing machine, even the tiniest particles of dirt may cause damage to the machine and the washing. To minimize the occurrence of such damage, it is advisable to install a fine filter of a type which is used in drinking water systems in the supply line upstream of the appliances. For use in wash basins or bath tubs, it is advisable to sterilise the water using a chlorine dosage pump.

1.9 Indian Examples of Rain Water Harvesting

Tamil Nadu was the first state to make rainwater harvesting compulsory for every building to avoid groundwater depletion. The scheme was launched in 2001 and has been implemented in all rural areas of Tamil Nadu. It gave excellent results within five years, and slowly every state took it as role model. Since its implementation, Chennai saw a 50% rise in the water table in five years and the water quality significantly improved.

In the city of Bangalore in Karnataka it is mandatory to install roof-top rainwater harvesting structures for every owner or the occupier of a building having a roof area of 2400 sq. ft. For newly constructed building the area is $\geq$1200 sq. ft. In this regard Bangalore Water Supply and Sewerage Board (BWSSB) has initiated and constructed "Rainwater Harvesting Theme Park" in the name of Sir M. Visvesvaraya in 4900 m^2 land situated at Jayanagar, Bangalore. In this park 26 different type of rainwater harvesting models have been demonstrated along with water conservation tips. In the auditorium on the first floor video clips about rainwater harvesting are shown to the students as well as to the general public.

In Rajasthan, rainwater harvesting has traditionally been practiced by the people of the Thar Desert. There are many ancient water harvesting systems in Rajasthan, which have now been revived. Water harvesting systems are also found in other areas of Rajasthan, for example the 'chauka' system of Jaipur district (www.rainwaterharvesting.org/Rural/Improvised.htm, accessed on 21.03.2017). A 'chauka' is a rectangular enclosure surrounded on three sides by earthen bunds or embankments. In this system, series of these rectangles are constructed in a checker-board pattern across a natural slope and connected with shallow canals. The embankments intercept the runoff rainwater which is collected at the down end of the 'bunded' field. During heavy rainfall, the water moves gradually from one 'chauka' to another, which gives it more time to infiltrate into the ground and recharge the groundwater.

1.10 Further Development of Technology

Rainwater harvesting appears to be one of the most promising alternatives for supplying freshwater in the face of increasing water scarcity and escalating demand. The pressure on rural water supplies, greater environmental impacts associated with new projects, and increased opposition from NGOs to the development of new surface water sources, as well as deteriorating water quality in surface reservoirs already constructed, constrain the ability of communities to meet the demand for freshwater from traditional sources, and present an opportunity for augmentation of water supplies using this technology.

2 Artificial Recharge of Groundwater

2.1 Introduction

Groundwater is recharged naturally by rain and snow melt and to a smaller extent by surface water (Fig. 18.7). Recharge may be impeded somewhat by human activities including paving, development, or logging. These activities can result in loss of topsoil resulting in reduced water infiltration, enhanced surface runoff, reduction in

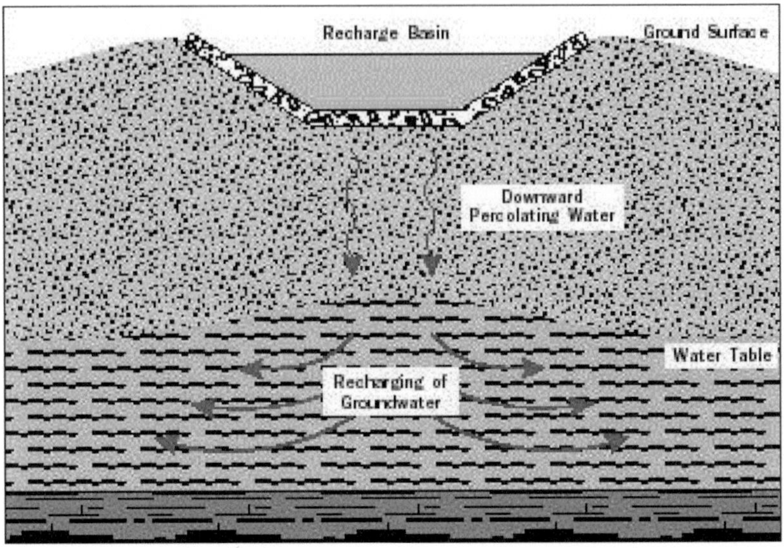

Fig. 18.7 Artificial recharge to groundwater

recharge and fall in water table. Use of groundwater, especially for irrigation, may also lower the water table.

Groundwater recharge is an important process for sustainable groundwater management, since the volume-rate abstracted from an aquifer in the long term should be less than or equal to the volume-rate that is recharged. Recharge can help move excess salts that accumulate in the root zone to deeper soil layers, or into the groundwater system. Tree roots increase water saturation into groundwater reducing water runoff. Flooding temporarily increases river bed permeability by moving clay soils downstream, and this increases aquifer recharge.

Artificial groundwater recharge is becoming increasingly important not only in India but also in other places of the world as a cost effective method to augment groundwater resources in areas where continued over-exploitation without due regard to their recharging options has resulted in various undesirable environmental consequences.

Artificial recharge (also known as aquifer re-injection) is the process of injecting (or recharging) water into the ground in a controlled way, by means of special recharge wells. The water is pumped from the dewatering system and then piped to the recharge location, which may be at a considerable distance away, where the water is injected back into the ground. Water may have to be treated prior to recharge to reduce the risk of clogging of recharge wells.

The concept of artificial recharge has been known for a long time. The practice began in Europe during the early nineteenth century. However, the practice has rarely been adopted on a large scale, with most large-scale applications being found in Netherlands, Germany, and USA. Israel transports 300 million cubic metres of water annually from north to south through the National Water Carrier System and stores two-thirds of it underground (Ambroggi, 1977). The water is used to meet high summer demands and offers a reliable source of supply during dry years. On the North Plain of China, which is prone to droughts, water from nearby streams is diverted into underground storage areas with capacities of about 500 million cubic metres. In India, subsurface storage is becoming popular to provide reliable source of irrigation water. A number of artificial recharge projects have been carried out in India (CGWB, 1994).

The area involved in artificial recharge projects may range from watershed, a limited area covering an urban, rural or industrial centre or administrative unit like mondal/block to large basins or larger administrative unit like districts/states.

Proper planning is essential for the successful outcome of any artificial recharge scheme. Planning of artificial recharge schemes involves the formulation of a suitable plan, under a given set of natural conditions, to augment the natural groundwater recharge. An artificial recharge scheme may be aimed at recharge augmentation in a specific area for making up the shortage of groundwater recharge compared to the groundwater draft either fully or partially (CGWB, 2007).

Artificial recharge may be carried out for various purposes (Asano, 1985; Oaksford, 1985). They are:

- To avoid depletion of water resources when dewatering is carried out in aquifers used for water supply.
- To ensure that ground settlements caused by drawdown are small, thus reducing the risk of damage to nearby structures.
- To reduce environmental impacts on sensitive water-dependent features such as wetlands.
- To control seawater intrusion in coastal aquifers.
- To maintain base flow in some streams.
- To raise water levels to reduce the cost of groundwater pumping.
- To conserve or dispose floodwater.
- To regulate groundwater abstraction temporarily.
- To improve water quality by removal of suspended solids by filtration through the ground or by dilution by mixing with naturally occurring groundwater.

2.2 *Methods of Artificial Recharge*

A variety of methods have been developed and applied to artificially recharge groundwater reservoirs in various parts of the world. Details of these methods, as well as related topics, can be found in the literature (e.g., Todd, 1980; Huisman and Olsthoorn, 1983; Asano, 1985; CGWB, 1994). The methods may be generally classified into the following four categories (Oaksford, 1985):

- Direct surface recharge technique
- Direct sub-surface recharge technique

Fig. 18.8 Direct surface recharge through check dam

- Combination of surface and subsurface methods
- Indirect recharge technique

2.2.1 Direct Surface Recharge Technique

This is the simplest and most widely applied method. In this method, water moves from the land surface to the aquifer by means of infiltration and percolation through the soil (Figs. 18.8 and 18.9). To achieve optimal infiltration rates, a number of features need to be considered in the design process, including clogging, water depth, groundwater level and water quality.

A major operational feature of infiltration systems for artificial recharge of groundwater is soil clogging caused by accumulation of suspended solids on the bottom and banks of the infiltration facility as they settle or are strained out on the soil surface. The suspended solids can be inorganic (e.g., clays, silts, fine sands) or organic (e.g., algae, bacterial flocks, sludge particles). As the clogging layer forms and its hydraulic resistance increases, the clogging layer becomes the controlling factor of the infiltration process, and infiltration rates decrease.

The water depth in infiltration basins is selected carefully. While high hydraulic heads produced by deep water result in high infiltration rates, they also tend to

Fig. 18.9 Direct surface recharge through percolation tanks in Satpura Mountainfront area in Maharashtra

compress clogging layers. Thus, contrary to intuitive expectations, deep basins can produce lower infiltration rates than shallow basins. Algae causes additional clogging of the soil as the biomass is strained out by infiltration. Another undesirable effect is that calcium carbonate may precipitate because of increase in the pH of the water as photosynthesizing algae take up dissolved carbon from the water, further aggravating the clogging.

Another design criterion is that the water table must be deep enough below the infiltration system so that it does not interfere with the infiltration process. This requirement applies to the mounding of the permanent water table caused by recharging, as well as to perched groundwater mounds that may form over restricting layers in the vadose zone. Where infiltration rates are controlled by the clogging layer (which is the rule rather than the exception for basins and ponds), the water table must be at least 0.5 m (1.6 ft) below the bottom of the basin.

Impaired quality water sources can vary in quality. For relatively unpolluted water, the most important quality parameters applicable when considering groundwater recharge are suspended solids (SS) content, total dissolved solids (TDS) content, and the concentrations of major cations such as calcium, magnesium and sodium.

2.2.2 Direct Sub-surface Recharge Technique

Direct sub-surface recharge methods access deeper aquifers and require less land than the direct surface recharge methods, but are more expensive to construct and maintain. Recharge wells, commonly called injection wells, are generally used to replenish groundwater when aquifers are deep and separated from the land surface by materials of low permeability (Fig. 18.10). These wells are similar to regular pumping wells.

All the sub-surface methods are susceptible to clogging by suspended solids, biological activity or chemical impurities. Recharge wells have been used to dispose off treated industrial wastewaters, to add freshwater to coastal aquifers experiencing

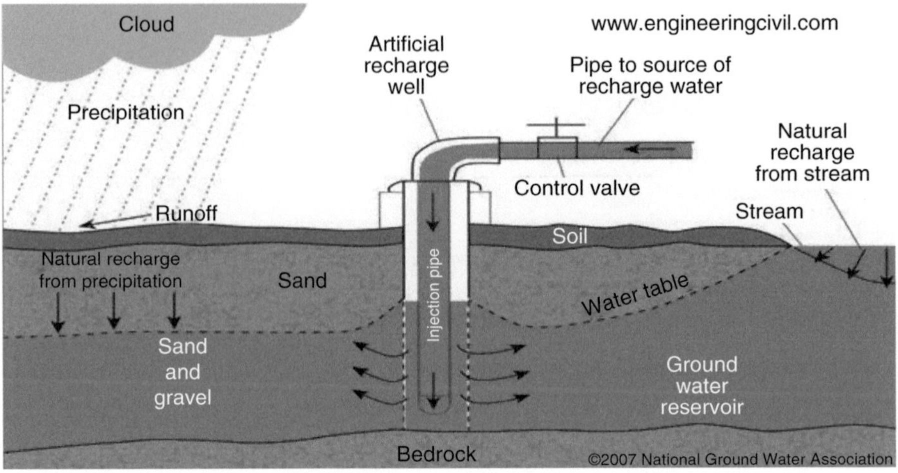

Fig. 18.10 Artificial recharge to groundwater by direct sub-surface recharge technique

saltwater intrusion, and to force water under pressure into permeable bedrock aquifers to arrest land subsidence resulting from extensive withdrawals of groundwater, although with variable success (CGWB, 1994).

2.2.3 Combination of Surface and Subsurface Method

This method includes subsurface drainage (collectors with wells), basins with pits, shafts, and wells.

2.2.4 Indirect Recharge Technique

This method includes the installation of groundwater pumping facilities or infiltration galleries near hydraulically-connected surface water bodies (such as streams or lakes) to lower groundwater levels and induce infiltration elsewhere in the drainage basin, and modification of aquifers or construction of new aquifers to enhance or create groundwater reserves.

2.3 Aquifer Storage and Recovery Wells

A rapidly growing practice in artificial recharge is the use of aquifer storage recovery (ASR) wells, which combine recharge and pumping functions. They are used for recharge when surplus water is available, and are pumped when the water is needed.

2.4 Sources of Water for Artificial Recharge

Generally three types of source waters having very different characteristics are used for artificial recharge. They are treated municipal wastewater, storm water runoff, and irrigation return flow. Normally, each of these source waters needs to be subjected to some kind of pre-treatment before being introduced into the soil or aquifer. The exact pre-treatment operations required depend on the type of source water and the nature of the recharge.

2.5 Environmental Effects of Artificial Recharge

The environmental effects of groundwater recharge vary from site to site, and there can be both beneficial and harmful impacts. In general, however, the types of environmental effects that should be considered when planning recharge facilities range from ecological effects on soil, hydrologic and aquatic ecosystems, to effects on species dependent on riparian habitats, and to possible effects on people's use of the water resources for recreation. As with any use of water, planners must take care to recognize that the impacts of their actions will affect not only local conditions, but also conditions downstream (third-party effects).

2.6 Operation and Maintenance

To ensure the effective and efficient operation of an artificial recharge system, a thorough and detailed hydrogeological study must be conducted before selecting the site and method of recharge. In particular, the following basic factors should be considered: the locations of geologic and hydraulic boundaries, the transmissivity of the aquifer, depth to the aquifer and lithology, storage capacity, porosity, hydraulic conductivity, and natural inflow and outflow of water to the aquifer, the availability of land, surrounding land use and topography, the quality and quantity of water to be recharged, the economic and legal aspects governing recharge, and the level of community acceptance.

2.7 Level of Involvement

Because of the technical complexity involved in siting and regulating artificial recharge, this technology is generally implemented at the governmental level.

2.8 Suitability

Groundwater recharge methods are suitable for use in areas where aquifers exist. Typically, unconfined aquifers are recharged by surface recharge methods, whereas confined aquifers are generally recharged through sub-surface injection. Surface recharge methods require relatively flat or gently sloping lands, while topography has little effect on subsurface recharge methods.

In temperate humid climates, the alluvial areas which are best suited to artificial recharge are areas of ancient alluvium, the buried fossil river-beds and interlinked alluvial fans of their main valley and tributaries. In the arid zone, recent river alluvium may be more favourable than in humid zones. In these areas, the water table is subject to pronounced natural fluctuations.

In hard rock areas injection wells are less expensive since they tend to be shallower and have a lesser risk of clogging. Guidelines for the socio-economic and financial appraisal of artificial recharge projects in developing countries, necessary to assess these net benefits, are provided by CGWB (1994).

Toxic substances must not be present in the recharge water, and if present they must be removed by pre-treatment or chemically decomposed by a suitable land or aquifer treatment system. Similarly, biological agents, such as algae and bacteria, can cause clogging of infiltration surfaces and wells, limiting the subsequent use of the recharged water.

2.9 Effectiveness of the Technology

Various artificial recharge experiments have been carried out in India by different organizations, and have established the technical feasibility of the artificial recharge of unconfined, semi-confined and confined aquifer systems. However, the most important, and somewhat elusive, issue in determining the utility of this technology is the economic and institutional aspects of artificial groundwater recharge (Gould, 1992). The cost is a function of availability of source water, conveyance facilities, civil constructions, land, and groundwater pumping and monitoring facilities (CGWB, 1994). Assuming that rainwater harvesting has been determined to be feasible, two kinds of techniques, that is, statistical and graphical methods have been developed to aid in determining the size of the storage tanks. These methods are applicable for roof-top catchment systems only, and detail guidelines for design of these storage tanks can be found in Gould (1992) and Pacey and Cullis (1989).

As surface water augmentation methods, such as dams and diversions, have become more expensive and less promising in terms of environmental considerations, the prospects of storing surplus surface water underground and abstracting it whenever and wherever necessary appears to be more effective technology. In urban areas, artificial recharge can maintain groundwater levels in situations where natural recharge has become severely reduced.

2.10 Disadvantages

There are a number of problems associated with the use of artificial recharge techniques. These include aspects related to recovery efficiency (e.g., not all of the added water may be recoverable), cost effectiveness, contamination risks due to injection of recharge water of poor quality, clogging of aquifers, and a lack of knowledge about the long-term implications of the recharge process. Hence, careful consideration should be given to the selection of an appropriate site for artificial recharge in a specific area.

2.11 Cultural Acceptability

Cultural considerations, stemming from socio-economic concerns, often enter into the selection of a recharge method and site. The availability of land, land uses in adjacent areas, public attitudes, and legal requirements generally play a role in defining the acceptability of artificial recharge in a given setting. In urban areas, where land availability, costs and uses in adjacent areas may pose restrictions, injection wells, shafts or small pits requiring highly controlled water supplies and little land area may be preferable to larger scale, surface spreading recharge methods (Widstrand, 1978).

2.12 Case Study: Mehsana Aquifer, Gujarat, India

The most common method of extraction of the groundwater in India is the tube wells dug into the aquifer. Due to the increasing demand for water these tube wells were drilled deeper and deeper reaching the deep seated aquifers. This has led to the over-exploitation of these aquifers. This heavy exploitation generally takes place in the areas where the drawn water is utilized for irrigational purposes (Abdulaziz, 1991). The 'Mehsana alluvial' aquifer in the western India is an excellent example of over-exploitation of the aquifer for irrigation. The earlier dug wells were substituted with tube wells for drawing more water. This initially improved the yield from the aquifer, but resulted in a steady decline in the piezometric levels of the aquifer (Phatdare 1992).

Direct approach of the artificial recharge was adopted by making use of the spreading channels and percolation tanks and recharge from the losses of the canals. Various experiments were conducted by the Central Ground Water Board in collaboration with the UNDP and State Groundwater Agency in order to artificially recharge the aquifer in and around the central Mehsana area of Gujarat (Rushton and Phadtare, 1989).

Groundwater level was intended to be restored in three ways: injection wells, spreading method, and recharge pit (CGWB, 1994). In the experiment conducted using the injection wells, the waters from the phreatic aquifer below the Saraswati River were utilized. A quantity of 225 cubic metres per day was injected continuously for 250 days, through pressure injection technique. At the end of this recharge cycle, an average rise of 5 m was observed in the injection well along with a rise of 0.6-1 m in the wells 150 m away from the injection well.

3 Conclusion

Rainwater harvesting and artificial recharge appear to be one of the most promising alternatives for supplying freshwater in the face of increasing water scarcity and escalating demand. The pressures on rural water supplies, greater environmental impacts associated with new surface water projects, and increased opposition from NGOs to the development of new surface water sources, as well as deteriorating water quality in surface reservoirs already constructed, constrain the ability of communities to meet the demand for freshwater from traditional sources, and present an opportunity for augmentation of water supplies using rainwater harvesting and artificial recharge.

References

Abdulaziz, A.T. (1991). Effectiveness of recharge from a surface reservoir to an underlying unconfined aquifer. *In*: Hydrology of Natural and Manmade Lakes. Proceedings of the Vienna Symposium, August 1991, IAHS Publ. No. 206.

Ambroggi, R.P. (1977). Underground Reservoirs to Control the Water Cycle. *Scientific American*, **236(5):** 21-27.

Asano, T. (ed.) (1985). Artificial Recharge of Groundwater. Boston, Mass.: Butterworth. 767 p.

Behzadian, K. and Kapelan, Z. (2015). Advantages of integrated and sustainability based assessment for metabolism based strategic planning of urban water systems. *Sc. of the Tot Environ.* Elsevier, **527-528:** 220-231:10.2016.

CGWB (2007). Manual on Artificial Recharge of Ground Water, Central Ground Water Board, 185 p.

CGWB (Central Ground Water Board) (1994). Manual on Artificial Recharge of Ground Water. Technical Series-M, No. 3. Ministry of Water Resources, Government of India. 215 p.

Dolman, B. and Lundquist, K. (2008). Roof Water Harvesting for a low Impact Water Supply: Featuring the Brazilian Ball Pre-Filter System—A Case Study. Occidental: The WATER Institute. Occidental Arts and Ecology Center (OAEC).

Gould, J.E. (1992). Rainwater Catchment Systems for Household Water Supply. Environmental Sanitation Reviews, No. 32. ENSIC, Asian Institute of Technology, Bangkok.

Huisman, L. and Olsthoorn, T.N. (1983). Artificial Groundwater Recharge. Pitman Publishing, Massachusetts.

Nag Choudhury, K.K., Misra, A.K., Sinha, S. and Mukherjee, P. (2016). Conjunctive use of Groundwater and Surface water for enhancement of irrigated area in parts of South 24 Parganas district, West Bengal. *Indian Groundwater*, **VII:** 65–74.

Oaksford, E.T. (1985). Artificial Recharge: Methods, Hydraulics and Monitoring. *In:* Artificial Recharge of Groundwater. T. Asamo (ed.). Butterworth Publishers, Boston.

Pacey, A. and Cullis, A. (1989) Rainwater Harvesting: The Collection of Rainfall and Runoff in Rural Areas, WBC Print Ltd., London.

Phatdare, P.N. (1992). Artificial Recharge Studies in Mehsana and Coastal Saurashtra Areas, Gujarat State – Interim Report on Concept, Methodology and Status of Work. CGWB Pilot Project for Artificial Recharge.

Rushton, K.R. and Phadtare, P.N. (1989). Artificial Recharge Pilot Projects in Gujarat, India. *In:* Groundwater Management: Quantity and Quality. IAHS Publication No. 188.

Todd, D.K. (1980). Groundwater Hydrology. Second Edition. John Wiley & Sons, New York.

Wall, B.H. and McCown, R.L. (1989). Designing Roof Catchment Water Supply Systems Using Water Budgeting Methods. *Water Resources Development*, **5:** 11-18.

Widstrand, C. (Ed.) (1978). The Social and Ecological Effects of Water Development in Developing Countries. Pergamon Press, New York.

Chapter 19
Hydrogeological Assessment for Development and Management of Baseflow for Public Water Supply in Semi-arid and Fluoride Affected Hard Rock Areas

Pradip K. Sikdar, Shouri Dutta, Saheb Das, Surajit Chakraborty, and Utsab Ghosal

1 Introduction

Worldwide, water resources in arid and semi-arid regions are finite; water tables are very low and run off is very high. Many of these areas have rapid growth of population resulting in increasing demand for water, deteriorating water quality, increasing environmental degradation and impending climate change. This situation requires more effort to assess water resources for local, regional and national planning and management in order to sustain development. Most of the economically viable development of water resources has already been implemented (Hamdy et al. 1995). The gap between demand and supply of water is increasing steadily in these regions. Therefore, it has now become imperative to develop innovative management options to close the gap between demand and supply of water which should take into consideration environmental concerns (Shadeed et al. 2007). In these areas groundwater is commonly the most important water resource. In many areas the groundwater has been over-exploited leading to the rapid fall in the water table or increase in the cost of exploitation because of the nature and disposition of the aquifer system. Baseflow which is a component of stream flow that is attributed to shallow groundwater discharge has generally been left unutilized. Therefore, development of baseflow for small to medium water supplies can be an important component of water management strategies in semi-arid areas of the world, in general and India, in particular.

P. K. Sikdar (✉) · S. Dutta · S. Chakraborty · U. Ghosal
Department of Environment Management, Indian Institute of Social Welfare and Business Management, Kolkata, India
e-mail: p_sikdar@hotmail.com

S. Das
Centre for Ground Water Studies, Kolkata, India

P. K. Sikdar (ed.), *Groundwater Development and Management*,
https://doi.org/10.1007/978-3-319-75115-3_19

Bankura district is one of such semi-arid areas in the western part of West Bengal where the groundwater at places contains high concentration of fluoride. Geographically this area is situated on the eastern slope of Chota Nagpur Plateau and located between 22° 38′N and 23° 38′N latitude and between 86° 36′E and 87° 46′E longitude, covering an area of about 6882 sq. km (DHDR 2007). The district is characterized by a dry tropical climate with a winter and a very hot summer. Southwest monsoon is the principal source of rainfall in the district. However, it is erratic in nature and there is occasional drought condition in between rain spells which may last for several days (Ray and Shekhar 2009). The area receives a mean annual rainfall varying from a little more than 1430 mm to the south-east to less than 1200 mm to the north-west which comes within 53–77 rainy days and about 70% of which comes during the four monsoon months of June to September. The sources of the water supply, for example, streams, small reservoirs etc., begin to dry up with the approaching summer when people have to depend entirely upon groundwater. But even the large diameter dug wells which are usually the source of groundwater in this region become almost ineffective by this time.

The drainage of the district is mainly controlled by the rivers Damodar, Dwarakeswar and the Kangsabati along with their network of tributaries (Fig. 19.1). They have in general a south easterly flow. The courses of the principal rivers are approximately parallel to each other. The river Dwarakeswar, the target river of this study, is the second important river in the district after river Damodar. It flows approximately through the middle of the district and divides it into two halves. It rises in the adjoining Purulia district, flows in a south easterly course and enters Bankura district in Chhatna block. Below its confluence with the river Silabati in the lower course, it is known as the Rupnarayan, which debouches into the river Bhagirathi near Diamond Harbour in 24 Parganas district of West Bengal. The Silabati, popularly known as Silai, is the largest tributary of the Dwarakeswar. The Joypanda is the principal tributary of the Silabati, and the only other important tributary of the Dwarakeswar is the Amodar, which rises in Joypur police station and leaves the district in Kotulpur Police Station to enter the Arambagh sub-division of the Hoogly district.

The geology of the area (Fig. 19.2) exhibits four different geological formations exposed from west to east (Saha et al. 1995). The area lying west of the line joining Barjora (23°25′:87°18′), Beliator (23°19′:87°13′), Bankura (23°14′:87°08′) and Bibarda (23°03′:87°02′) is covered by crystalline rocks of Precambrian age. Granite gneiss, mica schist, anorthosite, amphibolite, hornblende schist and gneiss are the principal rocks exposed in this part of the area. These rocks stand out prominently as mounds and ridges. They are well foliated and have well-defined joints. The foliation plane in the granite gneiss has usually a northwesterly trend with moderate to high easterly dips. Mainly three sets of joints are present in these rocks. They are (i) parallel to foliation, (ii) perpendicular to the foliation and (iii) another set trending NNW-SSE. All these joints are dipping at a very high angle (70°–90°). Other than these three prominent joint planes, there are numerous poorly developed joints. Towards north the foliation plane curves eastward and around Gangajalghati it trends roughly ENE-WSW dipping at a low angle towards north. In the extreme

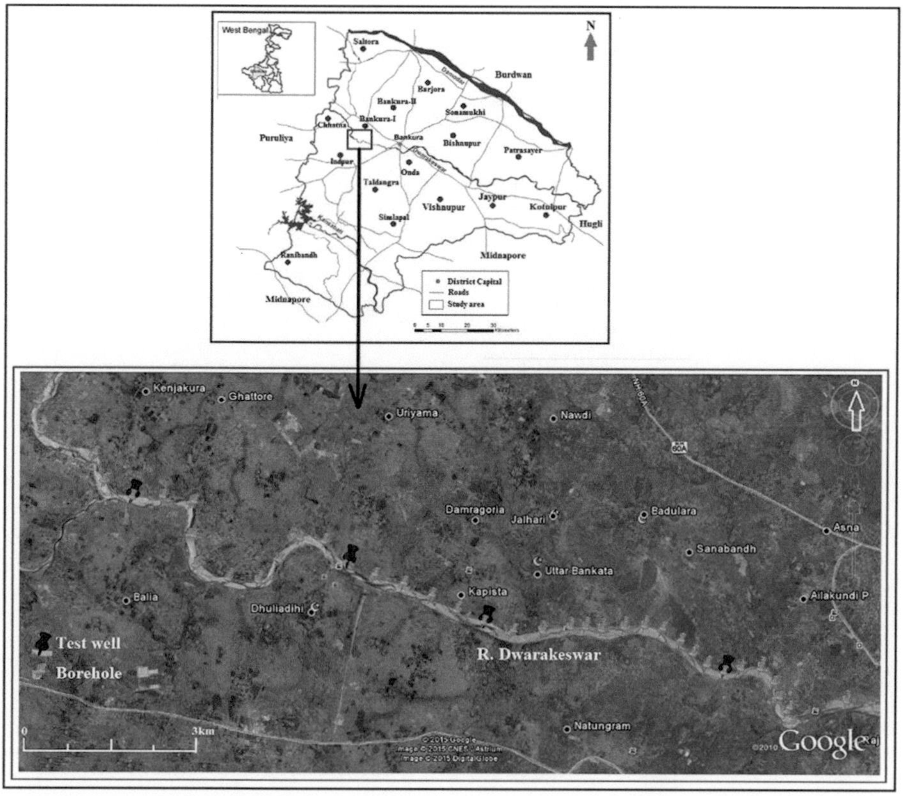

Fig. 19.1 Location of field area

north-western part of the district anorthosite rocks are exposed. The anorthosite is emplaced along the core of a doubly plunging (east-west) antiformal structure developed on the lithologic contacts of the associated metasediments and metanorites (Roy and Saha 1975; Roy 1977). The foliation planes in the anorthosites trend E-W with vertical joints parallel to the foliation plane. The joint planes in these rocks are important because the occurrence and movement of groundwater is mainly controlled by these planes.

To the immediate east of the Precambrian area a thick mantle of laterite and older alluvium occupies the highlands between the Damodar and Jaipanda-Silai rivers. They are believed to be of Pleistocene age. The laterite occurs as hard massive blocks or as nodules, being partly cemented in situ by weathered product of crystalline rocks. The older alluvium represents transported weathered product of the Precambrian rocks. The thickness of this laterite capping increases gradually from 6 m to 25 m from west and northwest to east and southeast until they are covered by recent alluvium which occurs in the river valleys and the flat land in the eastern part of the area (Saha et al. 1995).

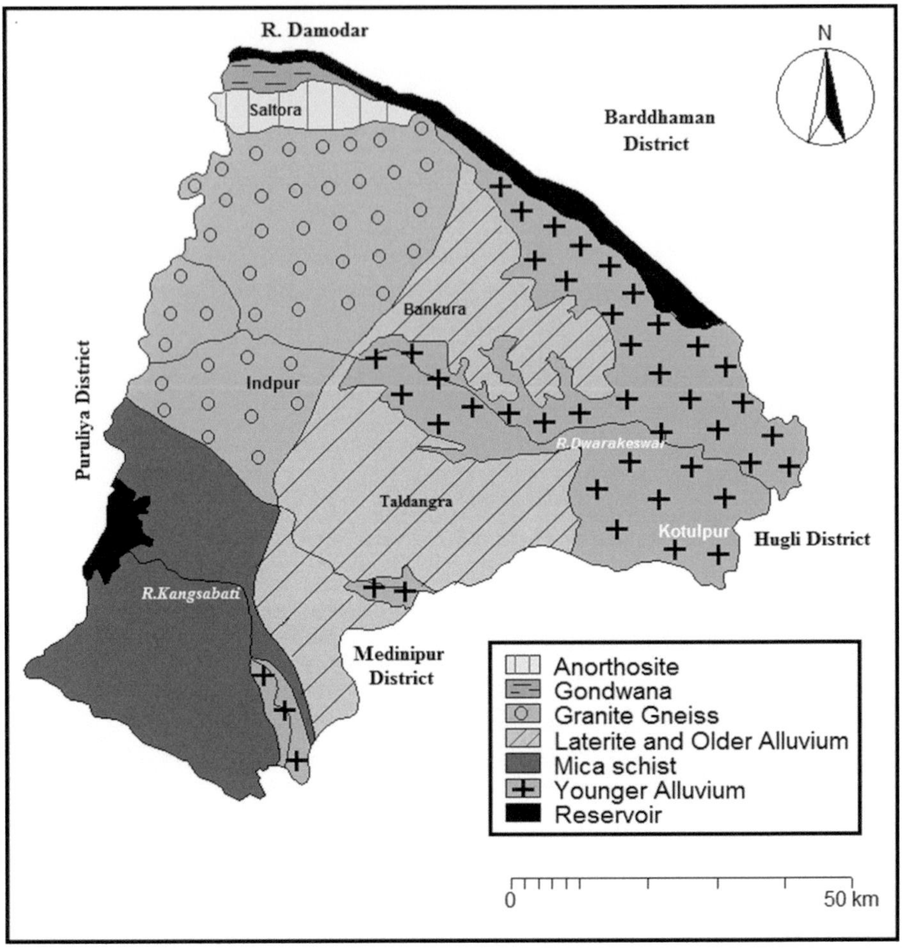

Fig. 19.2 Geological map of Bankura District

The lower Gondwana rocks of Raniganj Coalfield extend to the northern extremity of the district between Mejia and Borjora. These are mainly coarse feldspathic sandstone with interbedded grey shales containing plant fossils. A few seams have also been discovered within this formation.

Groundwater occurs under unconfined condition in the upper weathered residuum of the consolidated to semi-consolidated rocks. The thickness of the weathered mantle is <15 m and is being developed through dug wells (Saha et al. 1995). Groundwater also occurs in the deeper part below the weathered mantle in the fractures within granite gneiss between 30 and 60 m depth and is being developed through bored wells within 100 m bgl having a yield of 5–30 m^3/hr (CGWB and SWID 2011). The heterogeneity of fractures has limited the scope of large-scale development of groundwater. In both old and recent alluvium areas the aquifer occurs between 30 and 270 m below ground level (bgl) having storativity and transmissivity ranging from 1.02×10^{-3} to 2.18×10^{-4} and 273–806 m^2/day respectively. The wells tapping this aquifer discharge at 8–123 m^3/hr (CGWB and SWID 2011).

Hydrogeochemical investigations (Chakrabarti et al. 2012) revealed that the groundwater of 13 out of 22 blocks in Bankura district is severely affected with fluoride. The blocks affected with fluoride are Bankura-I, Bankura-II, Barjora, Gangajalghati, Mejia, Saltora, Chhatna, Indpur, Hirabandh, Raipur, Sarenga, Simlapal and Taldangra. The maximum concentration of 10.8 mg/L is found in Hirabandh block followed by 5.10 mg/L in Bankura-II block (Chakrabarti et al. 2012).

The stage of groundwater development of the western hard rock areas of Bankura district is very low, especially Bankura-I and Indpur block where it is only 7.78% and 2.07% respectively (CGWB and SWID 2011). But large scale groundwater development is not feasible because of complex hydrogeological set up and occurrence of fluoride in groundwater. Therefore, this water scarce and fluoride affected blocks of Bankura district require special attention from the point of view of safe drinking water supply. Action should be taken to identify, exploit and utilize areas with higher potential of good quality water resources particularly the baseflow of nearby ephemeral rivers.

This paper, therefore, focuses on the availability and utilization of the baseflow of the river Dwarakeswar by understanding the subsurface hydrogeology, aquifer characteristics, rate of baseflow and quality of the water at selected locations, then identifies sites for baseflow abstraction and finally provides the designs for the abstraction structures that may be constructed to supply drinking water to the adjacent fluoride affected villages.

2 Field Area

The field area is a 16.5 km stretch on the river Dwarakeswar between 86°54′12-″-87°2′13″E and 23°13′12″-23°16′2″N in Bankura-I and Chhatna blocks of Bankura district covering an area of 136 sq. km (Fig. 19.1).

3 Methods

3.1 Drilling and Well Construction

To investigate the subsurface lithology and to obtain sediment for investigation, 117 boreholes of 50 mm diameter, four test wells of 150 mm diameter and 16 observation wells of 50 mm diameter were drilled on the river beds of river Dwarakeswar during April–May 2015 (Fig. 19.3). The boreholes and observation wells were

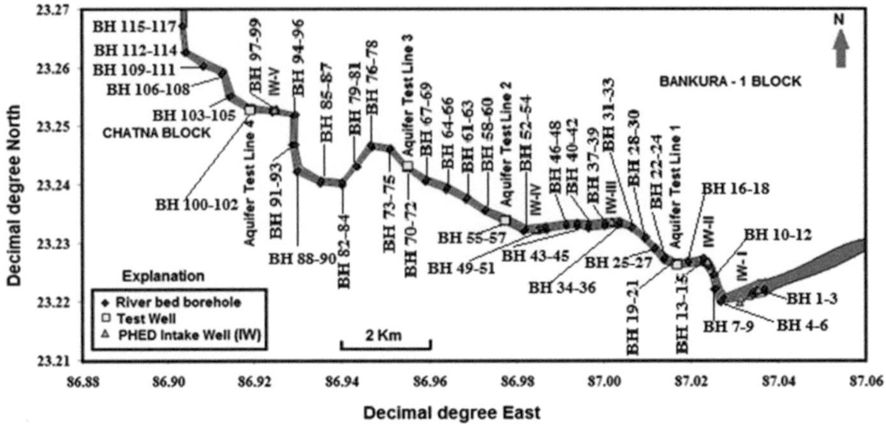

Fig. 19.3 Distribution of boreholes, test wells and observation wells on the bed of river Dwarakeswar

drilled using hand-sludging method of drilling (Ball and Danert 1999; Ali 2003) and the test wells were drilled using manually-operated direct rotary drilling method. Drill cuttings were collected at 1 m interval and logged, photographed, and sampled. The visualization package RockWorks® 15 was used to construct longitudinal and transverse sub-surface lithological cross-sections based on the logged boreholes.

After lowering of the casing assembly in the test and observation wells the annular space between the casing and the borehole was packed with clean, coarse sand. After this the test wells were developed by mechanical surging using a diesel operated submersible motor pump and observation wells by hand-pump to remove the fine materials from the screen apertures and aquifer around the wells.

3.2 *Pumping Test*

To calculate the cardinal aquifer parameters such as hydraulic conductivity, transmissivity and storage coefficient 72-hour pumping test was conducted on four test wells. Test wells and observation wells were constructed at suitable locations based on the drilling results.

During the test, observations include measurement of water levels during pumping in the test well and observation wells, discharge rate and measurement of recuperation of water level in the test well and observation wells. All water level measurements were done with reference to fixed measuring points in each test well and observation well. For this, steel tape graduated up to 1 mm interval was used. For discharge measurements, volumetric method and 'V' notch methods were used. The static water level was measured for the test wells and observation wells prior to commencement of pumping.

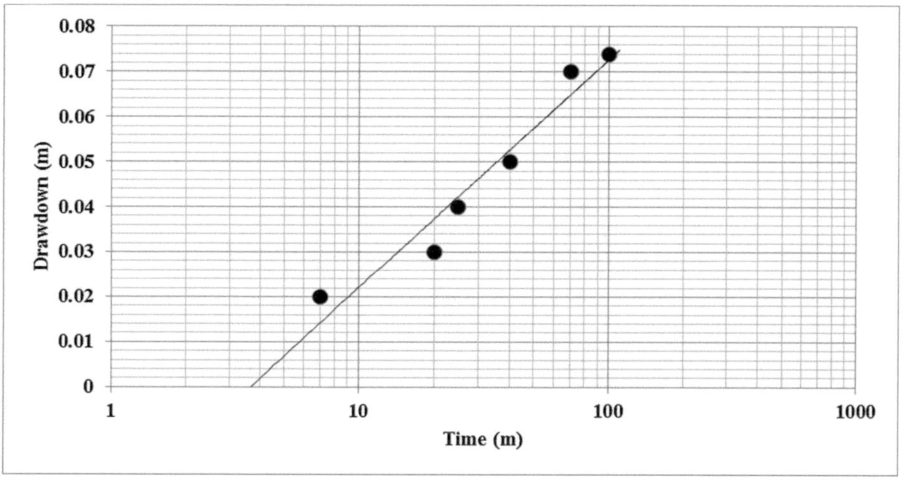

Fig. 19.4 Plots of drawdown against time of observation well 1.4 of test well 1

The test consists of pumping of a test well at a constant discharge and shutting off the pump after 72 h and recording of water levels in the test well and observation wells, both during pumping and after stoppage of pump. Immediately after starting of pump or its stoppage the water levels fall or rise very rapidly and hence water level measurements were rapid during this period in order to define the drawdown or recovery curves precisely.

Jacob's straight line method (Cooper and Jacob 1946) based on Theis (1935) equation has been used to estimate the cardinal aquifer parameters. In this method, the 'third' segment of 'time-drawdown' curve of an observation well has been used to determine transmissivity (T) and specific yield (SY) values. An example of the third segment of plots of drawdown against time in observation well 1.4 of test well 1 is given in Fig. 19.4.

The safe distance ($2r_o$) which is double the radius of the cone of depression has been determined using the non-steady relation of Theis (1935). An example of calculated drawdown curve of test well 1 is given in Fig. 19.5. Wells located at distance more than this will not interfere with each other when the two wells are pumped simultaneously.

3.3 *Water Level and Groundwater Flow*

To understand the groundwater regime of the field area, hydrological data were collected from dugwells from the near vicinity villages of the river Dwarakeswar in May 2015. For this, 52 network stations (dug wells) were selected from both sides of the river bank for water level monitoring in order to understand the behaviour of the

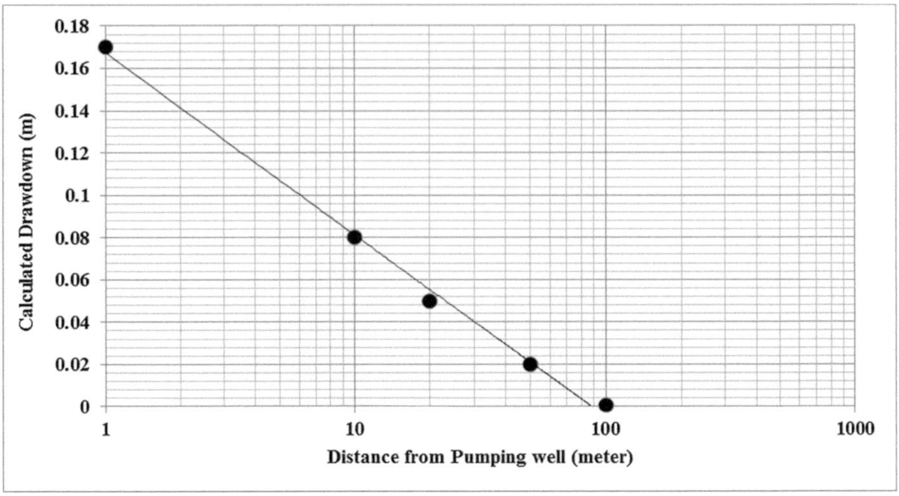

Fig. 19.5 Calculated drawdown curve of test well 1

water table. For water level measurements, steel tape graduated up to 1 mm interval and 169 Solinst Model 101 Water Level Meter were used.

The geographical location of each network station was determined using a GARMIN 72H GPS with WGS 84 as datum. The elevation of each borehole location in metres was extracted from the Shuttle Radar Topography Mission (SRTM) digital elevation data (EROS 2002) and CARTOSAT DEM, with a spatial and altitude resolution of 90 m and 30 m respectively, by bilinear interpolation. The elevation data were stored in an attribute file for contour generation using Surfer10. Local polynomial interpolation method was used for contour generation.

Using the water level measurements of the test wells and observation wells the quantity of groundwater flowing through the given cross-section of the river has been calculated using the Darcy's equation $Q = TiL$ where, $Q =$ quantity of water flowing through a given cross-section, $T =$ transmissivity of the aquifer, $i =$ hydraulic gradient, and $L =$ length of the cross-section.

3.4 *Grain Size Analysis*

The sediment samples of the test and observation wells were subjected to mechanical analysis by sieving with sieves of the American Society for Testing Material (ASTM) with standard sieve size 5–325 using 100–150 gm of samples (Folk 1968). The grain size retained in the sieve is plotted against cumulative percentage. An example of the grain size distribution curve of the sediments of test well 1 is given in Fig. 19.6. The grain size distribution curves were used to obtain the d_{10}, d_{40}, d_{60}, d_{70} and Uniformity Coefficient (UC) values. Based on these values the screen

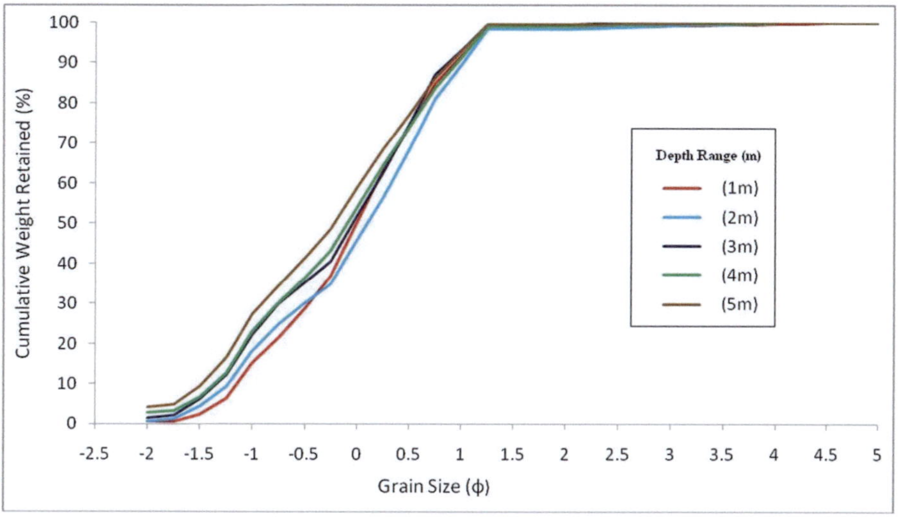

Fig. 19.6 Grain size distribution curves of sediment samples of test well 1

slot size and recommendation for construction of natural-pack infiltration gallery or gravel-pack gallery were suggested.

3.5 Design of Abstraction Structure

In geological environments beneath the river where the aquifer is thin and in an unconfined condition, infiltration galleries or wells with horizontal and vertical screens may be constructed. Infiltration gallery is a horizontal perforated or porous pipe with open joints surrounded by a gravel filter envelop laid in a permeable material with high water table and a continuous recharge. The design requirements for infiltration gallery are grain size, sediment thickness, hydraulic conductivity, water table depth and radius of influence for specified discharge. The maximum length with 100% safety factor of the infiltration gallery is calculated using the following equation (Driscoll 1986):

$$\frac{Q}{L} = q = \frac{K}{2r_o}\left(H^2 - h^2\right)$$

where Q = Discharge (m^3/day), q = Discharge per unit length (m^3/day/m), K = Hydraulic conductivity (m/day), r_o = Radius of influence (m), H = Height of static water level above gallery bottom (m), h = Height of pumping water level above gallery bottom under steady state condition (m) and L = Calculated length of gallery (m).

3.6 Water Quality

To understand the quality of the groundwater below the river beds, water samples were collected from the test wells and analysed for chemical, bacteriological and pesticides concentrations. The samples were collected in clean polyethylene bottles. Prior to collection the sampling bottles were washed three times with the well water before filling with the sample water. For bacteriological analysis samples were collected in sterilized dark glass bottles. The samples were taken to the laboratory within the shortest possible time and during transportation due care was taken to protect the water samples from direct sunlight. In the laboratory the samples were filtered using 0.45 mm Millipore filter paper and acidified with ultra-pure nitric acid for cation analyses. For anion analyses, these samples were refrigerated at 4 °C. The ions in the groundwater samples were determined as per the standard procedures. Each groundwater sample was analyzed in the laboratory for sixteen chemical parameters, two bacteriological parameters and 18 pesticide parameters as per the standard methods of APHA (1995). The analytical precision of chemical analysis of water samples was checked by calculating the charge balance error (Freeze and Cherry 1979). If the error is within ±10%, the analysis was assumed to be good.

4 Results

4.1 Lithology and Stratigraphy

Three sub-surface lithological cross-sections have been constructed using the borehole lithological data along the northern, middle and southern limits of the river bed. The depth of the boreholes ranges between 1 and 6 m within a stretch of 16.5 km. The subsurface lithological profiles (Fig. 19.7) reveal that the upper part of the lithological column up to a depth of 6 m consists of medium to coarse sand with pebbles. Below the depth of 6 m hard rock is encountered. The 4–6 m thick medium to coarse sand is the upper unconfined aquifer below the river bed through which the baseflow of the river takes place.

4.2 Aquifer Test

Based on the subsurface lithology below the river bed, four test wells and 16 observation wells were constructed on the bed of river Dwarakeswar (Fig. 19.3). Each of the aquifer test line contains one test well and four observation wells. The

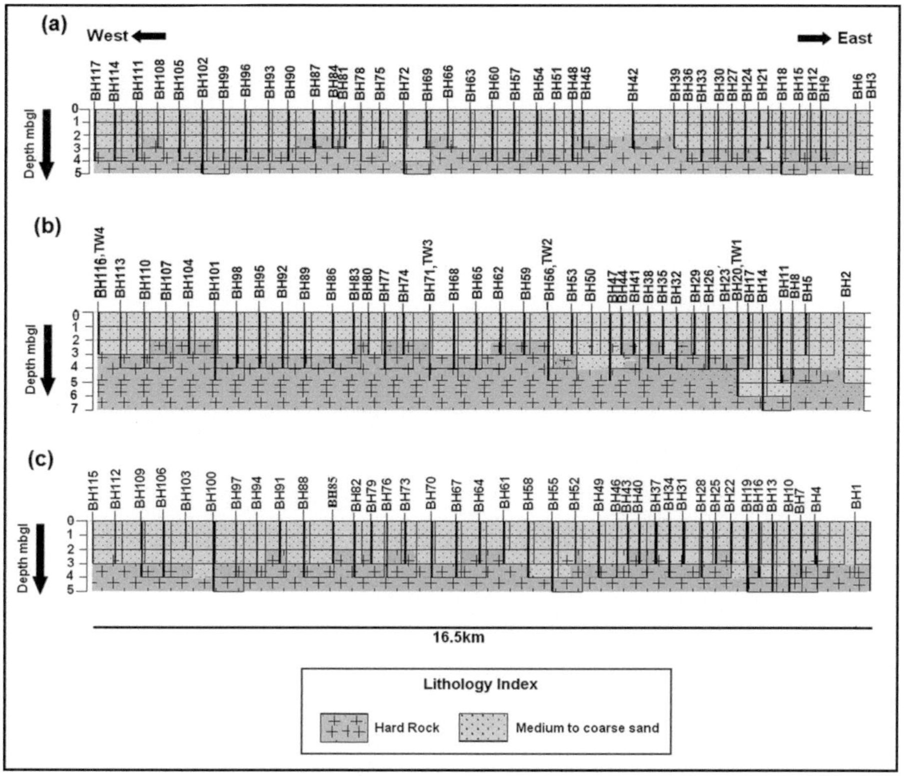

Fig. 19.7 Sub-surface lithological section along (**a**) northern, (**b**) middle and (**c**) southern limits of the bed of River Dwarakeswar

distribution of the test wells and the observation wells are shown in Fig. 19.8. The subsurface lithology and the aquifer test line along the direction of flow of the river is given in Fig. 19.9.

Water level measurements were carried out in each test well and observation well prior to pumping. Repeat measurements of the water level in the available wells were carried out towards the end of the field work in the month of May 2015 to understand the lowest water level. The water level data are given in Table 19.1. Pumping test was carried out in each of the test well for 72 h. The rate of discharge of the test well, maximum drawdown of the test wells and the observation wells and specific capacity of the test well are given in Table 19.2.

The water level below river Dwarakeswar is very shallow and rests between 0.11 and 1.01 m below the bed of river (Table 19.1; Fig. 19.9). The hydraulic parameter of the aquifer is given in Table 19.3. The storage coefficient value of the aquifer ranges between 0.14 and 0.35 indicating that the aquifer is unconfined in nature. The transmissivity of the aquifer is highest along aquifer test line 2. The radius of influence of the test wells ranges between 99 and 103 m.

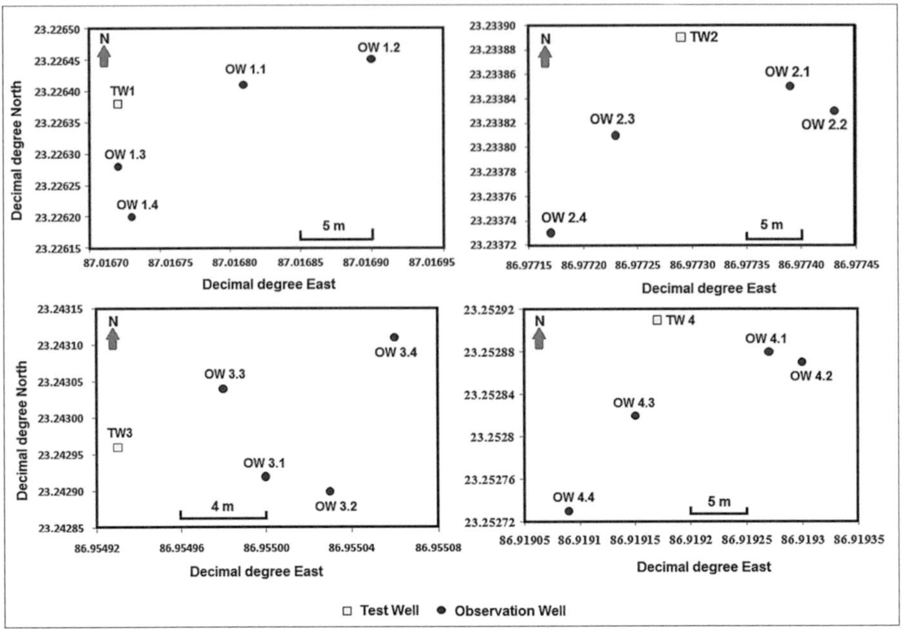

Fig. 19.8 Distribution of observation wells and test wells on River Dwarakeswar

4.3 Water Table Configuration

The lowest elevation of the water table in the study area is towards the river Dwarakeswar. Therefore, the groundwater flows towards the river from both sides. On the northern side of the river the groundwater flow is towards south and on the southern side flow is northerly (Fig. 19.10). Hence, river Dwarakeswar is an effluent or gaining river. Major lineaments in the study area were identified from the satellite data (Fig. 19.10). The lineaments were categorized into water-bearing and non-water-bearing depending on alignment of the lineaments and groundwater flow direction. The water-bearing lineaments trend towards NNW-SSE, NE-SW and NW-SE directions. These water bearing lineaments act as conduit for supplying the baseflow of the river.

4.4 Flow of Groundwater across Aquifer Test Line

The flow rates at the cross-sections of river Dwarakeswar along the aquifer test line have been computed using Darcy's equation and given in Table 19.4. The groundwater flow varies between 0.97 and 2.0 million gallons per day (MGD), being highest in test well 2.

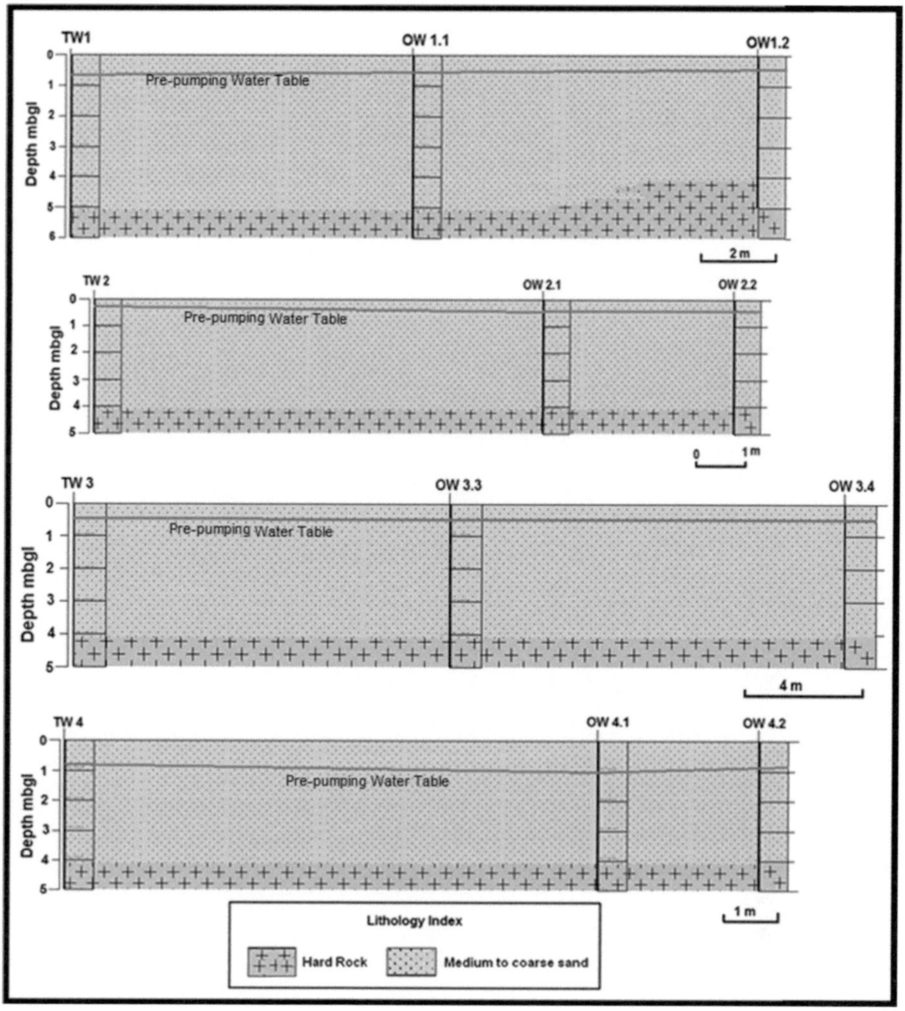

Fig. 19.9 Aquifer test line 1, 2, 3 and 4 section on river Dwarakeswar

4.5 Sediment Texture

The results of the grain size analysis of the test wells are given in Table 19.5. The sediments which constitute the aquifer material of TW1 are sand particles with size varying from coarse silt to pebble up to a depth of 5 m below river bed level (brbl). The mean grain diameter ranges between −0.06 ϕ (very coarse sand) and 0.18 ϕ (coarse sand) with an average of −0.10 ϕ (very coarse sand). The aquifer material of TW2 are sand particles which vary in size from very fine sand to granule up to a depth of 4 m brbl. The mean diameter of the particles ranges between 0.02 ϕ (coarse sand) and −0.05 ϕ (very coarse sand) with an average of −0.02 ϕ (very coarse

Table 19.1 Water level data in test wells and observation wells

Block	River	Well No.	MP (m)	WL BMP (m)	WL below river bed (m) May
Bankura –I	*Dwarakeswar*	TW1	0.70	1.43	0.73
		OW 1.1	0.20	0.78	0.58
		OW 1.2	0.14	0.71	0.57
		OW 1.3	0.32	0.65	0.33
		OW1.4	0.37	0.68	0.31
		TW 2	0.78	1.02	0.24
		OW 2.1	0.27	0.60	0.33
		OW 2.2	0.24	0.58	0.34
		OW 2.3	0.36	0.47	0.11
		OW 2.4	0.30	0.60	0.30
		TW 3	0.71	1.06	0.35
		OW 3.1	0.18	0.52	0.34
		OW 3.2	0.23	0.62	0.39
		OW 3.3	0.46	0.93	0.47
		OW 3.4	0.30	0.78	0.48
		TW 4	0.66	1.58	0.92
		OW 4.1	0.36	1.37	1.01
		OW 4.2	0.34	1.22	0.88
		OW 4.3	0.33	0.98	0.65
		OW 4.4	0.36	1.03	0.67

sand). The sediments of TW3 are sand particles with grain size varying from very fine sand to granule up to a depth of 4 m brbl. The mean diameter of the grains ranges between −0.32 ϕ (very coarse sand) and −0.48 ϕ (very coarse sand) with an average of −0.38 ϕ (very coarse sand).

The sediments which constitute the aquifer material of TW4 are sand with sizes from very fine sand to granule up to a depth of 4 m brbl. The mean diameter of the grains ranges between −0.17 ϕ (very coarse sand) and −0.29 ϕ (very coarse sand) with an average of −0.25 ϕ (very coarse sand).

4.6 Hydrogeochemistry

The results of the chemical, bacteriological and pesticide analyses are given in Table 19.6. The groundwater quality with respect to chemical parameters is, in

Table 19.2 Discharge-drawdown data of test wells and observation wells

Block	River	Test well no	Discharge (m^3/day)	Max drawdown (m)	Specific capacity (m^3/day/m)	Max drawdown in observation wells (m)			
						1	2	3	4
Bankura-I	Dwarakeswar	TW-1	1555	0.25	6220	0.03	0.025	0.04	0.01
		TW-2	1414	0.805	1757	0.065	0.02	0.07	0.03
		TW-3	1469	0.59	2490	0.09	0.075	0.12	0.09
		TW-4	389	0.25	1556	0.03	0.03	0.035	0.03

Table 19.3 Hydraulic parameters of aquifers below river bed

Well No.	Aquifer thickness (m)	Measuring point (m above river bed level)	S.W.L. (m below river bed level)	Pump capacity (H.P.)	Discharge (m^3/day) (Q)	Max drawdown (m)	Transmissivity (T) (m^2/day)	Hydraulic conductivity (K) (m/d)	Specific yield (S_y)	Radius of influence (r_o) m	Safe distance ($2r_o$) m
OW-1.3	4.5	0.32	0.33	3.5	972	0.04	6358	1413	0.19	99	198
OW-1.1	5	0.20	0.58	5	1555	0.03	5932	1186	0.40		
OW-1.3	4	0.32	0.33			0.04	3849	855	0.16		
OW-1.4	4.5	0.37	0.31			0.01	4593	1531	0.10		
Average							*5183*	*1246*	*0.21*		
OW-2.1	4	0.27	0.33	5	1414	0.065	7563	1891	0.22	101	202
OW-2.3	4	0.36	0.11			0.07	5756	1439	0.33		
Average							*6660*	*1665*	*0.28*		
OW-3.1	4	0.18	0.34	5	1469	0.09	4969	1242	0.38		
OW-3.3	4	0.46	0.47			0.12	4224	1056	0.49		
OW-3.4	4	0.30	0.48			0.09	4566	1142	0.18	100	200
Average							*4586*	*1147*	*0.35*		
OW-4.1	4	0.36	1.01	2.5	389	0.03	4667	1167	0.16		
OW-4.2	4	0.34	0.88			0.03	3389	847.3	0.13	103	206
OW-4.3	4	0.33	0.65			0.035	3195	799	0.13		
Average							*3750*	*938*	*0.14*		

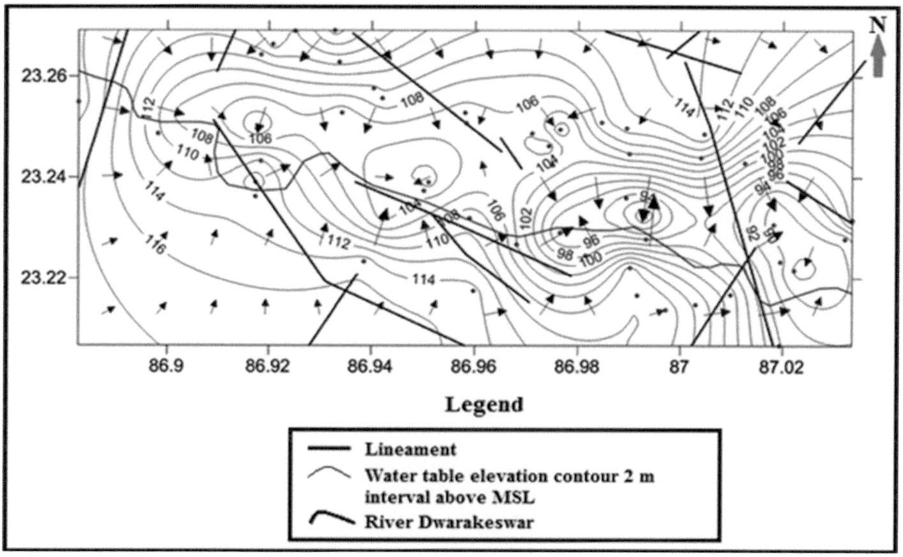

Fig. 19.10 Map showing groundwater elevation contour, groundwater flow direction and lineaments

general, within the permissible limit of IS 10500:2012. Bacteriologically, the groundwater samples are contaminated with coliform bacteria. Pesticides were not found in any of the samples (Table 19.7).

5 Discussions

5.1 Geology

In the study area, river Dwarakeswar drains through granite gneiss terrain. The subsurface lithological sequence of the river bed reveals that the upper part of the lithological column up to a maximum depth of 6 m consists of medium to coarse sand with pebbles which is the weathered residuum and below which hard gneissic rock is encountered. The thickness of the weathered residuum increases away from the river on both the banks. The weathered residuum is the unconfined aquifer which is tapped by dug wells. There are many lineaments which crisscross on either side of the river trending NNW-SSE, NW-SE, NE-SW and E-W directions dipping at high angles.

The grain sizes of sediments which constitute the aquifer materials below the river bed vary from very coarse silt to pebble. The mean diameter of the grains varies from very coarse sand to granules. Riverbed tubewell with horizontal and vertical screens or infiltration gallery can be constructed at appropriate locations.

Table 19.4 Groundwater discharge across aquifer test line

River	Location	Test well no.	Transmissivity	Cross-sectional length of river	Hydraulic gradient	Ground water flow rate		
			(m^2/day)	(m)		m^3/day	MGD	MLD
Dwarakeswar	Bankura-I	TW 1	5183	145	7.5×10^{-3}	5636	1.49	5.64
		TW 2	6660	103	11×10^{-3}	7546	2.0	7.57
		TW 3	4586	106	15.2×10^{-3}	6514	1.72	6.51
		TW 4	3750	132	7.43×10^{-3}	3678	0.97	3.67

Table 19.5 Results of grain size analysis of sediment samples of test well (TW1) on river Dwarakeswar, Bankura-I Block

	Depth range (m)	Uniformity coefficient (UC)	Screen slot size (mm)	Median (ϕ)	Mean diameter (ϕ)	Lithology	Porosity (%)	Screen pack
TW 1	0–1	2	1.14	−0.09	0.18	Coarse Sand	39	Natural
	1–2	2.3	1.1	0.1	−0.06	Very Coarse Sand	39	Natural
	2–3	2.28	1.23	−0.04	−0.18	Very Coarse Sand	39	Natural
	3–4	2.64	1.32	−0.1	−0.18	Very Coarse Sand	39	Natural
	4–5	2.69	1.45	−0.21	−0.27	Very Coarse Sand	39	Natural
	Average	2.38	1.25	−0.07	−0.1	Very Coarse Sand	39	Natural
TW 2	0–1	2.30	1.08	0.12	−0.01	Very Coarse Sand	39	Natural
	1–2	2.40	1.13	0.1	−0.05	Very Coarse Sand	39	Natural
	2–3	2.37	1.14	0.09	−0.05	Very Coarse Sand	39	Natural
	3–4	2.32	1.09	0.19	0.02	Very Coarse Sand	39	Natural
	Average	2.35	1.11	0.13	−0.02	Very Coarse Sand	39	Natural

(continued)

Table 19.5 (continued)

	Depth range (m)	Uniformity coefficient (UC)	Screen slot size (mm)	Median (ϕ)	Mean diameter (ϕ)	Lithology	Porosity (%)	Screen pack
TW 3	0–1	3.02	1.71	−0.4	−0.4	Very Coarse Sand	39	Natural
	1–2	3.11	1.65	−0.3	−0.32	Very Coarse Sand	39	Natural
	2–3	2.89	1.57	−0.3	−0.32	Very Coarse Sand	39	Natural
	3–4	3.05	1.74	−0.44	−0.48	Very Coarse Sand	39	Natural
	Average	3.02	1.67	−0.36	−0.38	Very Coarse sand	39	Natural
TW 4	0–1	2.69	1.48	−0.28	−0.27	Very Coarse Sand	39	Natural
	1–2	2.35	1.27	−0.12	−0.17	Very Coarse Sand	39	Natural
	2–3	2.51	1.43	−0.23	−0.25	Very Coarse Sand	39	Natural
	3–4	2.64	1.45	−0.25	−0.29	Very Coarse Sand	39	Natural
	Average	2.55	1.41	−0.22	−0.25	Very Coarse Sand	39	Natural

Table 19.6 Results of the chemical and bacteriological analyses

	Concentration				Groundwater quality with respect to IS 10500:2012
	Dwarakeswar rivers				
Parameters	TW1	TW2	TW3	TW4	(Acceptable Limit)
pH	7.5	7.2	7.6	7.4	6.5–8.5
Turbidity (NTU)	0.43	0.56	0.35	0.58	1
Alkalinity (as $CaCO_3$) (mg/L)	88	92	72	96	200
TDS (mg/L)	146	164	156.0	186.0	500
Hardness (mg/L)	108	120	112	120	200
Bicarbonate (mg/L)	130.4	144.3	130.4	137.7	200
Chloride (mg/L)	18.3	18.3	14.6	25.6	250
Sulphate (mg/L)	6.5	5.75	5.25	7.5	200
Nitrate (mg/L)	1.95	2.3	2.75	3.99	45
Fluoride (mg/L)	0.419	0.496	0.606	0.677	1.0
Arsenic (mg/L)	1.0	1.0	2.0	0.0	50
Sodium (mg/L)	17.6	20.5	18.3	20.5	–
Potassium (mg/L)	1.2	1.3	1.3	1.3	–
Calcium (mg/L)	24.4	28.3	25.1	25.9	75
Magnesium (mg/L)	8.6	10.5	10.0	10.9	30
Manganese (mg/L)	<0.025	<0.025	<0.025	0.05	0.1
Iron (mg/L)	0.13	0.11	0.10	0.12	0.3
Chromium (+3) (mg/L)	<0.05	<0.05	<0.05	<0.05	0.05
Total coliform (MPN/100 ml)	>1600	>1600	400	370	0
Faecal coliform (MPN/100 ml)	80	60	37	50	0

5.2 *Hydrogeology*

Groundwater in the study area occurs within both the moderately thick to thin aquifer under unconfined to semi-confined condition. The water table generally declines with varying gradients from west, north-west to east and south-east direction and broadly conforms to the topographical slope. In this region, the groundwater is mainly abstracted through open dugwell with limited number of low-duty tubewells. Heterogeneous character of the water-bearing formation with complex aquifer geometry prevails in the area and is feasible for open dugwells of 10 m to 15 m depth having 3 m diameter. The water-bearing formations are discontinuous and at places groundwater is held under pressure in the fractured conduits. Some water is also retained in the thin cover of soil and alluvium mantling the stream channel. The net groundwater availability of Bankura-I block is 35.56 MCM and the stage of groundwater development is 7.78%. The water bearing lineaments trends towards NE-SW, NNW-SSE and NW-SE directions. These water bearing lineaments act as

Table 19.7 Results of pesticide analysis

	Concentration				Groundwater quality with respect to IS 10500:2012 (Acceptable Limit)
	Dwarakeswar river				
Pesticides residue	TW1	TW2	TW3	TW4	
Alachor (μg/L)	<0.01	<0.01	<0.01	<0.01	20
Atrazine (μg/L)	<0.01	<0.01	<0.01	<0.01	2
Aldrin/Dieldrin (μg/L)	<0.01	<0.01	<0.01	<0.01	0.03
Alpha HCH (μg/L)	<0.01	<0.01	<0.01	<0.01	0.01
Beta HCH (μg/L)	<0.01	<0.01	<0.01	<0.01	0.04
Butachor (μg/L)	<0.01	<0.01	<0.01	<0.01	125
Chlorpyriphos (μg/L)	<0.01	<0.01	<0.01	<0.01	30
Delta HCH (μg/L)	<0.01	<0.01	<0.01	<0.01	0.04
2.4 Dichlorophenoxyacetic acid (μg/L)	<0.01	<0.01	<0.01	<0.01	30
DDT(o.p and p.p Isomers of DDT.DDE &DDD) (μg/L)	<0.01	<0.01	<0.01	<0.01	1
Endosulphan (alpha, beta and sulphate) (μg/L)	<0.01	<0.01	<0.01	<0.01	0.4
Ethion (μg/L)	<0.01	<0.01	<0.01	<0.01	3
Gamma-HCH (Lindane) (μg/L)	<0.01	<0.01	<0.01	<0.01	2
Isoproturon (μg/L)	<0.01	<0.01	<0.01	<0.01	9
Malathion (μg/L)	<0.01	<0.01	<0.01	<0.01	190
Methyl parathion (μg/L)	<0.01	<0.01	<0.01	<0.01	0.3
Monocrotophos (μg/L)	<0.01	<0.01	<0.01	<0.01	1
Phorate (μg/L)	<0.01	<0.01	<0.01	<0.01	2

conduits for baseflow of the river. The groundwater flow direction suggests that the river is effluent in nature in the pre-monsoon period when there is acute scarcity of drinking water. The transmissivity of the river bed aquifer is high which suggest that a good amount of baseflow can be transmitted through the riverbed aquifer. The storage coefficient value varies between 0.14 and 0.35 suggesting that the aquifer is unconfined in nature. The flow rates along the test lines on the river bed vary between 3678 and 7546 m^3/day. Therefore, the baseflow of the river can be tapped to make drinking water available to villages where groundwater is contaminated with fluoride.

5.3 *Groundwater Chemistry*

The overall groundwater quality with respect to chemical parameters is within the permissible limits (IS 10500:2012). Pesticides were also not found in any of the samples.

Bicarbonate in groundwater is generally derived from weathering of silicate minerals. Processes related to anthropogenic activities can be understood by identifying pollution by Cl^{-1}, NO_3^{-1} and SO_4^{-2} in groundwater (Sengupta et al. 2008;

McArthur et al. 2012). In the shallow aquifer of the Bengal Basin, pristine groundwater should be dilute in nature as the concentration of Cl^{-1} in rain is typically <3 mg/L (Sengupta et al. 2008). In Bengal Basin most dilute groundwater contains <5 mg/L Cl^{-1} that are not affected by salt water intrusion (McArthur et al. 2012). The natural baseline concentrations of both NO_3^{-1} and SO_4^{-2} in rainfall are very small. Groundwater in Bengal Basin is anoxic, so the small concentrations of natural NO_3^{-1} and SO_4^{-2} in recharge are rapidly removed, the latter into trace amount of diagenetic pyrite (Nickson et al. 2000). As a consequence, unpolluted groundwater contains a concentration of NO_3^{-1} that is undetectable and of SO_4^{-2} that is detectable but typically <0.1 mg/L (McArthur et al. 2012). The high concentration of NO_3^{-1} and SO_4^{-2} in groundwater samples of the study area are indicative of anthropogenic pollution from waste water from pit latrines, septic tanks of the nearby villages and open defecation. This impact is also corroborated by high total and faecal colliform count in the test wells. Hence the water should be adequately treated before supplying to the local people for drinking purpose.

5.4 Groundwater Abstraction Structures

Along the four aquifer test lines the sediments up to a depth of 4–5 m brbl consists of medium to coarse sand. The water level is very shallow and rests within 0.56 m below the bed of the river. The transmissivity of the aquifer varies between 3750 and 6660 m^2/day with an average of 5476 m^2/day. The average discharge of the four test wells is 1207 m^3/day with the average maximum drawdown of 0.47 m. The average groundwater flow below the river bed is 6565 m^3/day (1.73 MGD). Considering all these factors two types of abstraction structures can be constructed on the river bed. They are cluster of river bed tubewells and infiltration gallery.

For the first alternative, three well fields may be developed on aquifer test lines 1, 2 and 3 (Fig. 19.3). Aquifer test line 4 has not been considered for developing a new well field due to (i) relatively low hydraulic conductivity of the aquifer compared to the other three aquifer test lines (Table 19.3), (ii) failure of the aquifer to give sustainable yield while pumping the test well with 3 and 5 HP pump sets, (iii) the possibility that the thickness of the aquifer will decrease further upstream, and (iv) the relatively poor groundwater flow across that section of the river (Table 19.4). At each well field there should be three wells with both vertical and horizontal screens. The design yield of each well may be about 37.85 m^3/hour (10,000 GPH). If the wells are pumped for 16 hours per day then the three wells in a well field will pump out 1817 m^3/day (480,000 GPD or 0.48 MGD). Therefore, the total yield of three well fields will be 5451 m^3/day (1.44 MGD) which is 83% of the total baseflow.

In Well Field 1 (Fig. 19.11) the first well should be constructed at the TW1 site (N 23.226380, E 87.016720), the second 200 m downstream of the first well and the third 200 m downstream of the second well. Each well should have 1.85 m vertical

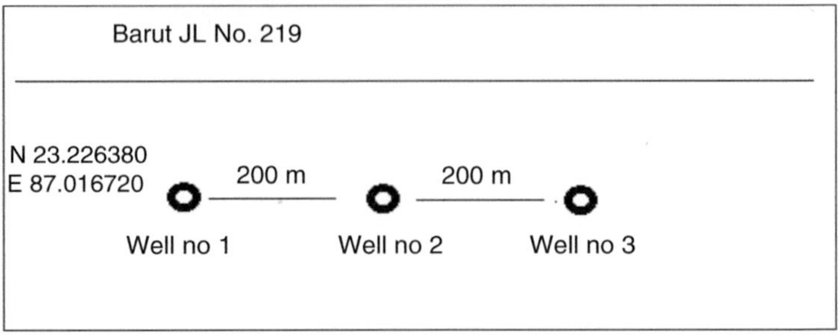

Fig. 19.11 Location of the proposed river bed tube wells in Well Field 1

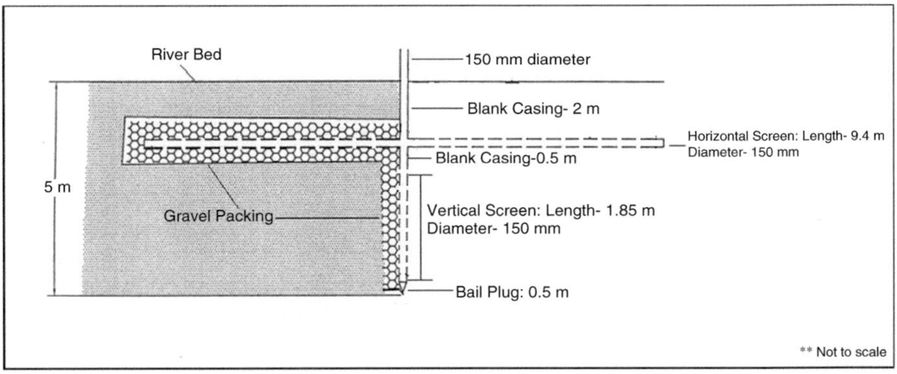

Fig. 19.12 Design of proposed horizontal and vertical tube well in Well Field 1

screen and 9.4 m horizontal screen (4.7 m on each side). The design of the well field 1 is given in Fig. 19.12. The slot size of the screens should be 1.33 mm.

In Well Field 2 (Fig. 19.13) the first well should be constructed at the TW2 site (N 23.233890, E 86.977290), the second 200 m downstream of the first well and the third 200 m downstream of the second well. Each well should have 1.60 m vertical screen and 8.4 m horizontal screen (4.2 m on each side). The design of the well field 2 is given in Fig. 19.14. The slot size of the screens should be 1.12 mm.

In Well Field 3 (Fig. 19.15) the first well should be constructed at the TW3 site (N 23.242960, E 86.954930), the second 200 m downstream of the first well and the third 200 m downstream of the second well. Each well should have 1.60 m vertical screen and 8.4 m horizontal screen (4.2 m on each side). The design of the well field 3 is given in Fig. 19.16. The slot size of the screens should be 1.66 mm.

The second alternative is the construction of infiltration galleries in the three well fields instead of tubewells. The design of the infiltration galleries is given in Table 19.8. Since the river is narrow and the minimum length of the gallery is 208 m the infiltration galleries should be constructed in the middle of the river along the river bed. Although sediment analysis suggests that artificial packing may not be required for the wells and infiltration galleries, but it may be prudent to have

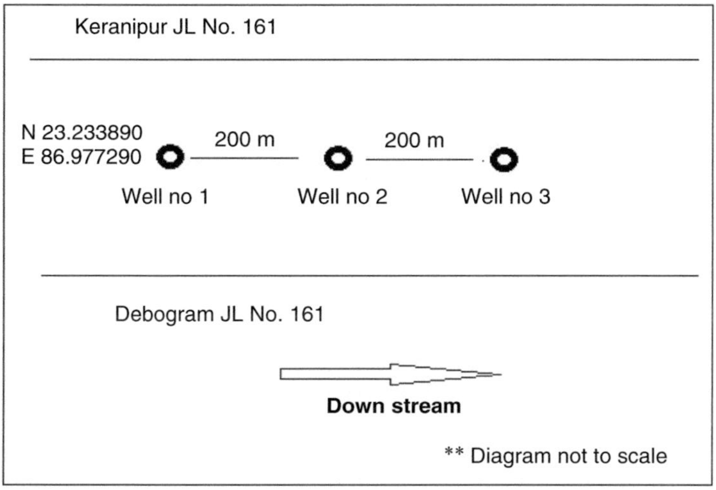

Fig. 19.13 Location of the proposed river bed tube wells in Well Field 2

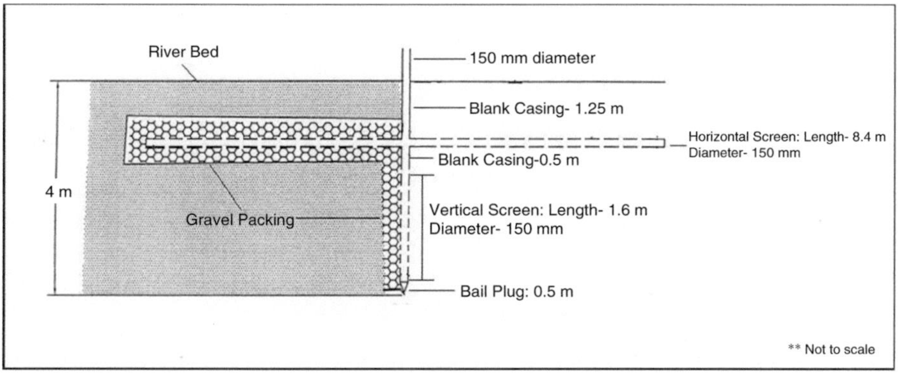

Fig. 19.14 Design of proposed horizontal and vertical tube well in Well Field 2

artificially packed screens. Gravels for the pack should be 5–6 times the screen slot size.

5.5 *Implications of the Design Yield*

A cautious approach should be adopted before planning for construction of abstraction structures on river bed based on the overall economics of the withdrawal of water from the site. Infiltration galleries may be a costlier proposal compared to river bed tubewells. Therefore, it is recommended to construct river bed tubewells with vertical and horizontal screens.

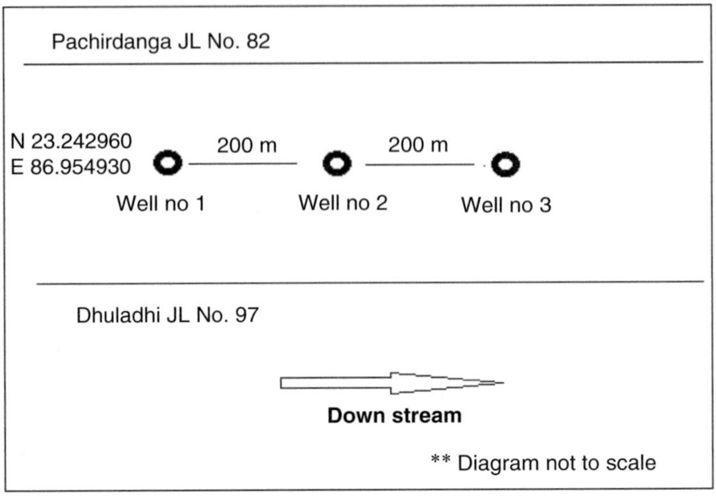

Fig. 19.15 Location of the proposed river bed tube wells in Well Field 3

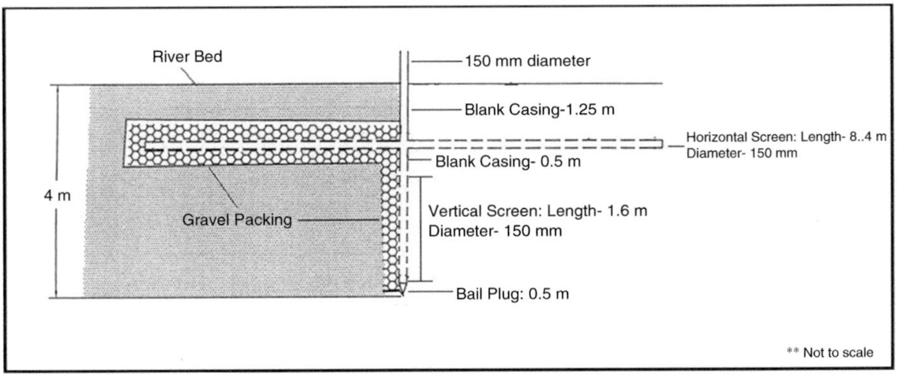

Fig. 19.16 Design of proposed horizontal and vertical tube well in Well Field 3

The river bed tubewells will draw water principally from discharging groundwater. The net groundwater available for future development in Dwarakeswar basin in Bankura-I block is 32.7×10^6 m^3/year (CGWB and SWID 2011). The total groundwater that will be withdrawn through the river bed wells is 1.99×10^6 m^3/year. This means about 6% of the net groundwater available will be abstracted from these three well fields. Hence the small amount of reduction in the baseflow of the river that may occur due to abstraction will not have any significant adverse effect on the river regime.

The three well fields should be operated for at least one year with 3785 m^3/day (1 MGD) capacity and monitored throughout the year. Based on the results obtained, further enhancement of the well yields can be thought of. A sub-surface dyke may be

Table 19.8 Design of the proposed infiltration gallery

Location	Latitude / Longitude	WF-1**	23.22638 / 87.01672	WF-2**	23.23389 / 86.97729	WF-3**	23.24296 / 86.95493
Pre-pumping water level (PWL) (m below river bed)		*0.73*		*0.24*		*0.35*	
Steady state water level (SWL) (m below river bed)		0.98		1.00		0.94	
Bottom of infiltration gallery (m below river bed)		4.5		4		4	
Height of pre-pumping water level above gallery bottom (H) (m)		3.77		3.76		3.65	
Height of water level above gallery bottom under steady state condition (h) (m)		3.52		3		3.06	
Allowable discharge (m^3/day)		5451 (1.44 MGD)					
K = T/b (m/day)		1246		1665		1147	
Safe distance ($2r_o$)		198		200		202	
Max. length of gallery (m)		478		416		417	
Min. length of gallery (m)		239		208		208.5	
Diameter of gallery (inch)		12		12		12	
Screen slot size (mm)		1.25		1.11		1.67	
Gravel packing		Naturally packed screen. Gravels for the pack should be 5–6 times the screen slot size		Naturally packed screen. Gravels for the pack should be 5–6 times the screen slot size		Naturally packed screen. Gravels for the pack should be 5–6 times the screen slot size	

**WF = Well field

constructed downstream of Well Field 1 which will provide further sustainability of well yield round the year.

6 Conclusions

Apprehension about the current supplies of drinking water in semi-arid and fluoride affected areas of Bankura district of West Bengal has led to the present systematic hydrogeologic investigation of river bed aquifer system to understand the quantity and quality of groundwater flowing below the bed of the river Dwarakeswar which is an effluent river at least during the pre-monsoon period. Based on the study, three well fields have been identified at the best possible locations on the river bed and abstraction structures have been designed to supply safe drinking water. Both river bed tubewells and infiltration galleries may be constructed, but river bed tubewells will be much more cost effective. The three well fields will supply 5451 m^3/day (1.44 MGD) of groundwater. In each well field three wells should be constructed at a distance of twice the radius of influence of each well to avoid interference effect. Each well should have both vertical and horizontal strainers which will provide more sustainability to the well yield. A sub-surface dyke downstream of well field 1 will provide further sustainability to water supply. The water needs to be treated for total and faecal colliform bacteria before supplying to the villages for drinking purpose. This chapter, therefore, illustrates the potentiality of river bed aquifer of ephemeral rivers of semi-arid and fluoride affected areas. The chapter also provides a framework for the engineers and planners to design groundwater abstraction structures for sustainable supply of safe drinking water to the vulnerable section of the community in the conurbations adjacent to ephemeral rivers in other semi-arid areas of India and elsewhere.

Acknowledgements The work was supported by PHED grant no. 4178/79/H/2011 dated 25.03.2015 to the first author. The authors thank Goutam Roy Chowdhury, B.R. Poddar, Ajay Kundu, Tanay Das, Santanu Chowdhury, Arindam Mazumder, Pareshnath Roy and Sudip Ghosh of PHED, Government of West Bengal; Director, IISWBM for providing necessary infrastructure for the research; Jyotirmoy Majumdar, Scientific Research Laboratory for assistance in chemical analysis and Apurba Mondal and his team for assistance in drilling.

References

Ali, M. (2003). Review of drilling and tubewell technology for groundwater irrigation. *In:* Rahman, A.A. Ravenscroft, P. (eds), Groundwater resources and development in Bangladesh—background to be arsenic crisis, agricultural potential and the environment. Bangladesh Centre for Advanced Studies, University Press Ltd., Dhaka.

APHA (American Public Health Association) (1995). Standard Methods for the Examination of Water and Wastewater, 19th edition. American Public Health Association, American Water Works Association and Water Pollution Control Federation, Washington, DC.

Ball, P. and Danert, K. (1999). Hand and Sludging: A Report from Northwest Bengal. Report of DFID KAR Project R7126 "Private Sector Participation in Low Cost Water Well Drilling", Cranfield University. ISBN 1861940.

IS 10500 (2012) Indian Standard Drinking Water—Specification (Second Revision). Bureau of Indian Standards, New Delhi.

CGWB (Central Ground Water Board) and SWID (State Water Investigation Directorate) (2011). Report on the dynamic ground water resources of West Bengal as on 31.03.2009, technical report series 'B' No. 234.

Chakrabarti, S., Majumder, A. and Bhattacharya, H.N. (2012). Fluoride contamination in groundwater of Bankura District, West Bengal: A Hydrogeo-Environmental Appraisal. *J Water Works Association*, April–June 2012, 134–142.

Cooper, H.H. and Jacob, C.E. (1946). A generalized graphical method for evaluating formation constants and summarizing well field history. *American Geophysical Union Transactions*, **27:** 526–534.

DHDR (District Human Development Report, Bankura) (2007). Development & Planning Department, Government of West Bengal. Published by HDRCC Development & Planning Department Government of West Bengal.

Driscoll, F.G. (1986). Groundwater and Wells. 2nd ed. Johnson Division, St. Paul, MN.

EROS (2002) Shuttle Radar Topography Mission (SRTM) Elevation Data Set. National Aeronautics and Space Administration (NASA), German Aerospace Center (DLR), Italian Space Agency (ASI). From: the National Center for Earth Resources Observations and Science, US Geological Survey, Sioux Falls.

Folk, R.L. (1968). Petrology of Sedimentary Rocks. University of Texas Publication, Austin. 170 p.

Freeze, R.A. and Cherry, J.A. (1979). Groundwater. Prentice-Hall, Englewood Cliffs, NJ. 604 p.

Hamdy, A., Abu-Zeid, M. and Lacirignola, C. (1995). Water Resource Management in the Mediterranean Basin. *Water Resources Development*, **11(4):** 24–29.

McArthur, J.M., Sikdar, P.K., Hoque, M.A. and Ghosal, U. (2012). Waste water impacts on groundwater: Cl/Br ratios and implications for arsenic pollution of groundwater in the Bengal Basin and Red River Basin, Vietnam. *Sc of the Tot Env*, **437:** 390–402.

Nickson, R., McArthur, J.M., Ravenscroft, P., Burges, W.G. and Ahmed, K.M. (2000). Mechanism of arsenic release to groundwater, Bangladesh and West Bengal. *J Appl. Geochem*, **15:** 403–413.

Ray, A. and Shekhar, S. (2009). Groundwater issues and development strategies in West Bengal. Bhu-Jal News, **24(1):** 1–17.

Roy, A.K. and Saha, A.K. (1975). Trace element geochemistry of Bengal Anorthosite and Associated rocks. *N. Jb. Miner. Abh.*, **125(3):** 297–314.

Roy, A.K. (1977). Structural and Metamorphic Evolution of the Bengal Anorthosites and Associated Rocks. *J Geol Soc Ind*, **18(5):** 203–223.

Saha, A.K., Dasgupta, S.P., Saha, D., Pal, S. and Sikdar, P.K. (1995). An approach to management of the groundwater problems of Bankura District, West Bengal. *Indian Minerals* (now Indian Journal of Geosciences), **49(1–2):** 1–12.

Sengupta, S., McArthur, J.M., Sarkar, A.K., Leng, M., Ravenscroft, P., Howarth, R.J. et al. (2008). Do ponds cause arsenic-pollution of groundwater in the Bengal Basin? : An answer from West Bengal. *Env Sc Tech*, **42:** 5156–5164.

Shadeed Sameer, Shaheen Hafez and Jayyousi Anan (2007). Management options of wadi faria baseflow. Eleventh International Water Technology Conference, IWTC11.

Theis, C.V. (1935). The relation between the lowering of the piezometric surface and the rate and duration of discharge of well using groundwater storage. *American Geophysical Union Trans.*, **16:** 519–524.

Chapter 20
Groundwater Pricing and Groundwater Markets

Achiransu Acharyya

1 Introduction

There are several general principles involved in assessing the economic value of water and the costs associated with its provision. First, an understanding of the costs involved with the provision of water, both direct and indirect, is key. Second, from the use of water, one can derive a value, which can be affected by the reliability of supply, and by the quality of water. These costs and values may be determined either individually, as described in the following sections, or by analysis of the whole system. Regardless of the method of estimation, the ideal for the sustainable use of water requires that the values and the costs should balance each other; full cost must equal the sustainable value in use so that the full range of environmental and economic services of groundwater need to be accounted for in policy decisions. Non-recognition of these services imputes a lower value for the groundwater resource in establishing policies. In this chapter, an attempt has been made to assess the value of groundwater in terms of pricing and cost and to analyse the role of groundwater markets in terms of groundwater pricing and accessibility to groundwater, especially for irrigation purposes.

2 Value of Water: A Historical Overview

As water resources have become increasingly scarce in the last few decades, the perception of water has changed. The debate over the treatment of water as an economic good has been a prevalent part of water resource management discussions in the literature as well as in real world. The topic is quite complicated, and

A. Acharyya (✉)
Department of Economics, Visva Bharati (A Central University), Santiniketan, India
e-mail: achiransu@gmail.com

P. K. Sikdar (ed.), *Groundwater Development and Management*,
https://doi.org/10.1007/978-3-319-75115-3_20

a general overview is difficult. However, the following review attempts to present a short summary of some of the main issues related to this topic.

2.1 The Dublin Statement and United Nations Agenda 21

The Dublin Statement, issued from the International Conference of Water and the Environment (ICWE) held in Dublin, Ireland, in January of 1992, was a primary catalyst of the debate over treatment of water as an economic good (ICWE 1992). Resulting from the call from 500 participants from 100 nations for fundamental new approaches to the management of freshwater resources, the Dublin Statement included within it the principle that, "Water has an economic value in all its competing uses and should be recognized as an economic good" (ICWE 1992, Guiding Principle No. 4). This was the first explicit recognition of water as an economic good, and this principle is often found quoted in literature that has ensued since its establishment. Shortly thereafter, this same idea was adopted by the Plenary in Rio de Janeiro at the United Nations Conference on Environment and Development in June of 1992, with some additions to the statement. Agenda 21, emanating from that meeting states, "Integrated water resources management is based on the perception of water as an integral part of the ecosystem, a natural resource and a social and economic good, whose quantity and quality determine the nature of its utilization" (Agenda 21, Proceedings of the United Nations Conference on Environment and Development, United Nations 1992).

2.2 The Many Values of Water

Water is not strictly limited to the status of an economic good. It is also a social good, and it has cultural and religious value as well (Gleick et al. 2003).

Water as a Social Good Access to clean water is vital to people. Water quality affects public health in the short and the long term. Water supply management for populations involves the building of large infrastructure. Such works are best handled with public oversight.

Water as an Economic Good Water is a scarce resource with value in competing uses. Allocation of water resources could be optimized to maximize benefits to society.

Water Has Ecological Value Water is not only essential for humans, but also for all life. Changing the hydrology of ecosystems threatens populations of many species.

Water Has Religious, Moral and Cultural Value Water figures into cultural and religious identities as part of rituals and symbolism. Moral values may come into play with property rights issues, when people feel they morally have a right to water.

Globalization, Privatization and Commodification of Water Globalization, privatization, and commodification of water are all relatively new phenomena in recent times. Commodification is the transformation of a formerly non-market good to a market good. While water has on a smaller scale had a market value in the past, with the issue of the Dublin Statement on water and changes in global markets, the commodification of water has increased (ICWE 1992). Globalization is the process of integrating markets internationally. The uneven distribution of water across the globe, coupled with newly opened global markets, has made water an item to be traded on the global scale. Water can be traded as a bulk good or as a value-added product as bottled water. Bottled water sales have been increasing noticeably in the last decade. As the case studies in this document show, water trade as a bulk quantity is also occurring. Privatization of water involves transferring control of all or parts of water systems from public into private hands. Privatization of water resources has been promoted as a way to improve water systems. There is a belief that business control is more efficient than government control and that the private sector can mobilize capital more quickly. There are also concerns about privatization. Among many risks of privatization that Gleick et al. (2003) outlines, privatization may result in social inequities, public ownership of the water itself may be at risk, ecosystem impacts could be ignored, and water use efficiency and water quality may not be as valued.

2.3 *Complexities in the Economic Behaviour of Water*

The question whether or not water can actually be treated as a true economic good is debated. Looking at water resources from a big picture perspective, it appears that by treating water as an economic good, pricing will improve overall allocations and encourage sustainable use. Dinar and Subramanian (1997) state that on both individual and social levels, if price reflects the value of the resource, water use efficiency will improve. Some argue that water cannot be treated like other economic goods because of its unique characteristics. Savenije (2001) outlines several characteristics of water that, together, illuminate how it is not an ordinary economic good. These characteristics of water lead it to behave differently from ordinary economic goods. To be effective, water pricing schemes need to be able to handle these complexities.

2.4 *Water as a Human Right*

As a response to the Dublin Statement identifying water as an economic good, there has been much outcry about the need to treat water as a human right (Baillat 2010). Because water is essential to life, and there are no substitutes for it, there is concern that treating it as an economic good will leave certain people without access to much needed freshwater resources. Scanlon et al. (2004) provide a review of this topic that covers many of the arguments found in literature. In their review of international laws, conventions and judicial decisions, they find that the human right to water has not been clearly defined by international instruments. It is implicit in existing fundamental human rights laws, and explicitly included only in non-binding instruments. Defining water as a human right would provide more protection to people and would obligate governments to ensure water to all people. A human right to water could help to set priorities for water policy and may help to focus attention to resolve conflicts over shared waters. It also could help to safeguard other human rights and environmental principles.

3 Measuring the Price of Water

In common terminology, water price is a volumetric price placed on metered water. A water rate is often the same thing as a water price. The term water rate, expressed plurally, typically refers to the entire package of charges applied by a water supplier. Indeed, any given supplier may simultaneously apply an extensive array of charges, with good reason. To begin with, water rates almost always include two categories:

(a) Charges that depend on the amount of water used, where the per-unit charges may vary according to the type of use, the amount of use, the time of use, and so on.
(b) Charges that are not based on water consumption such as new connection fees, "meter" charges, or irrigated acreage charges.

The fact that rates include water and non-water charges, and that the prices vary with an assortment of factors is an immediate complication of the issue at hand. Ideally, to foster good scarcity signalling the water charges will be independent of the non-water charges. Because the adequacy of revenue to cover the supplier's costs is an important concern, elements of the rate package are interdependent. Increase in one charge may allow another charge to be lowered. As a consequence, any study of the "best" water price is obligated to consider other elements of the rate structure. Just as importantly, the pursuit of efficiency should take full advantage of all available pricing tools.

3.1 Water-Based Charges

In many places water rates are sometimes called water tariffs. Although governments may be responsible for setting both taxes and water rates, there is an important distinction to be respected. Taxes are revenue-collecting mechanisms that enable governments to perform varied functions (maintain streets, build schools, operate the government, defend the borders, fund welfare programmes, etc.). Water rates are charges for the measured delivery of a valued commodity. This is not a tax. It is the cost of a service, and it is good to encourage an appreciation of this fact through one's choice of terminology. The term rate structure may address whether the per-unit price of water decreases, stays the same, or increases with the amount of water consumed. Figure 20.1 portrays the three available rate structures. The uppermost rate structure depicts decreasing block rates. For each customer, price is constant within every "block," but as metered consumption increases into the next higher block, price falls. The first block in this schedule exists from w_1 to w_2 units of water, and each water unit in this block costs the consumer p_1 rupees. While it is often true that $w_1 = 0$, some suppliers grant each consumer a small amount of water consumption, free of any volumetric price. If water consumption lies within a higher block, all units of water are still billed at the rate applicable for their block. Hence, the metered water bill for w units of water is not $p_2.w$. It is $p\,(w_2 - w_1) + p_2\,(w - w_2)$. It is also notable that the "marginal price" faced by this consumer is p_2. Different consumers served by this system may then face different marginal prices.

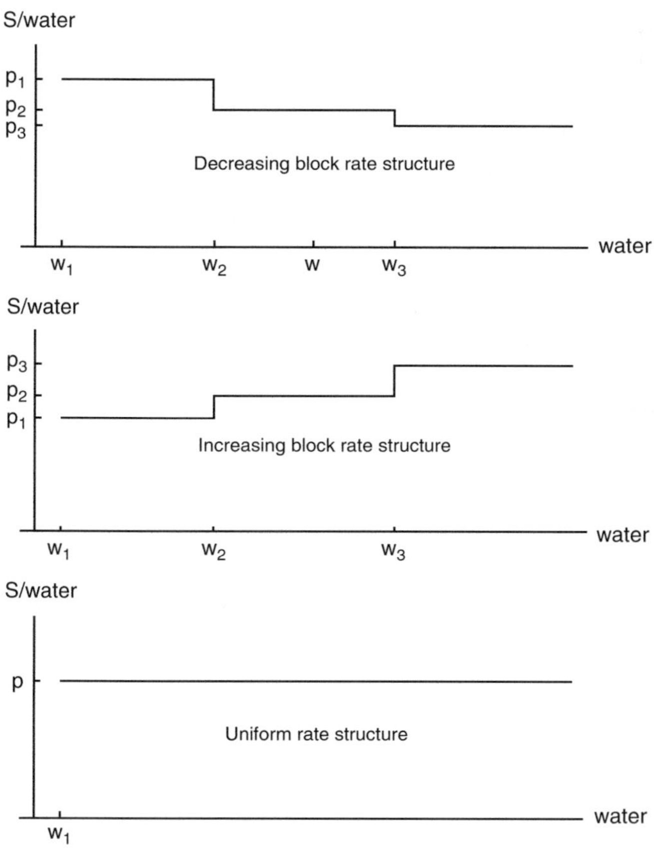

Fig. 20.1 Three types of water rates

Historically, decreasing block rates were favoured; although this has been changing as the economic circumstances of utilities evolve (Organisation for Economic Cooperation and Development 1999).

Three reasons explain the long-standing preference for decreasing block rates. The natural monopoly status of suppliers is due to the declining average costs of providing water. Said another way, greater system-wide deliveries lower the per-unit costs for everyone, so stimulating consumption with a lower price for large water consumers might seem appealing. Second, it is widely assumed that large water users such as businesses and industries are steadier in their water use in that their peak-hour and peak-day water use is not dramatically greater than their average water use. In contrast, it is typically presumed that small water users such as households contribute more to peak water usage. Because system capacity is both expensive and constructed to meet peak demands, it is arguable that residential users are causing higher average and marginal costs for the utility. Third, decreasing block rates are favourably viewed by suppliers because they stabilize revenue in the presence of climate-impacted demand. With decreasing block rates, a greater proportion of revenue is derived from the initial units of consumed water, and these units are less likely to be affected by climate.

The opposing rate structure is naturally termed increasing block rates, although inverted block rates is also an encountered term. Motivation for the adoption of increasing block rates comes from two sources. First, increasing block rates are often claimed to enhance water conservation because large water users are "penalized" for their behaviour. Second, because larger water users tend to be wealthier water users in residential settings, there may be a perceived degree of "fairness" associated with increasing block rates. In developing countries, increasing block rates may enjoy considerable support because the basic water uses undertaken by the poor are internally subsidized by this rate structure (Boland and Whittington 1998).

One method of time-dependent pricing is supported by contemporary metering practices. Monthly meter reading allows water prices to vary by month. Thus, as a utility moves through the year, encountering low-to-high water supply conditions relative to demand, it is feasible to apply month-specific prices. This is called 'time of year pricing'. While such a system has not gained complete favour, it is more efficient than keeping prices fixed for an entire year. Many urban suppliers now employ a simplified variant known as seasonal pricing in which separate winter and summer rates are applied. Winter rates apply for part of the year, and summer rates makeup the rest. Summer rates are justifiably higher because much of the supply system is only used during the summer. Given that there is idle system capacity during winter periods, it is clear that the purpose of the idle capacity is to provide summer service. It is therefore economically appropriate to assign these costs to the summer period, resulting in higher summer rates. The summer value of natural water is also higher in most regions.

Two other charges to water are also important. Both of these charges are rationalized by the capital intensity of the water supply industry, which has even greater capital requirements per dollar of product than the electricity, telephone, or railroad industries (Beecher et al. 1991). Both of these charges are focused on the many *points of use* at the end points of the water delivery system. Water managers refer to these end points as the number of *connections* or *meters* in their system.

The first of these fees is the meter charge, which is usually paid in every billing period. This fee can also be called the minimum charge or the service charge. When irrigators are charged on the basis of irrigated area (acreage), this fee functions much like a meter charge for each acre. Because it is not based on water consumption, the meter charge serves as a flat rate if it is not accompanied by a volumetric charge. Modern rate systems, however, incorporate both the meter charge and a water price. Historically, the meter charge component was employed in the absence of a volumetric charge. Irrigation districts have a strong propensity to rely on the acreage charge for revenue generation (Michelsen et al. 1999). Suppliers enjoy the revenue stability resulting from meter charges, and overall costs are lowered because meters do not have to be installed or regularly read. Yet the presence of a zero price for water provides a perverse incentive for consumers in light of the value of processed and possibly scarce water, variable operational costs (e.g., energy, treatment chemicals), and the value of the physical capital needed to obtain, store, treat and deliver this water. For these reasons, both meter installation and meter-reading efforts have been accepted as worthwhile undertakings in most modern systems (Organisation for Economic Co-operation and Development 2003).

The combined application of a water charge and a non-water charge also coincides with economic recommendations for declining-average-cost industries. In the technical economic literature concerning "two-part tariffs," the dual application of a meter charge and a volumetric charge enjoys extensive theoretical support (Brown et al. 1992; Kahn 1988; Ng and Weisser 1974).

The second significant non-water charge is the connection charge that modern utilities place on new connections to the delivery system which is a one-time fee for each new point of water use, such as a new home (Herrington 1987).

3.2 Equity, Efficiency and Sustainability in Groundwater Prices

There are many different ways to promote equity, efficiency and sustainability in the water sector and water pricing is probably the simplest conceptually, but maybe the most difficult to implement politically. For example, the typical command and control approach taken in most countries with respect to water management leads to large government involvement because of its needs for detailed hands-on monitoring and measurement. Using price policies, however, still requires significant government intervention to ensure that equity and public goods issues are adequately covered.

Economic theory has long ago explained how correct pricing of private and public goods can lead to gains in economic efficiency. Three generally accepted effects of price policy—demand reduction, efficient reallocation of the resource, and

Table 20.1 Effects of price policy on groundwater

Three well-known and three lesser known effects of price policy
(a) *Increased price reduces demand*
(i) Substitutes become cheaper
(ii) Conservation becomes affordable
(iii) Change consumption preferences
(b) Increased prices increase supply
(i) Marginal projects become affordable
(ii) Provides economic incentives to reduce water losses
(c) Increased price facilitates re-allocation between sectors
(i) From irrigation to domestic and industrial
(ii) From off-stream to in-stream uses
(d) Increased prices improve managerial efficiency due to increased revenues by
(i) Improving maintenance
(ii) Improving staff training and education
(iii) Making modern monitoring techniques affordable
(iv) Making modern management techniques affordable
(e) Increased prices lead to sustainability
(i) Reduces demands on resource base
(ii) Reduces pollution loads due to recycling of industrial water
(f) Increased prices reduce the per unit cost of water to poor people
(i) Increases coverage of poor urban and peri-urban populations because additional water is available for extending the system
(ii) Reduces reliance by the poor on water vendors

increasing the supply—together with three effects which are not generally associated with price policy, namely, improved equity, improved managerial efficiency, and improved sustainability of the resource are listed in Table 20.1. Here "water resources" encompasses surface water, groundwater and wastewater. We show that if water resources are managed in an integrated fashion where the economics, legal and environmental aspects complement each other, increased prices do improve equity, efficiency and sustainability of the resource.

3.3 *Full Cost Pricing of Water*

The problem faced by the water sector is that prices and tariffs are almost universally below the full cost of supply. This means that almost everywhere there are large inefficiencies in the water sector and that water prices need to be raised. The World Water Commission strongly endorsed the need for full-cost pricing of water services: Commission members agreed that the single most immediate and important measure that we can recommend is the systematic

Table 20.2 Important concept of groundwater for setting price

COST:	O&M costs, capital costs, opportunity costs, costs of economics and environmental externalities
VALUE:	Benefits to users, benefits from returned flows, indirect benefits, and intrinsic values
PRICE:	Amount set by the political and social system to ensure cost recovery, equity and sustainability. The price may or may not include subsidies. Prices for water are not determined solely by cost

adoption of full-cost pricing of water services (World Water Commission, 2000). Three important concepts from water economics is shown in Table 20.2.

4 Groundwater Markets

The term water markets connotes a localized, village level informal institutional arrangement through which owners of a modern water extraction mechanism (WEM) sell water to other farmers at a price. The poor farmer in the absence of a sound economic base and resource rich and big farmers due to the high degree of farm fragmentation enter into water markets as a buyer.

4.1 Important Features of Water Markets

(a) *Spontaneity:* Even though the WEMs are not installed primarily to sell water, water markets come into existence by spontaneous action initiated by individual farmers to exploit a mutually beneficial opportunity.
(b) *Informal:* The sole basis of the whole transaction is the mutuality of need between the buyers and sellers. There is no formal legal sanction behind the transactions in these water markets.
(c) *Unregulated:* These are unregulated and the state government or state electricity board does not exercise any direct or indirect control.
(d) *Localized:* Markets are mostly limited to a part of a village's fields.
(e) *Fragmented:* The option of one seller does not depend on the action of other sellers, but it depends upon the number of buyers and their respective area.
(f) *Non-seasonality:* Water markets operate in all the three crop seasons, namely, rabi (winter), kharif (monsoon) and boro (summer).
(g) *Impersonal:* Water markets are impersonal in the sense that sellers generally do not distinguish between various buyers in term of selling or quality of service provided.

4.2 Major Forms of Water Markets

(a) *Purely buyers:* This form of water market arises mainly because of small size of holding. Buyers are generally resource-poor farmers and they do not get a suitable partner to pool their resources to install a WEM.
(b) *Self users and buyers:* This form of water markets exists generally because of fragmentation of holdings.
(c) *Self users and buyers and sellers:* Existence and operation of this form of water markets is also due to high degree of farm fragmentation.
(d) *Self users and sellers:* These farmers are owner of WEMs and their land holdings are consolidated. They sell surplus water to other farmers because their land holdings are small to utilize a WEM at full capacity.
(e) *Purely self users:* Water markets do not exist in this category of farmers because they have WEMs to irrigate only their fields. Land holdings are generally consolidated requirement.

4.3 Experiences with Water Markets

Water markets can be broadly divided into formal and informal markets.

4.3.1 Formal Groundwater Markets

Formal water markets specify the volume and share of water to be sold, either for a set period of time or permanently. Informal markets usually involve the sale of unmeasured flows of surface water from a canal for a set period of time or of water pumped from a well for a set number of hours. Although the units sold in informal markets may not be metered, both the buyer and the seller have good information about the volume transferred. The key difference between the two markets is the way in which the trade is enforced. If the users must self-enforce trades because no formal water rights exist that can be enforced through the legal or administrative system, the market is informal. Formal water markets are usually found in North and South America, whereas informal markets are prevalent in the irrigated areas of South Asia.

4.3.2 Informal Groundwater Markets

Most of the groundwater markets are important for agricultural production and the distribution of water throughout the irrigated areas of South Asia. Saleth (1998) estimates that 20% of the owners of the 14.2 million pump-sets in India are likely to be involved in water trading. This means that water markets are providing water for about six million hectares, or 15% of the total area irrigated by groundwater. In Pakistan a survey reported that 21% of well owners sold water (NESPAK 1991). In areas where dependable precipitation recharges the groundwater, the benefits of buying and selling water from tubewells have increased farmers' income and production. The economic gains from groundwater markets reflect improved efficiency in pump management, in reducing conveyance losses, and in farm-level water use. These markets also increase access to irrigation, especially for smaller-scale farmers who do not own tubewells and cannot afford to invest in a well without a market for their water. Meinzen-Dick (1998), in one of the few studies estimating the economic returns from access to water markets, found that water markets increased the availability and reliability of water supplies. Both yields and income rose for those who purchased water, particularly for those who also had access to canal water supplies. The highest yields and income, however, were still found among farmers who owned their own tubewells and had access to canal water.

4.4 Problems with Groundwater Markets

4.4.1 Problems with Informal Groundwater Markets

There are several problems with informal groundwater markets. These include:

Preventing Overdrafts Given that markets for the sale of groundwater draw on an open-access resource (that is, one that is available for capture to anyone who has access), it is not surprising that problems arise in areas with high demands and limited supplies. Farmers have an incentive to ignore the scarcity and buffer stock value of the groundwater and pump until their cost of pumping equals the market price of water (Ramasamy 1996). Over time, the cost of pumping and the price of water rise as the groundwater level declines.

For example, the overdraft (that is, water use in excess of recharge) in the Coimbatore District of India is almost 5000 cubic metres a year. Ramasamy (1996) estimates that if the over-pumping continues, it will mean a drop in total net returns to farmers of between \$42 million and \$69 million, a result of the increased costs of power necessitated by increased pumping and additional investment to deepen wells. Here is a case where informal markets may exacerbate the problem, and formal markets may not work any better unless water rights can be established and enforced in strict quantity terms. The problem is not the water markets but the lack of exclusive property rights for groundwater. To establish such rights, the number of wells and the amount of water to be pumped would have to be agreed on and restricted. Such restrictions are probably unrealistic without strong support in the irrigation community. If exclusive water rights can be established, however, the water market should reflect the scarcity value of water and help restrain over-pumping.

Blomquist (1995) reports on one case where the demand for water is increasing and the community of water users has been able to stop the overdraft. In the dry Los Angeles metropolitan area in southern California, pumping is metered and taxed so that users have an incentive to shift from local groundwater to more expensive but more plentiful imported water. Surface and imported water are stored and used to recharge the groundwater in the basin. One result has been a halt in saltwater intrusion from the ocean in the area's coastal groundwater basins. In some of these basins, pumping rights have been defined, limited to the basin's average recharge, and made transferable to other users through sales. A more typical case, reported by Shah (1993), is in coastal Gujarat, India. Here, the overdraft of coastal aquifers has caused a decline in groundwater supplies in some areas and saltwater intrusion in others. Shah (1993) argues that any effective reduction in this overdraft is unlikely without good local leadership and the involvement of water user groups. He argues that legal, quasi-legal, and organizational instruments of public policy will not, on their own, succeed in securing the compliance of farmers unless they are accompanied by measures aimed at affecting private returns to irrigation or unless the structure of property rights on the water resource itself is drastically reformed. Similarly in Pakistan, Meinzen-Dick (1998:218) doubts whether government would have the institutional capacity to regulate sales among hundreds of thousands of private tube-well owners, and if it had such capacity, it is unclear what such direct intervention could achieve.

Yet in both India and Pakistan, any effect that water markets might have on the over drafting of groundwater is much less than the effect of subsidized electricity. The zero or near-zero marginal cost of pumping means that farmers have an incentive to use groundwater to the point where the marginal value of production is close to zero. This, of course, encourages farmers who can sell water to use their wells at close to full capacity. The low power rates not only create over-drafting problems but also waste electricity in countries without adequate power. As noted above, water markets can actually help solve the over-drafting problem by increasing the incentives for efficient water use and making it possible to purchase water from areas where water is abundant. The ability to find another source of water, but at a higher marginal cost, can help promote community action for self-regulation and demand management. Shah (1993) cites a case in coastal Gujarat where self-regulation became possible when additional new supplies were piped into the area.

Over-drafting tends to be concentrated in coastal areas of India and Pakistan and in the hard rock areas of southern India. In many of the northern areas, pumping actually improves growing conditions by lowering the water table below the root zone (Shah 1993; Meinzen-Dick 1998). In cases where water tables are high or recharge rates are rapid, water markets are not likely to cause negative externalities except possibly temporarily if neighbouring wells are too close or deep tubewells interfere with shallow wells. Where these externalities are small, personal trust and reputation may be enough to foster competitive informal water markets. This is particularly true where farmers own a number of separate plots that cannot be served by the same well. In such cases, most water sellers are also buyers because most farmers who own a well are able to irrigate only their large plots and must purchase water to irrigate other plots (Shah 1993; Meinzen-Dick 1998; Saleth 1998). In addition, their wells are likely to be underutilized unless they can sell water. Yet because of the costs of conveying water and the need for cooperation from neighbouring farmers when water is to be conveyed any distance, high transaction costs can restrict trades in areas with only a few wells and prevent water markets from being competitive.

Countering Monopoly Pricing This raises the other concern about water markets, the potential for monopoly pricing and discrimination. Groundwater markets are somewhat confined by the physical limits of the location and supply of groundwater. Still, pipelines can extend markets, and the investment costs of new wells should put a limit on monopoly power. An abusive monopolist who raises prices too high will find others investing in wells and undercutting the price. Shah (1993) notes a lack of balance between the numbers of buyers and sellers in areas with high capacity wells, where one seller may serve as many as 70 or 80 buyers. He fails to say how many sellers the average individual buyer can access. Monopoly pricing may be avoided if the buyers can purchase water from three or four sellers—so long as the sellers do not collude. The evidence on monopoly pricing is mixed. In a 1991–92 survey in Pakistan, Meinzen-Dick (1998) found that sellers were pricing water at little more than the cost of pumping. The two most common ways of charging for groundwater are a flat charge per hour of pumping (ranging from \$0.57 to \$3.27 an hour, depending on the pump type, capacity, and location) and arrangements whereby the buyer supplies the diesel and motor oil for the pump and pays an additional fee of \$0.16 to \$0.24 an hour to the well owner to cover the wear and tear on the engine. Sellers with diesel pumps were just recovering their own costs under either type of contract. In contrast, Saleth (1998) suggests that in some areas of India, monopoly rents may be extractive. He cites as evidence the variation in water charges compared with pumping costs in different areas. For example, water charges are 1.3–2 times higher than operating costs in the Indo-Gangetic region but 2.5–3.5 times higher in the water-scarce hard rock regions of southern India. The difference in rates, however, might be explained in part by the difference in water scarcity and in the value of water in those two regions. The degree of monopoly power may also be related to the terms of the transaction or contract for water. Not surprisingly, some of the contracts for water are quite similar to contracts for land. Water contracts include crop sharing, crop and input sharing, and labour arrangements. If the payment is cash-based, buyers have more freedom to take their business to another well owner anytime during the season. When the transaction is a contract in kind, especially one based on crop sharing or on crop and input sharing, the buyer is tied to the seller for at least one season, if not longer. Similarly, if buyers contract to pay for the water with their labour, they may find it difficult to change suppliers until they have fulfilled the contract. Yet in the villages, informal markets do not appear to face extreme cases of monopoly rents.

In fact, monopoly power that restrains trading in areas with serious problems of declining groundwater levels may help reduce over-extraction. In contrast, when suppliers are taking advantage of their monopoly position and there are adequate groundwater supplies, the best strategy is to encourage (legalize) trading and increase competition through community and private well development (Palanisami and William Easter 1991).

Thus informal water markets can improve water use and incomes in irrigated areas where water rights are not well defined or recorded. They also may be a good option if formal water markets are likely to produce third-party challenges and result in excessively high transaction costs. Finally, informal markets would work well in traditional irrigation systems where the farmers manage the irrigation system and would be able to maintain a relatively modest level of transaction costs.

4.4.2 Problems with Formal Groundwater Markets

In situations where informal markets can work well, it may not be necessary to incur the extra expense of establishing formal water markets. Formal markets will be required, however, to provide the certainty necessary for permanent water transfers or transactions between different sectors and jurisdictions. Because the need for permanent trades and inter-jurisdictional water exchanges is likely to become more important as non-agricultural demands for water grow, formal water markets are likely to become more common. The growing demand in water-scarce regions has been one of the driving forces behind the new interest in water markets. Several studies have illustrated the benefits that are possible from inter-jurisdictional trading in permanent water rights for short-term use.

In Texas, USA, 99% of the water traded has been transferred out of the agricultural sector in the Rio Grande Valley to non-agricultural users (Griffin 1998). Of the municipal water rights in the valley that existed in 1990, 45 percent had been purchased since 1970. Although water markets are not active in other areas in Texas, Griffin (1998) notes that the surface water law has evolved to the stage where trading will be more widespread in the future. In contrast, the groundwater law is just beginning to evolve.

Economic Gains In a study of the Guadalquivir Basin of southern Spain, Garrido (1998a) finds that the economic gains of trading within an individual water district or community may be relatively modest. In contrast, if permitted, trades among communities subject to different supply constraints and drought conditions could produce substantial gains. Garrido (1998a) estimates the total welfare gain at no more than 10% over the current water allocation for four communities where trades were only intra-community. Inter-community trading, however, could produce estimated economic gains in one of the older irrigation communities of almost 50%. Garrido (1998a) also shows that both types of trades are very sensitive to the level of transaction costs. If those costs exceed 8–12% of the market price, trading and the gains from trading would be too small to justify the expense of establishing formal markets. Yet Garrido (1998b) may underestimate the potential gains because he considers only the crops traditionally grown in the region (cotton, wheat, corn, oilseed and sugar beets) and excludes any transfers to non-irrigation uses. Evidence from Chile found significant changes in cropping as a result of water trading (Hearne and Easter 1997).

In contrast, Horbulyk and Lo (1998) found that most potential gains from introducing water markets in Canada's Alberta Province were likely to come from trades within a sub-basin. They considered four sub-basins and compared the current water allocation situation with the allocation under four separate markets (one in each sub-basin), as well as with a market encompassing the total basin. The four separate market scenarios created 90% of the welfare gains that were obtained when unrestricted trading was allowed among the four sub-basins. The urban sectors purchased most of the water, except on the River South Saskatchewan, where the agricultural sector purchased additional water when market trading was allowed among the sub-basins.

Trading Patterns and Transaction Costs In their analysis of selected water markets in Chile, Hearne and Easter (1997) found trading both within and between sectors. In the case of permanent transactions either within or between sectors, well-established water use rights that were recorded and recognized by the government were critical in fostering trade.

Several trades between farmers and the city of LaSerena were not consummated because of uncertainty regarding ownership of the water rights. La Serena is a growing vacation destination located on the coast in a dry region some 400 km north of Santiago. Rapid growth in demand has strained the city's water supply, particularly during the summer tourist season. The opening of water markets allowed the city to purchase water and delay development of new water sources. Starting in 1992, the city's water company purchased enough water to increase its water supply by 28%. Additional purchases were made by upstream households for domestic uses and by farmers.

Similarly, Archibald and Renwick (1998) found that high transaction costs in California limited a large number of potentially profitable trades. Two types of transaction costs were identified: administrative-induced costs, which are explicit and included in the price of water sold through the California Water Bank, and policy-induced transaction costs, which stem from existing legal requirements designed to avoid injuring owners of water rights, damaging fish and wildlife, and creating negative third-party effects in exporting areas. Policy-induced transaction costs in this range would be as much as or more than the potential gains from trading in the California Water Bank (Archibald and Renwick 1998).

Because of high transaction costs in Colorado, Howe (1998) recommends shifting the administrative responsibility for water transfers from the water courts to the State Engineer's Office. He also recommends reserving or acquiring water for "public good" uses such as recreation, as well as making other changes to allow water to be marketed as freely in Colorado as it is in the neighbouring states.

Colby (1990) suggests that the claims of Native Americans have the effect of imposing high transaction costs on water trading in many western rivers. She argues that even though markets do not work well with high transaction costs, when those costs are compared with the costs of litigated solutions, water markets look like a much better alternative.

Howitt (1998) reports that spot and options markets performed well during California's droughts in the 1990s. Even though these markets are a fairly recent phenomenon, he thinks they are a promising option for stabilizing available water supplies in California and other similar areas. Permanent shifts in demand, however, require a much more active formal market for water rights.

5 Conclusion

Contrary to the claims of many critics, water markets have worked and are likely to be a better mechanism for reallocating water than the alternative methods. There are both formal and informal water markets at work today. In addition, there are spot market sales, sales of permanent water rights, and leasing arrangements that are similar to those used for land, including crop sharing and cash rents.

Where water is scarce and large amounts of the available water supplies were committed to particular uses a long time ago, the economic benefits from water markets are likely to be large. In contrast, if the allocation was made fairly recently, based on the most highly valued uses of water and new opportunities are not available, then the gains will be much more modest.

For markets to be effective, transaction costs must be kept low. To keep these costs low, the appropriate institutional and organizational arrangements need to be in place, as well as flexible infrastructure and management. As pointed out earlier, the critical first step is to establish tradable water rights or water use rights separate from land, as well as the mechanisms to deal with third-party effects.

If it is difficult to establish legally enforceable, permanent water rights, a "thick" spot market may provide almost the same security as ownership of permanent water rights. In other words, the ability to buy the water needed at a reasonable price may provide enough security so that firms are willing to invest in enterprises that are dependent on this purchased water. A contingent water market can provide additional security so that firms can be assured of a given volume of water at a set price. With only a spot market and no contingent markets, firms may be subject to wide fluctuations in prices.

For those users needing certain supplies, spot water markets are probably cheaper alternatives than having to buy enough senior water rights so that one is guaranteed adequate supplies even in the worst drought. Owners of senior water rights have the right to whatever water is available, before the more junior water rights owners. In Pakistan, for example, the markets for groundwater have greatly improved the security of water supply, particularly in government irrigation projects. This security has allowed increased investment and increased production. It will be important to see if spot and contingent markets have similar effects on the productivity of water.

Finally, the evidence indicates that appropriately designed water markets, supported by sound institutions, are an effective mechanism for reallocating scarce water among sectors. Carefully designed water markets make it possible to meet the growing urban and industrial water demands without derailing growth in crop production.

Market transfers among sectors may make it possible to significantly scale back investments in new water supply projects. Government inaction, ineffective institutions for water management, and high transaction costs, however, are likely to prevent water markets from reaching their full potential for reallocating scarce water resources.

References

Agenda 21 (1992) United Nations Conference on Environment and Development, Rio D Janeiro.

Archibald, S. and Renwick, O.E. (1998). Expected Transaction Costs and Incentives for Water Market Development. *In:* Easter, K. William, Mark W. Rosegrant and Ariel Dinar (eds), Markets for Water: Potential and Performance. Kluwer Academic Publishers, Boston.

Baillat, A. (2010). International Trade in Water Rights. IWA Publishing, New York, USA.

Beecher, J.A., Mann, P.C. and Landers, J.R. (1991). Cost Allocation and Rate Design for Water Utilities. National Regulatory Research Institute, Columbus, OH, USA.

Blomquist, W. (1995). Institutions for Managing Groundwater Basins in Southern California. *In:* Ariel Dinar and Edna T. Loehman (eds), Water Quantity/Quality Management and Conflict Resolution: Institutions, Processes, and Economic Analysis., Praeger Publications, Westport, Conn.

Boland, J.J. and Whittington, D. (1998). The Political Economy of Increasing Block Tariffs in Developing Countries. Inter Develop Res Centre, Canada.

Brown, D.J., Heller, W.P. and Starr, R.M. (1992). Two-Part Marginal Cost Pricing Equilibria: Existence and Efficiency. *J Econ Theory*, **57:** 52-72.

Colby, B. (1990). Transaction Costs and Efficiency in Western Water. *Ameri J Agri Econo*, **5(72):** 1184-1192.

Dinar, A. and Subramanian, A. (1997). Water Pricing Experience: An International Perspective. World Bank Paper No. 386. The World Bank. Washington D.C.

Garrido, A. (1998a). An Economic Analysis of Water Markets within the Spanish Agricultural Sector: Can They Provide Substantial Benefits? *In:* Easter, K. William, Mark W. Rosegrant and Ariel Dinar (eds), Markets for Water: Potential and Performance. Kluwer Academic Publishers, Boston.

Garrido, A. (1998b). Economics of Water Allocation and the Feasibility of Water Markets in Agriculture. *In:* Organisation for Economic Co-operation and Development. Sustai Manag of Water in Agri: Issues and Policies. Paris.

Gleick, H.P., Burns, W.C.G., Chalecki, E.L., Michael, C., Cushing, K.C., Mann, A., Reyes, R., Wolff, G.H. and Wong, A. (2003). The World's Water 2002-2003: The Biennial Report on Freshwater Resources. Island Press: Washington DC.

Griffin, R.C. (1998). The Application of Water Market Doctrine in Texas. *In:* Easter, K. William, Mark W. Rosegrant and Ariel Dinar (eds), Markets for Water: Potential and Performance. Kluwer Academic Publishers, Boston.

Hearne, R.R. and Easter, W.K. (1997). The Economic and Financial Gains from Water Markets in Chile. *Agri Econ*, **15:** 187-199.

Herrington, P. (1987). Pricing of Water Services. Organisation for Economic Cooperation and Development, Paris.

Horbulyk, T.M. and Lo, L.J. (1998). Welfare Gains from Potential Water Markets in Alberta, Canada. *In:* Easter, K. William, Mark W. Rosegrant and Ariel Dinar (eds), Markets for Water: Potential and Performance. Kluwer Academic Publishers, Boston.

Howe, C.W. (1998). Water Markets in Colorado: Past Performance and Needed Changes. *In:* Easter, K. William, Mark W. Rosegrant and Ariel Dinar (eds), Markets for Water: Potentialand Performance. Kluwer Academic Publishers, Boston.

Howitt, R. E. (1998). Spot prices, option prices, and water markets: An analysis of emerging markets in California. In Easter, et al. (Ed.), Markets for water: Potential and performance. Boston: Kluwer.

ICWE (1992). International Conference on Water and Environment: Development Issues for the 21st Century, The Dublin Statement and Report of the Conference. Hydrology and Water Resources Department, Water Meteorological Association, Geneva.

Kahn, A.E. (1988). The Economics of Regulation: Principles and Institutions. Two vols. MIT Press, Cambridge.

Meinzen-Dick, R.S. (1998). Groundwater Markets in Pakistan: Institutional Development and Productivity Impacts. *The World Bank Research Observer*, **14(1)** (February 1999).

Michelsen, A.M., Taylor, R.G., Ray, G.H. and Thomas McGuckin, J. (1999). Emerging Agricultural Water Conservation and Price Incentives. *J Agri and Resou Econo*, **24:** 222-238.

NESPAK (National Engineering Services of Pakistan) (1991). Contributions of Private Tubewells in the Development of Water Potential. Field Report. Government of Pakistan, Ministry of Planning and Development, Islamabad.

Ng, Yew-Kwang and Weisser, M. (1974). Optimal Pricing with a Budget Constraint—The Case of the Two-Part Tariff. *Review of Economic Studies*, **41:** 337-345.

Organisation for Economic Co-operation and Development (1999). The Price of Water: Trends in OECD Countries. Organisation for Economic Co-operation and Development, Paris.

Organisation for Economic Co-operation and Development (2003). Improving Water Management: Recent OECD Experience. Organisation for Economic Co-operation and Development, Paris.

Palanisami, K. and William Easter, K. (1991). Hydro-Economic Interaction between Tank Storage and Groundwater Recharge. *Ind J Agri Econ*, **46(2):** 174-179.

Ramasamy, C. (1996). Intertemporal Allocation of Groundwater Resources with Stochastic Surface Supplies: The Case of Southern India. Ph.D. thesis. University of Minneapolis, Department of Applied Economics, St. Paul.

Saleth, M.R. (1998). Water Markets in India: Economic and Institutional Aspects. *In:* Easter, K. William, Mark W. Rosegrant and Ariel Dinar (eds), Markets for Water: Potential and Performance. Kluwer Academic Publishers, Boston.

Savenije, H.H.G. (2001). Why water is not an ordinary economic good. IHE Delft Value of Water Research Report Series No. 9.

Scanlon, J., Cassar, A. and Nemes, N. (2004). Water as a Human Right? IUCN Environmental Policy and Law Paper No. 51.

Shah, T. (1993). Groundwater Markets and Irrigation Development: Political Economy and Practical Policy. Oxford University Press, Bombay.

World Water Commission Report (2000). Earthscan Publications Ltd, London.

Chapter 21
Legislation for Effective Groundwater Management in South Asian Countries

Tapan K. Ray

1 Introduction

It is propounded by many policymakers that the future global economy will be dominated by the nations with high level of "*food security*". Agriculture is the mainstay of the South Asian countries. Agriculture and its allied sectors contribute to the gross domestic product (GDP) substantially. Economy of rural India, Bangladesh and Pakistan depends on agricultural growth.

In the year 1960, the term "*green revolution*" was introduced by William Gaud, the then Director of the US Agency for International Development. Green revolution was an important breakthrough in food production that emerged from scientific and technological developments. "*Grow more food*" was the slogan of that time. It is no denial that India has transformed to a food grain surplus country due to impressive success of the green revolution. Groundwater sources had played the cardinal role to meet the demand of intensive irrigation. Groundwater irrigation surpassed canal irrigation because of.

(i) Abstracting groundwater by sinking tube well (shallow depth and low capacity) is neither costly nor technologically difficult.
(ii) Cost is within the affordable limit of small and marginal farmers.
(iii) Reliable and under absolute control over the assured irrigation.
(iv) Groundwater falls upon open access regime.
(v) Ownership right of underground water beneath the land was enjoyed by the farmers.

Besides above, highly subsidized electricity tariffs and soft loan of banks have attracted farmers for groundwater irrigation. Later on rise of informal groundwater

T. K. Ray (✉)
State Water Investigation Directorate, Department of Water Resources Investigation & Development, Government of West Bengal, Kolkata, West Bengal, India
e-mail: tapankumarray.geologist@gmail.com

P. K. Sikdar (ed.), *Groundwater Development and Management*,
https://doi.org/10.1007/978-3-319-75115-3_21

market, untrammeled by any authority boosted up the "sale of groundwater". However, groundwater irrigation was the elixir of rural poverty alleviation. Let us take the example of India. In India, for districts where less than 10% of gross area is irrigated, 69% of the population is below poverty line (BPL), whereas in districts where irrigation covers more than 50% crop area poverty line is less than 30%. So economic growth performance is directly linked with water access. Not only in India, same scenario prevails in Bangladesh, Sri Lanka, Vietnam and Thailand. In India, Bangladesh and Pakistan groundwater irrigation has increased exponentially. But equitable access to groundwater is challenged due to deep pumping, resulting to fall of groundwater level below the capacity of low cost pumping devices (centrifugal pumps). Poor farmers are compelled to buy irrigation water from rich farmers. Thus landlords became *waterlords* in *rural society*.

Adverse impact on groundwater resources is essentially linked with the massive expansion of tubewell irrigation. In India, 60% of irrigation use comes from groundwater and it is the only source of irrigation for the poor farmers. Groundwater is also source of about 80% drinking water needs. As a result, India has become world's biggest user of groundwater (UNESCO 2012).

Let us look into the highlights of the 4th Minor Irrigation Census of India. There are 2.1 crore minor irrigation schemes in 641,062 villages. Within a span of 3 years (2001–2004) around ten lakh twenty thousand tubewells have been sunk. Groundwater based schemes are predominantly under private ownership (≈95%). But the share of shallow tubewells for the marginal and small farmers has gone down. There is also a decline in the number of irrigation dugwells. Deep tube wells are generally owned by big farmers, as a result big farmers have encroached the access to groundwater of small and marginal farmers.

The frenzied rush to attain food security has made the present groundwater scenario very uncomfortable in the context of sustainability and environmental concerns. So, the general wisdom for Indian subcontinent (India, Bangladesh and Pakistan) particularly supports the need for regulatory measures for groundwater abstraction. It has also been pointed out by the scholarly researchers that in the last few decades during the period of *'green revolution'* the elite meritocracy and interest group have failed to translate the understanding of groundwater sector problems into effective groundwater policy.

However, as laid down in the Section VII of National Water Policy (NWP) of India, periodical assessment of groundwater potential is done by a joint working team constituted by the members from the Central Groundwater Board, Government of India and the State Groundwater Organization (SGO). The main objectives of the task are: (a) to estimate the available dynamic/replenishable groundwater resource, (b) to infer the stage of development (SOD) as a ratio of net available groundwater and gross draft of groundwater expressed in percent, and (c) categorization of assessment unit (block, taluk etc.) as safe, semi critical, critical and over exploited.

As on March 2011, out of 6607 assessed unit 4530 are safe, 697 are semi critical, 217 are critical and 1071 are over exploited. Comparing with the previous assessment year it is noted that there is decline in the annual replenishable groundwater but increase in the annual groundwater draft manifested by declining trend of groundwater in many assessment units. Again, geogenic arsenic and fluoride contamination in groundwater is rapidly spreading in Indian subcontinent. Multi faceted problems are embroiling the groundwater development. So groundwater governance is emerging as one of the important issues in developmental discourse. However, in general '*water sector reforms*' are underway in most of the South Asian countries.

These reforms lead to achieve the governance over the groundwater resource. The concept of governance varies widely. The World Bank defines governance as "the manner in which power is exercised in the management of country's economic and social resources for development". As stated by Iyer (2003) the understanding of governance of water should begin with defining the crisis and dilemma of water resources development. The dilemma linked with groundwater resources are the right to groundwater, groundwater as an economic good, socio-ecological cost, tradable groundwater, competitive use of groundwater, trans-boundary aquifers, inter-state aquifers and many other factors. Besides, there are many social concerns that are attached with groundwater governance. Categorically, those are access to groundwater by the poor community, availability of groundwater in water stressed area and allocation there from controlling the quality of groundwater, ethical issues on burden of cost for non-availability of groundwater—mainly in drinking water sector. In Fig. 21.1, the framework for understanding water governance is depicted. However the detailed discussion of the framework of governance is beyond the scope of this chapter. The scope of this chapter is kept within the limit of discussion on Groundwater Act only.

2 Initiatives to Groundwater Act in India

Groundwater in water resource is a *'common pool resource'* (CPR), i.e. a resource made available to all by consumption (a progressive wasting away) and to which access can be limited at high cost. Groundwater being a CPR is prone to "tragedies of the commons". It is susceptible to overuse and/or misuse, thus initiating group interest and conflicts. All CPR's face problems of congestion and over-use. So groundwater should be typically nurtured in order to protect the resource. The management of CPR is a very complex process because: (i) it is difficult to select the institute which will be appropriate to govern the groundwater use, (ii) there is no linearity in the sector of groundwater use, and (iii) there is no panacea to reach a solution about the absolute ownership of the groundwater – a common pool resource.

The opinions are divided among the scholars. Some scholars advocated state ownership and some private ownership. Some water experts believe that simple solutions like government ownerships are neither possible nor appropriate. It is

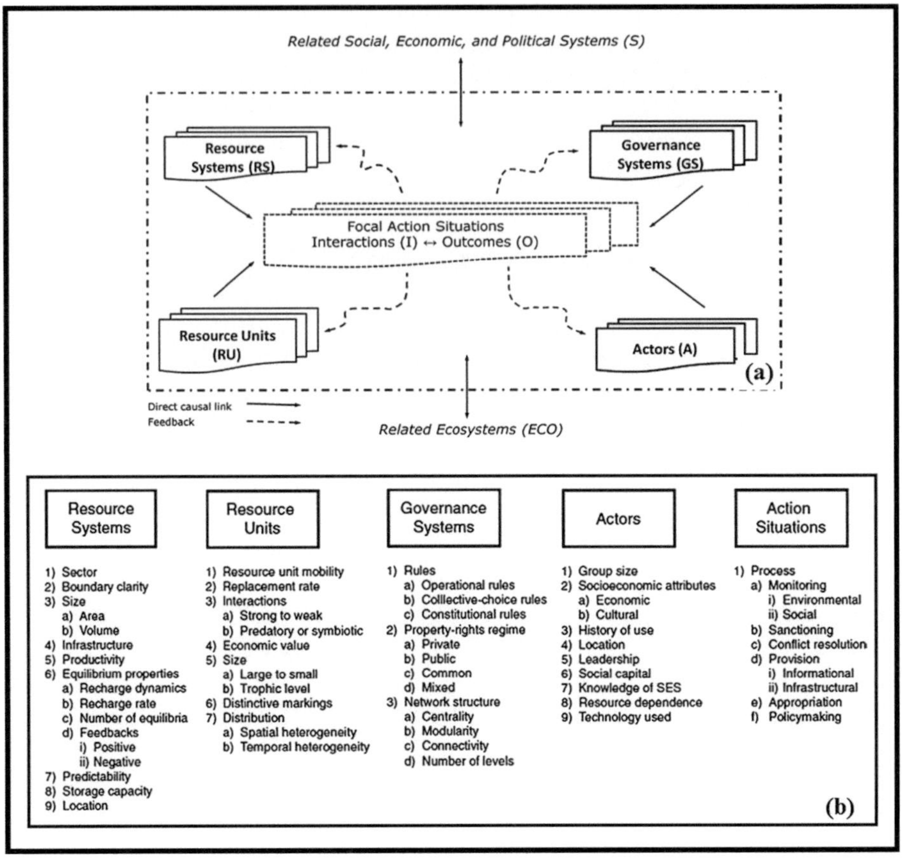

Fig. 21.1 (**a**) Social-ecological systems' framework for groundwater governance: first tier components and (**b**) second tier variables. (*Source:* Ostrom and Cox 2010)

opined that state ownership over the groundwater may prone the social expectation. Primarily groundwater is a life support system and only secondarily anything else like economic good, social good etc. So to protect the groundwater resource is a primary responsibility of the government.

From the historical records it is noted that water regulation is an important issue in the age of Vedas. There is discussion about the regulation of water even in Manusamvita. A chronology of water rules is given in Fig. 21.2.

Over the past few years renewed interest has been given to groundwater law by the Union Government of India as well as State Governments. Unfortunately the various layers of law are not very exhaustive and the statutory instruments are not strong enough for implementation of the Act.

In the context of "*Centre-State*" powers under Article 246 of the Indian Constitution (subject-matter of laws made by Parliament and by Legislation of State), water

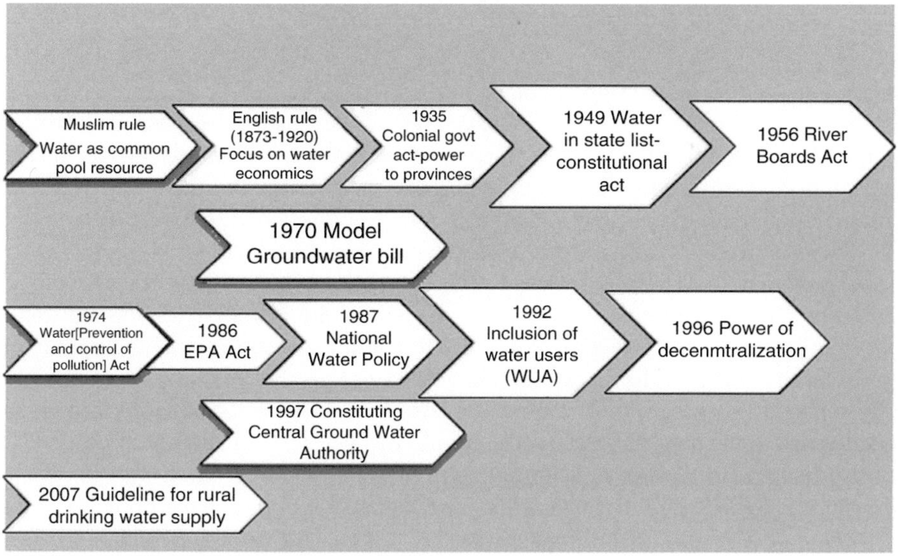

Fig. 21.2 Chronology of main water acts

is included in VIIth Schedule List II (State List). So the Legislature of any State enjoys full power to make laws. In the matter of groundwater resource state can enforce any law within its boundary or a part within its administrative jurisdiction.

In accordance to the Article 243G (powers of decentralized bodies of democratic governance—powers, authorities and responsibilities of Panchayats), the legislature of a State may endow the Panchayats to exercise powers, authority and responsibility in the matter of minor-irrigation, water management and water-shed development, rural drinking water supply, rural sanitation, fisheries etc. By law the State Legislature may also endow such powers to local urban bodies/municipalities.

In India, there is increase in recognition that water crisis are mainly governance and management crisis. Inadequacy of institutional infrastructure and ineffectiveness of governance are major reasons of groundwater crisis in many parts of the country. It is admitted that different water supply system is highly decentralized due to dominant role of groundwater not only for irrigation but also for drinking water supply.

Thus the necessity to control groundwater abstraction under institutional administration was recognized long back in the year 1970 by the Ministry of Water Resources, Government of India. In the same year the Ministry had drafted a Model Bill to regulate and control the development and management of groundwater. The Model Bill was circulated to state governments for adopting laws to regulate and control groundwater abstraction by different beneficiary sectors. As it was felt necessary, the said Bill had been revised from time to time and the latest version was published in the year 2011.

2.1 Model Bill for Groundwater Legislation

The Model Bill contains four chapters with a brief recital on the purposes. Though the importance of the Model Groundwater Bill was accepted by the States but it was a big challenge for the States to frame and promulgate Groundwater Act in the backdrop of agro-economic growth, consensus of farmers, political consequences and traditional agrarian social structure. However, following the Model Bill, so far 11 States and four Union Territories have adopted and enforced groundwater legislation. The Bill framed by the Union Government has wide scope to modify, based on local hydrogeologic condition, available administrative set up and traditional agro-practices. Model Bill for conservation, protection and regulation of groundwater 2011 is the latest version placed before the Planning Commission. The report of the Steering Committee on Water Resource and Sanitation has given cognizance to the revised version. The Model Bill was also an attempt to put an end of usufructory or riparian rights *of the land owner over the groundwater in accordance to Easement Act of 1882*. This important Bill seeks to make groundwater a '*common pool resource*' (Cullet and Koonan 2011). Thus Model Bill and National Water Policy address the governance of groundwater under *public trust doctrine.* State government being the trustee has the responsibility to protect and preserve this resource with all efforts to review its pollution and degradation. It is imperative to note that right to water is a part of "*Right to Life*" under Article 21 of the Indian Constitution. The full text of the Model Bill is available at www.ielrc.org/contente0506.pdf.

2.2 Brief Discussion on Model Groundwater Bill

The first chapter discusses preliminaries like short title, extent, commencement and definitions. The second part consists of definitions wherein "drinking water" is defined which includes both human beings and livestock. But it is not clearly stated about the consumption of livestock produced for commercial purposes. For commercial purposes groundwater is used as an economic commodity. On the more, definition of drinking water becomes ambiguous when it includes bathing, washing and other activities. Let us take the example of India. India is a humid tropical country. Groundwater estimation of Ministry of Water Resources has suggested 40 litres per capita per day (lpcd) for human consumption. Certainly it is a low figure in the context of the major climatic region of the country. It may be rational if it is defined as '*safe drinking water'* which happens to be a fundamental right of every citizen. National Rural Drinking Water Programme 2009/2010 ensures drinking water security for all in the community (available at www.ielrc.org/content/e1002.pdf in Sect. 4). However, National Rural Drinking Water Mission keeps a provision of 70 lpcd for on-going programmes of rural drinking water schemes. It may be agreed upon that for bathing, washing, cleansing and other purposes instead of groundwater other alternatives may be chosen.

To define "groundwater" the recent concept of hydrologic cycle may have been considered. In hydrologic cycle there is an area of overlap of surface water and groundwater. However, for resource estimation purposes, the dynamic groundwater resource is considered, wherein "input – output = change in storage" is the guiding principle. The control and regulation is essentially required for dynamic groundwater resource because abstracting water from static storage will be a compelling invitation to degrade the natural reservoir of groundwater.

The definition of "user group" should have been little elaborative because it includes both groups who will use water for life support and water for profit i.e. water as commodity. The purpose of use is certainly a critical rider to frame laws for different sectors like drinking, irrigation, industry, environment and others. In chapter 2 of the Model Bill, guidelines are given on the establishment of Groundwater Authority. Interestingly a bureaucratic frame of the Authority is suggested keeping aside the major issues of water law reforms. Broadly, the major water reforms seek reduction in the influence of the States. Most of the water law reforms suggest about the decentralization and participation of water users in the matter of regulating groundwater use. Even if the decentralization is not adopted the representation/nomination from the user's society would have been suggested. So, it may be concluded that the drawback of the Model Bill is that it employs '*top down approach*' instead of '*participatory approach*'. The Model Bill has proposed to empower the Authority to declare an area under the overall control for development and management by notification, demarcating the area as '*Notified Area*'. The Model Bill also proposes to grant *'permit'* for new abstraction structure and *'registration'* of all existing structures. The Model Bill has prescribed guidelines for granting permit and registration for building up of decisions in respect of each case. The guidelines help to prepare application forms, both for permit and registration. Information which are important for evaluating the permit are to be furnished in the application form e.g. information of the purpose of groundwater use, groundwater lifting device and capacity, hours of operation, total quantity of water to be abstracted per day etc. However, these proposals are to be examined in the context of groundwater availability, long term trend of the local groundwater level, quality of groundwater, the existence of other competitive users and other factors relevant thereto (e.g. spacing between operating tubewells etc.). In respect to registration, exhaustive information are recorded from the application forms. However, the Model Bill does not prescribe any guideline to refuse registration in places where over-use of groundwater has resulted in concernable critical problems including quality issues.

In Section 9 of Chapter IX, the Bill proposes the registration of drilling agencies on payment of fees to the respective Authority to be constituted under the Act. But no modality is suggested for such registration. In Sections 12, 13, 14 and 18 of Chapter XI, the powers of the Authority are highlighted. Surprisingly the powers bestowed upon the Authority are utmost favorable to any decision of the Authority leaving little or no space for the stakeholders. So in a word Model Bill is an authoritative bill instead of a people's friendly one.

Section 19 of Chapter III provides a brief discussion of *rainwater harvesting for groundwater recharge.* The bill proposes to impose stipulated conditions for providing roof top rainwater harvesting structures in the building plan in an area of 100 m^2 or more. Chapter IV of the bill deals on violation of the act and role of authority in the matter of offences, penalties and appeals. In brief these are the contents of the model groundwater bill. Most of the State groundwater laws follow the model groundwater bill. However, it cannot be denied that the Model Groundwater Bill is neither exhaustive nor explicit. It appears almost as an *ex cathedra* pronouncement ignoring the users and stakeholders views.

2.3 Groundwater in National Water Policy

After the first meeting of National Water Resource Council in 1985, the National Water Policy (NWP) was adopted in the year 1987. The NWP happens to be the most important document to manage the water resources as a whole. The policy document begins with the introduction on the need for a national water policy. It is indicated that around 432 billion cubic meter (bcm) groundwater may be available in India. But this availability is unevenly distributed over the country because of topographical, hydrogeological and climatic conditions in different parts of India. The status of information systems for water resources is discussed in Section 2 of the policy document. Other sections deal on water resource planning, institutional mechanism, water allocation priorities, project planning, groundwater development, drinking water provision, irrigation water use, resettlement and rehabilitation, financial and physical sustainability, participatory approach to water resource management, private sector participation, water quality, conservation of water, flood control and management, land erosion by sea or river, monitoring of water projects, water sharing/distribution amongst the states, performance appraisal, improvements, maintenance and modernization, safety of structures, science and technology for safety, economical managements and training on various domain on hydrology. The last section concludes the National Water Policy documents emphasizing on maintaining national consensus and commitment to the principles and objectives of NWP. It has also being proposed for periodical revision of the document.

In Section 7 of the NWP four issues are highlighted.

(i) Periodical reassessment of groundwater resource.
(ii) Exploitation of groundwater resource should be so regulated as not to exceed the recharging possibilities.
(iii) Integrated and coordinated development of surface water and groundwater resources and their conjunctive use.
(iv) Over-exploitation of groundwater should be avoided especially near the coast to prevent the ingress of sea water into fresh water aquifers.

Indian National Water Policy, 2002 is an exhaustive policy guideline to oversee the state water policies but cannot override any decision or policy adopted by the state governments.

2.4 *Role of Indian Judiciary to Develop Groundwater Law*

Indian judiciary has perhaps played the most important role to develop groundwater laws not only at national level but also at state levels. The nation will remain ever grateful to the Honourable Supreme Court of India for the land-mark judgement in the matter of MC Mehta v Union of India, 1996. The full document of the matter and judgement is available at www.ielre.org/content/e9619.pdf.

The keenness of judgement had compelled the Union Government to constitute the Central Groundwater Authority (CGWA). Verbatim excerpt of that historical judgement is "The Central Government in the Ministry of Environment and Forest shall constitute the Central Ground Water Board as an Authority under Section 3 (3) of the Act. The Authority so constituted shall exercise all the powers under the Act necessary for the purpose of regulation and control of groundwater management and development". The Act referred herein is the Environment (Protection) Act (EPA), 1986. The notification constituting the Central Ground Water Authority was issued on 14th January 1997 (amended on 13th January 1998, 5th January 1999 and 6th November 2000).

It is relevant to mention about a milestone judgement of Honourable Supreme Court in the matter of *MC Mehta v Kamal Nath*, 1996. Perhaps that was the end of century old Indian Easement Act, 1882 (Landowners usufructuary right). Important excerpts of the judgement are given below:

(i) The public trust doctrine primarily rests on the principle that certain resources like air, sea, waters and forests have such a great importance to the people as a whole that it would be wholly unjustified to make them a subject of private ownership. The said resources being a gift of nature, they should be made freely available to everyone irrespective of the status in life.

(ii) Our legal system based on English common law include the public trust doctrine as a part of its jurisprudence.

The state is the trustee of all natural resources which are by nature meant for public use and enjoyment.... These resources meant for public use cannot be converted into private ownership.

(iii) The public trust doctrine is a part of law of the land.

In the matter of state of West Bengal v. Kesoram Industries 2004, Honourable Supreme Court of India stated that "Deep ground water belongs to the state in the sense that doctrine of public trusts extends threats. Holder of a land may have only right of user and cannot take any action or do any deeds as a result whereof the right of others is affected".

Over the years, there has been immense amount of judicial activism on the subject of protecting the security of groundwater. The issues of declining / falling groundwater level and conservation of groundwater had been raised by the Apex court. While extending order in the case MC Mehta v Union of India wherein the rapidly falling levels of groundwater in Delhi were cited, the Supreme Court directed National Environmental Engineering Research (NEERI) to investigate the matter. Not only the directives of the Apex Court but verdicts of Delhi, Kerala, Karnataka, Calcutta, Haryana, Andhra Pradesh, Punjab, Uttar Pradesh and Gujarat have strengthened the groundwater law and policies.

It is worthy to mention that EPA, 1986 was passed under Article 253 of the Indian Constitution in order to implement the decisions taken at the United Nations Conference on the Human Environment, 1972 (Stockholm Declaration). This Act is considered as umbrella legislation to cover different environment laws including water/groundwater law. Under this law Union Government has constituted The Water Quality Assessment Authority (WQAA). The Central Ground Water Authority is another authority constituted under this Act. So within federal structure of the Union Government, the Central Government may assume the powers of regulating groundwater in the state realizing the local conditions.

2.5 Central Ground Water Authority

Following the directions of the Supreme Court in *MC Mehta v Union of India*, Central Groundwater Authority (CGWA) was constituted on 14th January 1997. Later it was amended on 13.01.1998, 05.01.1999 and 06.11.2000. CGWA is constituted under Subsection (3) of section 3 of the EPA, 1986. The main functions of the CGWA include the identification of critical/over-exploited areas and giving direction to state government authorities to take appropriate measures for the regulation and development of groundwater. So far 162 areas are notified by CGWA in Andhra Pradesh, Gujarat, Diu, Haryana, Punjab, Karnataka, Madhya Pradesh, Rajasthan, Puducherry, Tamilnadu, Telengana, Uttar Pradesh and West Bengal. Within the central government, the Ministry of Water Resources, River Development and Ganga Rejuvenation is responsible for the conservation and management of groundwater in the country. Ministry of Rural Development also takes up programmes related to groundwater management.

The members of the Authority are Chairman, Central Ground Water Board (CGWB, GoI) and four members from CGWB; Joint Secretary and Financial Adviser, MoWR, GoI; Joint Secretary, Ministry of Environment and Forests GoI; one Chief Engineer from Central Water Commission (CWC); and Director/General Manager (Exploration), ONGC. There are also invitee members from Agriculture Department, Urban Development Department, Rural Development Department, National Institute of Hydrology (NIH) Roorkee; and National Geophysical Research

Institute (NGRI), Hyderabad. The jurisdiction of the Authority is whole of India with the headquarters at Delhi. CGWA is constituted to "control, management and development of groundwater in the country and to issue necessary regulatory directions for this purpose". It is also mentioned that the Authority shall exercise powers and perform the following functions, namely,

(i) to exercise powers u/s 5 of the EPA, 1986 for issuing direction and taking such measures in respect of all the matters referred to in Sub-section (3) of the said Act;
(ii) to resort to the penal provisions in Sections 15(5) to 21 of the EPA, 1986;
(iii) to exercise powers under section 4 of the EPA, 1986 and above all; and
(iv) to regulate and control groundwater development in the country.

Thus, Central Ground Water Board, a pre-existing scientific and technical body under the Ministry of Water Resources, was designated as the CGWA under Section 3(3) of the Environment (Protection) Act 1986.

CGWA is the apex body to regulate, control and manage groundwater. For a federal government, good groundwater governance needs an apex administration because the problems which overlap between two or more states/countries can be accommodated. Some catchments of river basin may extent into more than one state. Again, groundwater catchment does not necessarily coincide with the surface water catchment. CGWA has the power to intervene in case of dispute between the states over the use of groundwater. CGWA has the power to intervene in the matter of legislation of a State where the interest of the other states is not protected. In case of any legal doubt on inter-governmental agreement and parallel legislation the judiciary may intervene.

However, the role of CGWA is not very significant in public domain. The Authority has very little or no contact with the users. The presence of Union Government through CGWS is little felt by a groundwater user. Even CGWA has taken little initiative to bring all the user sectors (drinking, irrigation and industry) under one umbrella. There is no representation of users group in the constituted Authority. It is an Authority constituted by the members from administrative services. CGWA has always taken an administrative role instead of participatory approach. On the contrary, the Indian Judiciary has the role to protect the interest of the beneficiaries also.

CGWA is yet to direct the States to take up the census of urban and industrial abstraction of groundwater. In the matter of judicious allocation of groundwater for major sectors (drinking, irrigation and industry), there is no guidelines/norms from CGWA. There is no instruction, guideline or recommendation from CGWA about the course of action to be adopted by the States in cases of geogenic arsenic and fluoride toxicity in groundwater.

So it is very difficult to understand the role of CGWA in public domain. It may be a role of advisor, an arbitrator, a body of experts, a legal authority or an ensign of the union government.

3 Groundwater Act in Other Countries

The economy of the South Asian countries i.e. India, Pakistan, Sri Lanka, Bangladesh, Bhutan, Afghanistan and Maldives by and large depend on irrigated agriculture. India, Pakistan, Bangladesh and North China use 380–400 km^3 of groundwater annually, over half of the world's total annual use. All of India, Northern Sri Lanka, Punjab and Sind of Pakistan and Bangladesh depend on groundwater irrigation for agricultural growth. Groundwater supports the irrigation available to small and marginal farmers. Asian groundwater is mainly used for agriculture whereas the rest of the world's groundwater is used mainly for urban and industrial use.

In India, the total number of mechanized wells and tubewells rose from less than a million in 1960 to an estimated 19 million in 2000. In Pakistan Punjab it increased from a few thousand in 1960 to 0.5 million in 2000 (Shah 2007). Before 1960 groundwater irrigation in Bangladesh was very insignificant. Groundwater irrigation took the boom from 1980 and onwards. It is interesting to note that between 1985 and 2004 the number of deep tubewells increased around 61.5% whereas in the sector of private owned shallow tubewells the increase is in the tune of 600% (Zahid and Ahmed 2006). It is to note that out of 270 to 300 million hectares of global irrigation economy, more than one third ($\approx$110 million hectares) comprises groundwater irrigated areas in the Indian sub-continent alone. Booming groundwater use in irrigation sector has brought a significant economic growth from the view points of livelihood and food security. But at the same time this groundwater abstraction boom has put an excessive pressure on the aquifers. Thus the sustainability of groundwater irrigation has become under risk. Eventually a common question arises how to manage this precious resource by implementing legislation. So the institutional administrations of India, Pakistan and Bangladesh have put their sincere endeavours to regulate the groundwater abstraction by promulgating acts and ordinances.

3.1 Groundwater Act in Pakistan

Around 35% agricultural water requirements are supplied from groundwater sources. The number of tubewells under private ownership is increasing at the rate of 20,000 per year A big investment is involved for tubewell irrigation. More than 30 billion rupees have been invested for tubewell irrigation in private sector. Groundwater usage contributes around 1.3 billion rupees to national economy per year.

The extensive groundwater abstraction under private control in and around groundwater-rich areas has compelled the Pakistan Government to form acts categorically for Indus Plain to protect the potential aquifers from over-exploitation and quality deterioration.

In the year 1958 an Act was promulgated in Pakistan styled as "The Pakistan Water and Power Development Authority Act, 1958". A federal agency was constituted under this act. Another Soil Reclamation Act was enforced, but it was not implemented by any institution. However a "Tubewell Promotion Policy" for agricultural development had been framed in the year 1960 which was subsequently revised in the years 1970 and 1980 for the areas outside the canal irrigation command of the Indus Plain.

In 1980 the licensing system was introduced to restrict installation of private tubewells in critical areas where groundwater was depleting at a faster rate and/or groundwater quality was deteriorating. At the provincial level the groundwater regulatory framework for Punjab was prepared in the year 1990 with the assistance of World Bank. Under Provincial Irrigation and Drainage Authority (PIDA 1999–2000) groundwater management rules were drafted. Baluchistan provincial government failed to implement the 'Balochistan Groundwater Rights Administration Ordinance' in 1978 and again in 2001.

Pakistan Council of Research in Water Resource (PCRWR), a government institute, had prepared a groundwater atlas for upper Indus Basin in the year 2014. The institute has recommended formulating a regulatory framework for efficient and sustainable use of groundwater.

3.2 *Regulation of Groundwater in Sri Lanka*

Government of Sri Lanka has constituted two Boards to manage water resources. They are the Water Resource Board (WRB) and National Water Supply and Drainage Board (NWS & DB), NWS & DB produces 310 million m^3/year treated water to cater a population 5.3 million.

3.3 *Groundwater Act in Bangladesh*

On 9th June 1985, the Groundwtar Management Ordinance (Ordinance No XXVII) was promulgated. The salient features of the ordinance are:

(i) Constituting upazillaparishad (Upazilla Irrigation Committee) - a constituted body under section IV of the Ordinance.
(ii) No tubewell shall be installed at any place without license granted by the Committee.
(iii) An application fee is fixed for obtaining the license.
(iv) The merits of the application are evaluated by local enquiry or aquifer characteristics.
(v) License has to be obtained within a period of 6 months for the existing tubewells.

(vi) In section VII of the Ordinance, the rules for suspension of license and rules for vacating the order of suspension are stated.
(vii) Section VII is on cancellation of the license.
(viii) Section X and section XI are on the offences and contraventions.
(ix) Section XII is on powers to make rules.
(x) Section XIII is on power to exempt any project or any specified area from this ambit of ordinance.

The Ordinance was the first step for the groundwater management in Bangladesh. Later on in 1999 Bangladesh adopted the first "National Water Policy (NWP)" in context of decentralization, environmental impacts and private sector investments. In 2004, Bangladesh also adopted the National Water Policy for "Arsenic mitigation". Recently, Water Act 2013 has been published by the Bangladesh Government. As per Water Act 2013 "all forms of water belong to the government". However, it is not mentioned in the Act about the maximum allocation of water for domestic and agricultural use. Bangladesh government is now serious to include legal, administrative and financial measure vis-à-vis awareness creation and capacity building for facing the challenges of groundwater crisis.

4 Legislation and Informal Groundwater Market

According to Mukherji (2007), "informal groundwater based pump irrigation services markets are all-pervasive agrarian institute in South Asia but have been criticized for bringing out less than equitable outcomes and causing groundwater overexploitation". The author also states that "groundwater markets in South Asia are a classic example of water markets since groundwater rights are inalienable from land rights". In Bangladesh, West Bengal (India) and Nepal 88%, 48% and 60% respectively of pump owners sell groundwater. The groundwater market is also expanding in the Indian States of Bihar and UP. In this way *'shallow tubewell owners'* are becoming *'shallow tubewell businessman'*. In village groundwater market the payment for water is in cash for hourly rates indicating that the groundwater market has reached a mature stage in India and Bangladesh. Thus it is opined by many experts in groundwater sector.

In all south Asian countries individual ownership of tubewell is preferred over community ownership (Wood and Jones 1991). Thus the spread of water market and its scale of operation are huge. So this phenomenon requires more attention than what has been accorded so far. Moench and Janakarajan (2006) point out that groundwater markets are spreading fast not only in eastern Indian states but also in parts of southern India especially in hard rock areas. So "the need for close study on factors underlying rental markets of groundwater and their implications for public policy especially in the realm of investment in irrigation assets cannot be gainsaid (Saleth 2004).

On the more, Indian agriculture is becoming '*small farm agriculture*'. Small holding accounts for overwhelming majority. Small farmers supply 62% of paddy, 54% wheat and 54% of coarse cereals. However there is a trend to shift from grain based cropping to high value crops like fruits, vegetables, spices etc.

Analysis of water markets in Karnataka and Gujarat reveals that the lack of well defined groundwater right increases monopoly power of the rich tubewell owners. The same situation is experienced also in West Bengal. Thus the '*bubble economy*' rises. Monopolistic water prices have put on negative impact on groundwater buyers. Hence economic value of groundwater has become a criterion to identify an area as 'water rich' or 'water scarce'. A water right system should have some key attributes, such as a requirement for effective and beneficial use of water, security of water use tenure and flexibility to reallocate water to more beneficial social, economical or ecological uses after periodical review.

The National Water Policy of India defines priorities for different water user sectors. NWP of India (1987) treats water as an '*economic good*' and proposes the use of water pricing in a manner that would cover the cost of investment, operation and maintenance. But nothing is indicated about the limit of profit. Many social activists argue that considering water as an economic good may lead to the concernable *commodification*. They fear that economic globalization may push down the South Asian countries to a path of greater inequities. In a special report, International Forum of Globalization (Barlow 2000) had mentioned categorically that, "When water is privatized, prices are set on the open market. As a result billions of poor people have been cut."

However both public and private institutions have failed to recognize the groundwater users as a "true stakeholders" that is, they are entitled to get all information about the available resource, present draft, projected draft, quality and sustainability. The public institutions are yet to decide on "better pricing" of groundwater used for irrigation. The rational is that groundwater has an incentive to use water efficiently when it has a price.

5 Conclusion

It was opined by great utilitarian sociologist Harber Spencer (1820–1903) "that government is the best government which interfere the least".

It is a difficult and challenging task for any public institution to interfere into the traditional system prevailing in water sector. But it becomes a compelling invitation for the government to interfere when inefficient uses of water resource cause to make the resource unsustainable, polluted, degraded and scarce. Groundwater use rights are often misused. For example, groundwater usage rights conventionally need not be tradable. Essentially the use rights should be embedded in the rules that can be designed, changed and adopted to different situations for good governance.

In India, Bangladesh and Pakistan, regulatory acts are enforced but implementation of acts depend on many factors like capacity building of nodal agency for implementing the act, stakeholders spontaneous participation, infrastructure strengthening, co-operation of the civil society, political support and above all a strong commitment from administration and politician. It appears that the administration is in doubt about enforceability of acts.

However, promulgation of act in south Asian countries has offered two major experiences such as: big farmers are more benefitted than small and marginal farmers, and present acts are not sufficient to control informal groundwater market. National Water Policy and groundwater acts have no meaning unless those are under-pinned at local level. In general farmers did not accept the acts because of limited faith on public institutions and they are also aggrieved for the promulgation of the act. Competitive uses are a serious issue which is the root of conflicts at village level. A large numbers of 'writ petitions' have been filed by the farmers to challenge the decision of the authority. The Indian judiciary has stood up to protect the security of groundwater.

It is understood that institutional changes and regulatory measures only will not be sufficient to protect the groundwater. The awareness of the stakeholders can only maximize the sustainable use of the precious groundwater resource.

References

Barlow, M. (2000). Blue Gold: The Global Water Crisis and The Commodification of The World's Water Supply. Diane Pub Co.

Cullet, P. and Koonan, S. (2011). Water Law in India: An Introduction to Legal Instruments. Second Edition. Oxford University Press.

Iyer, R.R. (2003). The Dilemmas of Water Resources Development. (Eds) Chopra, K., Rao, C.H.H. and Sengupta, R. Indian Society for Ecological Economics. Concept Publishing Company, New Delhi.

Moench, M. and Janakarajan, S. (2006). Are Wells a Potential Threat to Farmers' Well-being? *Economic & Political Weekly*, 41(37): 16–22.

Mukherji, A. (2007). Implications of Alternative Institutional Arrangements in Groundwater Sharing Evidence from West Bengal. Economic and Political Weekly, 2543–2551.

Ostrom, E. and Michael Cox. (2010). Moving beyond panaceas: A multi-tiered diagnostic approach for social-ecological analysis. Environmental Conservation, 37(04): 451–463.

Saleth, R.M. (2004). Strategic Analysis of Water Institutions in India: Application of a New Research Paradigm. Research Report 79. International Water Management Institute, Colombo, Sri Lanka.

Shah, T. (2007). The Groundwater Economy of South Asia: An Assessment of Size, Significance and Socio-ecological Impacts. *In:* The Agricultural Groundwater Revolution: Opportunities and Threats to Development. (Eds) Giordano, M. and Villholth, K.G. CAB International.

UNESCO (2012). Managing water under uncertainty and risk. The United Nation World Water Development Report 4, Vol. 1. United Nations Education Scientific and Cultural Organisation, France.

Wood, G.D. and Palmer Jones, R. (1991). The Water Sellers: A Cooperative Venture by the Rural Poor. Intermediate Technology Publications Ltd., London.

Zahid, A. and Ahmed, S.R. (2006). Groundwater resources development in Bangladesh: Contribution to irrigation for food security and constraints to sustainability. *Groundwater Gov Asia Ser*, 1: 27–46.

Chapter 22
Catalyzing Peoples' Participation for Groundwater Management

Ayan Biswas

1 Introduction

India is a groundwater civilization. About 90% of India's agriculture and 85% of India's domestic water sectors are dependent on groundwater. On an average, about 50% of water supplied to the cities is sourced from groundwater (Suhag 2016). Close to 30 million wells dot the Indian landscape. Indiscriminate pumping from these wells has led to "groundwater anarchy" (Shah 2009). Access to groundwater is almost unrestricted and individual land rights are inextricably linked to groundwater. At the same time, groundwater is mobile and cannot be stored in one location in its pristine state. This makes enforcement of rights difficult. Some of these complexities in resource characteristics are important to understand before we begin to analyze the different dimensions of participation in India's groundwater management.

2 Anomalies in Resource Characterization

Two contradicting schools of thought guide the resource discourse on groundwater. One group of practitioners and researchers acknowledges and theorizes groundwater as a Common Pool Resource (CPR). So this group's understanding of community's centrality to groundwater management and their own role with respect to the community tends to be different from the other group. The other group posits that groundwater is not CPR. Rather they focus on the systems understanding from a sustainability framework and focus on the conjunctive use of surface water and groundwater. A majority of the examples cited in this chapter make a case for

A. Biswas (✉)
Asia Hub, Akvo, Amsterdam, The Netherlands
e-mail: ayanbiswas2003@gmail.com

P. K. Sikdar (ed.), *Groundwater Development and Management*,
https://doi.org/10.1007/978-3-319-75115-3_22

"catalyzing peoples' participation for groundwater management". Before settling on to the argument that groundwater is CPR and this is how communities can participate in its management, it is important to understand the alternate point of view about the resource.

The alternate point of view espoused by the second group argues that downstream and ecological impacts of groundwater use are not well understood by both the hydrogeologists and the society. They also argue that these impacts are not linear. Hence the role of experts becomes necessary to understand the different kinds of impacts in different local contexts. In case of a groundwater dependent wetland, the wetland conservation should be part of aquifer planning. But when ecological interests are not at stake, elements of aquifer planning and the role of catalysts (either government or non-government) will be different.

The second group argues that aquifers as groundwater stocks, water budgeting and recharge based sustainability criteria may seem commonsensical but may not be a representation of the reality in its totality (Dumont 2013). They suggest that groundwater should be managed through its integration within a river basin and as a vehicle to sustain dependent ecosystems and ecological flows. In this construct, role of communities assumes a different proportion (sheer size of a river basin and the heterogeneity of stakeholders make it difficult for the communities to communicate and monitor, more on that later in the chapter) and role of different state and non-state stakeholders also needs a different kind of consideration.

But this chapter refrains from dealing much with the second group. Rather it goes with the assumptions of the first group (groundwater as CPR) and examines community's role, vis-à-vis the catalyst's role in some degree of detail.

3 Different Dimensions of Participation

Agriculture uses most of India's groundwater. Irrigated agriculture via tube wells and pumps led to an overdraft of groundwater in rural and peri-urban areas of India. Despite democratizing irrigation economy, drought proofing, increasing yields and income of the farmer, rampant withdrawal of groundwater has led to scarcity and quality issues across India—forcing farmers to migrate, move out of agriculture and even commit suicides.

Most of the management responses to this problem deal with how human beings have impacted the aquifers. It is equally important to understand how human beings respond to different aquifer situations across India. The perceptions of users of a particular aquifer determine how human beings would behave individually and collectively and whether participation is indeed possible in such a setting.

British Geological Survey (2004) defines an aquifer community or groundwater user group as people who are "mutually vulnerable and dependent because of the centrality of the resource in supporting livelihoods". An aquifer community's awareness regarding this inter-dependency is a measure of how strong/weak they are. While explaining the community/institutional dynamics in an aquifer, Shah (2008) uses a framework to describe five possible scenarios.

A. *Atomistic individualism:* A farmer is an insignificant user in a highly recharged water table. An individual farmer's groundwater use does not impact others in any significant manner. No interdependence is necessary and a community/user group is almost absent.
B. *Collusive opportunism:* Groundwater development has raised the cost of further abstraction. Hardly any effect on water scarcity and quality is visible. Wealthy farmers collude to control the resource. A weak community is established.
C. *Rivalrous gaming:* Groundwater development has not only raised the cost of further abstraction but has also limited water availability. Rivalry and competition among users begin. Strong interdependence among users is visible but the community is dysfunctional.
D. *Co-operative gaming:* Rivalrous gaming transforms into co-operative gaming under certain catalytic conditions (these conditions are discussed later in the chapter). Cost of water abstraction is reduced and water availability is improved. Strong sense of interdependence among users is palpable and a functional aquifer community is ready for self-governance.
E. *Exit:* Groundwater development results in quality issues without affecting its availability. Costs and risks of groundwater use become high, fatalism prevails over the sense of the interdependence and users begin to quit farming.

These scenarios can easily be linked to India's contextual realities. Atomistic individualism perhaps is best demonstrated in the alluvial plains of Ganga-Meghna-Brahmaputra, collusive opportunism is predominant in Western Rajasthan and North Gujarat, whereas hard rock areas of India demonstrate rivalrous gaming with a possibility to be organised in the cooperative gaming process. In coastal aquifers, farmers have begun to move out of agriculture. Shah's framework of pragmatic realism helps to understand how aquifer situations lead people to act in different ways, either to collaborate or not. But the rights discourse to participation takes the discussion to another level.

Property rights over groundwater are fundamental to improving rural livelihoods. Groundwater is anyway linked to land rights. Rural poor neither have the land, nor is their right to resource well established. Even if the poor have existing rights to land/groundwater, the political economy subsumes these rights in a manner that ostracizes the poor. Participation may create opportunities for rural poor to assert their rights and challenge the political economy of the system in the longer term. But these changes are contingent upon the strength of the group in preventing an 'elite capture' of the agenda. Some authors (Agrawal and Ostrom 2001) argue that introducing equity in property rights discourse will be key to the effectiveness of collective action. But introducing elements of collective action for managing groundwater is a difficult task, considering conflicting stakes and expectations of rich and poor farmers.

4 Theoretical Framework of Participation

Effective participation in groundwater management is generally dependent on commitment rather than coercion. It is also important that participants define participation in terms of extent and quality, as well as questions about who should participate.

Arnstein (1969) identified different forms of "citizen participation" as a ladder—moving from tokenistic forms of participation (manipulation) to more meaningful forms of involvement (citizen control). Sriskandarajah et al. (1996) suggested that at higher levels of participation (please refer Box), local people made decisions about the management of resource.

Participation and Collective Action
Participation and collective action are used interchangeably by many scholars and practitioners. Participation could vary from normal, passive, consultative, active, interactive and informed. All these forms of participation do not result in collective action. Participation and collective action could mean the same if "individual costs of participation are more than that of individual benefits"

While at the lower levels of participation, resource management decisions were made by bureaucrats and sector experts who are usually disconnected from the ground realities. In the tokenistic form of participation, communities were only involved either as voluntary or paid labour. At the higher order of participation, communities defined the end goals, or at least contribute substantially to developing and achieving these goals. Following sub-sections explain the elements that form the backbone of a theoretical framework to explain 'participation' in groundwater management.

4.1 Social Norms and/or Formal Legislation

Participation is key to managing groundwater. If citizens/primary stakeholders participate in managing groundwater, effective outcomes can be achieved and sustained over a longer term. Groundwater governance, a more matured yet ambitious theory for managing groundwater, attempts to combine participation with formal legislation—regulatory instruments that integrate social norms with conventional legislation.

Social norms to regulate groundwater usage hold the key to managing the highly decentralized and disaggregated nature of groundwater use in India. Since, social norms can be customized to a location and/or a situation these norms are often an outcome of participatory process that combine science and technology. Hence they hold greater potential to influence social behaviour. Social norms require communities to accept a certain kind of practice and change their behaviour, both long term ordeals; hence benefits may take a long time to manifest. But when they do, they are lasting and impactful. A validation of this hypothesis is available in the section "practical examples of catalyzing participation".

However, experience suggests that despite good social norms, 'free-riding' in regional aquifers can be an issue—particularly in alluvial and sedimentary aquifers. A combination of social norms and formal legislation becomes important in these contexts.

Prima facie, social norms and formal legislation may seem at odds with each other. But that is hardly the case. A command and control type of legislation is difficult to implement and replicate. Highly decentralized and 'anarchic' patterns of groundwater use are not compatible with centralized groundwater legislation either, even if developed and executed at the State level. However, if the groundwater legislation aims to protect social norms, it will confer a more 'legal' status to social processes. On the other hand, norms alone may not be adequate—especially for regional aquifers due to its sheer size and number of stakeholders. Hence, legislation and social processes need to compliment each other.

4.2 *Data Driving Participation*

Participation requires building trust among stakeholders. Generating data collectively, opening up and sharing databases drive the process of participation (Llamas 2005). While research highlights the importance of participation in supply-side interventions such as managed aquifer recharge (Shah 2014; Gale et al. 2006), participation remains crucial for managing demand too. Apparently commonsensical, yet paradox to some, the participatory groundwater management approach/ programme (refer the section on "practical examples of catalyzing participation") demonstrate varying degrees of success in its different projects—especially how data based knowledge systems can drive community participation. In these examples, knowledge about the aquifer situation and possible management options catalyze participation. The process is sustained by forming institutions that manage groundwater, especially at the scales of villages and small towns.

4.3 *Disparity in Scales*

An aggregated picture of India's groundwater situation (at the risk of generalizing and summing incomparable parts) cannot form the basis for community action. Community action in groundwater management has to be at an appropriate scale. Aquifers, in most cases, have to be understood locally. In addition, millions of farmers in India use groundwater in an 'atomistic' manner—prompting a decentralized action agenda. To make matters complicated, regional aquifers have a large span, go well beyond administrative boundaries and consist of many 'communities'—a scale that does not lend itself to participation and community action.

All these factors lead to a concept called 'groundwater typology' (Kulkarni and Vijay Shankar 2009). The theoretical framework of groundwater management needs to be reconciled with a given groundwater typology, the participation element also need to evolve differently with varying scales.

4.4 Nature of Resource

Groundwater's peculiar resource characteristics make it hard for people to participate. It is invisible and hence its state is difficult to assess. In some cases, physical extent of aquifers is too large to create a user community. The complexities of hydrogeology make common understanding of resource dynamics difficult. Given the resource characteristics and the response from the common pool resource (CPR) theorists (refer the next section), it seems evident that the design of a community based groundwater management approach needs to go beyond the familiar domains of hydrology, regulation and legislation. It needs to incorporate the elements of social capital formation, behavioral economics, vocational education, communication and leadership development. Details of some of these attributes will be discussed in the section on "practical examples of catalyzing participation".

5 The CPR Response

Acknowledging the CPR nature of groundwater, Kulkarni et al. (2015) puts forth COMMAN's (2005) 'design principles' based on Ostrom (1990), which helps people to gauge opportunities and constraints for managing groundwater as a community. They also help to develop a set of 'first' and 'second' order conditions for community based groundwater management.

First order conditions:

1. Interface between resource and management group (influences who receives benefits and who pays costs of group action).
2. Management group characteristics (affects ability to define groups of interest, management objectives and criteria for 'success').
3. Nested institutions (ensure that large scale problems are addressed; also helps absorb some of the transaction costs of group organization).
4. External environment – policies, institutions and processes (defines the wider influences and constraints on group management).

Second order conditions (applies only to existing group management schemes):

1. Rules defining groundwater access and/or use entitlements.

2. Monitoring and sanction arrangements exist for checking and enforcing compliance.
3. Mechanisms exist for modifying rules.

Comparing various community based approaches and considering the diversity of contexts, COMMAN suggested that a mix of first and second order conditions (formal legislation and locally developed social norms) is perhaps best suited to involve local communities to manage their own groundwater resources. Several conditions need to be met first, in order to make the CPR based groundwater management regime successful. Organizing and arranging communities, within the appropriate institutional framework, is one of them.

5.1 Institutional Arrangements

To begin with, using existing social structures to organize communities for groundwater management would make sense. The advantage is that these social structures usually have more or less accepted political institutions. In addition, groundwater users may not be prepared to invest in specialized groundwater organizations/aquifer communities till the gains from improved groundwater management are clear to them. However, the indivisible nature of groundwater makes it difficult to partition the resource and exclude other users, in case they choose not to follow extraction rules set by the aquifer communities. Also, aquifer systems are connected and groundwater users organized solely on the basis of existing social/political structures may be forced to bear the effects of depletion by neighbouring communities. Furthermore, there is much evidence (Lindberg et al. 2011) of traditional social/political organizations disintegrating while evolving as groundwater management institutions. Therefore, existing social structures may not be appropriate units for organizing groundwater users in the long term, if they are not accompanied by other institutional measures. Size, socio-economic and physical characteristics of the communities are important criteria for setting up groundwater collectives.

Classic CPR management theory suggests that all other things being equal, small community of users will manage groundwater the best. Generally, as the size of the community increases, coordination among groundwater users become difficult and the transaction costs of coordination also increase. Small mountain valleys perhaps present best conditions for setting up small groundwater management groups. On densely populated large river/groundwater basins, it will be difficult to organize groundwater users along similar lines. In case of large groundwater basins, the basin could be divided in various physically and or socio-economically homogeneous zones. Groundwater users could be organized in zones/collectives—right size to address short term phenomenon like drawdown. These groundwater collectives will also be effective for dealing with medium and long term effects of extraction, if they are represented at the level of the aquifer and can coordinate among each other. However, even within small natural communities or newly formed small groundwater collectives, the socio-economic and physical differentiation may still be substantial.

Different social and physical settings will require different institutional solutions— involving different costs of coordination. In homogeneous communities with little asset and income differentiation, groundwater users are likely to have similar interests. In this case, a few individuals (elders, leaders) may decide on the groundwater management issues. These decisions may satisfy most if not all users. The solution to a groundwater problem could be achieved with a relatively small proportion of the community engaged in decision making—low cost of coordination. In heterogeneous societies, with sharp socio-economic and physical differences, decision making cannot be left to a few individuals. Left to them, the elite may not even bother to involve the groundwater collective in decision making—use their superior socio-economic situation to their own benefit. However, if the elite is dependent on some forms of support from other members of the community, that might create a bargaining opportunity for other members of community. This bargaining may lead to a more restrained and optimal form of collective action (Van de Laar 1990).

Differences in physical settings make it difficult for groundwater collectives to cooperate. People living in low-lying areas with higher groundwater tables, or areas with access to surface water, will not feel the necessity to contribute towards any management solution to solve the problems in another locality (Kumar 1994). Yet, if any kind of dependencies exist between those affected by groundwater problems and those who are not, a sub-optimal form of collective action could take place through bargaining.

5.2 *Transaction Costs*

Bargaining (defining rules, monitoring and enforcement) entails transaction costs, under the CPR regime or otherwise. Someone has to bear these costs of transaction. In case of homogeneous communities, groundwater users are expected to share the transaction costs equally. In case of diverse interests and varying capacities for paying transaction costs, the elite may be willing to pay the most, considering their greater interest in controlling the management regime.

The state can also reduce transaction costs for communities involved in groundwater management. The state may reduce the cost of exclusion, e.g., by legally banning new sources. The state may also support local groundwater collectives to reduce coordination costs. In general, the state may secure property rights, legalize the groundwater collectives and preserve law and order. On the contrary, the state may also increase transaction costs for groundwater communities by withdrawing one of the above forms of support. However, the state's intervention may reduce the communities' flexibility in decision making. It may also impede a learning process for the community involved—especially to deal with uncertain resource conditions.

5.3 Monitoring and Rule Enforcement

Monitoring is an indispensable component of managing groundwater as CPR. Two types of monitoring assume importance in this context: monitoring groundwater as a resource and its extraction by the users. Monitoring incurs transaction costs, but effective monitoring may reduce the overall transaction and exploitation costs. Monitoring individual groundwater extraction rates are difficult. Monitoring cumulative impacts of extraction is easier.

Apart from the measurement issues, there is also a scale issue involved in monitoring. Monitoring extraction from large numbers of individually owned wells may not be feasible (Mosley and Lincklaen Arriëns 1995). Self-discipline and trust form the backbone of monitoring. If the users trust each other, there is no need for individual monitoring. Similarly, if the users trust that the groundwater monitoring system will detect the 'defectors' then the users will exercise self-discipline.

In general, people compete for groundwater and have a tendency for breaking rules. Once the rules are set following negotiations, the groundwater community should negotiate rules for penalizing the 'defectors'. This will ensure that no external coercion is necessary for rule enforcement. As a rule of thumb, the defection should be monitored at a reasonable transaction cost. In case the elite (economically, socially, politically powerful) get away with defection, the groundwater community should have recourse to higher authorities. Similarly, local rules require legitimacy and should therefore be embedded in national legislation. Hence, the formal groundwater legislation needs to create enough flexibility within its framework to embed local rules and adapt itself to local conditions.

5.4 Catalysts and Transformational Change

At an early stage, groundwater communities need support from an external agency to generate and share information, to make rules, to take management decisions, and to solve conflicts. Leadership needs to be nurtured. And an external agency could be helpful in identifying and nurturing the leaders. Other forms of support include developing a CPR based groundwater management regime, establishing user organizations or strengthening existing management arrangements. This will reduce transformation costs for the resource users. The State may subsidize part of the transformation costs by encouraging the collection of information, enabling complex institutional arrangements, and endorsing local rules through legislation. At the secondary management level, community ownership and use of groundwater sources usually involve substantial transformation costs—especially to reduce the social tension and conflicts, developing and maintaining the infrastructure. Poor people find it difficult to bear these costs. Private organizations or NGOs have typically absorbed a major part of these transformation costs, and have thus helped these groundwater communities to overcome these difficulties.

Under a CPR regime, external agencies are necessary to help communities overcome initial hurdles of managing groundwater and bear a substantive portion of the transformation costs. But the same regime dictates that some portion of the costs is borne by the users themselves. Hence at an early stage, the external agencies/NGOs should clarify their roles vis-à-vis the groundwater users. The agency must also explain that although they may bear substantive portions of some of these early transformation costs, this may not be the case with future transaction costs. The agency needs to clarify that communities need not participate in groundwater management if their expectation is to receive material incentive from the agency. The true test of transformational change arises, when the perception of such incentives cease and communities begin to manage the groundwater themselves and bear a substantive portion of the transformation costs. The catalytic role played by the external agency assumes importance in this context, especially how they develop and nurture communities and build their capacities to manage groundwater on their own.

6 Practical Examples of Catalyzing Peoples' Participation

Despite the complexities involved in ensuring people's participation in its highest level and derive maximum benefits, several organisations have catalyzed 'successful' models of groundwater management in India. Based on the role played by the catalyst and the central theme of participation, these examples can be classified into different categories. This section elaborates on these practical examples and attempts to understand their theoretical underpinnings.

6.1 Self-Regulation Through Social Norms—Rajasthan and Gujarat

Self–regulatory systems for managing groundwater have been developed through community participation and building social norms, particularly in areas with shallow, semi-confined aquifers. These collective management systems are quite location specific and rudimentary at times. But more importantly, self-regulated groundwater management systems operate at a scale that matches its extent of overuse. Groundwater community formation and self-regulation becomes easier where the impact of recharge or pumping is immediate and dramatic.

6.1.1 Rajasthan

Eastern Rajasthan is a semi-arid location. Some parts of Eastern Rajasthan suffer from extensive groundwater overdraft. In Eastern Rajasthan, NGOs such as Tarun Bharat Sangh (TBS) and Professional Assistance for Development Action (PRADAN) have catalysed community action in rainwater harvesting and groundwater recharge.

PRADAN began working in the Alwar District (East Rajasthan) in the 1980s to improve the implementation of anti-poverty programmes. Following this beginning, they developed a water conservation programme based on rainwater harvesting. PRADAN built water harvesting cum recharge structures within the watershed planning framework. Community organisations and a democratic decision making process were set up prior to building the physical infrastructure.

TBS worked in roughly 550 villages of the Alwar district. TBS worked on a variety of water harvesting structures in these villages. Beginning slowly, TBS's work gathered momentum since the mid 1990s.

An interesting element emerged from PRADAN and TBS's experiences in Rajasthan. PRADAN emphasized on building sustainable local institutions, which improved the quality of their work but compromised the speed and scale. TBS focused on building *ad hoc* local organisations that scaled quickly.

In case of TBS, once the benefits became visible, demand for similar work began on its own—catalyzing community participation across many other villages.

6.1.2 Gujarat

Hindu religious sect *Swadhyaya Pariwar* catalyzed the Saurashtra groundwater recharge movement. Subsequently, other sects of Hinduism, NGOs and grassroots organisations joined the movement especially in the wake of the drought during 1985–1987.

Following instructions from Pandurang Athvale, a *Swadhyayee* teacher, many farmers within the sect began harvesting rainwater and recharging groundwater. Around 1992, the recharge experiments almost became a movement. Most farmers (*Swadhyayee* or otherwise) began harvesting rainwater and channel the water towards a recharge structure. NGOs also joined the movement around the same time. Diamond merchants of Saurashtra contributed significantly to the recharge movement too. They put in money and helped organize the movement in different locations.

No study on the actual scale of the recharge is available. However, references (Van Steenbergen and Shah 2003) suggest that between 1992 and 1996, close to 100,000 wells were recharged in Saurashtra and about 300 *Nirmal Neer* (farm ponds) were constructed.

Two aspects about the Saurashtra recharge movement are significant. First, the selection of appropriate technology for groundwater recharge. Second, attracting people's participation across different social segments. In both cases, simplicity of technology and communication played an important role. *Swadhyaya Pariwar* communicated the practice widely and rapidly. Some of them even went from one village to another to ensure mass adoption of the recharge technique. As a result, what started out as a movement organized and propagated by one religious sect, became a self-perpetuating groundwater recharge movement. A combination of reasons contributed to this evolution.

- *Swadhyayees* had a strong sense of affiliation to Pandurang Athvale.
- Underplaying the economics, Athvale marketed the groundwater recharge message within the fold of instrumental devotion. That seemed to have worked better.
- *Swadhyayees* were dealing with one of the most pressing problems of people; groundwater scarcity in Saurashtra was palpable.
- Communities/groups of people became part of the movement as opposed to discrete individuals. An entire village or substantial portions of it turned to harvesting and recharge. This allowed community to internalize the positive externality of the recharge, showed dramatic results in quick time and led to a snowballing effect.
- As more farmers beyond the *Swadhyayees* joined the movement, it evolved and a new narrative began to emerge around 1994. The central narrative of the movement shifted from that of an instrumental devotion to an economically rationale act.

Among the honourable mentions in this category, Utthan, a Gujarat based NGO also had a successful experience in Rajula. Based on their catalytic role, people in several villages accepted the norm of not allowing tubewells deeper than 65 m. People also followed the well-spacing specifications issued by the groundwater community.

6.2 Accelerating Regulation—Erstwhile Undivided Andhra Pradesh

NGO/Government agencies formed and nurtured farmer groups through a catalytic process. Social regulations were developed through a consensus building approach and shared widely. Farmers generated scientific information at an appropriate hydrological scale to support the regulatory process, either systematically or in an ad-hoc manner.

6.2.1 Self-Regulation Through Systemic Knowledge

6.2.1.1 Andhra Pradesh Borewell Irrigation Scheme (APWELL)

APWELL project (Das 2000) is a unique experiment in community based groundwater management. Under APWELL, farmers were trained to carry out participatory hydrological monitoring. Farmer groups reported their monitoring findings to a field hydrologist, who helped to analyse the results and suggested possible recharge measures. Combining hydrological monitoring to agricultural extension helped manage both demand and supply side issues.

APWELL demonstrated that external regulation was clearly not enough to have communities participate in groundwater management—strengthening local management systems, developing a systematic data based decision support system for recharge and scaling up the lessons from experimental pilots can lead to self-perpetuating groundwater recharge movements.

6.2.1.2 Andhra Pradesh Farmer Managed Groundwater Systems (APFAMGS)

The APFAMGS initiative was borne out of APWELL. APFAMGS is based on the principle of demystifying science for common citizens. APFAMGS achieved this demystification by training communities on hydrogeology and improving their knowledge of groundwater management. The focus was on facilitating communities to assess the groundwater potential at the village level and estimating the available water before each crop season. These estimates were integrated at the hydrological unit level.

APFAMGS also attempted to find the appropriate granularity for measuring groundwater levels; observation wells were set up at the village level. Communities also prepared crop-water budgets at the village level and developed a seasonal cropping pattern based on groundwater availability. These details were shared across villages within the hydrological unit. The hands-on approach with external scientific or technical inputs helped to improve the awareness of farmers regarding groundwater and cropping pattern. The initiative grabbed national and international policy attention. It was especially lauded for its efforts in demystifying science, promoting community awareness and forging institutional linkages for aggregating information at the village level.

APFAMGS raised awareness about the availability of water at the aquifer level, which helped to arrest rampant development of groundwater sources. But APFAMGS did not encourage investments in recharge structures. It hardly addressed the equity issues—sharing water with un-irrigated farmers. In fact, most importantly APFAMGS did not focus on enforcing social regulations related to water conservation/sharing. There were no economic incentives to follow these regulations either. In fact, some argue that APFAMGS plays an advisory role without incentivizing or dis-incentivizing individual/community practices to follow the advisories.

6.2.2 Social Regulation Approach

Two other community based groundwater management models from erstwhile undivided Andhra Pradesh, viz. Social Regulations in Water Management (SRWM) and Andhra Pradesh Drought Adaptation Initiative (APDAI), adopted social regulation to manage groundwater. Unlike APFAMGS, awareness building about water availability in aquifers and data generation by the village communities was not integral to these models (Garduño et al. 2009). They were important constituents of the model but the process was not as systematic as APFAMGS. Bringing consensus among the communities to share water, between well owners and others, was the most important component of this model. A variety of economic incentives were put in place to incentivize 'positive' farmer behaviour and community action. These incentives ranged from subsidies for micro irrigation, provision for protective irrigation and irrigation support in case of well failure etc. In addition, water harvesting structures were promoted to increase recharge, distribution losses were reduced by supplying water directly to fields via pipelines and drip irrigation was promoted to improve water use efficiency.

Enforcing social regulation led to community action. No new borewells were drilled in these locations. Small and marginal farmers also received benefits along with well owners. Equity issues were addressed and overall social welfare improved. However, researchers (Srinivasa Reddy et al. 2012) argue that both SRWM and APDAI were not equally effective in enforcing social regulation and subsequent community action for groundwater management. They suggest that SRWM was more effective than APDAI. Researchers argue that Watershed Support Services and Action Network (WASSAN) and some of the other NGO partners played an important catalytic role in SRWM. They had a longer history of association in the areas where SRWM was practiced. NGOs also followed a more informal, localized and hence more intensive approach to community building, while APDAI adopted a wider and more formal approach involving government departments.

6.2.3 Lessons from the Andhra Models

APFAMGS approach was knowledge intensive and was not designed to address equity. In the absence of any social regulation, formal or informal, farmers do not have an incentive to follow the good practices. SRWM and APDAI showed that encouraging water sharing between well owners and others could result in achieving the twin objectives of conservation and improved access with equity. But it seems that localized and informal approaches to water sharing works better. NGOs seem to do a better job of catalyzing the process. But having acknowledged that one is still uncertain whether the SRWM approach can/should be mainstreamed in government programming. If mainstreamed, how the process would play out in a larger scale is a question which still remains to be answered.

Sustainability is a major concern across all the approaches. None of the approaches have a well-defined exit protocol. APDAI involves a number of formal institutions and is perhaps better placed to develop a protocol. But at the same time, strong leadership at the village level is necessary to steer the community and take the initiative forward. These two apparently conflicting requirements make this a tough balancing act. NGOs played an important role in the success of SRWM, but financial contribution from farmers remained a problem in SRWM too. Despite the readiness among community, sustaining the action after the funding period is over remained a challenge. External funders and fund flows appear to be critical for the early success of the initiatives. And the initiatives may continue at a smaller scale in some villages due to strong local leadership and commitment of the NGOs even without funding. But scaling up these initiatives will require a more strategic approach.

6.3 *Pioneering Leadership—Hiware Bazar, Maharashtra*

Hiware Bazar is a village in Ahmednagar district, Maharashtra. The village underwent a groundwater led economic transformation in the '90s. Treating groundwater as CPR was the basic premise of this transformation. Poor rainfall, steady degradation of forest land, lack of education opportunities in Hiware Bazar prompted several people to migrate into the nearby cities. But the situation changed soon.

Led by Shri Popatrao Powar, villagers came together for a development plan. The plan accorded highest priority to safe drinking water, followed by irrigation and other productive or non-productive uses.

Due to watershed development measures, the groundwater situation improved and irrigated area increased. From traditional millets, farmers switched to cash crops and horticulture. Farmer incomes grew, with increased cropping intensity and different cropping pattern. Farmers began to invest in water storage structures too. Increased incomes allowed farmers to take loans for cattle and invest in vegetative measures, which increased milk yields and improve incomes further. Hiware Bazar is an example of inclusive growth, almost all families gained from this transformation, either directly or indirectly.

6.3.1 Community Participation in Groundwater Management

Community led the transformation in Hiware Bazar. They collectively chose the development priorities, contributed labour, and managed their natural resources (including groundwater) by regulating and enforcing norms. Most community decisions were taken at Gram Sabha meetings. But it takes time and effort to build a strong community. Shri Popatrao Powar showed exemplary leadership and catalyzed the community building process. He helped people understand the extent of the problems through exposure visits. Most importantly, Shri Powar focused on improving transparency of the decision making, made the process more inclusive and held the Panchayat office bearers accountable to their responsibilities. Groundwater conservation and recharge efforts rode on the back of these initiatives. Community's involvement in Hiware Bazar had spread to all segments of the population. School children read rain gauges and measured groundwater levels. Women collected and managed the monthly water tariff for individual connections. Water budgeting and crop planning decisions were taken in the Gram Sabha meetings (IDFC 2012).

6.3.2 Lessons from Hiware Bazar's Success

Hiware Bazar's success hinged on Shri Popatrao Powar's leadership and the subsequent involvement of the entire community. He nurtured and empowered community that collectively planned and implemented their development priorities. Hiware Bazar's success was also achieved due to convergence of objectives, funds and actors like government, beneficiaries and NGOs. Shri Powar's leadership helped achieve the convergence early into the watershed development programme and allowed community to own the process over longer term. Some lessons from Hiware's Bazar's success are as follows:

- Delink land rights from groundwater use rights: Irrespective of who owned the land, village community in Hiware Bazar managed groundwater as CPR.
- Water budgeting and prioritizing drinking water: Community was involved in the water budgeting process. Local use norms and rules were set based on the water available in the aquifer. Among all the uses of groundwater, drinking water was accorded highest priority. Crop plans were developed based on the amount of groundwater available in the aquifer after allocating water for drinking.
- Aquifer based watershed development programming: Community in Hiware Bazar considered aquifer characteristics while making the water use allocations. Village community was trained on different characteristics of an aquifer and how water budgeting can be carried out based on these characteristics.
- Low transaction costs of coordination: Hiware Bazar was relatively homogenous in terms of caste and religion. Without undermining Shri Powar's efforts, perhaps it is safe to suggest that mobilizing community to share resources and achieving consensus in decisions was relatively easier in Hiware Bazar. Co-ordination may have been an issue if the watershed community would have been heterogeneous.

6.4 *Citizens and Supportive Governments—Kerala and Madhya Pradesh*

6.4.1 Mazhapolima, Kerala

Mazhapolima ("Bounty of rain") is an open well recharge programme based on rainwater harvesting. Thrisur district administration initiated Mazhapolima in collaboration with the Panchayati Raj institutions during 2009. The programme was designed to ensure water security to households in Thrissur district.

Mazhapolima works with a simple logic. Harvest rainwater, channelize the harvested rainwater either directly to the open well (after passing via sand and gravel filter) or put it into the recharge pits constructed next to an open well. It also has a very Kerala specific context. Within a span of 6–7 years, Mazhapolima helped to build between 15,000 and 20,000 wells in Thrisur and other neighbouring coastal districts in Kerala (Biswas 2015).

Mazhapolima is a citizen led movement. One man's zeal (Jos Raphael) combined with a conscientious media house (Malayalam Manorama) and a pro-active bureaucrat (Kurien Baby) helped Mazhapolima's succeed. Mazhapolima's numbers speak for its achievements but the story merits further analysis.

6.4.1.1 Non-linear Approach to Scale

The word 'scale' is clichéd and used loosely in the development parlance. There is hardly a clear and objective definition. But Mazhapolima's achievements resemble the kind of imagery 'scale' evokes. It is interesting to understand what contributed to the scale. Partly it has to do with the simplicity of the idea. And partly it has to do with the approach. Catalysed by Mr. Raphael, common citizens picked up a simple idea (open well recharge) and pursued passionately. In addition, the district administration helped to diffuse the idea at different levels: schools, colleges and media houses. Arghyam (a donor agency based out of Bengaluru) helped to set up an administrative team, within the Thrissur district collector's office. The team trained a large number of masons, Panchyati Raj Institution (PRI) members on open well recharge. Slowly, the idea captured popular imagination. In the process, several champions were created within and outside the government system. Once a critical mass of actors got involved, the idea became self-sustaining.

6.4.1.2 The Kerala Context

Homesteads are hallmark of staggered settlement patterns in Kerala. Open wells are a common feature in the homesteads. Together, homesteads and common wells form part of a unique heritage in Kerala that influences the living standards and shapes its socio-cultural identity. Hence, an open well recharge movement found easy resonance. In addition, Kerala boasts of a strong grassroots' democracy. A deliberative and empowered grassroots democracy has helped devolution of functions and resources to the Gram Panchayats. This has helped all developmental activities in Kerala, and Mazhapolima is no exception.

6.4.1.3 The Downsides

In general, people are satisfied with Mazhapolima and report that groundwater levels have improved in their wells. But some suggest that effects of recharging open wells in different topography have been varying. Lack of scientific understanding about the recharge sites has led to such variations, they suggest.

In addition, there is a competition lurking at the background. Groundwater abstraction is around 50% in Thrissur. Borewells are also making their way to Thrissur. Neither there is an understanding nor there is any preparedness to deal with this emerging issue.

6.4.1.4 Summing Up

To understand and explain Mazhapolima is an enterprise fraught with difficulty. There are many facets to this story. And a short sub-section will definitely not be able to do justice. Yet, some conclusions need to be drawn.

Mazhapolima is a bold experiment. Flawed as some would argue, still it has achieved scale and captured popular imagination. Perhaps it is best suited for Kerala's context, perhaps with some tweaking it could have performed better. But it has truly created opportunities for people to participate in a movement. It is a journey of common citizens with a simple idea that not only managed to convince the rough and tumble of the bureaucracy, but also created conditions for a large-scale social change.

6.4.2 Bhagirath Krishak Abhiyan

Dewas is a district in Madhya Pradesh. Since 2005, about 6000 farm ponds of sizes varying from a quarter to 5 ha have been dug in Dewas (Pallavi 2012). They changed the agricultural landscape in Dewas. Water scarcity became a phenomenon of the past. Groundwater levels have improved too.

6.4.2.1 The Programme

Around 2005, some progressive farmers had dug ponds in different villages of Dewas district. These farmers were coping better with water scarcity. The District Collector of Dewas noticed the developments. There was neither a government policy to allocate funds for digging ponds nor did any bank had a provision for giving loan. However, the agriculture department, at the collector's behest, decided to help farmers, especially, the ones who owned land. In fact, the agricultural department prepared a list of 7,000 landowners who owned tractors necessary for digging operations. Follow-up awareness and training sessions were held. The campaign was named *Bhagirath Krishak Abhiyan*. Farmers who dug ponds were called "*Bhagirath Krishak*" and ponds were called *Rewa Sagar*. Farmers were advised to set aside 10 per cent of their land for ponds. They were also provided assistance in accessing necessities like tractors at a reasonable cost.

The efforts bore fruit. About 600 *Rewa Sagar* ponds were dug in 2006. In response, the state announced a subsidy of Rs. 16,350 per pond. When the paltry sum failed to encourage the farmers, a subsidy of Rs. 80,000 per pond was announced, irrespective of its size. An additional Rs. 20,000 for scheduled castes and tribes was announced. Many ponds were constructed and groundwater levels improved in these villages. Incidental benefits of improved groundwater situation accrued to the landless too.

6.4.2.2 Analysis and Summing Up

Bhagairath Krishak Abhiyan is an example of government catalyzing citizen participation. Some enlightened farmers took to constructing ponds on their own and reaped benefits. Other farmers followed suit. The district government took cognizance. Not only they took cognizance, they furthered the cause through awareness building, follow up training and monetary incentives. This is an example how opportunistic and nimble policy making can take advantage from proven successes and create conditions for replication at scale.

6.5 Testing the Model of Participatory Groundwater Management (PGWM)

The PGWM model borrows from experiences of managing groundwater as CPR. Supported by Arghyam (a water and sanitation focused donor agency based out of Bengaluru, Karnataka), Advanced Centre for Water Resources Development and Management (ACWADAM), Arid Communities and Technologies (ACT), Watershed Support Services and Action Network (WASSAN) and People's Science Institute (PSI) piloted the PGWM idea in different locations/aquifer typologies. Different agencies and their respective pilot sites were as follows:

- ACWADAM in Pune and Satara (Maharashtra),
- ACT in Bhuj (Gujarat), WASSAN in Ranga Reddy (Telengana), and
- PSI in Sirmour (Himachal Pradesh).

Across different pilot sites, these agencies played a catalytic role. They worked with local communities and helped them adopt some basic principles for managing groundwater:

- Irrespective of the nature of land ownership, groundwater was treated as CPR.
- Groundwater problems were clearly defined.
- Drinking and cooking needs of the community were met first before putting groundwater to other uses.
- Principles and processes of management would cut across different uses, e.g., drinking water, irrigation etc.
- While the individuals and communities can own groundwater structures, the water was shared with others in the water user group, along with operational costs.
- Aquifer was the unit for managing groundwater. Minimum unit of management was local aquifer and maximum unit was regional aquifer.
- Long-term engagement (at least 8 years) in each location was necessary.
- The local community was in charge of planning, management and monitoring of groundwater. Local community received support from external agencies while carrying out their management actions.
- A combination of local knowledge and formal science was prioritized for managing groundwater.
- No overriding was allowed.

6.5.1 Community Participation

By definition and practice, participation remained an integral component of PGWM. Following are the ways in which communities were involved in different aspects of PGWM:

- All the external agencies were essentially NGOs. They learnt about the traditional practices on groundwater management and incorporated them while developing management plans and use protocols with the communities.
- Communities were trained on the science of hydrogeology and were encouraged to use hydrogeology in combination with traditional resource understanding to make appropriate groundwater use decisions.
- *Gram Sabha* endorsed and legitimized the groundwater management plans prepared by the communities.

6.5.2 Analysis and Summing up

People's participation was central to the PGWM construct. Most of these NGOs had been working in these pilot sites for years. Prior to working on PGWM, the NGOs had invested in trust building exercises with the community. They had also helped build local leadership. The history of association, mutual sense of trust and reciprocity and local leadership helped the NGOs play a catalytic role in the PGWM process (Kasturi Rangan 2016). Communities were involved in different aspects of PGWM: hydro-geological and socio-economic surveys, groundwater budgeting, setting up a weather monitoring station, training and orientation sessions on basic concepts of hydrogeology etc. Once the community understood the importance of managing the resource and the role of collective action in this endeavour, it became easier to inculcate the idea of PGWM.

Despite some challenges related to replicating the success at scale and delinking land ownership from owning the groundwater underneath (which has implications especially for landless farmers), PGWM proponents successfully empowered communities in the pilot sites. The PGWM examples also demonstrate that NGOs (working in close collaboration with government) can catalyze collective action among the local communities. In this context, the role of catalysts becomes important. NGOs can act as catalysts to build competencies, impart awareness and support decisions for designing locally relevant groundwater management programmes.

In addition to the PGWM model, some other examples of community-based groundwater management also deserve an honourable mention: Water budgeting experiences from Barefoot College (Tilonia, Rajasthan), Foundation for Ecological Security's (FES) experiences of working with communities on water balance and groundwater use in Rajasthan, Madhya Pradesh (MP) and Andhra Pradesh (AP), ACWADAM's experiences with *Samaj Pragati Sahayog* (SPS) in Bagli (MP) and *Pani Panchayats* of Maharashtra. As a matter of fact, ACWADAM's past experiences in knowledge based, typology driven aquifer management strategies served as backgrounders for developing the PGWM programme.

7 Conclusions and Way Forward

Several conclusions can be drawn from the analysis of theories and practices related to catalyzing participation in groundwater management. Some of these are direct conclusions, others are derived as interpretations. Conclusions also lead to the possible ways forward. Details are as follows:

- Creating **economic incentives** for **participation** is important for catalyzing community participation in groundwater management. In most cases, people don't participate since a clear economic case is not made by the catalysts/facilitators. It is neither the sense of righteousness nor the sense of higher morality but an economic rationale that drives participation.
- If the **problem is palpable**, people participate. In a number of cases, groundwater scarcity or quality issues have become key drivers of participation. This assumes importance if people are dependent on groundwater for life and livelihoods.
- Making the '**invisible**' resource '**visible**' also helps. Training the community on water budgeting, getting them to play simulated games and imparting awareness

about the nature of aquifers help people develop a common resource understanding and make groundwater use choices.

- **Nature of aquifer** and its **spatial extent** play an important role. The spatial extent of aquifer should allow dependent users to engage in meaningful community action. Hence, catalyzing community action in large aquifers (e.g. alluvial) is difficult.
- In case users expect significant **future dependence** on groundwater, an external agency can get people to rally around the future perceptions and help propel community action.
- A history of relationship between the external agency (NGO in most cases) and local community helps to build **trust and a sense of mutual reciprocity.** This facilitates community action on groundwater management.
- **Local norms/social regulations** tend to work better than legal rights in relation to groundwater management. Also, communities internalize the norms better than rights enshrined legally—perhaps of their proximity to and/or involvement in the norm making process.
- **Formal legislation** will work best when it legitimizes/legalizes the local regulations. For instance, delinking land and groundwater use rights is an important aspect of building social norms. Once the norm achieves the legality, it will be easier to enforce and monitor its implementation.
- **Institutionalizing the norms** is important for communities to manage groundwater in the longer run. Loosely enforced norms in several situations are a powerful alternative, but there are limitations to what management by norms can achieve and hence mainstreaming the norms via appropriate institutions is important.
- **Data and information** are key. The whole process of collecting data/information, analyzing and making decisions is essential elements of groundwater management. Each of these elements is capable of engaging communities in the management process.
- **Universality** is important i.e., not excluding any potential user in the regulations. None of the successful real-life examples barred a new entrant from having access to groundwater or defined the quantitative right of one well owner over another.
- **Homogeneity of communities and their use interests** are important to understand before facilitating any community action. In fact, the facilitation process becomes relatively easier in case of homogenous communities. Community participation and its facilitation become increasing difficult for communities with heterogeneous interests.
- **State engagement is necessary.** Groundwater over-exploitation is a widespread problem in India. Considering the enormity and spread of the problem, one feels that the state is best suited to address the issue with systemic action. In addition to the challenges of scale, communities need time and continued investments in capacity building before they begin to manage groundwater sustainably. The State agencies have and need to ensure that community-based groundwater management models do not fail because of lack of support and formalization.

They need to ensure that experiences from successful interventions remain available for replication.

- **Simple messaging and calls to action** work the best. People understand simple action-orientated messages better. In most cases, simple messaging has prompted people to take action at scale.

Finally, people need to manage their own resource. Considering the historicity of resource management in India and its current political, economic and social underpinnings, people managing their own resource seem the most appropriate narrative. Groundwater is no exception. There are enough theoretical, empirical and analytical evidences to suggest that people can actively participate in the groundwater management. But their participation will be dependent upon the nature of the community, their relationship with the resource and the characteristics of the resource itself. The state and the non-state actors can play an important catalytic role in improving the community-resource relationship in groundwater and bring about a transformative change in the society.

References

Agrawal, A. and Ostrom, E. (2001). Collective Action, Property Rights, and Decentralization in Resource Use in India and Nepal. *Politics & Society*, **29(4):** 485–514.

Arnstein Sherry, R. (1969). A Ladder of Citizen Participation. *J Amer Plann Assoc*, **35(4):** 216–224.

Biswas, A 2015. Personal Communication, Mazhapolima: Recharging open wells in Kerala, www, indiawaterportal.org.

British Geological Survey (2004). Community Management of Groundwater Resources: An Appropriate Response to Groundwater Overdraft in India? London, UK: British Geological Survey.

COMMAN (2005). Managing Groundwater Resources in Rural India: The Community and Beyond. British Geological Survey Commissioned Report CR/05/354.

Das, S.V. (2000). Participatory hydrological monitoring (PHM): An effective tool for community managed groundwater system. Paper at International Conference on Land Resource Management for Food. Employment and Environmental Security, Delhi, India, pp. 9–13.

Dumont Aurelien (2013). Groundwater is not a Common-Pool resource: Ordering sustainability issues of groundwater use. 3rd International IWA Conference on Water Economics, Statistics and Finance, Marbella, Spain: 2–10.

Gale, I.N., Macdonald, D.M.J., Calow, R.C., Neumann, I., Moench, M., Kulkarni, H., Mudrakartha, S. and Palanisami, K. (2006). Managed Aquifer Recharge: An assessment of its role and effectiveness in watershed management. British Geological Survey Commissioned Report, CR/06/107N. p 80.

Garduño, H., Foster, S., Raj, P. and Van Steenbergen, F. (2009). Addressing Groundwater Depletion Through Community-based Management Actions in the Weathered Granitic Basement Aquifer of Drought-prone Andhra Pradesh (India), Sustainable Groundwater Management. Lessons from Practice, Case Profile Collection, Number 19. The World Bank.

IDFC Policy Group (2012). Hiware Bazar: A water-led transformation of a village. *Quarterly Research Note*, **16:** 1–4.

Kasturi Rangan, A. (2016). Participatory Groundwater Management: Lessons from Programmes Across India. *IIM Kozhikode Society & Management Review, Perspective*, **5(1):** 8–15.

Kulkarni, H. and Vijay Shankar, P.S. (2009). Groundwater: Towards an aquifer management framework. *Economic and Political Weekly*, Commentary, February 7.

Kulkarni, H., Shah, M. and Vijay Shankar, P.S. (2015). Shaping the contours of groundwater governance in India. *J Hydrol: Regional Studies*, **4:** 172–192.

Kumar, H.D. (1994). Water Issues in Mehsana. *In:* M. Moench and M.D. Kumar (eds), Local Water Management Initiatives: NGO activities in Gujarat. Vikram Sarabhai Centre for Development Interaction, Ahmedabad, pp. 27–43.

Lindberg, S., Rajagopal, A., Djurfeldt, G., Venkatesh, B.A. and Vidyasagar, R. (2011). Designing Collective Action: Problems of Local Water Management in Tiruchi District. *Review of Agarian Studies*, **1(2):** 157–176.

Llamas, Martínez-Santos (2005). Intensive Groundwater Use: Silent Revolution and Potential Source of Social Conflicts. J Watr Resour Plann Manag © ASCE/September/October.

Mosley, M.P. and Lincklaen Arriëns, W.T. (1995). Towards a Policy for Water Resources Development and Management in the Asian and Pacific Region: Issues and Opportunities. Discussion paper prepared for the Interdepartmental Water Resources policy group. Asian Development Bank, Manila, Philippines.

Ostrom, E. (1990). Governing the Commons: The Evolution of Institutions for Collective Action. Cambridge University Press, New York.

Pallavi, A. (2012). Penny Foolish, Pond Wise. Down To Earth, January 31, 2012.

Shah, T. (2008). The Challenge of balancing livelihoods and environment. *In:* Handbook on Environmental Economics in India. Chopra, K. and Dayal, V. (Eds). Oxford University Press, New Delhi, India. 11–12.

Shah, T. (2009). Taming the Anarchy: Groundwater Governance in South Asia. Routledge: New Delhi, India: 34–36.

Shah, T. (2014). Towards a Managed Aquifer Recharge strategy for Gujarat, India: An economist's dialogue with hydro-geologists. *J Hydrol*, **518(Part A):** 94–107.

Srinivasa Reddy, V., Reddy, R. and Rama Mohan, R.V. (2012). Institutionalising Groundwater Management: A Tale of Three Participatory Models in Andhra Pradesh, Working Paper No. 114. RULNR Working Paper No. 15.

Sriskandarajah, N., Fisher, R.J. and Packham, R.G. (1996). Community Participation in Natural Resource Management: Lessons from Field Experience. *In:* Proceedings of the Regional Seminar "Ecotone V" on Community Participation in Conservation, Sustainable Use and Rehabilitation of Mangroves in Southeast Asia, Ho Chi Minh City, Vietnam.

Suhag, R. (2016). Overview of Groundwater in India. PRS Legislative: 1–2.

Van de Laar, A. (1990). A Framework for the Analysis of Common Pool Natural Resources, Working paper no. 77. Institute of Social Studies, The Hague.

Van Steenbergen, F. and Shah, T. (2003). Rules rather than rights: Self-regulation in intensively used groundwater systems. *In:* Llamas R. and Custodio, E. (eds), Intensive use of groundwater challenges and opportunities. Taylor and Francis, pp. 241–254.

Chapter 23
Way Forward and Future Strategies

Pradip K. Sikdar

1 Development Strategy to Management Strategy

The average rainfall in South Asia is plentiful with respect to global standards. Most of this rain falls in relatively brief period during the monsoon, which also differs from region to region. The vagaries of the climate coupled with man-made climate change have driven the farmers, households and industries to increasingly depend on groundwater rather than surface water. This dependence has led to deterioration in the South Asia's groundwater (Chap. 1), in general, and India's groundwater (Chap. 2), in particular, in terms of over-exploitation, water quality deterioration, saline water intrusion and threat of land subsidence.

This book has dealt with diverse subjects – from groundwater distribution, groundwater science, water well drilling, construction and development, groundwater conservation, treatment technologies, environmental hazards of contaminated groundwater to social, economic and legal aspects associated with sustainable groundwater management. Each chapter in the book examines the current situation, provides recommendations or suggestions for future actions and emphasizes on a paradigm shift from development strategy to management strategy based on a better appreciation for the social, legal and economic drivers of groundwater use for ensuring security of food and energy in South Asian region. This book thus provides an outline of challenges that are in dire need of efficient management, based on sustainability principles, quantity, quality, ecosystem support, socio-economic benefit and good governance. The book also depicts a road-map to evolve an entirely different paradigm of ensuring sustainable groundwater management which would ensure safety and security of groundwater-based water supply in South Asia.

P. K. Sikdar (✉)
Department of Environment Management, Indian Institute of Social Welfare and Business Management, Kolkata, India
e-mail: p_sikdar@hotmail.com

P. K. Sikdar (ed.), *Groundwater Development and Management*,
https://doi.org/10.1007/978-3-319-75115-3_23

2 Change of Groundwater Management Paradigm

The authors of this book bring out the need for a change of groundwater management paradigm. Aquifer management is politically complex. Therefore, management models should be inclusive and give opportunities for innovations in managing typical problems in various hydrogeological, hydrogeochemical and physiographical framework such as arid, semi-arid, hilly, alluvial and coastal areas. While developing groundwater in these different frameworks, great cautions should be observed in estimating available resources precisely—both annually and periodically (season-wise)—to avoid over-exploitation resulting in environmental imbalances. The reservoir configuration and their carrying capacity should be well understood by mapping aquifers with the help of geological, geophysical, geostatistical and remote sensing techniques (Chaps. 3, 5 and 6) followed by appropriate drilling technique (Chap. 10) and pumping test (Chap. 11). Another initiative for better groundwater management is to develop numerical models (Chaps. 7 and 8) coupled with isotope studies (Chap. 4) to understand the dynamics of groundwater flow at regional, intermediate and local scales and for predictive assessment. Pricing of groundwater (water is not a free commodity), although politically very sensitive, must be implemented for effective groundwater management. Pricing of groundwater could be based on the resource itself or the energy required to pump groundwater. Reallocating scarce water among various sectors by appropriately designed water markets supported by sound institutions will make it possible to meet the growing urban and industrial water demands without jeopardizing growth in crop production (Chap. 20). Groundwater managers should therefore advocate the importance of changing the established use patterns and managing water at catchment or river basin scales and not on political and administrative boundaries. Development at stages backed by well-conceived monitoring, surveillance, numerical models and pricing instead of jumping to rapid growth should be the key approach for future groundwater development and management.

3 Development of Scientific Database of Groundwater

Groundwater system is complex and limits of abstraction are not well established. Therefore, hydrogeological and hydrodynamical structure of aquifer systems should be properly understood to promulgate a comprehensive and adaptive policy. In most cases, hydrogeological survey is not adequate and even where there are regular surveys it is incomplete. This makes groundwater to be an unseen resource. Quality and abstraction data is not well monitored or compiled and is generally unavailable in the public domain. Governments in many countries conduct groundwater surveillance when confronted with groundwater problem, but it is discontinued after the problem is mitigated. In India groundwater resource is categorised into safe, semi-critical, critical and over-exploited based on stage of groundwater development and long-term decline of groundwater level (Chap. 2). The quality is not taken into consideration. Depreciation of groundwater quality can limit the value of aquifers as source of water for different uses. So blocks where groundwater is contaminated with arsenic or fluoride are at times categorised as 'safe' for drinking purpose. Therefore, governments should develop scientific database for policy making through strengthened groundwater quantity and quality monitoring, geostatistical analysis, modelling and data sharing mechanism, supported by universities, research institutes and community-based institutions.

4 Augmentation of Groundwater Through Baseflow Management, Water Conservation and Artificial Recharge

The impact of climate change on the hydrological cycle including groundwater is largely uncertain. While water stress is likely to increase over many regions of South Asia, in several areas, there will be surplus (Chap. 17). In semi-arid areas development of baseflow for small to medium water supplies can augment drinking water supply (Chap. 19). Traditional water conservation practices and groundwater recharging through re-excavation of old, silted up ponds, below the average groundwater level during pre-monsoon, will provide water security to the communities in the lean period (Chap. 18). Desiltation and rejuvenation of 1.5 million village tanks and ponds in India, will ensure 6,60,000 villages against recurrent droughts along with enhancement of groundwater recharge and water level recovery in shallow aquifers. For recovery of water level in deep aquifers injection of alternate high quality water is essential. A sustainable groundwater management should take rainwater as the starting point for integrated catchment management which may give rise to new opportunities. Rainwater harvesting including roof-top rainwater harvesting, artificial recharging, sprinkler and drip irrigation, and conjunctive use should be given more importance in water deficit and groundwater contaminated areas.

5 Management of Poor Quality Aquifers

There are large areas where groundwater is contaminated from point and non-point sources (Chaps. 12 and 14), which increases health risks (Chap. 15) due to geogenic arsenic (India, Bangladesh and Pakistan) and fluoride (India, Sri Lanka and Pakistan). Arsenic pollution assumes particular importance in view of the fact that from a point-source of contamination (contaminated wells for drinking water) it changes to a diffuse-source of contamination (when it enters the human food web) through the use of arsenic-contaminated water in irrigated agriculture, that may increase the number of people at risk from arsenic contamination. Mitigation options to reduce diffuse-source of contamination should be given immediate attention by the Governments. They are (i) development/identification of suitable low As-accumulating high yielding crops/varieties, (ii) identification/development of varieties/crops which accumulate less As in the consumable parts and where the ratio of inorganic to organic forms of As is low, (iii) developing cost-effective phyto- and bio-remediation options, (iv) increased use of FYM and other manures + green manure crops, as well as application of appropriate inorganic amendments etc. (Chap. 13).

For managing point-source of contamination it is important to first comprehend the sub-surface generic sedimentology and then delineate wells which may remain uncontaminated for the foreseeable future. For the contaminated wells, alternative source of water from deep aquifer or surface water should be considered. If alternative source is not available then cost-effective innovative treatment technologies for point-sources should also be implemented for better utilization of these contaminated aquifers (Chap. 16). Two-stage arsenic removal process, oxidation-co-precipitation followed by adsorption, is found to be the best method to reduce the concentration of arsenic to <10 μg/L. For fluoride removal from contaminated groundwater effective sorbent-based treatment technologies may be used. HIX–NanoZr, in which nano particles of zirconium oxide are loaded within a polymeric support of ion-exchange resin, has been found to be effective for selective fluoride removal from a background of other competing ions (Chap. 16). Sea water intrusion into coastal plains has become evident especially in coastal cities. This needs to be addressed through controlled groundwater use and managed aquifer recharge. Again, large brackish water resources available in the shallow aquifers of coastal and inland areas have to be utilized for agriculture, industry and other uses after dilution with fresh water and after economic treatment (Chap. 8).

6 Springshed Development

Springs are important source of water in the hilly regions throughout South Asia. In India, about five million springs support about 200 million people in mountainous regions from the Nilgiris to the Himalayas. But spring discharges are in decline due to ecological degradation, climate change and increasing demand for water. Therefore, improved management of springs is required to ensure water security for people in mountain regions, as well as downstream communities in the plains. The steps involved are: first, rigorous and comprehensive surveys to derive spring density numbers and the population dependent on springs; second, for springshed development the catchment area of the spring or springshed area should be mapped using the combination of science of groundwater, local knowledge and participation of the local community; and third, targeted treatments of the springshed such as restoration or protection should be carried out. For success of springshed development capacity building of the local community is imperative. Springs should be a national priority to be addressed to ensure water security under increasing demand, changing land use and climate variability (Chap. 9).

7 Groundwater Legislation

Groundwater legislation is still weak and fragmented in South Asia. In Pakistan, only Baluchistan Provincial Government issued legislation to control groundwater mining. Bangladesh also has some acts and rules for management of groundwater resource. But their implementation and enforcement is not enough to reverse negative consequence of groundwater development. Groundwater legislation in these and other South Asian countries as it stands today should be modified to focus on community ownership, with a community–management approach associating beneficiaries.

In India, groundwater legislation exists in many states but is based on the common law doctrine which allows rights to landowners to access groundwater by viewing it as an attachment to the land they own. As a result well users simultaneously pump groundwater causing interference effect, more drawdown, change in groundwater flow pattern and drying up of wells without submersible pumps. The present legislation does not consider the water needs of poor people without land ownership and only considers the need of the affluent landowners. Hence, groundwater legislation should include resource protection incorporating the role of groundwater in safeguarding the environment (Chap. 21). The Constitution of India lists minor irrigation, water management and drinking water as items to be handled by Panchayats. Hence, it should be made mandatory to involve Panchayats in groundwater resources management in India. Each Indian state needs to formulate State Water Policies, based on National Water Policy (2012), keeping in view the region's human, social and economic needs.

8 Community Participation

Community participation is the key to the success of groundwater management, especially in the context of groundwater use by small users (Chap. 22). Community participation can take place at various levels ranging from individual waterwells to an aquifer system and even to the river basin or national level. Participation should be encouraged at all levels since it can make an important contribution to groundwater conservation, management and protection. The governing principles of participatory management include building trust among stakeholders, creating economic incentives for participation, equal rights, long-term commitment, community management (embracing all users), extensive promoting activity under CSR programme and reliable baseline data. Delinking land and groundwater use rights by formal legislation may also help in community participation especially for communities with heterogeneous interests. Capacity building of community especially farmers and local well drillers is very important so that they can manage their own systems. This includes education in hydrogeology, groundwater monitoring, water testing and surveillance. Groundwater being a common pool resource, a community-based management approach should also incorporate the fundamentals of social capital formation, behavioural economics, vocational education, communication and leadership development.

Community participation in the context of groundwater management should promote equity amongst the users, avoid groundwater access being dominated by the big landholders, avoid resource depletion, optimize pumping costs, better estimate of groundwater balance, change of groundwater consumption pattern in the long-term communal interest, increase productivity of groundwater (by using appropriate soil moisture management, improved crop selection and irrigation techniques), and enhance aquifer recharge.

9 Future Groundwater Managers

Present groundwater managers are either hydrogeologists or civil engineers who depend on rationality, calculation and technology, and follow ordered structures and approaches. These managers mostly are unable to understand the importance of environmental and social dimensions of groundwater development. The future groundwater managers must develop communication skills to deal with the various stakeholders including politicians. Therefore, what we urgently need is to develop a group of 'social hydrogeologists/engineers' or 'ecological hydrogeologists/engineers' who have the skills to better understand not only the technical, but also financial, social and environmental aspects and their impacts. Therefore, universities and management institutions should develop new curriculum that can produce new groundwater managers who can look at groundwater from a holistic point of view.

Index

P. K. Sikdar (ed.), *Groundwater Development and Management*,
https://doi.org/10.1007/978-3-319-75115-3

MIX
Papier aus verantwortungsvollen Quellen
Paper from responsible sources
FSC® C105338

If you have any concerns about our products,
you can contact us on
ProductSafety@springernature.com

In case Publisher is established outside the EU,
the EU authorized representative is:
Springer Nature Customer Service Center GmbH
Europaplatz 3, 69115 Heidelberg, Germany

Printed by Libri Plureos GmbH
in Hamburg, Germany